The Quantum Universe

Essays on Quantum Mechanics, Quantum Cosmology, and Physics in General

The Quantum Universe

Essays on Quantum Mechanics, Quantum Cosmology, and Physics in General

James B Hartle

University of California at Santa Barbara, USA

World Scientific

NEW JERSEY • LONDON • SINGAPORE • BEIJING • SHANGHAI • HONG KONG • TAIPEI • CHENNAI • TOKYO

Published by

World Scientific Publishing Co. Pte. Ltd.

5 Toh Tuck Link, Singapore 596224

USA office: 27 Warren Street, Suite 401-402, Hackensack, NJ 07601

UK office: 57 Shelton Street, Covent Garden, London WC2H 9HE

Library of Congress Cataloging-in-Publication Data

Names: Hartle, J. B. (James B.), author.

Title: The quantum universe : essays on quantum mechanics, quantum cosmology and physics in general / James B Hartle.

Description: New Jersey : World Scientific, [2021] | Includes bibliographical references.

Identifiers: LCCN 2020046398 | ISBN 9789811216398 (hardcover) | ISBN 9789811216404 (ebook for institutions) | ISBN 9789811216343 (ebook for individuals)

Subjects: LCSH: Quantum theory. | Quantum cosmology.

Classification: LCC QC174.12 .H378 2021 | DDC 530.12--dc23

LC record available at https://lccn.loc.gov/2020046398

British Library Cataloguing-in-Publication Data

A catalogue record for this book is available from the British Library.

The author and publisher would like to thank the publishers of various journals and books who have been approached for permission to include the reprints found in this volume.

For any available supplementary material, please visit
https://www.worldscientific.com/worldscibooks/10.1142/11716#t=suppl

Contents

Preface

Understanding the regularities of nature is the goal of science. Physics is concerned with understanding the regularities of specific classes of subsystems — stars, atoms, the elementary particles, black holes, superconductors, gravitational waves, etc. But beyond this, physics has a special charge — to understand the universal regularities exhibited by all subsystems, without exception, without qualification, and without approximation. These are usually called the fundamental laws of physics.

As physics has progressed, its most fundamental theories have become more distant from the physics of the every day. That distance has posed challenges for understanding — notably with quantum mechanics.

From time to time, the author has had an occasion or an impulse to write a short piece of an essentially expository nature to address those challenges and improve understanding. A number of these essays are collected here.

The essays are written at a level that should be accessible to many physicists, but they are not popular articles aimed at a general public. And they do not replace the author's longer efforts at exposition, for example, those in his Jerusalem Lectures (arXiv.org/abs/1805.12246) and Les Houches Lectures (arXiv.org/abs/gr-qc/9304006).

The essays were written over a period of some thirty years and vary significantly in length, style, subject, quality, level, overlap, point of view, and perhaps even consistency among them.

The essays are divided into four rough categories:

- *General:* Having something to do with physical theories generally.
- *Quantum Mechanics:* Dealing with issues in the quantum mechanics of a closed system in a fixed background spacetime.
- *Cosmology:* Dealing with aspects of the quantum universe on the largest scales of space and time, in particular, with the generalizations of Copenhagen quantum mechanics necessary for semiclassical quantum spacetime geometry (quantum gravity) and quantum cosmology.
- *Personal Recollections:* Short personal recollections of two distinguished scientists the author had the privilege to work with.

These categories are only rough guides. Many of the essays touch on several of the categories and are concerned with the connections among them.

The rest of this preface consists of short descriptions of each essay — what issues they address and how those are resolved. With these short paragraphs, the author aims at providing a measure of unity for the collection that will also enable the reader to select among them without thumbing through the whole volume.

General

1. *Computability and Physical Theories*

(with Robert Geroch) What would physics be like if its theories predicted measurable numbers that were non-computable in the mathematical sense? That is, numbers for which no single computer program exists to compute them to any accuracy that may be specified. Not that much would be changed, but the existence of algorithms for approximate computation would be an important question for such theories.

2. *Excess Baggage*

In many advances in physics, some previously accepted general idea was found to be unnecessary and dispensable. The idea was not truly a general feature of the world, but only perceived to be general because of our special place in the universe and the limited range of our experience. The idea was *excess baggage* which had to be jettisoned to reach a more general and productive perspective. This article discusses excess baggage from the perspective of quantum cosmology, and asks, "What excess baggage is in our present theoretical framework?"

3. *Sources of Predictability*

Sources of predictability from the basic laws of physics are described in the most general theoretical context — the quantum theory of the universe.

4. *Theories of Everything and Hawking's Wave Function of the Universe*

If a cat, a cannon ball, and an economic textbook are dropped from a building, they all fall to the ground with roughly the same acceleration. That is an example of a universal law of physics. But that law tells us little about cats, cannon balls, or economics. This essay describes how a theory of these universal laws — what in physics is called 'a theory of everything' — must include a theory of a quantum state of the universe. A 'theory of everything' does not predict everything that we observe, but only certain features of what we observe that are related to the universal regularities exhibited by all physical systems. (Hawking's wave function of the universe is today mostly called "the no-boundary wave function of the universe" and is the joint work of Stephen Hawking and the author. See Chapter 25 for references).

5. *Scientific Knowledge from the Perspective of Quantum Cosmology*

Scientific knowledge is limited for at least three reasons: (1) Physical theories predict only the regularities of our experience and not every detail of it. (2) Predictions may require intractable computation. (3) The process of induction and test is limited. Quantum cosmology displays all three kinds of limits. This essay briefly describes these limits and the place of the other sciences in this most comprehensive of physical frameworks.

6. *The Physics of 'Now'*

The world is four-dimensional according to fundamental physics, governed by basic laws that operate in a spacetime that has no unique division into space and time. Yet our subjective experience is divided into past, present and future. This essay discusses the origin of this division in terms of simple models of information gathering and utilizing systems (IGUSes). Past, present, and future are not properties of four-dimensional spacetime but rather notions describing how individual IGUSes process temporal information. Their origin is to be found in how these IGUSes evolved or were constructed. The past, present, and future of an IGUS are consistent with the four-dimensional laws of physics and can be described in four-dimensional terms. The present is not a moment of time in the sense of a spacelike surface extending over all spacetime. Rather, there is a localized notion of present at each point along the world line of an IGUS. The present time of a group of localized IGUSes like ourselves is an approximate notion appropriate when members of the group are sufficiently close to each other and have relative velocities much less than that of light. Modes of temporal organization that are different from past, present and future can be imagined that are consistent with the physical laws. We speculate that the past, present and future modes of organization might be favored by biological evolution for exploiting the laws of physics and therefore a cognitive universal to be found widely here on Earth and other places in the universe.

Quantum Mechanics

7. *The Quantum Mechanics of Closed Systems*

A pedagogical introduction is given to the quantum mechanics of closed systems, most generally the universe as a whole. This quantum mechanics aims at predicting the probabilities of alternative coarse-grained time histories of a closed system. But, probabilities consistent with the rules of probability theory cannot be assigned when there is quantum mechanical interference between individual histories of the set.

In Copenhagen quantum mechanics, probabilities can be assigned to histories of a subsystem that have been measured. In the quantum mechanics of closed systems, containing both observer and observed, probabilities are assigned to those sets of alternative histories for which there is negligible quantum interference between individual histories in the set as a consequence of the system's quantum state and dynamics. Such sets of histories are said to decohere.

Copenhagen quantum mechanics is an approximation to the more general quantum framework of closed systems. It is an approximation that is appropriate when there is an approximately isolated subsystem that is a participant in a measurement situation in which the decoherence of alternative registrations of the apparatus can be idealized as exact.

8. *The Spacetime Approach to Quantum Mechanics*

Feynman's sum-over-histories formulation of quantum mechanics is reviewed as an independent formulation of quantum theory in spacetime form. It differs from the Copenhagen formulation that utilizes states on spacelike surfaces because it assigns probabilities to different kinds of sets of alternative histories. The general notion is a partition (coarse-graining) of the histories into an exhaustive set of exclusive classes. With this generalization, the sum-over-histories formulation can be said to be in fully spacetime form with the dynamics represented by path integrals over spacetime histories and alternatives defined as partitions of spacetime histories into exclusive classes. When restricted to alternatives at definite moments of times, this generalization is equivalent to Copenhagen quantum mechanics. However, the quantum mechanics of more general spacetime alternatives that extend over time does not have an equivalent Copenhagen formulation. We suggest that, in the quantum theory of gravity, the general notion of observable is supplied by diffeomorphism invariant partitions of spacetime metrics and matter field configurations. By generalizing the usual alternatives so as to put quantum theory in fully spacetime form, we are led to a generalized quantum mechanics of spacetime free from the problem of time.

9. *Quantum Physics and Human Language*

Human languages tacitly assume specific properties of the limited world which they have evolved to describe. These properties are true features of that limited context, but may not be general or precise properties in fundamental physics. Human languages must therefore be qualified, discarded, or otherwise reformed to give a clear account from fundamental physics of even the phenomena that the languages have evolved to describe. The surest route to clarity is to express the constructions of human languages in the language of fundamental physical theory, not the other way around. These ideas are illustrated by an analysis of the verb 'to happen' and the word 'reality' in special relativity and the modern quantum mechanics of closed systems. This essay contains the author's views on what is 'real'.

10. *Quantum Pasts and the Utility of History*

Why are we interested in the past? It is over and done with. This article describes the process of retrodicting the past in quantum cosmology with quantum probabilities conditioned on present data. There is not just one unique past that is over and done with, but rather, many different possible ones corresponding to different decohering sets of alternative coarse-grained past histories. Retrodicting a past is useful because it helps in predicting the future as in cosmology.

11. *The Quasiclassical Realms of this Quantum Universe*

The most striking feature of our indeterministic quantum universe is the wide range of time, place, and scale on which the deterministic laws of classical physics hold to an excellent approximation. This essay describes how this domain of classical predictability emerges from a quantum theory of the universe's state and dynamics that is characterized generally by indeterminacy and distributed probabilities.

12. *Living in a Superposition*

This essay considers a model quantum universe consisting of a very large box containing a screen with two slits and an observer (us) that can pass through the slits. The modern quantum mechanics of closed systems is applied to calculate the probabilities for alternative histories of how we as an observer would move through this universe and what we would see. After passing through the screen with the slits, the quantum state of the universe is a superposition of classically distinguishable histories. We would then be living in a superposition. Some frequently asked questions about such situations are answered using this model. The model's relationship to more realistic quantum cosmologies is briefly discussed.

13. *The Quantum Mechanical Arrows of Time*

The familiar textbook quantum mechanics of laboratory measurements incorporates a quantum mechanical arrow of time — the direction in time in which state vector reduction operates. This arrow is usually assumed to coincide with the direction of the thermodynamic arrow of the quasiclassical realm of everyday experience. But, in the more general context of cosmology, we seek an explanation of all observed arrows, and the relations between them, in terms of the conditions that specify our particular universe. This essay investigates quantum mechanical and thermodynamic arrows in a time-neutral formulation of quantum mechanics for a number of model cosmologies in fixed background spacetimes. We find that a general universe may not have well defined arrows of either kind. When arrows are emergent they need not point in the same direction over the whole of spacetime. Rather, they may be local, pointing in different directions in different spacetime regions. Local arrows can therefore be consistent with global time symmetry.

14. *Why Our Universe Is Comprehensible*

Einstein wrote memorably that "The eternally incomprehensible thing about the world is its comprehensibility". This essay argues that the universe must be comprehensible for information gathering and utilizing systems such as human observers to evolve and function.

Cosmology

15. *Quantum Cosmology: Problems for the 21st Century*

Two fundamental laws are needed for prediction in the universe: (1) a basic dynamical law and (2) a law for the quantum state of the universe. Quantum cosmology is the area of basic research concerned with the search for a theory of the cosmological quantum state. The issues involved in this search are presented in the form of eight problems.

16. *Anthropic Reasoning and Quantum Cosmology*

In quantum cosmology, anthropic reasoning uses probabilities that are conditioned on the existence of us as physical observing systems within the universe. This essay discusses how anthropic reasoning depends on the quantum state of the universe. Every prediction of our observations of the universe involves anthropic considerations.

17. *Are We Typical?*

(with Mark Srednicki) We have no observational evidence that as human observers we are typical of any class of objects in the universe, and there is no reason to believe that the laws of physics have to be such as to make our observations typical of other observations that might be made elsewhere in the universe. Indeed, an assumption that we are atypical is a testable hypothesis.

18. *Science in a Very Large Universe*

(with Mark Srednicki) Inflation can make the universe large enough that there is significant probability that we are replicated as physical systems at other locations in the universe. Predictions of our future observations then require an assumed probability distribution for our location among the possible ones (the xerographic distribution) in addition to the probabilities arising from the quantum state. It is the combination of fundamental theory plus the xerographic distribution that can be predictive and testable by further observations.

19. *The Observer Strikes Back*

(with Thomas Hertog) In the modern quantum mechanics of cosmology, observers are physical systems within the universe. They have no preferred role in the formulation of quantum theory, nor in its predictions of third person probabilities of what occurs. However, observers return to the importance of the prediction of first person probabilities for what we observe of the universe: What is most probable to be observed is not necessarily what is most probable to occur. This essay reviews the basic framework for the computation of first person probabilities in quantum cosmology starting with an analysis of very simple models. It is shown that anthropic selection is automatic in this framework. First person probabilities generally favor larger universes resulting from inflation where there are more places for us to be. In very large universes it is probable that our observational situation is duplicated elsewhere in spacetime. The calculation of first person probabilities then requires a specification of whether our particular observational situation is assumed to be typical of all the others. It is the combination of the model of the observational situation, including this typicality assumption, and the third person theory which is tested by observation.

20. *Quantum Multiverses*

A theory of the quantum universe may present us with an ensemble of possible situations, but only one of which is observed by us. Such ensembles constitute a quantum multiverse. Quantum multiverses occur naturally in quantum cosmology in the form of decoherent sets of alternative histories describing the evolution of our universe's spacetime geometry and matter content. Quantum multiverses are not a choice or an assumption, but are consequences (or not) of the theories of the universe's quantum state and dynamics. Quantum multiverses can provide mechanisms for the variation of constants in effective theories (like the cosmological constant). These conclusions are illustrated in simple cosmological models. In an FAQ, we argue that the quantum multiverses are scientific, real, testable, falsifiable, and similar to those in other areas of science even if they are not directly observable on arbitrarily large scales.

21. *Quantum Mechanics at the Planck Scale*

Copenhagen quantum mechanics requires a fixed, background, spacetime geometry and its associated causal structure. A generalization of the usual theory is needed at the Planck scale for quantum theories of gravity in which spacetime geometry is a quantum variable. The elements of generalized quantum theory are briefly reviewed and illustrated by generalizations of usual quantum theory that incorporate spacetime alternatives, gauge degrees of freedom, and histories that move forward and backward in time. A generalized quantum framework for cosmological spacetime geometry is sketched. This theory is in fully four-dimensional form and free from the need for a fixed causal structure. Usual quantum mechanics is recovered as an approximation to this more general framework that is appropriate in those situations where spacetime geometry behaves classically.

22. *The State of the Universe*

What is the quantum state of the universe? That is the central question of quantum cosmology. This essay describes the place of that quantum state in a final theory governing the regularities exhibited universally by all physical systems in the universe. It is possible that this final theory consists of two separate parts: (1) a dynamical theory such as superstring theory, and (2) a quantum state of the universe such as the no-boundary wave function. Both are necessary because prediction in quantum mechanics requires both a Hamiltonian and a state. The simplicity revealed by observations of the early universe gives hope that there is a simple, discoverable quantum state of the universe. It may be that, like the dynamical theory, the predictions of the quantum state for late-time, low-energy observations can be summarized by an effective cosmological theory. That should not obscure the need to provide a fundamental basis for such an effective theory which gives a unified explanation of its features and is applicable without restrictive assumptions. It could be that there is one principle that determines both the dynamical theory and the quantum state. That would be a truly unified final theory.

23. *The Impact of Cosmology on Quantum Mechanics*

When quantum mechanics was developed in the '20s of the last century, another revolution in physics was just about to start. It began with the discovery that the universe is expanding. For a long time, quantum mechanics and cosmology developed independently of one another. Yet, the very discovery of the expansion would eventually draw the two subjects together because it implied the big bang where quantum mechanics is important for cosmology and for understanding and predicting our observations of the universe today. Textbook (Copenhagen) formulations of quantum mechanics are inadequate for cosmology for at least four reasons: (1) They predict the outcomes of measurements made by observers. But in the very early universe, no measurements were being made and no observers were around to make them. (2) In Copenhagen quantum mechanics, observers were outside of the system being measured. But we are interested in a theory of the whole universe where everything, including observers, are inside. (3) Copenhagen quantum mechanics could not retrodict the past. But retrodicting the past to understand how the universe began is the main task of cosmology. (4) Copenhagen quantum mechanics requires a fixed classical spacetime geometry not least to give meaning to the time in the Schrödinger equation. But in the very early universe, spacetime is fluctuating quantum-mechanically (quantum gravity) and without definite value. A formulation of quantum mechanics general enough for cosmology was started by Everett and developed by many. That effort has given us a more general framework that is adequate for cosmology — decoherent (or consistent) histories quantum theory in the context of semiclassical quantum gravity. Copenhagen quantum theory is not an alternative to decoherent histories. It is rather, an approximation to this more general quantum framework that is appropriate for measurement situations. We discuss whether further generalization may still be required.

Recollections

24. *Murray Gell-Mann: A Short Appreciation*

25. *Working with Stephen Hawking*

At various times in his career, the author was privileged to collaborate with scientists who must be considered as among the greats in recent history. The last essays contain short personal appreciations for two of them: Murray Gell-Mann (24) and Stephen Hawking (25), that were written by the author for various occasions.

James B. Hartle
hartle@physics.ucsb.edu
November 2020

Acknowledgments

The author was encouraged to put this collection together in a book by a number of colleagues, but particular thanks should go to Lars Brink, Jenann Ismael, and Simon Saunders in this regard. The author thanks the National Science Foundation for many years of support of the research that led to these essays.

The author thanks his editor Lakshmi Narayanan and the rest of the World Scientific staff for their patience, understanding and competence in producing this collection. Thanks are also due to Kristy McCants who proofread all 622 pages of the page proofs.

CHAPTER 1

Computability and Physical Theories*

The familiar theories of physics share the feature that the application of the theory to make predictions in specific circumstances can be done by means of an algorithm. We propose a more precise formulation of this feature — one based on the issue of whether or not the physically measurable numbers predicted by the theory are computable in the mathematical sense. Applying this formulation to an approach to quantum theory of gravity, indications imply that there may exist no such algorithms in this case. Finally, we discuss the issue of whether the existence of an algorithm to implement a theory should be adopted as a criterion for acceptable physical theories.

"Can it be that there is ... something of use for unraveling the universe to be learned from the philosophy of computer design?"

J. A. Wheeler[1]

---

*With Robert Geroch (Enrico Fermi Institute, University of Chicago, Chicago, IL 60637, USA).
arXiv:1806.09237

1. Introduction

The process of making predictions in physics involves two steps. The first is the formulation of a theory covering a broad class of phenomena; the second is the application of that theory to the specific circumstances of interest. For making predictions about the motion of the planets of our solar system, for example, the first step is the formulation of Newton's theories of mechanics and gravitation; the second is finding the solution for a system of many dozens of ordinary differential equations. These two steps are rather different in character. The first involves the selection, from a wide variety of phenomena, of those that one judges to have a common cause, and then the formulation of a mechanism for how that cause is to operate. The test of a good theory is its simplicity and breadth. The second step involves the selection of a specific algorithm by which to extract the implications from the theory. The test of a good algorithm is its convenience.

What underlies the confidence that physics can be structured in this way? Perhaps it is primarily our experience with present-day physical theories. In both classical and quantum physics, the theory culminates in the formulation of differential equations, while the implementation of the theory consists of selecting some algorithm — of which there are many — for solving the equations. But not all theories of physics need to be of this type. We may someday in physics be confronted with a situation in which the line between the theory and its implementation is not so sharp — in which the activity of applying the theory is not so different from that of finding the theory in the first place. Specifically, we may reach a point at which there exist no algorithms whatsoever for applying a theory mechanically to specific circumstances. Indeed, there are suggestions that, with quantum gravity, we may already be at this point.

John Wheeler has stressed on several occasions[1,2] the relation between physical laws, computation and logic. Here, on the occasion of his 75th birthday, we discuss certain possible formulations of, and attitudes toward, these issues. Section 2 is a review of mathematics. We briefly summarize the notion of solvability of a mathematical problem by computer, and the closely related notion of a computable number. Section 3 involves physics. We propose, in parallel with the notion of a computable number in mathematics, that of a measurable number in a physical theory. The question of whether there exists an algorithm for implementing a theory may then be formulated more precisely as the question of whether the measurable numbers of the theory are computable. We argue that the measurable numbers are in fact computable in the familiar theories of physics, but there is no reason for this to be the case in order for a theory to have predictive power. Indeed, in some recent formulations of quantum gravity as a sum over histories, there are candidates for numbers that are measurable but not computable. In Sec. 4, we speculate the consequences for physics should there be an accepted theory having measurable numbers that are not computable. We suggest that such a theory should be no more unsettling in physics than the existence of well-posed problems in mathematics that are unsolvable by any algorithm.

But, in the spirit of many of John Wheeler's papers, our aim is more to raise issues and stimulate discussion than to state conclusions.

2. Computability

Computability theory deals with the issue of whether certain well-posed mathematical problems can be solved by a digital computer. The key result is a theorem to the effect that one particular problem cannot be solved in this way. From this theorem, one shows unsolvability of a variety of other problems, including that of finding rational approximations to certain real numbers. These are called the noncomputable numbers. We briefly summarize this subject below.[a]

We begin with a few ground rules. First, we wish to deal only with digital computers, and not with analog, for we are concerned for the moment only with mathematics, and not with physics. We are interested only in what is possible in principle with such computers, and not with such issues as efficiency or speed. Fix some finite character set, e.g. the standard one of 256 characters. By a string,

[a]For a more detailed and rigorous treatment of this material see, e.g. Ref. 3.

we mean a finite sequence of characters. Thus, while each string is in itself finite, there is no overall upper limit to the allowed lengths of strings. The computer is to have access to an infinite number of memory locations, each capable of holding a string. All these locations are to be blank initially, in order to avoid a situation in which "the answer to the problem" is already encoded, from the beginning, in the memory. The language of the computer is to consist of the specification of certain allowed strings, called the instructions in that language. Of course, not every possible string needs to be an instruction. A program in the language consists of a finite sequence of instructions. We may represent a program by placing its instructions one after another to form a single long string. Again, not every possible string needs to be a program. Finally, we demand of the language that it includes at least the following three instructions: (i) an input instruction, which temporarily interrupts the running of the program while the user reads in, for possible later use in the program, any string, (ii) a print instruction, which causes the computer to display to the user any string produced by the program, and (iii) a halt instruction, which causes the program to cease execution.

The above is an outline of the types of languages we have in mind. In practice, there are numerous specific languages in which computer programs are commonly written, e.g. Fortran, Basic, Pascal, Cobol, and various machine languages. One might think that whether a problem can be solved by a computer program will depend on the language in which that program is to be written. But, at least for these specific languages, this is not the case: A problem that can be solved in any one of these languages can be solved in any other. Consider, for example, Pascal and Basic. One could write a program in Basic that "imitates Pascal", i.e. that takes a sequence of Pascal instructions, determines for each what would be done in Pascal, and then does precisely that. Any problem solvable in Pascal is thus solvable in Basic, and similarly for other combinations of these standard languages.

Since, in the sense above, the language is irrelevant, let us fix once and for all a language, and call it "Fortran," with the understanding that this word can be regarded as standing for any of the others.

What problems can be "solved" by a Fortran program? The idea is to specify in some detail what one wants one's program to do, and then to determine whether or not there exists such a program.

Consider, as an example, Fermat's conjecture: that there exist no positive integers x, y, z, and n, with $n \geq 3$, such that $x^n + y^n = z^n$. Does there exist a Fortran program that prints "true" if Fermat's conjecture is true, or "false" if it is false, and then halts? The answer to the question, stated in this way, is yes. Indeed, consider the program, with just two instructions, that merely prints "true" and halts; and a second program that prints "false" and halts. Since Fermat's conjecture is either true or false, one of these is the desired program. So, there does indeed *exist* a program that does what we have asked. Similarly, any well-posed "yes-or-no" problem is solvable. One might be tempted to reformulate the question so that its answer requires the specification of *which* is the desired program. But it turns out to be difficult to do this in a precise way.

Clearly, the more interesting problems are those that are actually a sequence of problems. For example, does there exist a program having a single input instruction such that, when any integer n is read in, the program prints "prime" or "not prime" according to whether that integer is prime or not, and then halts? The answer, of course, is yes: Most computer users have written such a program. So, in this sense, the problem of determining whether or not a number is prime is solvable. Similarly, there exists a program that continues to print out the successive digits in the decimal expansion of π, never halting.

As a final example, we return to Fermat's conjecture, but this time with a different question. Does there exist a Fortran program having a single input instruction such that, when any integer $n \geq 3$ is read in, the program prints "yes" or "no" according to whether or not there are positive integers x, y, and z with $x^n + y^n = z^n$, and then halts? Since Fermat's conjecture is open, we certainly are not in possession of such a program. [It would not do, for example, to have one's program merely try various combinations of x, y, and z, for then, were n such that there are no suitable combinations, the program would never halt.] Indeed, it is even unclear whether or not there exists such a program.[b] There does, for example, if Fermat's conjecture is true (namely, the program that always prints "no" and halts), or if the conjecture is false for all but a finite number of n (namely, the program that compares the n read in with this finite number of candidates). But what if the n's for which the theorem fails are more "scattered"?

The key theorem in this subject displays an explicit example of a problem that cannot be solved by any program. It is convenient, for this discussion, to introduce the term *monoprogram* to mean a program having a single input instruction. The problem is to determine, given any monoprogram and the string we intend to read in for its input, whether that monoprogram will ultimately halt. In more detail, consider a program X, having just two input instructions and operating as follows. For the first input instruction, let there be the string representing a monoprogram P, and for the second, an arbitrary string S. Then X itself must always halt, having printed "halts" or "does not halt" according to whether or not the monoprogram P, with string S read in for *its* input, halts or not. [Should the first string read in to X not represent any monoprogram, then X can do anything.] It is instructive to attempt to write, say, for Fortran, such a program X. One first thinks of various simple arrangements by which P, with S read in, might halt, and so writes X to look for those arrangements. But one continually discovers ever more complicated ways by which P might halt, and so the program X becomes more and more complicated. Eventually, with no end in sight, one gives up in frustration.

There exists no such program X. The proof is simple, but subtle. Suppose, for contradiction, that there were such an X. Construct from it the following monoprogram P. When string S is read in for its input, P merely runs program X, reading in string S for *both* of X's inputs. Then P waits patiently for X to finish. If X would have printed "halts," then P goes into an infinite loop, never halting; while if X would have printed "does not halt," then P halts. Thus,

[b]For purposes of this illustration, we ignore the recent result that Fermat's conjecture holds for all but at most a finite member of n. See Ref. 4.

given program X one could, by merely changing a few instructions as indicated above, write this program P. Now run program P, but, for its input string, read in the string representing the program P itself. Will P so run halt or not halt? Either answer gives a contradiction. Suppose, for example, that P, with the string representing P read in, halts. This means, from the way P was constructed, that X, with the string representing P read in for both its inputs, will print "does not halt." But this in turn means, from the way X is supposed to operate, that P, with the string representing P read in, does not halt — a contradiction. Similarly for the alternative supposition. This contradiction establishes in turn that the desired program X does not exist. This result is called the *halting theorem*.

Note that the halting theorem asserts only that there is no general algorithm (i.e. no program X) for deciding whether or not programs halt. There need not be any specific program such that the issue of whether or not *that* program halts is undecidable. Indeed, in the formal sense, whether a specific program halts is not merely a "yes-or-no" question. Imagine that one's job consists of receiving from one's employer a monoprogram and a string, and then trying to decide whether or not that program, with that string read in, halts. The first few cases turn out to be quite easy. Then one day one receives a more difficult case; but, after a month's work, devises a most clever solution. This new method works for the next few cases. But eventually one will receive a case requiring a new, still more clever, solution. The halting theorem states in essence that this job will never become routine — that new and ever more clever solutions will always be required.

It is of interest to see the role played in the halting theorem by the language being used. Clearly, the particular language — provided only that it has the three basic instructions — is irrelevant. But it is crucial that the program X be written in the *same* language as the program P that X tests. Consider, for example, the simplest possible language — that having just the three instructions "input," "print," and "halt." Then the halting theorem in this case asserts that no program X written in this simple language can solve the halting problem for programs written in this same language. One sees directly that this is true, because programs in this simple language cannot "branch." Suppose, however, that one allows the program X to be written in Fortran. There does exist a Fortran program X that will solve the halting problem for a program P written in our simple language. [Program X merely checks whether or not program P includes a "halt" instruction.] But, by the halting theorem, no Fortran X will solve the halting problem for Fortran P's. Next, imagine a language that is essentially Fortran, but having an instruction of the form "Do, for I $= 1$ to ∞," with appropriate modifications to incorporate this into the rest of the language. One can write a program X in this "super-Fortran" that solves the halting problem for ordinary Fortran programs P. [Program X basically asks whether P has halted after I steps, for "I $= 1$ to ∞."] But again no such X will solve the halting problem for program P also written in "super-Fortran."

While one can imagine the action of the "super-Fortran" language, it is much more difficult to see how one might build a physical computer that actually runs programs in this language. To execute "Do, for I $= 1$ to ∞" in the traditional way would require performing an infinite number of steps in finite time. Indeed, to go one step further, it is difficult to see how any language that could actually be run on a physical computer could do more than Fortran can do. The idea that there is no such language is called Church's thesis.[c] Thus, under Church's thesis, there is a universal notion of an "algorithm" — of "the most that can be done on a physical computer" — and this notion is correctly captured by standard computer languages. The role of the thesis, then, is to give technical results of the form "... cannot be solved by a Fortran program" with the broader meaning "... cannot be solved by any algorithm."

Since we shall shortly be concerned with physics, it is convenient to re-express the above in terms of the objects with which physics deals: numbers.

A real number w is said to be *computable* if there exists a (Fortran) program with a single input instruction such that, when any

[c]For further discussion of the physical limitations on computing, see, e.g. Ref. 5.

positive rational number ϵ is read in, the program prints some rational number r with $|w - r| < \epsilon$ and then halts. Thus, a number is computable if there is an algorithm that generates rational approximations to the number. [Rationals are used in the definition because they can be "finitely displayed."] So, for example, the number π is computable. Indeed, it is easy to write out an appropriate program in detail, using, e.g. a power-series expansion for π. Similarly, $\pi^e \sin\sqrt{2}$ — and, in fact, any other number one can think of off-hand — is computable.

That there must exist noncomputable numbers is easily seen, for the cardinality of the set of Fortran programs (that of the integers) is smaller than the cardinality of the set of real numbers. In fact, one can give an "explicit example" of such a number. Fix once and for all an ordering (first, second, etc.) for pairs consisting of a monoprogram and a string. [One might, for example, first fix an ordering for the (finite) character set. Then apply dictionary ordering to those pairs for which the program and string combined have just one character, then to those having just two, etc.] Now consider the following number:

$$K = \sum_{n=1}^{\infty} \alpha_n 3^{-n} \tag{1}$$

where α_n is 1 if the nth monoprogram with string (in the ordering above) halts, and 0 if it does not. This sum, since it necessarily converges, certainly defines a number — in fact, one between 0 and 1. We now claim that this number K is not computable. Indeed, suppose that there did exist a program P to compute it, as above. Then we would immediately have a solution to the halting problem: If you wish to determine whether the nth monoprogram with string halts, merely read in to P the value $\epsilon = 3^{-n-1}$, and, from the rational approximation to K then printed by P, determine whether or not the appropriate term appears in the sum (1). It thus follows, in light of the halting theorem, that K is not computable. More generally, this strategy can always be applied to make the transition between "whether there exists a program to answer a sequence of yes-or-no questions" and "whether a certain number is computable."

It is curious that, while the number K above is not computable as we have defined that term, it is "determinable" in a somewhat weaker sense. We claim: There exists a program that successively prints out an increasing sequence of rational numbers, never halting, such that this sequence converges to K. Construct such a program as follows. Fix an integer, say 10. Let our program write out the first ten monoprograms and strings, run each of these programs for just ten steps, and then evaluate the sum of the first ten terms in (1), letting α_n be 1 if the corresponding program has halted in ten steps, 0 otherwise. This is the first rational approximation. The successive approximations are obtained by increasing the integer, say, to 100, then 1,000, etc., and repeating the process. There clearly results an increasing sequence of rational numbers converging to K. Note that this program does less than would be required to show K computable, for we do not know at any stage how close the "rational approximation" is to K. If, for example, the first monoprogram required 1,017 steps before halting, then our sequence of rationals would "jump upward" by about 1/3, very far along in the sequence. Note also that there can exist no program that prints out a *decreasing* sequence of rationals converging to K, for the existence of programs yielding both decreasing and increasing sequences would imply that K itself is computable.

We emphasize that computability is an attribute of the *number itself*, and not how that number is presented. Thus, for example, the number K given by Eq. (1) must be irrational, for every rational number is computable. As a further illustration of this point, consider the number

$$M = \sum_{n=1}^{\infty} 3^{-n}/L(n) \tag{2}$$

where $L(n)$ is the number of steps taken by the nth monoprogram and string before it halts (or "infinite," i.e. omit the corresponding term in (2), if the program does not halt). Is the number M computable? One might think of writing, along the following lines, a program to provide rational approximations to M. The program would first determine how many terms of the sum (2) would be required to approximate M to within the given allowed error ϵ, it would then determine the corresponding $L(n)$, and would finally take the sum of those terms. But this strategy must fail, for, by the halting theorem, there is no program that will determine the $L(n)$. Yet, despite this failure, the number M is actually computable! A program that shows this is the following. In order to approximate M to within error, say, $= 1/100$, it suffices to deal only with the first ten terms in the sum (2) and, even for these, only either to determine $L(n)$ or else ensure that it exceeds 1,000. So, given $\epsilon = 1/100$, our program merely runs the first ten monoprograms with strings for 1,000 steps each, letting $L(n)$ be infinite for any monoprogram that has not by then halted. Similarly for other ϵ. Thus, a number may appear to be noncomputable by virtue of the way it is presented, and yet turn out to be computable by appeal to some more sophisticated algorithm.

3. Measurability

Imagine a physical theory whose prediction, for the result of some particular experiment, is the noncomputable number K of Eq. (1). The theory has thus made a perfectly definite prediction, for K is merely some specific number. Imagine next that we wish to test this prediction. We do so in the usual way: Select some desired accuracy ϵ, carry out the experiment to within that accuracy, and then check to see whether the number that results from the experiment is within ϵ of K. But, in actually carrying out this test, we would be confronted by the following difficulty: There exists no algorithm to determine whether a given number is within ϵ of K — for this is the essence of K's being noncomputable. Of course, it is nonetheless possible to test the theory — and to do so to arbitrary accuracy. The point is only that the test cannot be carried out mechanically: Each new level of accuracy will require new ideas for the testing procedure. We regard this as the prototype of the situation in which a theory has no algorithm for its implementation. In this section, we formulate this issue more precisely. In the following section, we discuss possible consequences for physics should such a theory come to be accepted.

The key step is the identification of the experimentally determinable numbers that emerge from the theory — what we shall call the measurable numbers of the theory. This having been done, the issue of whether a theory can be mechanically applied can be stated thus: Are all measurable numbers of the theory computable? It turns out that "measurable number" must be formulated with some care, in particular to ensure that we thus distinguish only the final physical predictions of the theory, not numbers from intermediate calculations. As a consequence, this term must be given a meaning slightly more restrictive than its common one.

Regard number w as measurable if there exists a finite set of instructions for performing an experiment such that a technician, given an abundance of unprepared raw materials and an allowed error ϵ, is able by following those instructions to perform the experiment, yielding ultimately a rational number within ϵ of w. "Measurable" is analogous to, although of course much less precise than, "computable." The technician is analogous to the computer, the instructions to the computer program, the "abundance of unprepared raw materials" to the infinite number of memory locations, initially blank. Indeed, one can think of the measurable numbers as those that are "computable" using an analog, rather than a digital, computer.[d] A few examples will illustrate what we have in mind.

The number π is measurable. Suitable instructions for this case might direct that a metal disk be turned on a lathe, that measurements of its circumference and diameter be taken, and that their ratio be computed. Of course, the care with which the disk is to be constructed, and with which the measurements are to be taken, would be dictated by the allowed error ϵ.

The number $e^{1/2}$ is measurable. Suitable instructions might direct that a projectile be released in a viscous fluid, that measurements be taken of the distance d_1 that the projectile has traveled when it slows to half its initial speed as well as the ultimate distance traveled d_2, and that the combination $d_2/(d_2 - d_1)$ be computed. Why

[d]For other connections between digital computers and the natural "analog" world, see Ref. 6. The thesis raised in these papers is that limitations on the power of digital computers are universally applicable to their analog counterparts and so can be used to place limitations on the behavior of physical systems. We are not directly concerned with this issue here.

would not a similar argument show that e^K, where K is the number given by (1), is measurable? It is only necessary to replace in the instructions "half" by "a fraction K." But these modified instructions would not be regarded as acceptable, for, since K is not computable, it would not be possible for the technician to perform the experiment by merely "following instructions."

The ratio of the energies of two given levels in the helium atom is measurable. Suitable instructions might direct that there be constructed helium atoms, and that the transition energies be measured with appropriate care.

The fine structure constant is measurable. Suitable instructions[e] might direct that there be measured the wavelengths of the photons emerging from positronium decay as well as that of the lowest energy photon that will ionize hydrogen in its ground state, and that the square root of twice the ratio of these wavelengths be computed.

The meaning of "unprepared raw materials" is illustrated by the following example. Place two metal rods on a table: Do the obvious instructions show the ratio of their lengths to be measurable? We contend not, for two carefully calibrated rods are hardly "unprepared." Were we to allow such examples, then clearly any number would be measurable. Measurable numbers are to be those that flow from the operation of physical laws, not from mere initial conditions. The analogous requirement, in the computable case, is that all memory locations initially should be blank. Otherwise, any number would be computable. The lathe used to turn the disk in the first example above does not constitute "prepared raw materials," for one could build a lathe from scratch. The Earth–Moon mass ratio would not be regarded as measurable, since the Earth–Moon system is "prepared." But we would regard as measurable the proton–electron mass ratio.

All numbers in physics are essentially dimensionless, for, even for example, an elapsed time reported in seconds merely gives the ratio between that elapsed time and the elapsed time that is the standard second. But, for the issue of measurability, the choice of standard is important. If, for example, the standard second is 1/86,400 of

---

[e]See Ref. 7 for this and other methods of measuring α.

the mean solar day, then instructions reporting times in these units would not be acceptable, for the experiment involves "prepared" raw materials. But this problem does not arise if the standard second is the time of 9, 192, 631 and 700 vibrations of a certain hyperfine transition in Cs133. It is curious that most such "unprepared" standards arise via quantum mechanics.

It is intended that "measurable" be an attribute of the number itself, and not of how that number is presented. Thus, a given number is either measurable or it is not, the instructions merely serving (as does the computer program in the computable case) to show whether it is measurable. Consider, for example, instructions that direct the technician to work with an unstable system, so that the number finally reported varies wildly from one run to the next. There is no number that these instructions show to be measurable. This is analogous to a Fortran program that prints rationals that vary wildly from one ϵ to the next. There is no number that this program shows to be computable.

The notion "measurable" involves a mix of natural phenomena and the theory by which we describe those phenomena. Imagine that one had access to experiments in the physical world, but lacked any physical theory whatsoever. Then *no* number w could be shown to be measurable, for, to demonstrate experimentally that a given instruction set shows w measurable would require repeating the experiment an infinite number of times, for a succession of ϵ's approaching zero. One could not even demonstrate that a given instruction set shows measurability of any number at all, for it could turn out that, as ϵ is made smaller, the resulting sequence of experimentally determined rationals simply fails to converge. It is only a theory that can guarantee otherwise. The situation is analogous to that of trying to demonstrate that a given Fortran program shows some number to be computable. There is no general algorithm for deciding this. In particular, it would not do merely to run the program for a few selected values of ϵ.

So, "The number w is measurable" is basically a statement about the theory within which we are working, and only indirectly — only to the extent that the theory correctly describes the physical world —

a statement about natural phenomena. Thus, in the example of the measurement of π, the theory is that of classical continuous elastic solids in Euclidean space. Classical continuous elastic solids are required to ensure that arbitrarily accurate measurements can, in principle, be made on the disk (without interference, for example, from quantum mechanics or atomic structure). Euclidean geometry ensures that the ratio obtained will be independent of disk size, and also that π is actually the number to which the approximations converge. In the example of $e^{1/2}$, the theory is that of viscous fluids, and in particular, its prediction of exponential decay. In the example of the helium atom, the theory is quantum mechanics; in that of the fine-structure constant, quantum electrodynamics.

Every computable number is measurable. This is easy to see: Let the instructions direct that the raw materials be assembled into a computer, and that a certain Fortran program — one specified in the instructions — be run on that computer. That is, every digital computer is, at its heart, an analog computer.

We now ask whether, conversely, every measurable number is computable — or, in more detail, whether current physical theories are such that their measurable numbers are computable. This question must be asked with care. Consider, for example, the fine-structure constant α^{-1}. This number is measurable in quantum electrodynamics, as discussed above. Is α^{-1} computable? The problem is that in quantum electrodynamics the actual value of α^{-1} is not specified by the theory, but rather is left to be determined by experiment. There is no "number," mathematically specified, whose computability can be investigated. There is no meaning to be attached to the statement "α^{-1} is computable." [Of course, were quantum electrodynamics reformulated with α^{-1} specified, say, to be 137.036 (exactly!), then the computability of the fine-structure constant would be meaningful.] We are thus led to divide the measurable numbers into two classes: those assigned specific values by the theory, and those not. For example, π, as measured by the disk, and $e^{1/2}$, as measured by the viscous fluid, are clearly assigned values by the theories. The example of the helium atom is more subtle. If the theory is that of a helium atom with infinite mass, point-like nucleus and no relativistic effects, then the ratio of energy levels is assigned a specific value by that theory. But a finite mass brings in the electron-nucleon mass ratio, and relativistic effects of the fine-structure constant, neither of which is assigned a specific value in conventional theory. Of course, specific values may be assigned in some future theory.

We suggest that the familiar theories of physics are such that all the measurable numbers specified by these theories are in fact computable. The following example, representing a "typical" theory whose mechanism is a differential equation, will support this suggestion. Let the state of the physical system be characterized by a single variable x, evolving with time according to the equation.[f]

$$\frac{dx}{dt} = 5(x^2 - t^2)\sin(xt). \tag{3}$$

Let it be the case that a technician, given any allowed error ϵ, is able to (i) set the system initially in state x within ϵ of any given rational number, and (ii) wait for time t within ϵ of any given rational number, and then measure to within ϵ the final state of the system. Then, under these ground rules, the measurable numbers of the theory are clearly the following: all numbers of the form $x(t_0)$, where t_0 is any computable number and $x(t)$ is the solution of (3) with initial condition of any computable x_0. Are all of these numbers computable? This is a purely mathematical question, involving the character of the solutions of (3). It turns out that all these numbers are indeed computable, as one sees by noting that one could integrate (3) numerically on a computer. Thus, at least in this model theory, the measurable numbers are computable. Note that, were the coefficient 5 on the right in (3) replaced by a constant λ not further specified, then the measurable numbers would not be specified by the theory; and were it replaced by the noncomputable number of (1), then the measurable numbers would not be computable. In a more realistic treatment, the issue of what the technician can set and what he can measure would be determined by the interactions in

[f]The right side of (3) could be any smooth "explicit" function — or even some not so explicit. We here merely select one as an example in order to avoid the issue of what is the class of allowed functions.

the theory. We note, finally, that a similar argument applies to field theories governed by hyperbolic equations,[g] although the technician must for such examples be given considerable power to control the initial conditions.

Thus, we argue, that the conventional theories of physics do have the property that all measurable numbers specified by these theories are computable. Here, then, is at least one sense in which it is true that these familiar theories can be applied by means of algorithms.

There is a recent example[9] in which this issue comes very much to the fore. Consider one version of quantum gravity for closed cosmologies, formulated as a sum over histories. An observable in the theory is a functional, $A[\mathcal{G}]$, of geometries $\mathcal{G}$ on compact four-manifolds. The expectation value of such an observable in the analog of the ground state is given, formally, by

$$\langle A \rangle = \frac{\sum A[\mathcal{G}] \exp\left(-\mathcal{I}[\mathcal{G}]\right)}{\sum \exp\left(-I[\mathcal{G}]\right)} \tag{4}$$

where $I[\mathcal{G}]$ is the Euclidean gravitational action. The sums in Eq. (4) are over all compact four-geometries — including in particular all possible topologies for the four-manifolds. Now fix a "reasonable" observable A. Is the number given by Eq. (4) measurable? Is it computable?

As we have seen, even for simple theories, the issue of whether a given number is measurable can be subtle. But this theory is anything but simple, for the theory purports to provide the quantum description of the entire Universe. What instructions would allow a technician to "measure" $\langle A \rangle$? Our understanding of the physics of quantum gravity is not yet to the point at which we can answer this question. Nonetheless, one expects that there will be some measurable numbers in the final theory, and such expectation values would seem to be good candidates.

Is $\langle A \rangle$ computable? Unfortunately, the mathematics of quantum gravity has not yet been developed to the point that we have a complete mathematical formulation of what Eq. (4) is to mean. But one

---

[g]For examples of lack of computability in solutions of differential equations, see Ref. 8.

possible meaning that has been proposed involves the use of Regge calculus. Fix an integer n. Consider a simplicial four-manifold with exactly n vertices. That is, take n points as vertices, and specify which pairs of vertices are to be connected by an edge, which triples of adjoining edges are to be connected by a face, which quadruples of adjoining faces are to be connected by a tetrahedron, and which quintuplets of adjoining tetrahedra are to be connected by a four-simplex — all such that the resulting space is a topological four-manifold. Next, specify numerical values for the lengths of all the edges of this simplicial four-manifold. The result is a simplicial four-geometry $\mathcal{G}$. We can now evaluate $A[\mathcal{G}]$ and $I[\mathcal{G}]$, and thus one term in each of the sums in Eq. (4). To evaluate the sums themselves, we proceed as follows: first keep the simplicial four-manifold fixed and integrate over the values of the edge lengths; and then sum the result over all simplicial four-manifolds with exactly n vertices. There results the n-vertex simplicial approximation $\langle A \rangle$, a number we denote as $\langle A \rangle_n$. Finally, take the limit of $\langle A \rangle_n$, as n goes to infinity — assuming this limit exists! The result is our candidate for the expectation value $\langle A \rangle$.

This particular meaning for (4) has the following curious feature. Imagine an attempt to evaluate the $\langle A \rangle_n$ by computer. One would select and read in some value for the integer n. The computer would then consider n vertices, and try all combinations of which vertices are to be connected by edges, which edges by faces, etc. From the resulting list of simplicial complexes, the computer would then eliminate all those that are not four-manifolds. Then, from the resulting list of simplicial four-manifolds, the computer would eliminate all duplications — cases in which there are included more than one simplicial representation of a single four-manifold. The computer would then perform, for each simplicial four-manifold in this reduced list, the integrations over edge lengths. Finally, the computer would take the sums over the four-manifolds. The result would be reported as $\langle A \rangle_n$. The curious feature is that there exists no computer that can carry out this task! The key problem is with the elimination of duplications. It is known that the issue of whether two simplicial four-manifolds are topologically identical or not is undecidable.

The theorem, in more detail, is the following.[10] Let simplicial four-manifolds be specified by giving the number n of vertices, for which, vertices are to be connected by edges, edges by faces, etc. This information could be encoded, for example, into a string. Then, there exists no Fortran program that will accept as inputs two such strings, and ultimately print "same manifold" or "different manifolds" according to whether the four-manifolds represented by the strings are topologically the same or different, and then halt. Further, even the elimination of those simplicial complexes that are not manifolds may be problematic. It appears likely that the question of whether a simplicial complex is a manifold or not, is decidable in dimension four, but undecidable in higher dimensions.[h]

All this is not to say that quantum gravity will admit measurable numbers that are not computable. The mere fact that one particular strategy for writing a program to compute $\langle A\rangle_n$ fails does not mean that the $\langle A\rangle_n$ are noncomputable. We have seen an example in (2). Further, even if the $\langle A\rangle_n$ should turn out to be noncomputable, this does not guarantee that their limit, $\langle A\rangle$, is also noncomputable. It is easy to write down examples of sequences of noncomputable numbers having computable limits. [The noncomputable numbers are dense in the reals.] It could also turn out that the meaning ultimately assigned to (4) is different from that indicated here. For example, one might simply adopt the rule that duplications of four-manifolds are not to be excluded in the sums; or, alternatively, the rule that the sums are not to be over different four-manifolds. It could turn out that the numbers ultimately measurable in the theory are quite different from these expectation values. Or, finally, what is ultimately quantum gravity may bear little resemblance to a sum over simplicial histories.

Yet quantum gravity does seem to be a serious candidate for a physical theory for whose application there is no algorithm.[i]

---

[h]For a review of these results, see Ref. 10.

[i]For a discussion and other examples of undecidable problems in physics and mathematical physics, see Refs. 11 and 12.

4. Discussion

What must be the character of a physical theory in order that it be regarded as satisfactory? Surely, it must include some mechanism for making quantitative predictions — or producing, under specified experimental arrangements, quantities subject to experimental verification. Otherwise, we would hardly have a "theory." But this is by no means the only criterion. The familiar theories of physics possess, in addition to predictive power, such features as simplicity, generality, and elegance. Historically, these criteria have not always been incidental, for they have at times played a central role in the search for physical laws. General covariance is an example — one for which, remarkably, the "criterion" turned out to be empty!

We here raise the possibility of another type of criterion: that the theory be such that its predictions can be extracted not merely "in principle", possibly by ever higher levels of skill and sophistication, but also "in practice, mechanically." The criterion, in the present more precise version, is that all the measurable numbers specified by the theory be computable numbers. At issue is the status of this criterion.

First note that it may be difficult to decide whether or not a given theory satisfies this criterion. One mathematical formulation of the theory may provide no algorithm for implementing the theory, and yet another, equivalent, formulation does. The physical issue of whether a number is measurable, as well as the mathematical issue of whether it is computable, can be a subtle matter. Nevertheless, it should be far more straightforward to decide whether a theory satisfies this criterion than, for example, to decide whether a theory is "elegant." It is at least more a question of fact than of taste.

In present physical theories, the computability of the measurable numbers manifest, so this issue is rarely raised. But now the issue may be forced upon us, by quantum gravity. There are indications that this theory may have measurable numbers that are not computable, at least if its formulation as a sum over simplicial histories is a good guide. What, then, would be the consequences for physics should there be accepted a theory whose measurable numbers are not computable? Fix some experimental arrangement, and attempt to extract from such a theory its prediction for that experiment. To predict to within, say, 10%, one manipulates the mathematics of the theory for a while, arriving eventually at the predicted number. To predict to within 1%, it will be necessary to work much harder, and perhaps to invent a novel way to carry out certain mathematical steps. To predict to 0.1% would require still more new ideas. The process would be similar to that of trying to compute rational approximations to the noncomputable number given by (1). The theory certainly "makes definite predictions," in the sense that predictions are always in principle available. It is just that ever increasing degrees of sophistication would be necessary to extract those predictions. The prediction process would never become routine.

Clearly, this would not be a particularly desirable state of affairs. But it would seem to be merely an inconvenience — far from a disaster for physics.

The status of the other criteria for physical theories arose, not through general arguments, but rather through experience with specific examples. Whether physical theories will be required to possess some algorithms by which they can be implemented will presumably be settled by the same process.

Acknowledgments

Many colleagues gave helpful advice on a preliminary version of this paper. We are grateful in particular to B. DeWitt, J. Friedman, B. Horne, G. Horowitz, W. Kohn, A. Komar, G. Kreisel, D. Scalapino, R. Sorkin, M. Srednicki and S. Wolfram. This work was supported in part by the NSF.

References

1. J. A. Wheeler, *Int. J. Theor. Phys.* **21**, 557 (1982).
2. C. W. Misner, K. S. Thorne and J. A. Wheeler, *Gravitation* (Freeman, San Francisco, 1970).
3. A. Yasuhara, *Recursive Function Theory and Logic* (Academic Press, New York, 1971).
4. G. Faltings, *Invent. Math.* **73**, 349 (1983).

5. *Int. J. Theor. Phys.* **21**, Nos. 3–4 (1982).
6. S. Wolfram, *Phys. Rev. Lett.* **54**, 735 (1985); *Sci. Am.* **251**, 188 (1984).
7. B. N. Taylor, W. H. Parker and D. N. Langenberg, *Rev. Mod. Phys.* **41**, 375 (1969).
8. M. B. Pour-El and I. Richards, *Ann. Math. Logic* **17**, 61 (1979); *Adv. Math.* **39**, 215 (1981).
9. J. B. Hartle, to appear. [Note added later: This was a reference to J. B. Hartle, Simplicial minisuperspace I. General discussion, *J. Math. Phys.* **26**, 804–814 (1985), Section VI.]
10. W. Haken, in *Word Problems*, W. W. Boone, F. B. Cannonito and R. C. Lyndon, eds. (North-Holland, Amsterdam, 1973).
11. A. Komar, *Phys. Rev. B* **133**, 542 (1964).
12. G. Kreisel, *Synthese* **29**, 11 (1974).

CHAPTER 2

Excess Baggage*

Many advances in physics have in common that some idea which was previously accepted as fundamental, general and inescapable was subsequently seen to be consequent, special and dispensable. The idea was not truly a general feature of the world, but only *perceived* to be general because of our special place in the universe and the limited range of our experience. It was *excess baggage* which had to be jettisoned to reach a more general perspective. This article discusses excess baggage from the perspective of quantum cosmology which aims for a theory of the universe's quantum initial state. We seek to answer the question: "Which features of our current theoretical framework are fundamental and which reflect our special position in the universe or its special initial condition?"

(*continued overleaf*)

*This article appeared in *Elementary Particles and the Universe: Essays in Honor of Murray Gell-Mann*, ed. J. H. Schwarz (Cambridge University Press, Cambridge, 1991). The article has not been revised except to add an abstract, correct minor typos, update references which were "to be published" at the time, add two footnotes concerning terminology, and restore $\hbar$ to some equations.
arXiv:gr-qc/0508001

Past instances of cosmological excess baggage are reviewed, such as the idea that the Earth was at the center of the universe or that the second law of thermodynamics was fundamental. Examples of excess baggage in our current understanding are the notion that measurement is central to formulating quantum mechanics, a fundamental quantum mechanical arrow of time, and the idea that a preferred time is needed to formulate quantum theory. We speculate on candidates for future excess baggage.

1. Quantum Cosmology

It is an honor, of course, but also a pleasure for me to join in this celebration of Murray Gell-Mann's sixtieth birthday and to address such a distinguished audience. Murray was my teacher and, more recently, we have worked together in the search for a quantum framework within which to erect a fundamental description of the universe which would encompass all scales — from the microscopic scales of the elementary particle interactions to the most distant reaches of the realm of the galaxies — from the moment of the big bang to the most distant future that one can contemplate. Such a framework is needed if we accept, as we have every reason to, that at a basic level the laws of physics are quantum mechanical. Further, as I shall argue below, there are important features of our observations which require such a framework for their explanation. This application of quantum physics to the universe as a whole has come to be called the subject of quantum cosmology.

The assignment of the organizers was to speak on the topic "Where are our efforts leading?" I took this as an invitation to speculate, for I think that it is characteristic of the frontier areas of science that, while we may know what direction we are headed, we seldom know where we will wind up. Nevertheless, I shall not shrink from this task and endeavor, in the brief time available, to make a few remarks on the future of quantum cosmology. The point of view that I shall describe owes a great deal to my conversations with Murray.

One cannot contemplate the history of physics without becoming aware that many of its intellectual advances have in common that some idea which was previously accepted as fundamental, general and inescapable was subsequently seen to be consequent, special and dispensable. Further, this was often for the following reason: The idea was not truly a general feature of the world, but only *perceived* to be general because of our special place in the universe and the limited range of our experience. In fact, it arose from a true physical fact, but one which is a special situation in a yet more general theory. To quote Murray himself from a talk of several years ago: "In my field an important new idea ... almost always includes a negative statement, that some previously accepted principle is unnecessary and can be dispensed with. Some earlier correct idea ... was accompanied by unnecessary intellectual baggage and it is now necessary to jettison that baggage".[1]

In cosmology, it is not difficult to cite previous examples of such transitions.[2] The transition from Ptolemaic to Copernican astronomy is certainly one. The centrality of the earth was a basic assumption of Ptolemaic cosmology. After Copernicus, the earth was seen not to be fundamentally central but rather, one planet among others in the solar system. The earth was in fact distinguished, not by a law of nature, but rather by our own position as observers of the heavens. The idea of the central earth was excess baggage.

The laws of geometry of physical space in accessible regions obey the Euclidean laws typified by the Pythagorean theorem on right triangles. Laws of physics prior to 1915, for example those governing the propagation of light, incorporated Euclidean geometry as a fundamental assumption. After Einstein's 1915 general theory of relativity, we see Euclidean geometry not as fundamental, but rather as one possibility among many others. In Einstein's theory, the geometry in the neighborhood of any body with mass is non-Euclidean and this curvature gives a profound geometrical explanation for the phenomena we call gravity. On the very largest scales of cosmology, geometry is significantly curved and it is the dynamics of this geometry which describes the evolution of the universe. Euclidean geometry is the norm for us, not because it is fundamental, but only because our observations are mostly local and because we happen to be living far from objects like black holes or epochs like the big bang. The idea of a fixed geometry was excess baggage.

In each of these examples, there was a feature of the current theoretical framework which was perceived as fundamental but which in truth was a consequence of our particular position and our particular time in the universe in a more general theoretical framework. Our description was too special. There was excess baggage which had to be discarded to reach a more general and successful viewpoint. Thus, in our effort to predict the future of quantum cosmology, the question naturally arises: Which features of our *current* theoretical framework reflect our special position in the universe[a] and which are fundamental? Which are excess baggage?

We live at a special position in the universe, not so much in place, as in time. We are late, living some ten billion years after the big bang, a time when many interesting possibilities for physics could be realized which are not easily accessible now. Moreover, we live in a special universe. Ours is a universe which is fairly smooth and simple on the largest scales, and the evidence of the observations is that if we look earlier in time it is smoother and simpler yet. There are simple initial conditions which are only one very special possibility out of many we could imagine.

The question I posed above can therefore be generalized: Which features of our current theoretical framework are fundamental and which reflect our special position in the universe *or* our special initial conditions.[b] Are there natural candidates for elements of the framework which might be generalized? Is there excess baggage? This is the question that I would like to address today for quantum cosmology.

2. Quantum Mechanics

First, let us review the current quantum mechanical framework today as it was developed in the 20s', 30s' and 40s', and as it appears in most textbooks today.

[a]In a currently (2005) popular terminology, the regularities associated with these features would be called *emergent*.

[b]The current (2005) usage is *initial condition* (singular) since there is only one wave function of the universe.

It is familiar enough that there is no certainty in this world and that therefore we must deal in probabilities. When probabilities are sufficiently good, we act. For example: (1) It wasn't certain that my plane to Los Angeles wouldn't be blown up, but I came because I thought that the probability was sufficiently low, (2) experimentalists can't be certain that the results of their measurements are not in error, but they publish because they estimate the probability of this as low, and so on. It was the vision of classical physics that fundamentally the world *was* certain and that the use of probability was due to lack of precision. If one looked carefully enough to start with, one could with certainty predict the future. Since the discovery of quantum mechanics some sixty years ago, we have known that this vision is only an approximation and that probabilities are inevitable and fundamental.

Indeed, the transition from classical to quantum physics contains a good example of excess baggage. In this case it was classical determinism. Classical evolution was but one possibility out of many although it is by far the most probable evolution in the great majority of situations we were used to.

Quantum mechanics constructs probabilities according to characteristic rules. As a simple example let us consider the quantum mechanics of a slowly moving particle such as an electron in a solid. Its path in spacetime is called its history.

In classical mechanics, the time history of the electron is a definite path determined by the electron's initial conditions and its equation of motion. Thus, given sufficiently precise initial conditions, a later observation of position can yield only a single, certain, predictable result. In quantum mechanics, given the most precise possible initial conditions, all paths are possible and all possible results of the observation may occur. There is only a probability for any one of them. The probability of an observation of a position at time t yielding the value x, for example, is constructed as follows.

A complex number, called the amplitude, is assigned to each path — this number has the form $\exp[iS(\text{path})/\hbar]$ where the action S summarizes the inertial, the inertial properties of the electron together with its interactions, and $\hbar$ is Planck's constant. The

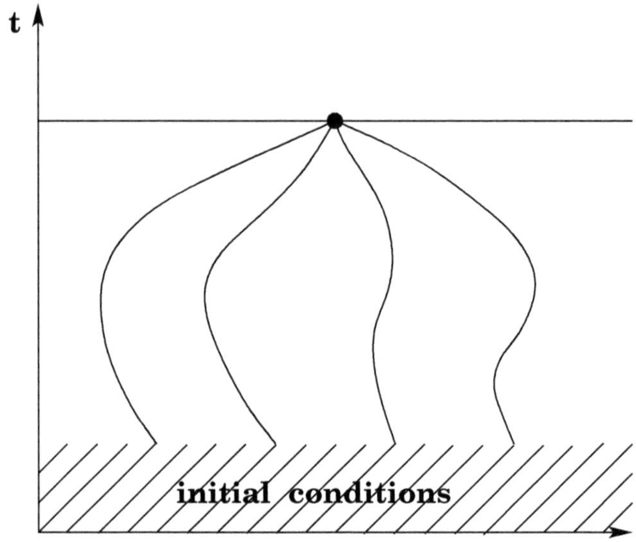

Fig. 1. In quantum mechanics, the probability for a particle to be observed at position x at time t is the absolute square of the amplitude to arrive there. In Feynman's sum-over-histories formulation, that amplitude is the sum over all paths which are consistent with the initial conditions and arrive at x and t weighted by the amplitude for each path $\exp[iS(\text{path})/\hbar]$. In the Hamiltonian formulation of quantum mechanics, this sum is called the wave function.

amplitude to arrive at x, and t is the sum of the path amplitudes over all paths which are consistent with the initial conditions and which end at x and t (Fig. 1).

$$\begin{pmatrix} \text{amplitude to} \\ \text{arrive at position } x \\ \text{at time } t \end{pmatrix} = \sum_{\text{paths}} \exp[iS(\text{path})/\hbar]\,. \tag{1}$$

The sum here is over all paths consistent with the initial conditions and ending at position x and time t. For example, if the position of the particle were actually measured at a previous time but not in between, one would sum over all paths which connect that position with x. The probability to arrive at x and t is the absolute square of this amplitude. In essence, this is Feynman's sum-over-histories formulation of the rules of quantum mechanics.

There is another, older, way of stating these rules called the Hamiltonian formulation of quantum mechanics. Here, the amplitude for observing the electron at x at time t is called the wave function $\psi(x,t)$. If, as above, we can construct the amplitude $\psi(x,t)$ at one time, we can also employ this construction at all other times. The time history of the wave function gives a kind of running summary of the probability to find the electron at x. The wave function is thus the closest analog to the classical notion of "state of a system at a moment of time". The wave function obeys a differential equation called the Schrödinger equation

$$i\hbar\frac{\partial\psi}{\partial t} = H\psi \tag{2}$$

where H is an operator, called the Hamiltonian, whose form can be derived from the action. It summarizes the dynamics in an equivalent way. Thus, although the electron's position does not evolve according to a deterministic rule, its wave function does.

Feynman's sum-over-histories formulation of quantum mechanics is fully equivalent to the older Hamiltonian formulation for this case of slowly moving particles. One can calculate the wave function from (1) for two nearby times and demonstrate that it satisfies the Schrödinger equation (2). One can use formulation of quantum mechanics in this and many other situations as well.

More generally than the probability of an observation of a position at one time, the probabilities of time *sequences* of observations are of interest. Just such a sequence is needed, for example, to check whether or not the electron is moving along a classical path between the initial conditions and time t. The amplitude for positive answers to checks of whether the electron is located inside a sequence of position intervals is the sum of the amplitude for a path over all paths which thread these intervals (Fig. 2). The joint probability for the sequence of positive answers is the square of this amplitude.

This means that the probability of finding the particle at x and t depends not only on the initial conditions, but also on what measurements were carried out between the initial conditions and t. Even if one does not know the results of these measurements, the probabilities at t will still depend on whether or not they were carried

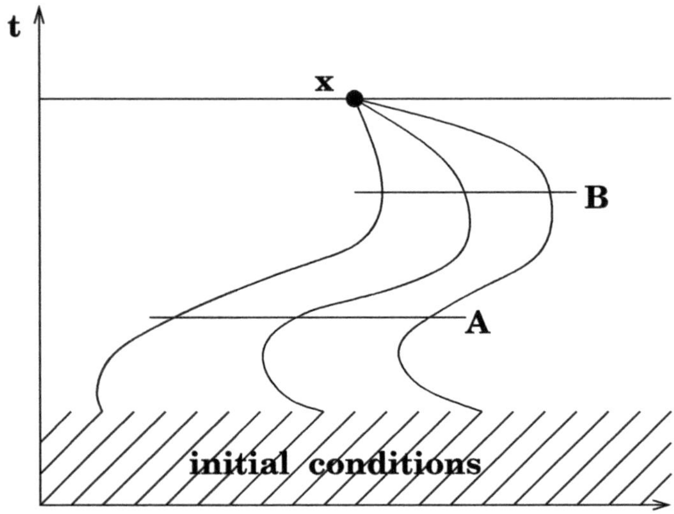

Fig. 2. Checks may be carried out as to whether the particle passed through position intervals $A, B, \ldots$ at times in between the initial conditions and t. The amplitude to arrive at position x and time t with a sequence of positive answers to these checks is the sum over all paths which meet the initial conditions and arrive at x and t having threaded the intervals $A, B, \ldots$. The probability to arrive at x and t will depend on whether these checks are carried out *even if we know nothing of the results.*

out. Thus, there are in quantum mechanics different rules for evolution depending on whether measurements occurred or not. In the Hamiltonian formulation this means that the wave function evolves by the Schrödinger equation only in between measurements. At a measurement it evolves by a different rule — the notorious "reduction of the wave packet".

In some circumstances it makes no difference to the probabilities whether prior measurements were carried out or not. For example, the classical limit of quantum mechanics occurs when the initial conditions are such that only a single path to x — the classical one — contributes to the sum over histories. Then, as long as sufficiently crude measurements are considered, it makes no difference to the probability of x and t whether they are made or not. The electron follows the classical evolution. In such cases the classical history is said to "decohere".

3. From Bohr, To Everett, To Post-Everett

The framework of quantum mechanics described in the previous section was the starting point for the "Copenhagen" interpretations of this subject. An idea characteristic of the Copenhagen interpretations was that there was something external to the framework of wave function and Schrödinger equation which was necessary to interpret the theory. Various expositors put this in different ways: Bohr[3] spoke of alternative descriptions in terms of classical language. Landau and Lifshitz[4] emphasized preferred classical observables. Heisenberg and others[5] stressed the importance of an external observer for whom the wave function was the most complete summary possible of information about an observed system. All singled out the measurement process for a special role in the theory. In various ways, these authors were taking as fundamental the manifest existence of the classical world that we see all about us. That is, they were taking as fundamental the existence of objects which have histories whose probabilities obey the rules of classical probability theory and, except for the occasional intervention of the quantum world as in a measurement, obey deterministic classical equations of motion. This existence of a classical world, however it was phrased, was an important part of the Copenhagen interpretations for it was the contact with the classical world which mandated the "reduction of the wave packet".

The Copenhagen pictures do not permit the application of quantum mechanics to the universe as a whole. In the Copenhagen interpretations the universe is always divided into two parts: To one part, quantum mechanical rules apply. To the other part, classical rules apply.[6]

It was Everett who, in 1957, first suggested how to generalize the Copenhagen framework so as to apply it to cosmology.[7] His idea (independently worked out by Murray in 1963) was to take quantum mechanics seriously and apply it to the universe as a whole. He started with the idea that there is one wave function for the universe that is always evolving by the Schrödinger equation and never reduced. Any observer would be part of the system described by quantum mechanics, not separate from it. Everett showed that, if the universe contains an observer, then its activities — measuring, recording, calculating probabilities, etc. — could be described in this generalized framework. Further, he showed how the Copenhagen structure followed from this generalization when the observer had a memory which behaved classically and the system under observation behaved quantum mechanically.

Yet, Everett's analysis was not complete. It did not explain the manifest existence of the classical world, much less the existence of something as sophisticated as an observer with a classically behaving memory. Classical behavior, after all, is not generally exhibited by quantum systems. The subsequent, post-Everett, analysis of this question involves a synthesis of the ideas of many people. Out of many, I might mention in particular the works of Joos and Zeh,[8] Zurek,[9] Griffiths,[10] and later, Murray, Val Telegdi, and myself. It would take us too far to attempt to review the mechanisms by which the classical world arises, but we can identify the theoretical feature to which its origin can, for the most part, be traced.

A classical world cannot be a general feature of the quantum mechanics of the universe, for the number of states which imply classical behavior in any sense is but a poor fraction of the total number of states available to the universe. Classical behavior, of course, can be an approximate property of a *particular* state, such as for a particle in a wave packet whose center moves according to the classical equations of motion. But, the universe *is* in a particular state (or a particular statistical mixture of states). More exactly, particular quantum initial conditions must be posed to make any prediction in quantum cosmology. It is to the particular features of the initial conditions of the universe, therefore, that we trace the origin of today's classical world and to explore the possibility of such information gathering and utilizing systems for ourselves.

In retrospect, the Copenhagen idea that a classical world, or an observer together with the act of measurement must occupy a fundamentally distinguished place in quantum theory can be seen to be excess baggage. These ideas arose naturally, in part from our position in the late universe where there *are* classically behaving objects and even observers. In part, they arose from the necessary focus on laboratory experiments as the most direct probe of quantum phenomena where there *is* a clear distinction between the observer and the observed. However, these features of the world, while being true physical facts in these situations, are not fundamental. Quantum mechanically, the classical world and observers are but some possible systems out of many and measurements are but one possible interaction of such systems out of many. Both are unlikely to exist at all in the very early universe. They are, in a more general post-Everett framework, *approximate* features of the late universe arising from its particular quantum state. The classical reality to which we have become so attached by evolution is but an approximation in an entirely quantum mechanical world made possible by specific initial conditions.

The originators of the Copenhagen interpretation were correct; something beyond the wave function and the Schrödinger equation *is* needed to interpret quantum mechanics. But, that addition is not an external restriction of the domain to which the theory applies. It is the initial conditions of the universe specified within the quantum theory itself.

4. Arrows of Time

Heat flows from hot to cold bodies, never from cold to hot. This is the essence of the second law of thermodynamics — that entropy always increases. In the nineteenth century this law was thought to be strict and fundamental.[11] A direction of time was distinguished by this law of nature. After the success of the molecular theory of matter, with its *time symmetric* laws of dynamics, the increase in entropy was seen to be *approximate* — one possibility out of many others, although the overwhelmingly most probable possibility in most situations. A direction of time was distinguished, not fundamentally, but rather by our position in time relative to initial conditions of simplicity and our inability to follow molecular motion in all accuracy. The idea of a fundamental thermodynamic arrow of time was excess baggage.

Exchanging the thermodynamic arrow of time for simple initial conditions might seem to be exchanging one asymmetry for another. "Why," one could ask, "was the universe simple in the past and not in the future?" In fact, this is not a question. There is no way of specifying that direction in time, which is the past, except by calling it the direction in which the universe is simple. It's not an arrow of time which is fundamental, but rather the fact that the universe is simple at one end and not at the other.[12]

The Hamiltonian formulation of quantum mechanics also possesses a similar distinction between past and future. From the knowledge of the state of a system at one time *alone*, one can predict the future but one cannot, in general, retrodict the past.[13] This is an expression of causality in quantum theory. Indeed, quantum mechanics in general prohibits the construction of history. The two-slit experiment is a famous example (Fig. 3). Unless there was a measurement, it is not just that one cannot say for certain which slit the electron went through; it is meaningless even to assign a probability. In this asymmetry between past and future, the notion of "state" in quantum mechanics is different from that of classical physics from which both future and past can be extrapolated. In Hamiltonian

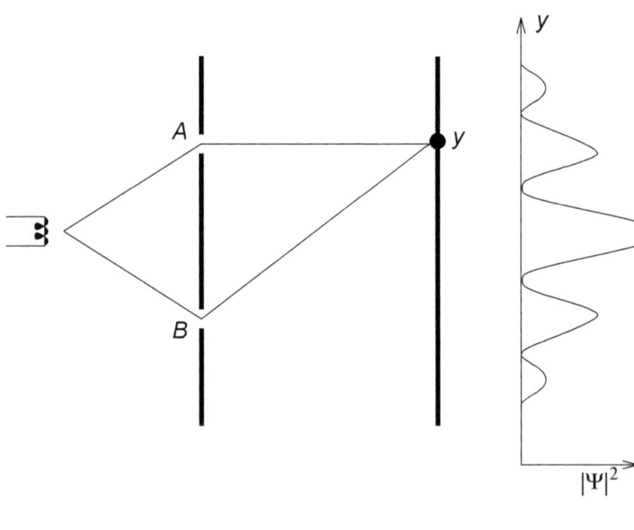

Fig. 3. The two-slit experiment. An electron gun on the left emits an electron traveling toward a screen with two slits, its progress in space recapitulating its evolution in time. From a knowledge of the wave function at the detecting screen alone, it is not possible to assign a probability to whether the electron went through the upper slit or the lower slit.

quantum mechanics, the past is the direction which is comprehensible by theory; the future is the direction which is generally predictable.

As long as the distinction between future and past in cosmology is fundamental, it is perhaps reasonable to formulate a quantum cosmology using a quantum mechanics which maintains this same distinction as fundamental. However, it would seem more natural if causality were an empirical conclusion about the universe rather than a prerequisite for formulating a theory of it. That is, it would seem more natural if causality were but one of many options available in quantum mechanics, but is the one appropriate for this particular universe. For example, one might imagine a universe where both initial *and* final conditions are set. The generalization of Hamiltonian quantum mechanics necessary to accommodate such situations is not difficult to find.[14] In fact, it is ready to hand in the sum-over-histories formulation described briefly in Sec. 2. The rules are the same, but both initial and final conditions are imposed on a contributing history. In the context of such a generalization, the arrow of time in quantum mechanics, and the associated notion of state[15] would not be fundamental. Rather, they would be features of the theory arising from the fact that those conditions which fix our particular universe are comprehensible at one of its ends, but not at the other.

5. Quantum Spacetime

Gravity governs the structure and evolution of the universe on the largest scales, both in space and time. To pose a quantum theory of cosmology, therefore, we need a quantum theory of gravity. The essence of the classical theory of gravity — Einstein's general relativity — is that gravity is curved spacetime. We thus need a quantum theory of space and time in which the geometry of spacetime will exhibit quantum fluctuations.

Finding a consistent, manageable quantum theory of gravity has been one of the goals of theoretical research for the last thirty years. The chief problem has been seen as finding a theory which puts out more predictions than there are input parameters. But, in cosmology there is also a conflict with the existing framework of quantum mechanics. These difficulties are called "the problem of time".[16]

Time plays a special and peculiar role in the familiar Hamiltonian formulation of quantum mechanics. All observations are assumed to be unambiguously characterized by a single moment of time and we calculate probabilities for "observations at one moment of time". Time is the only observable for which there are no interfering alternatives (as a measurement of momentum is an interfering alternative for a measurement of position). Time is the sole observable not represented in the theory as an operator but rather, it enters the Schrödinger equation as a parameter describing evolution.

If the geometry of spacetime is fixed, external and classical, it provides the preferred time of quantum mechanics. (More exactly, a fixed background spacetime provides a family of timelike directions equivalent in their quantum mechanics because of relativistic causality.) If the geometry of spacetime is quantum mechanical — fluctuating and without definite value — then it cannot supply a notion of time for quantum mechanics. There is thus a conflict between the familiar Hamiltonian quantum mechanics with a preferred time and a quantum theory of spacetime with no fixed notion of time. This is the "problem of time".

The usual response to this difficulty has been to keep Hamiltonian quantum mechanics but to give up on spacetime. There are several candidates for a replacement theory. Perhaps space and time are separated quantum-mechanically, with the beautiful synthesis of Minkowski and Einstein emerging only in the classical limit.[18] Perhaps, spacetime needs to be augmented by other, now hidden, variables which play the role of a preferred time in quantum mechanics.[19] Perhaps, spacetime is simply a totally inappropriate notion fundamentally and emerges only in the classical limit of some yet more subtle theory.

Each of these ideas could be right. I would like to suggest, however, that there is another alternative. This is the notion of a preferred time in quantum mechanics being another case of excess baggage, that the Hamiltonian formulation of quantum mechanics is simply not general enough to encompass a quantum theory of spacetime, and that the present formulation of quantum mechanics is but an approximation to a more general framework — an approximation that is appropriate because of our special position in the spacetime of our universe.

It is not difficult to identify the property of the late universe which would lead to the perception that there is a preferred time in quantum mechanics. Here, now, on all accessible scales, from the smallest ones of the most energetic accelerators to the largest ones of the furthest seeing telescopes, spacetime *is* classical. The preferred time of quantum mechanics reflects this true physical fact. However, it can only be an approximate fact if spacetime is quantum mechanical. Indeed, like all other aspects of the classical world, it must be a property of the particular quantum initial conditions of the universe. The present formulation of quantum mechanics thus may be only an approximation to a more general framework that is appropriate because of particular initial conditions which mandate classical spacetime in the late universe.

What is this more general quantum mechanics? Feynman's sum-over-histories supplies a natural candidate.[20–22] I shall now sketch why. We saw earlier that Feynman's approach was equivalent to the Hamiltonian one for particle quantum mechanics. It is also for field theory. A little thought about our example shows, however, that this equivalence arises from a special property of the histories — that they do not double back on themselves in time. Thus, the preferred time enters the sum-over-histories formulation as a restriction on the allowed histories.

In a quantum theory of spacetime, the histories are four-dimensional spacetime geometries $\mathcal{G}$ with matter fields $\phi(x)$ living upon them. The sums over histories defining quantum amplitudes therefore have the form:

$$\sum_{\mathcal{G}} \sum_{\phi(x)} \exp\left(i\, S[\mathcal{G}, \phi(x)]/\hbar\right) \tag{3}$$

where S is the action for gravity coupled to matter. The sums are restricted by the theory of initial conditions and by the observations whose probabilities one aims to compute.

The interior sum in (3)

$$\sum_{\phi(x)} \exp\left(i\, S[\mathcal{G}, \phi(x)]/\hbar\right) \tag{4}$$

defines a usual quantum field theory in a temporarily fixed background spacetime $\mathcal{G}$. The requirement that the fields be single-valued on spacetime is the analog of the requirement for particles that their paths do not double back in time. This quantum field theory thus possesses an equivalent Hamiltonian quantum mechanics with the preferred time directions being those of the background $\mathcal{G}$.

When the remaining sum over $\mathcal{G}$ is carried out, the equivalence with any Hamiltonian formulation disappears. There is no longer any fixed geometry, any external time, to define the preferred time of a Hamiltonian formulation. All that is summed over. There would be an *approximate* equivalence with a Hamiltonian formulation were the initial conditions to imply that for large scale questions in the late universe, only a single geometry $\hat{\mathcal{G}}$ (or an incoherent sum of geometries) contributes to the sum over geometries. For then,

$$\sum_{\mathcal{G}} \sum_{\phi(x)} \exp\left(i\, S[\mathcal{G}, \phi(x)]/\hbar\right) \approx \sum_{\phi(x)} \exp\left(i\, S[\hat{\mathcal{G}}, \phi(x)]/\hbar\right) \tag{5}$$

and the preferred time can be that of the geometry $\hat{\mathcal{G}}$. It would be in this way, that the familiar formulation of quantum mechanics emerges as an approximation to a more general sum-over-histories framework appropriate to specific initial conditions and our special position in the universe so far from the big bang and the centers of black holes. The preferred time of familiar quantum mechanics would then be another example of excess baggage.

In declaring the preferred time of familiar quantum mechanics as excess baggage, one is also giving up as fundamental, a number of treasured notions with which it is closely associated. These include the notion of causality, any notion of state at a moment of time, and any notion of unitary evolution of such states. Each of these ideas requires a background spacetime for its definition and in a quantum theory *of* spacetime there are none to be had. Such ideas, however, would have an approximate validity in the late universe on those scales where spacetime is classical. Many will regard discarding these notions as too radical a step, but I think it is no less radical than discarding spacetime — one of the most powerful organizing features of our experience.

6. For The Future

In this discussion of the excess baggage which has been discarded to arrive at a quantum mechanical framework general enough to apply to the universe as a whole, we have progressed from historical examples, through reasonable generalizations, to topics of current research and debate. Even in the last case, I was able to indicate to you candidates for the necessary generalizations. I would now like to turn to features of the theoretical framework where this is not the case; where there may be excess baggage but for which I have no clear candidates for the generalizations.

6.1. *"Initial" Conditions*

A candidate for excess baggage is the idea that there is anything "initial" about the conditions which are needed in addition to the laws of dynamics to make predictions in the universe. Our idea that conditions are initial comes from big bang cosmology, in which the universe has a history with a simple beginning ten and some billions of years ago. The global picture of spacetime is that there is a more or less uniform expansion everywhere from this moment. Perhaps the very large scale structure of spacetime is very different. Perhaps, as A. Linde has suggested,[23] there are many beginnings. Our local expanding universe might be just one inflationary bubble out of many others. Perhaps, therefore, the idea that conditions should reflect a simple state at one end of the universe is excess baggage arising from our confinement to one of these bubbles. Conditions in addition to dynamics would still be needed for prediction, but they might be of a very different character.[24]

6.2. *Conditions and Dynamics*

In the framework we have discussed so far, there are three kinds of information necessary for prediction in quantum cosmology: First, there are the laws of dynamics summarized by the action. Second, there is the specification of initial (or other) conditions. Third, there are our specific observations. Is it fundamental that the process of prediction be structured in this way?

Our focus on laboratory science, I believe, is the origin of the idea that there is a clean separation between dynamics and conditions. Dynamics are governed by the laws of nature; that which we are seeking, not that which we control. By contrast, the conditions represented in the experimental arrangement are up to us and therefore not part of the laws of nature. This ideal of control is not truly realized in practice in the laboratory, but in cosmology it is never realized. The initial conditions of the universe are most definitely not up to us, but must be specified in the theoretical framework in the same law-like way as the dynamics. A law of initial conditions has, therefore, become one of the central objectives of quantum cosmology.[26] However, is it not possible that the distinction between conditions and dynamics is excess baggage arising from the focus on laboratory science to which our limited resources have restricted us? Is it not possible that there are some more general principles in a more general framework which determine both the conditions and the dynamics? Is it not possible, as it were, that there is a principle which fixes both the state and the Hamiltonian?

Recent works by Hawking,[27] Coleman,[28] Giddings and Strominger[29] among others, gives some encouragement to this point of view. They show that in a quantum theory of spacetime which includes wormholes — small handles on spacetime connecting perhaps widely separated spacetime regions — that there is a closer connection between initial conditions and dynamics than had been previously thought. Specifically, the initial conditions determine the form of the Hamiltonian that we see at energies below those on which spacetime has quantum fluctuations. Indeed, they may determine a range of possibilities for the Hamiltonian, but only one of which is realized in our large scale universe. Dynamics and initial conditions are thereby entwined at a fundamental level.

6.3. *The Laws of Physics*

The last candidate for excess baggage that I want to discuss is the idea that the laws of physics, and in particular laws of initial conditions, are unique, apart from the universe, apart from the process of their construction, and apart from us.[30,31] Scientists, like mathematicians, proceed as though the truths of their subjects had an independent existence. We speak, for example, of "discovering" the laws of nature as though there were a single set of rules by which the universe is run with an actuality apart from this world they govern.

Most honestly, the laws of physics are properties of the collective data that we have about the universe. In the language of complexity theory, this data is compressible. There are computational algorithms by which the data can be stored in a shorter message. Take, for example, an observed history of motion of a system of classical particles. This history could be described to a given accuracy by giving the position and momentum of each particle at a suitably refined series of times. However, this message can be compressed to a statement of the positions and momenta at *one* time plus the equations of motion. Some of these initial values might be specified by observation but most, for a large system, will be specified by theory and statistically at that. For large systems, the result is a much shorter message. Further, we find for many different systems that the data can be compressed to the form of initial conditions plus the *same* equations of motion. It is the universal character of this extra information beyond the initial conditions which gives the equations of motion their law-like character. It is similar in quantum mechanics. Thus we have:

$$\begin{matrix}\text{All} \\ \text{Observations}\end{matrix} \rightarrow \begin{cases}\text{Some Observations} \\ + \\ \text{Laws of Dynamics} \\ + \\ \text{Laws of Initial Conditions}\end{cases} \tag{6}$$

The laws of physics, therefore, do not exist independently of the data. They are properties of our data much like "random" or "computable" is a property of a number, although probably (as we shall see below) in a less precise sense.[32]

This characterization of compression is incomplete. Any non-physicist knows that one can't just take a list of numbers and compress them in this way. One has to take a few courses in physics and mathematics to know what the symbols mean and how to compute with them. That is, to the list in (6) should be appended the algorithms for numerical computation for practical implementation of the theory, that part of mathematics which is needed to interpret the results, the rules of the language, etc. Yet more properly we should write:

$$\begin{array}{c}\text{All}\\ \text{Observations}\end{array} \rightarrow \left\{\begin{array}{l}\text{Some Observations}\\ +\\ \text{Laws of Dynamics}\\ +\\ \text{Laws of Initial Conditions}\\ +\\ \text{Algorithms for Calculation}\\ +\\ \text{Mathematics}\\ +\\ \text{Language}\\ +\\ \text{Culture}\\ +\\ \text{Specific Evolutionary History}\\ +\cdots\end{array}\right. \tag{7}$$

What confidence do we have that our data can always be compressed in this way? I have discussed the possibility that a clear division between initial conditions and equations of motion may be excess baggage. Bob Geroch and I have discussed similar questions for the division between algorithms and the rest of physical theory.[33]

However it is subdivided, what guarantee do we have that the resulting theory will be unique, independent of its process of construction, and independent of the specific data we have acquired? Very little, it seems. As we move down the list on the right we encounter more and more items which seem particular to our specific history as a collectivity of observers and to our specific data. It is an elementary observation that there are always many theories which will fit a given set of data, just as there are many curves which will interpolate between a finite set of points. Further, when the data is probabilistic as it is in quantum theory, there is always the possibility of arriving at different theories regardless of whatever criteria are used to distinguish them.

Beyond this, however, what confidence do we have that different groups of observers, with different histories, with growing but different sets of specific data, will, in the fullness of time, arrive at the same fundamental theory? I do not mean to suggest that the theories might vary because they are *consequences* of specific history, for that is not science. Rather the question is whether there are observationally indistinguishable theories which are different because of the processes of their discovery.

There may be specifically cosmological reasons to expect nonuniqueness in theories of initial conditions. A theory of initial conditions, for example, must be simple enough so that it can be stored within the universe. If the initial conditions amounted to some particular complex specification of the state of all matter, this would not be possible. The act of constructing theories may limit our ability to find them. The gravitational effect of moving a gram of matter on Sirius by one centimeter in an unknown direction is enough to ruin our ability to extrapolate classically the motions of a mole of gas particles beyond a few seconds into the past.[34] In view of this, there must be many theories of initial conditions rendered indistinguishable simply by our act of constructing them.

It would be interesting, I think, to have a framework which dispensed with the excess baggage that the laws of physics were separate from our observations of the universe, a framework in which the inductive process of constructing laws about the universe was described in it, and in which our theories were seen as but one possibility among many.

7. Conclusion

The assignment of the organizers was not just to speak on the subject "Where are our efforts leading?" They also wanted to know "In fifty or one hundred years time how do you think today's efforts will appear?" I have been bold enough to try their first question, I shall not be foolish enough to essay the second. I shall, however, offer a hope, and that is: In the future this might be seen as the time when scientists began to take seriously the idea that it was important to consider the universe as a whole and science as a unity, the time when they began to take seriously the search for a law of how the universe started, to work out its implications for science generally, and to discard the remainder of our excess baggage.

Acknowledgments

The author is grateful for the support of the John Simon Guggenheim Foundation and the National Science Foundation (NSF grant PHY 85-06686) during the preparation of this essay. He would also like to thank the Department of Applied Mathematics and Theoretical Physics, Cambridge University, for their hospitality while it was being written.

References

1. M. Gell-Mann, Talk given at the Symposium in memory of Arthur M. Sackler to celebrate the opening of the Arthur M. Sackler Gallery at the Smithsonian Institution, September 11, 1987.
2. A number of developments in physics are considered from this point of view by the philosopher C. A. Hooker in *Bell's Theorem, Quantum Theory, and Conceptions of the Universe*, ed. M. Kafatos (Kluwer, Boston, 1989).
3. See especially the essays "The Unity of Knowledge" and "Atoms and Human Knowledge" reprinted in N. Bohr, *Atomic Physics and Human Knowledge* (John Wiley, New York, 1958).
4. L. Landau and E. Lifshitz, *Quantum Mechanics* (Pergamon, London, 1958).

5. For clear statements of this point of view see F. London and E. Bauer, *La Théorie de L'observation en Mécanique Quantique* (Hermann, Paris, 1939); R. B. Peierls, in *Symposium on the Foundations of Modern Physics*, eds. P. Lahti and P. Mittelstaedt (World Scientific, Singapore, 1985).
6. Some have seen this division as motivation for restricting the domain of application of quantum mechanics and introducing new non-linear laws which apply in a broader domain. Examples are Wigner's discussion of consciousness [E. P. Wigner in *The Scientist Speculates*, ed. I. J. Good (Basic Books, New York, 1962)] and Penrose's discussion of gravitation [R. Penrose, in *Quantum Concepts of Space and Time*, eds. C. Isham and R. Penrose (Oxford University Press, Oxford, 1986)].
7. The original paper is H. Everett, *Rev. Mod. Phys.* **29**, 454 (1957). There is a useful collection of papers developing the Everett approach in B. DeWitt and N. Graham, *The Many Worlds Interpretation of Quantum Mechanics* (Princeton University Press, Princeton, 1973). A lucid exposition is in R. Geroch, *Noûs* **18**, 617 (1984). Even a casual inspection of these references reveals that there is considerable diversity among the various Everett points of view.
8. H. Zeh, *Found. of Phys.* **1**, 69 (1971) and especially E. Joos and H. D. Zeh, *Zeit. Phys. B* **59**, 223 (1985).
9. W. Zurek, *Phys. Rev. D* **24**, 1516 (1981); *Phys. Rev. D* **26**, 1862 (1982).
10. R. Griffiths, *J. Stat. Phys.* **36**, 219 (1984).
11. See, *e.g.* the discussion in A. Pais, *Subtle is The Lord* (Oxford University Press, Oxford, 1982).
12. Arrows of time in quantum cosmology are discussed in S. W. Hawking, *Phys. Rev. D* **32**, 2989 (1985); D. Page, *Phys. Rev. D* **32**, 2496 (1985); S. W. Hawking, *New Scient.*, July 9, p. 46 (1987).
13. The arrow of time in quantum mechanics is discussed in many places. See, *e.g.* Y. Aharonov, P. Bergmann and J. L. Lebovitz, *Phys. Rev. B* **134**, 1410, 1964 and R. Penrose in *General Relativity: An Einstein Centenary Survey*, eds. S. W. Hawking and W. Israel (Cambridge University Press, Cambridge, 1979).

14. M. Gell-Mann and J. B. Hartle, in *The Physical Origins of Time Asymmetry*, eds. J. Halliwell, J. Pérez-Mercader and W. Zurek (Cambridge University Press, Cambridge, 1994), gr-qc/9309012.
15. This was stressed by W. Unruh, Quantum measurement in new techniques and ideas in quantum measurement theory, *Ann. N. Y. Acad. Sci.* **480**, 242 (1986).
16. For reviews see J. A. Wheeler, in *Problemi dei fondamenti della fisica, Scuola internazionale di fisica "Enrico Fermi"*, Corso 52, ed. G. Toraldo di Francia (North Holland, Amsterdam, 1979); K. Kuchař, in *Quantum Gravity 2*, eds. C. Isham, R. Penrose and D. Sciama (Clarendon Press, Oxford, 1981), p. 329ff. For recent views see the articles and discussion in Ref. 17.
17. *Proceedings of the Osgood Hill Conference on the Conceptual Problems of Quantum Gravity*, eds. A. Ashtekar and J. Stachel (Birkhauser, Boston, 1989).
18. See, especially articles by K. Kuchař in Ref. 16 and A. Ashtekar in Ref. 17.
19. See, *e.g.* W. Unruh and R. Wald, *Phys. Rev. D* **40**, 2598 (1989).
20. C. Teitelboim, *Phys. Rev. D* **25**, 3159 (1983); *Phys. Rev. D* **28**, 297 (1983); *Phys. Rev. D* **28**, 310 (1983).
21. R. Sorkin, in *History of Modern Gauge Theories*, eds. M. Dresden and A. Rosenblum (Plenum Press, New York, 1989).
22. J. B. Hartle, in *Gravitation in Astrophysics*, eds. J. B. Hartle and B. Carter (Plenum Press, New York, 1987); *Phys. Rev. D* **37**, 2818 (1988); *Phys. Rev. D* **38**, 2985 (1988); and in Ref. 17.
23. See, *e.g.* A. Linde, *Mod. Phys. Lett. A* **1**, 81 (1986); *Physica Scripta* **T15**, 169 (1987).
24. Few of the proposals currently under investigation as candidates for the conditions of the universe genuinely have the character of "initial" conditions in the sense of posing conditions at a definite moment of time. Indeed, characteristically, as described in Sec. 5, the notion of a single time breaks down in the early universe. Rather these proposals fix something like the quantum state of the universe for all times. But also typically these proposals imply simplicity in the early universe and a global spatial structure which consists of a single inflationary bubble. The "no

boundary" proposal[25] is a case in point. The point of the remarks in Sec. 6.1 is that there may be other possibilities.

25. S. W. Hawking, in *Astrophysical Cosmology*, eds. H. A. Brück, G. V. Coyne and M. S. Longair (Pontifica Academia Scientarum, Vatican City, 1982); S. W. Hawking, *Nucl. Phys. B* **264**, 185 (1984).
26. For reviews of these efforts from the author's perspective, see J. B. Hartle in *Proceedings of the International Conference on General Relativity and Cosmology*, Goa, India, 1987 (Cambridge University Press, Cambridge, 1989); and in *Proceedings of the 12th International Conference on General Gravitation*, Boulder, Colorado, 1989 (Cambridge University Press, Cambridge, 1989). For a bibliography of papers on this subject as of 1991 see J. J. Halliwell at the end of his lectures in *Quantum Cosmology and Baby Universes: Proceedings of the 1989 Jerusalem Winter School for Theoretical Physics*, eds. S. Coleman, J. B. Hartle, T. Piran and S. Weinberg (World Scientific, Singapore, 1991), pp. 159–243.
27. S. W. Hawking, *Phys. Lett. B* **195**, 337 (1985).
28. S. Coleman, *Nucl. Phys. B* **310**, 643 (1988).
29. S. Giddings and A. Strominger, *Nucl. Phys. B* **307**, 854 (1988).
30. J. A. Wheeler has for many years stressed the "mutability" of the laws of physics from a viewpoint which would entail much more fundamental revisions in physics than those suggested here but yet which has something in common with them. See, *e.g.* J. A. Wheeler in *Foundational Problems in the Special Sciences*, eds. R. E. Butts and K. J. Hintikka, D. Reidel (Springer, Netherlands, 1977) and J. A. Wheeler, *IBM J. Res. Dev.* **32**, 4 (1988).
31. For a general discussion of the status of the laws of physics and references to other points of view see, P. C. W. Davies, What are the laws of nature, in *The Reality Club No. 2*, ed. J. Brockman (Lynx Communications, New York, 1989).
32. For an effort at making such ideas more precise see R. Sorkin, *Int. J. Theor. Phys.* **22**, 1091 (1983).
33. R. Geroch and J. B. Hartle, *Found. Phys.* **16**, 533 (1986).

34. As discussed by H. Zeh in *Die Physik der Zeitrichtung*, Springer Lecture Notes in Physics, Vol. 200, (Springer, Berlin, Heidelberg, 1984), [Translated and updated as *The Physical Basis of the Direction of Time*, 2nd ed. (Springer-Verlag, Berlin, Heidelberg, 1992)]. He attributes the remark to E. Borel, *Le Hasard* (Alcan, Paris, 1924).

CHAPTER 3

Sources of Predictability*

Sources of predictability in the basic laws of physics are described in the most general theoretical context — the quantum theory of the universe as a whole.

*Appeared in the proceedings of the conference on *Fundamental Sources of Unpredictability* held at the Santa Fe Institute, March 28–30, 1996. http://arxiv.org/abs/gr-qc/9701027v2

1. Introduction

The title of our conference, *Fundamental Sources of Unpredictability*, invites the question: "What are the obstacles and limitations to the predictive power of the laws of physics?" But in a quantum mechanical world characterized by indeterminacy and distributed probabilities, in a world where even classically describable phenomena can be intractable or impossible to compute, and in a finite universe with limited opportunities for observation and induction, we might more straightforwardly ask the question: "What are the fundamental origins of predictability?" More specifically, what features of the basic laws of physics lead to predictable regularities in the universe? This essay is devoted to a few general remarks about this question from the perspective of quantum cosmology — the quantum theory of the universe as a whole.

It seems inescapable that there are limits to the predictive power of the laws of physics. The world appears to be complex. But the basic laws that govern the regularities of the world must be simple to be discoverable, comprehensible and effectively applicable. It is a logical possibility that every feature of our experience — the shape of each galaxy, the number of planets circumnavigating each star, the character of each biological species, the course of human history, and the result of every experiment — is the output of some short computer program with no input. However, there is no evidence that our universe is so regular. Even the most deterministic classical theories did not claim this. They claimed only to predict future evolution when *given* initial conditions. Not everything that can be observed can be predicted — only the regularities in those observations are the province of science. The regularities of the universe are limited and therefore there are limits to the predictive power of physics. This kind of limit is therefore not a failure of the scientific enterprise. Limits are inherent to that enterprise, and their demarcation is an important scientific question.

To ask about the *fundamental* sources of predictability is to ask for the regularities implied by the basic laws that apply universally to all physical systems — without exception, qualification, or approximation — in the most general theoretical context. In contemporary physics, that is the subject of quantum cosmology — the quantum theory of the initial condition of the universe and its subsequent history. In present thinking this theory is based on two laws:

- The basic theory of dynamics (e.g. heterotic superstring theory),
- The theory of the initial condition of the universe (e.g. the "no-boundary" wave function of the universe).

There are no interesting predictions that do not depend on these two laws, even if only very weakly. In this essay, we shall describe some specific features of these two basic laws which give rise to predictable general regularities of the universe. We cannot pretend to discuss the origin of all the regularities considered by science in this brief compass, but we shall describe some of the more general ones. We begin in Sec. 2 with the general regularities exhibited by any quantum system, and then proceed to the regularities that depend on the particular properties of the laws of dynamics and the initial condition that hold in our universe. Specifically, we consider the origin of a realm of approximate classical determinism in Sec. 3 and the origin of individual, approximately isolated subsystems in Sec. 4. The mere existence of such regularities, however, is not enough for science. They must be accessible to inference and amenable to computation. We mention a few properties of the basic laws that allow for these qualities in Secs. 5 and 6.

2. A Quantum Universe

It is an inescapable inference from the physics of the last seventy years that we live in a quantum mechanical universe and that the basic laws of dynamics and initial condition are consistent with the general framework of quantum theory. Let us accept the inference and ask for its implications for predictability.

To keep the discussion manageable, let us neglect gross quantum fluctuations in the geometry of spacetime such as would be expected to occur in the first 10^{-43} s after the big bang. The quantum mechanics of a universe of matter fields moving in a fixed spacetime geometry exhibits the determinism of the Schrödinger equation:

$$i\hbar \frac{\partial \Psi}{\partial t} = H\Psi. \tag{1}$$

The basic theory of dynamics supplies the Hamiltonian H, the theory of the initial condition supplies the initial quantum state Ψ_0. The only regularities exhibited with certainty by all quantum systems, independent of the specific natures of H and Ψ_0, are paraphrases of the prediction that at any time t, the universe is in the state $\Psi(t)$ that evolved deterministically from the initial state Ψ_0 by the Schrödinger equation.

But it is not merely these certain and general regularities of the Schrödinger equation which are of interest. We are interested in the much broader class of regularities which are not certain but occur with high probability as a consequence of the *particular* properties of the laws of dynamics and the initial condition. The approximate regularities of classical physics are the most prominent example.

Most generally quantum mechanics predicts the probabilities of particular *histories* of the universe in sets of alternative possible histories. Examples of sets of alternative histories are the alternative histories of the abundances of the elements over time, the alternative histories of the motion of the earth around the sun, the alternative histories of outcomes of a laboratory experiment testing quantum mechanics, the alternative evolutionary tracks of biological species on earth, and the alternative histories of the S&P stock average. Probabilities for the members of each of these sets of alternative histories can in principle be calculated from the initial state of the universe and the basic dynamical law.

Probabilities of single events or single histories are not necessarily definite predictions. From a theory that a coin is unbiased, we calculate the probability of the outcome of a single toss as 1/2, but we cannot be said to have made a definite prediction of the outcome. The definite predictions of a theory of the initial state and dynamics are the sets of histories in which only one member has a probability very close to unity and the rest have probabilities close to zero. Among the examples of sets of alternative histories mentioned above, only the first — the alternative histories of the abundances of the elements — has this character. Current theories of the initial state and the dynamics predict a high probability of primordial abundances of approximately 75% hydrogen, 25% He^4, small but definite abundances of He^3, deuterium, lithium, and negligible percentages of the other elements. There are similarly definite predictions for some other large scale features of the universe, but the probabilities for the other sets of alternatives mentioned above are likely to be highly distributed and so are not definite predictions. We expect only a few definite predictions from the basic laws of dynamics and the initial condition alone, unaugmented by further information. The largest and most useful general regularities of this kind are those summarized by the deterministic laws of classical physics. It is to these we now turn.

3. The Quasiclassical Realm

Classical deterministic laws approximately govern the regularities in time, of a wide range of phenomena over a broad span of time, place and scale in the universe. The domain of applicability of these classical laws is the quasiclassical realm of everyday experience. In a quantum mechanical universe, these classical equations can be but approximations to unitary evolution by the Schrödinger equation and reduction of the state vector. To what do we owe the validity of this classical approximation over such an extensive realm of phenomena, time, place and scale?

The answer is certainly not merely the determinism of the Schrödinger equation. The regularities in time that it summarizes generally relate quantities that are nothing like those of classical physics. The simple example of Ehrenfest's theorem for the quantum mechanics of a single particle moving in one dimension suggests what is necessary for classical behavior. Ehrenfest's theorem relates the acceleration of the expected position to the expected value of the force:

$$m\,\frac{d^2\langle x\rangle}{dt^2} = -\left\langle\frac{\partial V(x)}{\partial x}\right\rangle. \tag{2}$$

This is an exact consequence of the determinism of the Schrödinger equation, but not a classical equation of motion because it is not a differential equation for $\langle x \rangle$. It becomes a deterministic equation in the approximation that the expected value of the force may be replaced by the force evaluated at the expected position

$$m\frac{d^2\langle x\rangle}{dt^2} \approx -\frac{\partial V(\langle x\rangle)}{\partial x}. \tag{3}$$

There are at least four requirements for the validity of this approximation. First, the Hamiltonian must be of the form that can give rise to an equation of motion. Second, the approximation is true *only for certain states*, typically narrow wave packets. Third, the right variables must be followed, in this case, position. Finally, there must be coarseness in how they are followed, here average rather than exact position.

Only certain states and certain coarse-grained descriptions exhibit patterns of classical correlations in this simple model. Similarly, but more generally, we cannot expect a quasiclassical realm in a universe with a generic initial condition and a generic Hamiltonian. The approximate quasiclassical realm that extends over most of cosmological space and time in this universe owes its existence to an appropriate coarse-grained description in terms of the variables of classical physics and the particular properties of the initial quantum state and dynamical law. Classical predictability is an emergent feature of the universe's particular initial condition and dynamical law.

4. Regularities of Subsystems

Most of science is not concerned with the regularities exhibited universally by all physical systems but rather with the particular regularities of classes of individual, and localized, subsystems. Stars, atoms, oceans, biological species and individual human behaviors are just a few of the myriad examples. In interesting circumstances such subsystems approximately preserve their identity over significant periods of time and can be treated as approximately isolated. Their properties and evolution can thus be studied individually rather than as a feature of the evolution of the whole universe. There is no particular reason to believe that a universe with a generic Hamiltonian and a generic initial state would exhibit such approximately isolated subsystems. To what features of our particular basic laws do we owe the existence of such subsystems and what is the origin of their regularities?

The locality of the effective interactions between the elementary particles that is a consequence of the basic dynamical law is certainly necessary for the approximate isolation of localized subsystems. However, properties of the law of initial condition also play a role. There were no individual subsystems in the early universe. The evidence of the observations is that the early universe was nearly smooth, featureless, homogeneous, isotropic, and far from equilibrium. Today's subsystems have evolved from the quantum fluctuations, away from this smoothness, that are properties of the initial quantum state primarily by the universal action of gravitational attraction. The existence of individual galaxies, for example, is a consequence of the gravitational collapse of primordial fluctuations in the uniform density of matter.

There are several possible sources for the regularities among members of certain classes of approximately isolated subsystems. The simplest is a common historical origin as with individual members of a biological species. But the laws of dynamics and initial condition can also be a source of regularities. Einstein's theory of gravity ensures that one black hole is similar to any other no matter how different were the situations in which they were formed. The regularities among galaxies arise from their origin in similar circumstances across the universe as mandated by an initial condition of close to perfect homogeneity and isotropy.

Thus the specific regularities of particular subsystems that are the subject of much of science outside of physics, arise in part from special properties of the basic laws of dynamics and the initial condition of the universe.

5. Inference and Computation

In the previous sections we have discussed how the fundamental laws of dynamics and the initial condition permit prediction in this quantum mechanical universe. But these laws are not presented to us by revelation. They are arrived at by a process of induction from observation, theoretical development and experimental test. Further, exhibiting the predictions of these laws is not simply a matter of displaying their form. Their consequences must be computed in particular circumstances. No treatment of the sources of predictability would be complete without discussion of those features of the two fundamental laws that permit the processes of induction and application. Space does not permit an examination of these questions that is worthy of their depth. We can only list a few features of the fundamental laws that help us in our task.

The locality of the fundamental law of dynamics, and the separation of phenomena by scale that it permits, have both aided its discovery. Locality enables us to extrapolate the regularities of familiar scales both in much larger and smaller domains. Assuming locality, from the gravitational phenomena on earth, we are able to infer the rules governing the evolution of the planets, galaxies, and eventually the universe itself. Assuming locality, from the electromagnetic forces between macroscopic bodies we are able to infer those between the constituents of atoms.

Separation of phenomena by scale has permitted a kind of progression in the discovery of the fundamental dynamical law — each stage being characterized by a discoverable effective theory of limited validity. Thus we moved from the motion of the planets in the solar system, to the motion of electrons in an atom, to the motion of protons and neutrons in the nucleus, to the motion of quarks in the proton and neutron, etc. Thus classical physics was discovered before quantum mechanics, Newtonian mechanics before special relativity, and Newtonian gravity before general relativity, etc. The similarity of phenomena on different scales has been advanced[1] as the reason for the "unreasonable effectiveness of mathematics in the natural sciences."

Locality of effective interactions and separation of phenomena by scale also reduce the computation of certain predictions to manageable tasks. Locality has already been mentioned as a prerequisite for approximately isolated subsystems. For these, the application of the dynamical law leads to computations which are more tractable than those arising in its application to the universe as a whole. The separation of phenomena by scale permits the computation of the evolution of a wide range of phenomena using simpler, approximate, and effective forms of the dynamical law.

More generally, it is important for practical predictability that the known laws of physics predict computable numbers that are extractable in particular circumstances by the application of standard algorithms. This is not a trivial statement because there are infinitely many more noncomputable numbers than computable ones.

The phenomenon of chaos is commonly mentioned as a source of unpredictability in physics. But in an essay on sources of predictability, it is more appropriate to focus on the existence of nonchaotic, integrable systems whose evolution can be tractably computed. The source of this integrability lies in the conservation laws of the dynamical theory and these arise from its symmetries — exact or approximate. These same conservation laws are at the heart of the effectiveness of statistical mechanics in separating out quantities that approach equilibrium slowly enough to be governed by phenomenological equations of motion. In these ways, symmetries of the dynamical law can be a source of predictability.

6. Conclusion

Predictable regularities in this specific quantum universe arise from particular features of its fundamental laws of dynamics and the initial condition. As observers of the universe we rely on the existence of a quasiclassical realm, the locality of effective interactions, the separation of phenomena by scale, and the existence and regularities of approximately isolated subsystems — all properties which depend on the form of the basic laws of dynamics and initial condition. One could ask: "Why do the fundamental laws exhibit these features which are of such use to us?" That, however, is the wrong way of putting the question. Individually and collectively, we are complex adaptive systems within the universe that evolved to exploit the emergent regularities that the universe presents. Ask rather, how, or whether such systems would have evolved if the fundamental laws did not display the features we have described. While much is beyond our power to answer at present, this question does show that our adaptation to the regularities of our specific universe is another source of predictability in physics.

7. Further Reading

Reference 2 is a *Scientific American* introductory article on quantum cosmology. Also, the author has expanded on many of the topics in this brief essay in Ref. 3 where a more extensive list of references may be found.

Acknowledgments

Preparation of this essay was supported in part by the US National Science Foundation under grants PHY95-07065 and PHY94-07194.

References

1. M. Gell-Mann, *Nature Conformable to Herself*, Banquet talk.
2. J. Halliwell, Quantum cosmology and the creation of the universe, *Scient. Ame.* **265**, 76 (1991).
3. J. B. Hartle, *Scientific Knowledge from the Perspective of Quantum Cosmology* in *Boundaries and Barriers: On the Limits to Scientific Knowledge*, eds. John L. Casti and A. Karlqvist (Addison-Wesley, Reading, Mass., 1996), arXiv:gr-qc/9601046.

CHAPTER 4

Theories of Everything and Hawking's Wave Function of the Universe*

1. Introduction

It is an honor, of course, to participate in this celebration of Stephen's 60th birthday and to address such a distinguished audience. But for me it is a special pleasure. I first met Stephen over thirty years ago, when we were both beginning in science at the Institute of Theoretical Astronomy here in Cambridge. I have been following his inspiration ever since and have had the privilege of working with him on several directions that he has pioneered. I want to talk to you about one of these today.

---

*Appeared in *The Future of Theoretical Physics and Cosmology: Stephen Hawking 60th Birthday Symposium* (Cambridge University Press) with two additional figures.
arXiv:gr-qc/0209047

2. Different Things Fall with the Same Acceleration in a Gravitational Field

If a cat, a cannon ball and an economics textbook are all dropped from the same height, they fall to the ground with the same acceleration (9.8 m/s^2) under the influence of gravity. This equality of the gravitational accelerations of different things is one of the most accurately tested laws of physics. The accelerations are known to be equal to an accuracy of a few parts in a thousand billion. That law, however, tells you very little about cats, cannon balls, or economics. That, in a nutshell, is the theme of my talk today (Fig. 1).

Were a cat, a cannon ball, and an economics textbook dropped from a real leaning tower, the accelerations would not be equal to an accuracy of a thousand billion. Air resistance, magnetic fields, and a

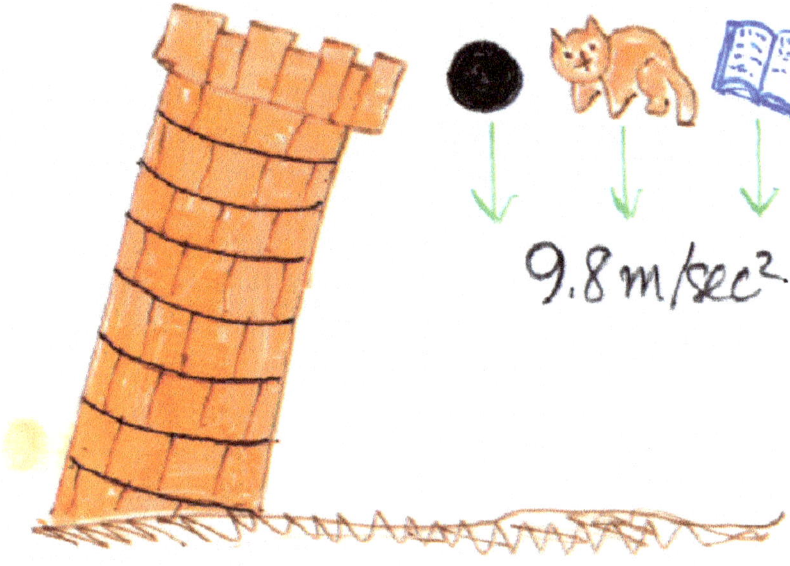

Fig. 1. The lecture in a nutshell: A cat, a cannon ball and an economics textbook all fall to the ground with the same acceleration if acted upon only by gravity. The equality of accelerations of different things in a gravitational field is an example of a fundamental law of physics. But that law does not tell you much about cats, cannon balls, or economics. [From a transparency used in the lecture.]

host of other effects mask the purely gravitational one. Rather, the equality of accelerations is revealed to that accuracy only in delicate laboratory experiments shielded from such extraneous effects, or in the study of the accelerations of astronomical bodies for which they are not important. The accuracy record holder at the moment is the lunar laser ranging demonstration that the Earth and the Moon fall with the same acceleration toward the Sun.[1]

The manned and unmanned missions to the Moon in the late '60s and early '70s left behind arrays of corner reflectors. A corner reflector is an arrangement of three mirrors making up the sides of a cube that meet at one corner. It has the property that any light ray incident on the reflector bounces from mirror to mirror and is reflected back in exactly the direction from whence it came. Since 1969, a systematic program to monitor the Moon's orbit using these reflectors has been carried out by the McDonald Observatory in Texas and the Observatoire de Côte d'Azur in Grasse, France.[1] High powered lasers send light pulses to the Moon that bounce off the reflectors. By measuring precisely the time it takes for the light pulse to return (about 1 s), and dividing that by the velocity of light, the distance to the Moon can be determined very accurately. About a billion photons are sent out ten times a second and one returning photon is detected every few seconds. The returning photons have been detected for over thirty years, and as a consequence of this effort, we know the position of the Moon relative to the Earth to an accuracy of about a few centimeters. Study of its orbit shows that it is falling towards the Sun with the same acceleration as the Earth to an accuracy of a few parts in a thousand billion.

This law of equality of accelerations of different things is a cornerstone of Einstein's general relativity — Stephen's subject. However, it is not general relativity specifically that I want to talk to you about today, but about such laws of physics more generally.

3. The Fundamental Laws of Physics

The equality of gravitational accelerations of different things is an example of a regularity of nature. Everything falls in exactly the same way. The regularity is universal. No exceptions! There are other regularities that may apply to specific systems in similar circumstances. For example, if we drop three cats, they might all go "meeeeow" in roughly the same way on the way down. Or perhaps there are only statistical regularities for cats — 8 out of 10 cats will go "meeeeow" in the average drop. Other systems will do something different according to their own particular regularities. Cannonballs, for example, do nothing on the way down. But the equality of accelerations of different things is a special kind of law because it applies *universally* to anything falling in a gravitational field. It is an example of a fundamental law of physics.

Identifying and explaining the regularities of nature is the goal of science. Physics, like other sciences, is concerned with the regularities exhibited by particular systems. Stars, atoms, fluid flows, high temperature super-conductors, black holes, and the elementary particles are just some of the many examples. Studies of these specific systems define the various subfields of physics — astrophysics, atomic physics, fluid mechanics, and so forth. But beyond the regularities exhibited by specific systems, physics has a special charge. This is to find the laws that govern the regularities that are exhibited by *all* physical systems — without exception, without qualification and without approximation. The equality of gravitational accelerations of different things is an example. These are usually called the *fundamental* laws of physics. Taken together, they are called informally a "theory of everything". Stephen has been a leader in the quest for these universal laws.[2] Today, I would like to ask the question: "How much do we know about the world if we have a theory of everything?"

Ideas for the nature of the fundamental laws have changed as experiments and observations have revealed new realms of phenomena and have reached new levels of precision. But since they have been studied, it has been thought that the fundamental laws consist of two parts

- *Dynamical laws* that govern regularities in *time*. Newton's laws of motion governing the orderly progression of the planets or the trajectory of a tennis ball are examples, as is the law that different things fall with the same acceleration in a gravitational

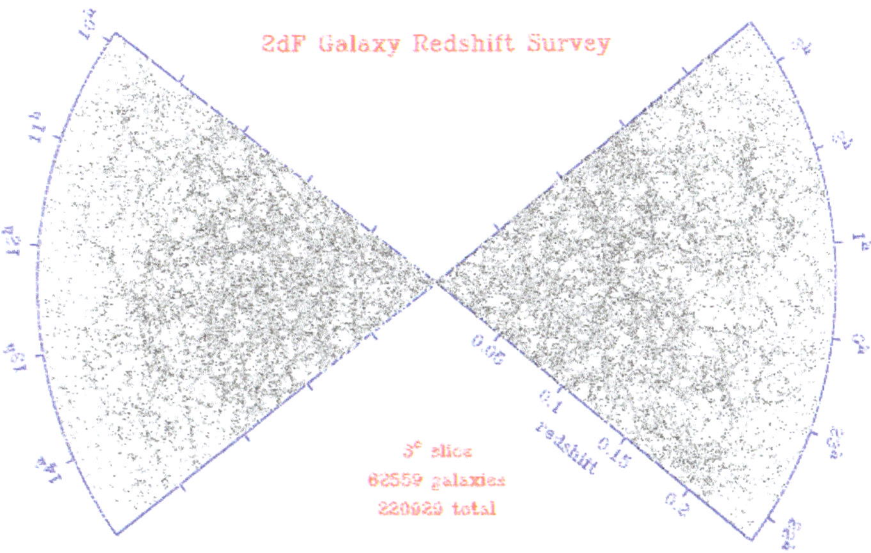

Fig. 2. This plot from the 2dF Galaxy Redshift Survey[3] shows the location of 62,559 distant galaxies in two angular slices of the universe approximately 90° wide and 3° thick as viewed from the Earth. A galaxy's redshift is used as a measure of its distance from our location at the center. The statistics of the distribution of knots, filaments and voids in this picture is an example of a large scale regularity of the universe that is directly linked to its initial condition. [J. Peacock and the 2dF Galaxy Redshift Survey Team.]

field and the Einstein equation governing the evolution of the universe.

- *Initial conditions* that govern how things started out and therefore most often specify regularities in *space*. The statistical regularities of the large scale distribution of the galaxies in the universe is a possible example (Fig. 2).

Laplace is a name associated with this view of a theory of everything. He wrote, famously,[4]

> *An intelligence knowing, at any given instant of time, all forces acting in nature, as well as the momentary positions of all things of which the universe consists, would be able to comprehend the motions of the largest bodies of the world and*

> *those of the smallest atoms in one single formula. ... To it, nothing would be uncertain, both future and past would be present before its eyes.*

That was the theory of everything circa 1820 — Newtonian determinism.

Both parts of a theory of everything are needed to make any predictions. Newton's dynamical laws by themselves don't predict the trajectory of a tennis ball you might throw. To predict where it goes, you must also specify the position from which you throw it, the direction, and how fast. In short, you must specify the ball's *initial condition.* One of Stephen's most famous achievements is just such an initial condition,[2] but not for tennis balls. Stephen's no-boundary initial condition is for the whole universe.

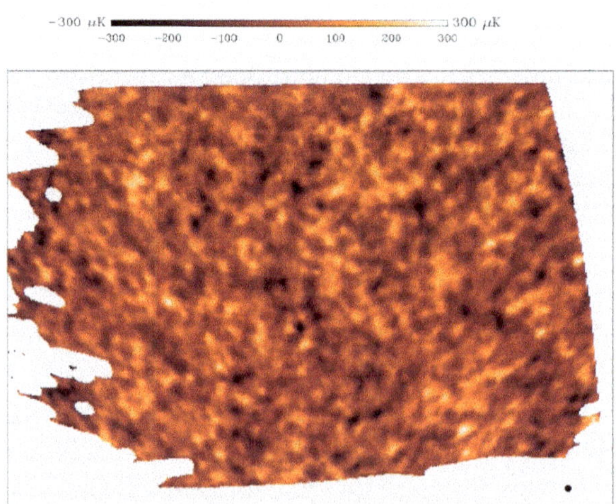

Fig. 3. This picture from the Boomerang experiment[5] shows the temperature distribution across a patch of sky of the cosmic background radiation. The light and dark regions differ in temperature by only a few ten millionths of a degree, that is about a mean of 2.73 K. This is as close as one can come to a picture of the early universe some hundreds of thousands of years after the big bang. The temperature variations are evidence for tiny concentrations of density, which then in the intervening 15 billion years grew by the action of gravitational attraction to become the galaxies we see today with the large scale distribution illustrated in Fig. 2.

The search for a theory of the dynamical laws has been seriously underway since the time of Newton. Classical mechanics, Newtonian gravity, Maxwell's electrodynamics, special and general relativity, quantum mechanics, quantum field theory and superstring theory are but some of the milestones in this search. But the search for a theory of the initial condition of the universe has been seriously underway for only the twenty years since Stephen's pioneering work on the subject. Why this difference? The examples used above to discuss regularities governed by the two parts of a theory of everything hint at the answer. The trajectory of a tennis ball was used to illustrate the regularities of dynamical laws, and the large scale distribution of galaxies was used to illustrate the regularities implied by the law of the initial condition. There is a difference in the *kind* and *scale* of regularities that the two laws predict.

Dynamical laws predict regularities in time. It is a fortunate empirical fact that the fundamental dynamical laws are local — both in space and time. The trajectory of a tennis ball depends only on conditions that are nearby both in space and time, and not, for example, either on what is going on in distant parts of the universe or a long time ago. This is fortunate because that means that dynamical laws can be discovered and studied in laboratories on Earth and extrapolated, assuming locality, to the rest of the universe. For example, because it is local, the law that different things fall with the same acceleration in a gravitational field can be discovered by experiments in laboratories here, and indeed all over the galaxy (Fig. 4). Without that simplicity of the dynamical laws in the here and now it is possible that we would never have discovered them.

By contrast, the regularities governed by the law of the initial condition of the universe occur on large, cosmological scales. The universe isn't simple on small spatial scales. Look at the disorder or complexity in the room we're in right now, for example. But the universe *is* simple on large, cosmological scales — more or less the same in one direction as in any other, more or less the same in one place as in any other.

The mottled map of the cosmic background radiation in Fig. 3 illustrates this. In the beginning the universe was opaque, hot, and

Fig. 4. The equality of accelerations of different things in a gravitational field is an example of a *local* fundamental dynamical law. A body's acceleration depends only on the gravitational field at its location and not, say, on the arrangement of distant matter or events in the past that could not be summarized in this way. This fundamental law can therefore be discovered by experiments done locally, by experimenters in their laboratories across the universe. [From a transparency used in the lecture.]

glowing. As it expanded it got cooler. A few hundred thousand years after the big bang, the matter cooled to a temperature at which it was transparent. Light from that time has been propagating to us ever since, cooled to a temperature now of only a few degrees above absolute zero. The cosmic background radiation and map of the temperature of the radiation in different directions, i.e. that in Fig. 3, is the earliest picture we have of the universe.

The mottled map in Fig. 3 may not look like evidence that the early universe was smooth, but the temperature difference between the light and dark regions is only 30 millionths of a degree. At resolution of the eye, the universe would look completely uniform at this time. The reason for which serious research on the initial condition at the big bang began only recently is because only lately have we had cosmological observations like this that are detailed enough to reveal the regularities that the laws govern.

4. Quantum Mechanics

There is another way in which our vision of the fundamental laws and the nature of a theory of everything has changed since the times of Newton and Laplace. That is quantum mechanics. We don't yet know the final form the fundamental laws will take, but the inference is inescapable from the physics of the last seventy-five years that they will conform to that subtle framework of prediction we call quantum mechanics.

In quantum mechanics, any system — the universe included — is described by a wave function Ψ. There is a dynamical law called the Schrödinger equation which governs how the wave function changes in time

$$i\hbar\frac{d\Psi(t)}{dt} = H\Psi(t) \qquad \text{(dynamical law)}.$$

Here the operator H, called the Hamiltonian, summarizes the dynamical theory. There are different forms of H for Maxwell's electrodynamics, for a theory of the strong interactions, etc., as well. Like Newton's laws of motion, the Schrödinger equation doesn't make any predictions by itself, it requires an initial condition. This is

$$\Psi(0), \qquad \text{(initial condition)}.$$

When we consider the universe as a quantum mechanical system, this initial condition is Hawking's wave function of the universe.[2]

Probabilities are the key difference between classical and quantum mechanics. Let's first think about probabilities in classical physics. If I say that there is a 60% chance of hitting an audience member if I toss a ball in this room, I am not expressing a lack of confidence that its trajectory will be governed by the deterministic laws of Newtonian mechanics. Rather, the 60% reflects my ignorance of the exact initial speed that I would be imparting to the ball, of the influence of air on its motion, and perhaps my ability to do an accurate calculation. If I practice to control the initial condition when it is thrown, the subsequent evolution of the tennis ball becomes more certain. Probabilities in classical physics result from ignorance.

But in quantum mechanics, probabilities are fundamental and uncertainty, inevitable. No amount of careful determination of the present state of the tennis ball will achieve certainty for its trajectory. In quantum mechanics there is some probability that a ball will take *any* trajectory as it leaves my hand. However, in classical situations, one trajectory — the one obeying Newton's laws — is much more probable than all the others. The determinism of classical physics is an approximation, but an approximation on which we can rely in many practical circumstances.

5. A Theory of Everything is Not a Theory of Everything

My colleague Murray Gell-Mann used to ask me, "If you know the wave function of the universe, why aren't you rich?" A quantum mechanical theory of the Hamiltonian and the initial state *does* predict probabilities for every event that might happen in the universe. That is, in one sense, what a "theory of everything" is. However, only a few things are predicted with near certainty. The vast majority of alternatives are predicted with 50%-50% probabilities giving no useful information. That's the sense in which a "theory of everything" is *not* a theory of everything.

We hope that the probability is high for the form of the effective dynamical theories that are sufficient to predict the outcomes of experiments done in every laboratory constructed so far. We also hope that a theory of everything predicts some large scale features of the universe with near certainty, such as its approximate smoothness on large scales, the statistics of the distribution of galaxies illustrated in Fig. 2, the vast disparity between the age of the universe, and the time scales characterizing the fundamental dynamical theory, and so forth.

But it is too much to hope that an interesting probability like that for the FTSE to go up tomorrow is in this category. Since a rise tomorrow is an event in the history of the universe, a quantum mechanical probability could, in principle, be calculated for it from a theory of everything. (Although I suspect that it is well beyond our present powers of computation.) But it's likely that, after all that work, the predicted probability would be 50% for an upward tick. That is why you can't get rich knowing the wave function of the universe, and that is why a theory of everything isn't a theory of everything.

But there is a deeper reason why a theory of everything doesn't explain everything — it's too short. Everyone with a PC knows that picture files are typically a lot longer than text files. The files for the undergraduate general relativity text that I am writing, for example, consists of roughly 1 Mb of text and 100 Mb of pictures. A very rough estimate is that it would take a billion, billion, billion, billion, billion, billion, billion, billion, billion (10^{81}) compact disks (CDs) to describe the visible universe at a reasonably coarse-grained level of classical description. And that is just at one moment in time! All the matter in the planets, stars, and galaxies in the visible universe would not be enough to make this number of CDs.

However, the description of the universe can be compressed because it exhibits regularities on large scales. Compression is an idea familiar from computation. When a large file exhibits regularities, its length can be compressed. For instance, a string of 1,000 zeros "0000000000000000000. . . 0000" in a file is a kind of regularity. It can be replaced by a shorter string "1,000 zeros". The regularities summarized by the laws of nature similarly permit compression. Rather than saying, "cannonballs fall at 9.8 m/s^2, cats fall at 9.8 m/s^2, economics textbooks fall at 9.8 m/s^2 etc.", the laws of nature permit us to say "*everything* falls at 9.8 m/s^2 near the surface of the Earth". That is shorter and therefore more useful. It is possible that everything we see, every detail of every leaf, every event in human history, every thought, is a long string which is compressible to a law implementable by a very short computer program. But there is no evidence that the universe exhibits such regularity, and even Laplace did not propose such a thing.

The initial condition of a deterministic classical theory that did not allow ignorance would have to be as detailed as a present description of the universe. It seems likely therefore that, were the world governed by classical laws, it might take so many CDs to write out the law of its initial condition that we could never make the CDs to do it.

Let's contrast this situation in classical physics with Stephen's no-boundary quantum wave function of the universe. It is given by the following simple formula[2]

$$\boxed{\Psi = \int \delta g \delta \phi e^{-I[g,\phi]}} \qquad \begin{pmatrix}\text{The no-boundary wave} \\ \text{function of the universe}\end{pmatrix}$$

That's short! — maybe 45 LATEX key strokes to write out. Furthermore, it is as complete a description of the initial condition as it is possible to have in quantum mechanics. It implies uncertainty but contains no ignorance. Presumably, the basic equations of the fundamental dynamical theory are similarly short.

You might think this a little misleading because I haven't yet said what Ψ means, what g means, and what $\int$ means, etc. But, even including the lengths of the ten or so texts that physics students read to understand what it all means, a maximum of 10 or 20 Mb would be needed to state the law — easily fitted on one CD. This means that the law is discoverable and implementable, applicable to every happening in the universe, but predicting the near certainty of only a *few* of its many regularities. A complete discoverable theory of everything is only possible in quantum mechanics where some things are predictable, but not everything.

Before I left Santa Barbara to come to Cambridge, my colleague Steve Giddings asked me what I was going to talk about in this lecture. I said that I would speak on the question, "What do we know if we have a theory of everything?" and to answer — "Not all that much." He replied something like, "I hope you're going to say something more hopeful than that!" But it *is* a hopeful message. It is only because so little of the complexity of the present universe is predicted by the fundamental universal laws that we can discover them.

6. Reduction

The earlier parts of this essay have emphasized the regularities of subsystems of the universe that follow from the universal laws of physics (Fig. 1) and apply to all subsystems. Specific classes of subsystems (like cats) have their own regularities in addition to the universal ones. Where do the these regularities come from in a quantum mechanical universe?

They are the results of chance events that occur naturally over the history of the universe — some classical, some quantum mechanical. As my former colleague Murray Gell-Mann put it,[6] they result from frozen accidents. "Chance events of which particular outcomes have a multiplicity of long-term consequences, all related by their common ancestry."

The regularities of cats probably do depend a little bit on the fundamental physical laws, for example, a quantum state that implies a spacetime that is locally smooth across the universe, has three spatial dimensions, etc. But the origin of most of the regularities of cats can be traced to the chance events of four billion years of biological evolution. Cats behave in similar ways because they have a common ancestry and development in similar environments. The mechanisms which produce those chance events that led to cats are very much dependent on fundamental biochemistry and ultimately atomic physics. But the particular outcomes of those chances have little to do with the theory of everything.

These specific regularities are different enough that they are studied in separate sciences like, biology, geology, economics, psychology, and human history.

Do psychology, economics, biology, etc., reduce to physics? The answer is YES, because everything considered in those subjects must obey the universal, fundamental laws of physics. Does every one of the subjects of study in these sciences — humans, market tables, historical documents, bacteria, cats, etc. — fall with the same acceleration in a gravitational field? The answer is Yes. But the other regularities central to subjects like psychology, economics, biology, etc., are not predicted by the universal laws alone, even in principle. They arise from frozen accidents. The answer to the question as to whether subjects like biology reduce to physics depends on what you mean by 'reduce'.

7. The Main Points Again

I have always liked the part of the BBC news called "The Main Points Again". Here are my main points again:

- The fundamental laws of physics constituting a "theory of everything" are those which specify the regularities exhibited by every physical system, without exception, without qualification, and without approximation.
- A theory of everything is not (and cannot be) a theory of everything in a quantum mechanical universe.
- If it is short enough to be discoverable, it is too short to predict everything.
- The regularities of human history, personal psychology, economics, biology, geology, etc., are consistent with the fundamental laws of physics, but do not follow from them.

But remember also, especially on this occasion, that all the beautiful regularities that we observe in the universe, whether certain or not, predictable or not, could be the result of quantum chances following from the fundamental dynamical theory and Hawking's no-boundary wave function of the universe.

Acknowledgment

The author's research was supported in part by the National Science Foundation under grant PHY00-70895.

References

1. *e.g.* J. O. Dickey *et al.*, Lunar laser ranging: A continuing legacy of the Apollo Program, *Science* **265**, 482 (1994); J. D. Anderson and J. G Williams, Long range tests of the equivalence principle, *Class. Quant. Grav.* **18**, 2447 (2001).
2. *e.g.* S. W. Hawking, The quantum state of the universe, *Nucl. Phys. B* **239**, 257 (1984).

3. *e.g.* M. Colless, G. Dalton, S. Maddox, W. Sutherland *et al.*, The 2dF galaxy redshift survey: spectra and redshifts, *MNRAS* **328**, 1039–1063 (2001).
4. P. S. Laplace, *A Philosophical Essay on Probabilities*, translated from the 6th French edition (Dover publications, New York, 1951).
5. P. de Bernardis *et al.*, A flat universe from high-resolution maps of the cosmic microwave background radiation, *Nature* **404**, 955 (2000).
6. M. Gell-Mann, *The Quark and the Jaguar* (W. Freeman, San Francisco, 1994).

CHAPTER 5

Scientific Knowledge from the Perspective of Quantum Cosmology*

Existing physical theories do not predict every feature of our experience, but only certain regularities of that experience. That difference between what could be observed and what can be predicted is one kind of limit on scientific knowledge. Such limits are inevitable if the world is complex and the laws governing the regularities of that world are simple. Another kind of limit on scientific knowledge arises because even simple theories may require intractable or impossible computations to yield specific predictions. A third kind of limit concerns our ability to know theories through the process of induction and test. Quantum cosmology — that part of science concerned with the quantum origin of the universe and its subsequent evolution — displays all three kinds of limits. This paper briefly describes quantum cosmology and discusses these limits. The place of the other sciences in this most comprehensive of physical frameworks is described.

*Contribution to the proceedings of the workshop *Limits to Scientific Knowledge* held at Abisko, Sweden, May 15–19, 1995.
arXiv:gr-qc/9601046

1. Introduction

The assignment of the organizers was to speak on the subject of "limits to scientific knowledge". This is not a topic on which I have been forced to reflect a great deal in the course of my efforts in astrophysics, but I shall try to offer a few thoughts on it from the perspective of cosmology. Like any assignment, the first task is to understand what it might mean. I shall say more about this later, but one thing is immediately clear: This is not simply an empirical question, but rather concerns the relationship between what we observe and our theories of what we observe. Limits therefore depend on theories and will vary from one scientific theory to another. The question of what are the *fundamental* limits to scientific knowledge must be examined in the most general theoretical context. In physics, this is the subject of quantum cosmology — the quantum mechanics of the universe as a whole and everything inside it. The nature of scientific knowledge in this most comprehensive of theories is the subject of this essay. I shall try to describe a little of what quantum cosmology is about, and address the question of limitations to scientific knowledge in this most general of contexts.

2. Three Kinds of Limits to Scientific Knowledge

In this section three different kinds of limits to scientific knowledge are identified. No claim is made that these are the only kinds of limits, but these three have a general character that is inherent in the nature of the scientific enterprise. Subsequent sections will illustrate these general kinds of limits with examples from quantum cosmology.

2.1. *Limits to What is Predicted*

The task of science, as Bohr said, is "to extend the range of our experience and to reduce it to order".[1] To reduce experience to order is to compress the length of a description of that experience. That compression is achieved when a computer program can be exhibited which, given a certain input, outputs a string describing some parts of our experience and the length of that program together with its input are shorter than the length of the output description. Theory supplies the program. For instance, a detailed description of the observations of the positions of the planets over the last 100 years might make up a very long table, but Newton's equations of motion can be used to compress all that information into two much shorter strings: a string stating Newton's theory and another string giving the positions and velocities of the planets at one time.

It is a logical possibility that *every* feature of our experience — the wave function of every quark, the velocity of every molecule, the position of every leaf, the character of each biological species, the action of every human, etc. — is just a very long output of a short computer program with *no* input. However, in the history of scientific inquiry there is no evidence that the universe is so regular. Even the most deterministic classical theories did not claim this. With Newtonian mechanics, Laplace proposed only to predict the future and retrodict the past *given* the present position and velocity of each particle in the universe. That list of initial data would be vastly longer than a few treatises on Newtonian mechanics. Existing theories predict a string describing our experience only given some other, shorter, string as input. Theories do not predict everything that is observed, but only certain regularities in what is observed. Some things are predicted, some are not, and that limit to what is predicted is one kind of limit to scientific knowledge.

Scientific laws must have some degree of simplicity to be discoverable, comprehensible and effectively applicable by human beings and other complex adaptive systems. If the complexity of the present universe is large, then this necessary simplicity of the laws implies that this kind of limit to scientific knowledge is inevitable. Not everything can be predicted, but only those regularities that are summarized in the laws of science. In the following, we shall describe what is predicted and what is not predicted in quantum cosmology.[a]

[a]For a lucid discussion in popular language of the notion of complexity and of prediction in quantum cosmology, as well as a summary of some of the author's work with M. Gell-Mann, see Ref. 2.

2.2. *Limits to Implementation*

To be tested, the predictions of an abstractly represented theory covering a broad class of phenomena must be implemented in particular circumstances. The theory must produce numbers, and that process involves computation. Even if the laws are precisely specified, even if the input to those laws is exactly stated, limitations of our ability to compute may limit our ability to predict. This is another kind of limit to scientific knowledge. The practical limitations of present computing machines are all too familiar. Computing the motion of every particle in a group of classical gas 10^{22} particles in less than its real evolution time is well beyond the powers of contemporary computers. However, beyond the limitations of contemporary machinery, we may ask whether there are fundamental limitations on what can be computed that are inherent in the form of the laws themselves. The phenomenon of chaos is the source of one kind of limitation. The precision required of initial data to extrapolate a given time into the future increases exponentially with that time for a wide variety of classical systems. Another kind of limit arises in cosmology where resources for computation, both in time and space, are limited. Further, as we shall see, there is some evidence that certain predictions of quantum cosmology may be noncomputable numbers.

It is not difficult to display predictions which are computationally intractable but which are measurably inaccessible. Given initial conditions, classical theory predicts the orbits of every molecule of gas in a room. The explicit computation of this prediction at the operating speeds of present computers would take much longer than the age of the universe because of the large number of particles involved. Yet, for the same reason, neither the initial condition nor the predicted orbits are measurably accessible quantities. Merely exhibiting phenomena which are impossible or intractable to compute is not much of a limit if the phenomena are impossible or extraordinarily difficult to measure. The most interesting limits concern phenomena that are easy to measure but difficult to compute.

2.3. *Limits to Verification*

The above discussion has assumed that we know the laws of physics. However, we arrive at those laws by a process of induction and test. Competing laws consistent with known regularities are winnowed by the process of checking their predictions with new observations. Are there fundamental limits to what we can test, and therefore fundamental limits to how well the theory can be known? Cosmology will provide examples.

2.4. *False Limits*

Beware of false limits that arise only from imprecise language or the comparison of a correct theory with an incorrect one. A classic example is provided by the uncertainty principle in quantum mechanics

$$\Delta x \, \Delta p \geq \hbar/2. \tag{1}$$

That relation is sometimes described as a limit on our ability to predict (or "measure" or "know") both the position and momentum of a particle at one time to accuracies better than those restricted by (1). However, the uncertainty principle is more accurately characterized as a limit on the use of classical language in a quantum mechanical situation.

There is no state of a quantum mechanical particle with a precisely defined position and momentum. That is the content of (1). The uncertainty principle, therefore, is not a limit to what observed properties of quantum particles are predicted by the theory. Since there is no quantum state with precisely defined position and momentum, quantum theory predicts that we shall never observe both simultaneously. Thus, as far as the position and momentum of a particle are concerned, there is no disparity within quantum theory between what can be predicted and what is observed arising from (1) as there would be in the case of a genuine limit of the type discussed in Sec. 2.1.

As mentioned earlier, limits to prediction are properties of the theories which specify what can be predicted. Of course, if we *compare* two theories, one may predict different phenomena from the other. In classical physics there are states in which the position and momentum of a particle *are* simultaneously specified. In quantum theory, there are not. But quantum theory is correct and classical theory incorrect for the domain of phenomena we have in mind. The uncertainty principle (1) may be viewed as a kind of limit to how far classical concepts and language can be applied in quantum theory, but, were we to strictly adhere to the language and concepts of quantum theory, there would be no limit at all.

3. Dynamical Laws and Initial Conditions

As we mentioned above, fundamental limits to what is predicted, to how predictions can be implemented, and to how theory can be verified depend on what the basic theory is. This section sketches some essential features of basic physical theory today. Of course, we are on dangerous ground here. The most basic laws are often the furthest from definitive experimental test. Nevertheless, it is interesting to see what kinds of limits might exist in the kind of basic theoretical framework that is under active investigation by physicists today.

The most general framework for prediction is quantum cosmology — the quantum theory of the universe as a whole and everything that goes on inside it. In the following, I shall describe a little of this theory.

Historically, physics for the most part has been concerned with finding dynamical laws — laws which compress the description of evolution over time to the description of an initial condition. Thus, these dynamical laws require boundary conditions to yield predictions. There are no particular laws governing these boundary conditions. They are specified by our observations of the part of the universe outside the subsystem whose dynamics is of interest. In a room, if we observe no incoming radiation, we solve Maxwell's equations there with no-incoming-radiation boundary conditions. If we prepare an atom in a certain atomic state, we solve Schrödinger's equation with that initial condition, etc.

But in cosmology we are confronted with a fundamentally different kind of problem. Whether classical or quantum, the dynamical laws governing the evolution of the universe require boundary conditions. But in cosmology there is no "rest of the universe" to pass their specification off to. The boundary conditions must be part of the laws of physics themselves. There is no other place to turn.

A present view, therefore, is that the most general laws of physics involve two elements:

- The laws of dynamics prescribing the evolution of matter and fields and consisting of a unified theory of the strong, electromagnetic, weak and gravitational forces.
- A law specifying the initial boundary condition of the universe.

There are no predictions of any kind which do not depend on these two laws, even if only very weakly, or even when expressed through phenomenological approximations to these laws (like classical physics) appropriate in particular and limited circumstances with forms that may be only distantly related to those of the basic theory.

The search for a fundamental theory of the dynamics of matter has been seriously under way since the time of Newton. Classical mechanics, Newtonian gravity, electrodynamics, special relativity, general relativity, quantum mechanics, quantum electrodynamics, the theory of the electroweak interactions, quantum chromodynamics, grand unified theories, and superstring theory are but some of the important milestones in this search. The search for a theory of the initial condition of the universe has been seriously under way for not much more than a decade. (See Ref. 3 for a review.) The reason for this difference can be traced to the scales on which the regularities summarized by these two laws emerge. The trajectory of a ball in the air, the flow of water in a pipe, or the motion of a planet in the solar system all exhibit the regularities implied by Newtonian mechanics. The regularities of the dynamical laws of atomic and particle physics can be exhibited in experiments carried out in laboratories or large accelerators. However, characteristic regularities implied by a theory of the initial condition of the universe emerge mostly on much larger, cosmological scales.

On any scale, the universe exhibits some regularities in space as distinct from regularities in time. Rocks on one part of the earth are related to rocks on another part. Similarly, there are relations between individual members of biological species, and human history in different locations. These regularities have their origins in the common origin of rocks on earth, the evolution of biological species, and the facts of human history. On cosmological scales, the universe is more regular in space than it is on smaller scales. The progress of observation in astronomy in recent decades has given us an increasingly detailed picture of the universe on ever larger scales of space and time. The remarkable inference from these observations is that the universe becomes increasingly simple as we move to larger scales in space and more distant times in the past. Galaxies are not very complicated objects, but still exhibit a variety of types and considerable individuality. On the larger scale of a tenth of the radius of the universe, the galaxies are no longer individual objects, but there is considerable structure in their distribution. Pictures of the distribution of galaxies on the sky, which probe out to greater distances show less structure. On the largest scales, the distribution of the cosmic background radiation temperature, which is as close as we can come to a picture of the universe three hundred thousand years after the big bang, reveals almost no structure at all. (See Fig. 1.) The deviations in this temperature from exact smoothness (exact isotropy) are measured in tens of *millionths* of a degree. However, those deviations are important! They are the origin of all the complexity in the universe we see today. As the universe evolves, these fluctuations grow, collapse, and fragment through gravitational attraction to become the galaxies, stars and planets which characterize the universe today. Initially very close to equilibrium, the matter in the universe is thereby driven further away from equilibrium. That disequilibrium is necessary for chemistry, geology, life, biology and human history.

The evidence of the observations then is that the universe was a simpler place earlier than it is now — more homogeneous, more isotropic, with matter more nearly in thermal equilibrium. The aim of quantum cosmology is a quantum theory of this simple initial condition.

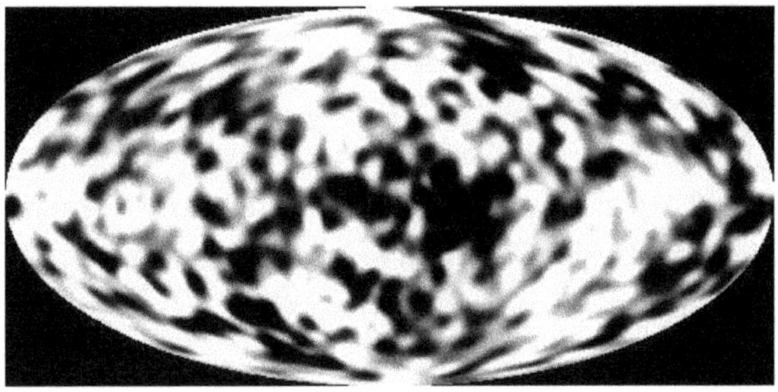

Fig. 1. A sky map of temperature fluctuations in the cosmic background radiation. This figure may be thought of as a picture of the universe approximately 300,000 years after the big bang. The hot mixture of matter and radiation that existed immediately after the big bang cools as the universe expands. About 300,000 years later the universe has cooled enough that matter and radiation no longer significantly interact. Photons from that time have been traveling freely towards us ever since. Their characteristic temperature now is only 2.7 degrees above absolute zero, yet they can be detected at microwave wavelengths by sensitive instruments. The figure above shows a sky map of the temperature of that radiation based on data taken with the COBE satellite. The dark spots are where the sky is cooler than the mean temperature and the white areas are where it is hotter. The differences in temperature between the darkest black and the whitest white is only a few hundred *micro* degrees Kelvin. The universe is thus essentially featureless at 300,000 years after the big bang except for these tiny fluctuations. These small fluctuations, however, are the origin of all the complexity in the universe that we see today. [Grayscale adaptation by J. Gundersen of the results of C. Bennett *et al.*, *Ap. J.* **436**, 423 (1994).]

4. Classical and Quantum Initial Conditions

It is an inescapable inference from the physics of the last sixty years that we live in a quantum mechanical universe — a world in which the basic laws of physics conform to that general framework for prediction we call quantum mechanics. We perhaps have little evidence for peculiarly quantum mechanical phenomena on large and even familiar scales, but there is no evidence that the phenomena that we do see cannot be described in quantum mechanical terms and explained by quantum mechanical laws. This is the first reason such that the search for a theory of the initial condition is carried out in the framework of *quantum* cosmology. There is, however, another reason: quantum indeterminacy is probably necessary for a comprehensible, basic, scientific theory of the initial condition.

To explain this necessity and also to understand a bit of the machinery of quantum cosmology, consider a model universe. Suppose the universe consists of a box the size of the visible universe containing a large number N of particles interacting by fixed potentials. To simulate the expansion of the universe, we could let the box expand. That is actually not a bad model for what goes on in more recent epochs of the universe.

Classically, a history of this model universe is a curve in a $6N$-dimensional phase space of the positions and momenta of all the particles in the box. Classical evolution is deterministic — if the point in phase space specifying the system's configuration is known at one time, the location at all other times is determined by the equations of motion. A classical theory of the initial condition of the model universe thus might specify the initial point in phase space at $t = 0$. However, such a theory would necessarily be hopelessly complex because it would have to encode all the complexity we see today. Its description would be too long to be comprehensible.

A *statistical* classical initial condition could be simpler. Such an initial condition would only give a *probability* for the initial point in phase space and therefore only a probability for the subsequent evolution. Present predictions of the future would then be probabilistic. For example, observers at any time in the history of the universe can only see galaxies within a distance close enough such that light from the galaxies could have reached them in the time since the big bang. This cosmological horizon expands as the universe ages. One new galaxy comes over this cosmological horizon approximately every 10 minutes. A statistical initial condition might not predict with near certainty, say, the specific locations of the individual new galaxies, but rather their statistical distribution on the sky. Similarly, with a classical initial condition in which matter was initially in thermal equilibrium, one might predict the overall intensity of the background radiation on the sky, but not the location of any particular fluctuation in its intensity.

Probabilities in classical physics reflect ignorance. A classical statistical law of the initial condition would mean that we have some information about how the universe started out, but not all. However, we learn from observation. With every observation we could refine our theory of the initial condition which would therefore become increasingly complex, reflecting the complexity of the present, and thus become increasingly less comprehensible.

Quantum mechanics is inherently indeterministic and probabilities are basic. The most complete specification of the initial state of our model box of particles would be a wave function on the configuration space of all their positions

$$\Psi\left(\vec{x}_1, \ldots, \vec{x}_N\right) \tag{2}$$

— a wave function of the universe for this model. Unlike classical physics, subsequent observation will not improve this initial condition, although the results of observation can be used to improve future predictions. Thus, in quantum mechanics, it is natural to have a simple, comprehensible law of the initial condition which is consistent with the complexity observed today.[b]

5. What is Predicted in Quantum Cosmology?

My colleague, Murray Gell-Mann, once asked me, "If you know the wave function of the universe, why aren't you rich?" The answer is that very little is predicted with certainty by such a quantum initial condition of the universe and it is thus certainly not of much use in generating wealth. What might be predicted by an initial condition for cosmology is the subject of this section.

Quantum mechanics predicts probabilities for sets of alternatives. In our model universe in a box, for example, it might predict the probabilities for alternative ranges of the position of a particle at a particular time, or the probabilities for alternative distributions of energy density in the box, and many other sets of alternatives. These are the probabilities for alternatives which are *single* events in a *single* closed system — the universe as a whole.

---

[b]For an early statement of this, see Ref. 4.

What do such probabilities of single events mean? Some may find it helpful to think of these probabilities as predictions of relative frequencies in an imaginary infinite ensemble of universes, but they are not frequencies in any accessible sense. Rather, to understand what the probabilities of single events mean, it is best to understand how they are used. Probabilities of single events can be useful guides to behavior even when they are distributed over a set of alternatives so that none is very close to zero or one. Examples are the probability that it will rain today or the probability of a successful marriage. However, because the probabilities are distributed, the event which occurs — rain or no rain, divorce or death before parting — does not test the theory that produced the probabilities. Tests of the theory occur when the probabilities are *near certain*, by which I mean sufficiently close to zero or one, such that the theory would be falsified if an event with probability sufficiently close to zero occurred, or an event with a probability sufficiently close to one did not occur.[c] Various strategies can be used to identify sets of alternatives for which probabilities are near zero or one. The most familiar is to study the frequencies of outcomes of repeated observations in an ensemble of a large number of identical situations. Such frequencies would be predicted with certainty in an infinite ensemble. However, since there are no genuinely infinite ensembles in the world, we are necessarily concerned with the probability of deviations of the frequency in a finite ensemble from the expected behavior of an infinite one. Those are probabilities for single properties (the deviations) of a single system (the whole ensemble) that become closer and closer to zero or one as the ensemble is made larger.

Another strategy to identify alternatives with probabilities near zero and one is to consider probabilities conditioned on other information besides that given in the theory of dynamics and the initial condition of the universe. Present theories of the initial condition do not predict the observed orbit of Mars about the sun with any

---

[c] *How* close to zero or one must probabilities be for near certain predictions depends on the circumstances in which they are used, as I have discussed elsewhere.[5]

significant probability. But they do predict that the *conditional* probability for the observed orbit is near one *given* a few previous observations of Mars' position. Such conditional probabilities are what are used in the rest of the sciences when they are viewed from the perspective of quantum cosmology, as we shall discuss in more detail in the subsequent sections.

In the following discussion it will be helpful to use just a little of the mathematics of quantum mechanics to discuss quantum cosmology.[d] For simplicity and definiteness, let us continue to discuss the model universe of N particles in a box. The quantum initial state of this model universe is represented by a state vector $|\Psi\rangle$ in a Hilbert space, or equivalently by a wave function of the coordinates of all the particles in the box:

$$\Psi\,(\vec{x}_1, \ldots, \vec{x}_N). \tag{3}$$

General alternatives at a moment of time whose probabilities we might want to consider can always be reduced to a set of "yes-no" alternatives. For instance, questions about the position of a particle can be reduced to questions of the form: "Is the particle in this region — yes or no?", "Is the particle in that region — yes or no?", etc. A set of "yes-no" alternatives at one moment of time, say $t = 0$, is represented by a set of orthogonal projection operators $\{P_\alpha\}$, $\alpha = 1, 2, \ldots$ — one projection operator for each alternative. (A projection operator is one whose square is equal to itself.) The projection operators satisfy

$$\sum\nolimits_\alpha P_\alpha = I, \qquad \text{and} \qquad P_\alpha P_\beta = 0, \quad \alpha \neq \beta, \tag{4}$$

showing mathematically that they represent an exhaustive set of exclusive alternatives. The same set of alternatives at a later time t is represented by a set of (Heisenberg picture) projection operators $\{P_\alpha(t)\}$. The time dependence of each $P_\alpha(t)$ is given by

$$P_\alpha(t) = e^{iHt} P_\alpha e^{-iHt} \tag{5}$$

---

[d]For more details at an elementary level see Ref. 6, and for greater depth see Ref. 7.

where H is the Hamiltonian encapsulating the basic dynamical theory. The probability predicted for alternative α at time t is

$$p(\alpha) = \|P_\alpha(t)|\Psi\rangle\|^2\,, \tag{6}$$

where $\|\cdot\|$ means the length of the Hilbert space vector inside. For this model, the Hamiltonian H specifies the first of the two elements of a basic physical theory described in Sec. 3 — the fundamental theory of dynamics. The state vector $|\Psi\rangle$ or equivalently the wave function $\Psi(\vec{x}_1, \ldots, \vec{x}_n)$ specifies the second element — the initial condition.

Probabilities for alternatives at a moment of time are not the most general predictions of quantum mechanics. More generally, one can ask for the probabilities of sequences of sets of alternatives at a series of different times $t_1 < t_2 < \cdots < t_n$ making up a set of alternative *histories* for the universe. Each history corresponds to a particular sequence of alternatives $(\alpha_1, \ldots, \alpha_n)$ and is represented by an operator that is the chain of projections corresponding to the sequence of alternatives

$$C_\alpha = P^n_{\alpha_n}(t_n)\cdots P^1_{\alpha_1}(t_1). \tag{7}$$

Here, the index α is shorthand for the whole sequence $(\alpha_n, \ldots, \alpha_1)$ and the superscripts on the P's indicate that different sets of alternatives can be considered at different times. When the operator C_α is applied to the initial state vector $|\Psi\rangle$, one obtains the *branch state vector* $C_\alpha|\Psi\rangle$ corresponding to the history α. The probability of the history α is the length of the history's branch state vector:

$$p(\alpha) = \|C_\alpha|\Psi\rangle\|^2. \tag{8}$$

Probabilities of histories are essential for predicting such everyday things as the orbit of the moon, which is a sequence of positions at a series of times.

We can now begin to analyze the question of what is predicted in quantum cosmology and what is not predicted. The most characteristically quantum mechanical limitation on what can be predicted is that not every set of alternative histories that may be described can be assigned probabilities by the theory because of quantum mechanical interference. That is very clearly exemplified in the two-slit thought experiment illustrated in Fig. 2. Electrons proceed from an electron gun through a barrier with two slits on their way to detection at a screen. Passing through slit A or slit B defines two alternative histories for the electrons arriving at a fixed point y on the screen. In the usual story, if we have not measured which slit an electron had passed through, then it would be inconsistent to predict probabilities for these alternative histories. It would be inconsistent because the probability to arrive at y would not be the sum of the probability to pass through A to y and the probability to pass through B to y:

$$p(y) \neq p_A(y) + p_B(y). \tag{9}$$

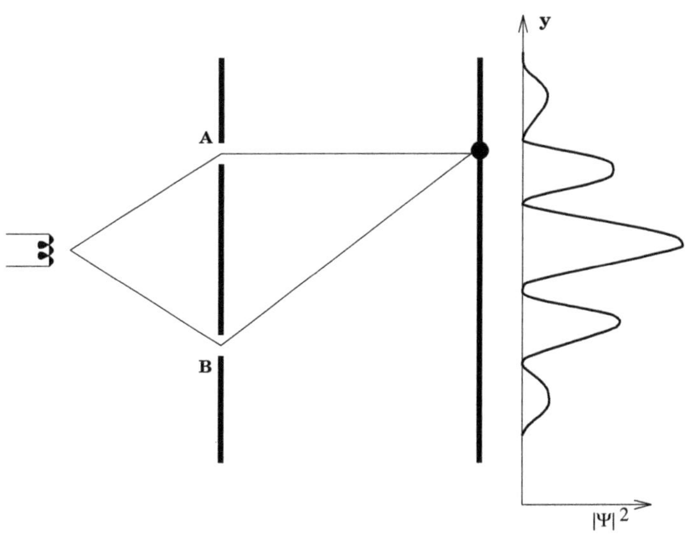

Fig. 2. The two-slit experiment. An electron gun at left emits an electron traveling towards detection at a screen at right, its progress in space recapitulating its evolution in time. In between, there is a barrier with two slits. Two possible histories of an electron arriving at a particular point on the screen are defined by whether it went through slit A or slit B. In quantum mechanics, probabilities cannot be consistently assigned to this set of two alternative histories because of quantum mechanical interference between them. However, if the electron interacts with an apparatus that measures which of the slits it passes through, then interference is destroyed, the alternative histories decohere, and probabilities can be assigned to the alternative histories.

That is because in quantum mechanics, probabilities are the squares of amplitudes and

$$|\psi_A(y) + \psi_B(y)|^2 \neq |\psi_A(y)|^2 + |\psi_B(y)|^2. \tag{10}$$

It is not that we are ignorant of which slit an electron passes through, so that the probabilities are 50–50. It is inconsistent to discuss probabilities at all.

Quantum mechanics, in any of its various levels of formulation, therefore contains a rule specifying which sets of alternative histories may be assigned probabilities and which may not. In the most general context of the quantum mechanics of the universe that rule is as follows[8–10]: Probabilities may be consistently assigned to just those sets of histories for which there is vanishing interference between the individual members of the set as a consequence of the universe's initial state $|\Psi\rangle$. Such sets of histories are said to decohere. The condition for a decoherent set of histories is that the branches of the initial state $C_\alpha|\Psi\rangle$ corresponding to individual histories be mutually orthogonal:

$$\langle\Psi|C_\alpha^\dagger \cdot C_\beta|\Psi\rangle \approx 0, \quad \alpha \neq \beta. \tag{11}$$

The most general probability sum rules are satisfied as a consequence. Consistency limits the predictions of quantum theory to the probabilities of *decoherent* sets of alternative histories.

As an example of how the decoherence of a set of histories comes about, think about a single millimeter-sized dust grain in a quantum state that is a superposition of two positions about a millimeter apart located deep in intergalactic space. Consider alternative histories of the position of this particle at a sequence of a few times. (The P's in (7) would then be projections onto ranges of this position.) Were the particle isolated, this situation would be analogous to the two-slit experiment, and histories of differing positions would not decohere. However, even deep in space this particle is not isolated. The all-pervasive light from the big bang illuminates the particle, and about 10^{11} cosmic background photons scatter from it every second. Through these interactions, this seemingly isolated dust grain becomes correlated with radiation in a part of the universe whose size is growing at the speed of light. The two states with different positions become correlated with two different, nearly orthogonal states of the radiation after a time of about a nanosecond. By this means, a branch of the initial state in which the grain is initially at one position becomes orthogonal to a branch in which the grain is a millimeter away. Decoherence of alternative histories of position has been achieved because the relative phase between states of different positions has been dissipated by feeble interactions with the background radiation. Mechanisms such as this are widespread in the universe and typical of those effecting the decoherence of histories of the kinds of classical variables we like to follow. (See, e.g. Refs. 11 and 12.)

In the above example, decoherence of alternative histories of the position of the dust grain is achieved at the cost of ignoring the photons that are effecting the decoherence. That is an example of *coarse-graining*. Were we to consider a set of alternative histories of states of the cosmic background radiation as well as the position of the grain we would be, in effect, following all possible phase information. Such a set of alternative histories would generally not decohere. Except for trivial cases, sets of histories must describe *coarse-grained* alternatives in order for probabilities to be predicted at all. This necessary imprecision is a genuine limit to what can be predicted in quantum cosmology, in contrast to the limits of the kind associated with the uncertainty principle, which are merely limits to the applicability of classical modes of description.[e]

---

[e]Decoherence also implies another kind of limit to classical predictability which should be mentioned, although we cannot discuss it in any depth here. As described, realistic mechanisms of decoherence involve the dispersal of phase information concerning a subsystem into an environment that interacts feebly with it. Those interactions produce noise which limits the classical predictability of the subsystem. Thus, for classical predictability, appropriate and sufficient coarse-graining is needed for the decoherence necessary to predict probabilities at all. But further coarse-graining is needed for the subsystem to have sufficient inertia to resist the noise that those mechanisms of decoherence produce and thereby become classically predictable. (For an introductory discussion, see Ref. 13. For a more detailed one, see Ref. 7.)

We thus have the picture of a vast class of all possible sets of alternative histories and a smaller subclass of decoherent sets of histories for which quantum theory predicts probabilities. For almost none of these decoherent sets there exists a history predicted with certainty on the basis of the initial state alone. If one history has probability one, then all alternatives to it must have probability zero. Suppose we have such a set, and let α_c be the label of the certain history, then from (8)

$$\|C_\alpha|\Psi\rangle\|^2 = 0, \quad \alpha \neq \alpha_c, \tag{12}$$

which implies

$$C_\alpha|\Psi\rangle = 0, \quad \alpha \neq \alpha_c. \tag{13a}$$

Then, since $\Sigma_\alpha C_\alpha = I$ as a consequence of (4), we also have

$$C_\alpha|\Psi\rangle = |\Psi\rangle, \quad \alpha = \alpha_c. \tag{13b}$$

Decoherence, see Eq. (11), is then automatic for such sets of histories in which one is certain.

Equation (13b) shows that operators of histories that are predicted with certainty act as projection operators on the initial state. An alternative predicted with probability one is thus mathematically equivalent to the alternative corresponding to the question "Is the universe in state $|\Psi\rangle$"? These are very special questions. Out of the class of sets of decoherent histories, almost none correspond to sets in which one history is a certain prediction of the initial condition and the theory of dynamics alone.

In quantum cosmology we might hope that some gross features of the universe might be among those that are predicted with near certainty from the initial condition and dynamics alone. These include features such as the approximate homogeneity and isotropy of the universe on scales above several hundred megaparsecs,[f] its vast age after the big bang when measured on elementary particle time scales, and certain features of the spectrum of density fluctuations that grew

[f] A megaparsec (Mpc) is a convenient unit for cosmology. One megaparsec = 3.3 million light years = 3.1×10^{24} cm. The size of the universe visible today is of the order of several thousand megaparsecs.

to produce the galaxies. On more familiar scales we may hope that the laws of the initial condition and dynamics would predict the homogeneity of the thermodynamic arrow of time and the wide range of scale and epoch on which the regularities of classical physics are exhibited. There has even been speculation that phenomena on very small scales, such as the dimensionality of spacetime or certain effective interactions of the elementary particles at accessible energy scales, may be near certain predictions of the initial condition and dynamics. But there is little reason to suspect that a simple theory of the initial condition and fundamental dynamics will predict anything about the behavior of the New York stock market with near certainty and about a great many other interesting phenomena as well. That is why you can't get rich knowing the wave function of the universe!

The situation is very different if information beyond laws of dynamics and the initial condition is supplied and probabilities conditioned on that information are considered. There are many sets of *conditional* probabilities in which one member of the set is near certain. These conditional probabilities are the basis of prediction in all the other sciences when viewed from the perspective of quantum cosmology, as will be described in the next section.

I have described various limitations on what can be predicted in quantum cosmology. Yet there is a sense in which we, as information gathering and utilizing physical systems, make use of only a small part of the possible predictions of quantum cosmology. That is because of our almost exclusive focus on alternatives defined in terms of the variables of classical physics — averages over suitable volumes of densities of energy and momentum, densities of nuclear and chemical species, average field strengths, etc. Such classical quantities are represented by quantum operators called quasiclassical operators. (They are termed *quasi*classical because they do not behave classically in all circumstances.) Certainly, our immediate experience can be described in terms of quasiclassical variables even when — as in the clicks of a Geiger counter — these variables do not obey deterministic classical laws.

Even in our theorizing about regions of space or epochs in time that are very distant from us, we often focus on histories of alternatives of quasiclassical operators. Only in the microscopic arena do we consider nonquasiclassical alternatives such as electron spin and coherent superpositions of position. Even then we typically consider such alternatives only when they are tightly correlated with a quasiclassical variable as in a measurement situation.

However, quantum field theory exhibits many more kinds of variables than the small set of quasiclassical ones. Decohering sets of histories can be constructed from alternative values of nonquasiclassical operators as well as from quasiclassical ones. Indeed, the quasiclassical sets of histories are but a small subset of the whole class of decohering histories. Quantum theory does not privilege one set of decohering histories over another. Probabilities are predicted for all such sets of alternatives. Histories of nonquasiclassical alternatives are not beyond reach. Suppose we were to make measurements of peculiarly quantum mechanical variables involving large numbers of particles in regions of macroscopic dimensions. The histories that would be relevant for the explanation of the outcomes of these measurements would not be histories of quasiclassical variables in these regions, but rather histories of the nonquasiclassical alternatives that were measured. The reason for our preference for quasiclassical sets of alternative histories, like all other questions concerning ourselves as particular physical systems, probably lies in our evolutionary history — not in the framework of quantum theory itself.

6. Differences between the Sciences

Using the conditional probabilities of quantum cosmology, a particular orbit of the earth about the sun could be predicted with near certainty given a few previous positions of the earth and a description of the earth and solar system in terms of the fundamental fields which are the language of quantum cosmology. The probabilities for the outcome of chemical reactions become near certain predictions of quantum cosmology given a description in terms of fundamental fields of the molecules involved and the conditions under which they interact. The probabilities for the behavior of sea turtles in particular environments could, in principle, become predictions of quantum cosmology given a description of sea turtles and their environments in the language of quantum cosmology. Even the probabilities for the different behaviors of human beings — both individually and collectively — could in principle be predicted given a sufficiently accurate description of the individuals, their history, their environment and their possible modes of behavior. In this way, *every* prediction in science could be viewed in terms of a conditional probability in quantum cosmology. Why, then, do we have separate sciences of astronomy, chemistry, biology, psychology, and so on? The answer, of course, is that it is neither especially interesting nor practical to reduce the predictions of these sciences to a computation in quantum cosmology.

One measure of the difference between the sciences, is how sensitive the regularities they study are to the forms of the initial condition of the universe and the fundamental theory of dynamics. The phenomena studied in chemistry, fluid mechanics, geology, biology, psychology and human history, depend only very little on the particular form of the initial condition. All of these sciences, especially chemistry, depend on the form of the theory of dynamics in some approximation, but as we move through the list we are moving towards the direction of the study of the regularities of *increasingly specific subsystems of the universe.* Specific subsystems can exhibit more regularities than are implied generally by the laws of dynamics and the initial condition. The explanation of these regularities lies in the origin and evolution of the specific subsystems in question. Naturally these regularities are more sensitive to this specific history than they are to the form of the initial condition and dynamics. That is especially clear in a science like biology. Of course, living systems conform to the laws of physics and chemistry, but their detailed form and behavior depend much more on the frozen accidents of several billion years of evolutionary history on a particular planet moving around a particular star than they do on the details of superstring theory or the "no-boundary" initial condition of the universe.[g] Conversely, the phenomena studied by these sciences do not help much

[g]See, e.g. Ref. 2 for more discussion in greater depth and examples from this point of view.

Table 1. Some differences between the sciences.

	Length of Coarse-grained Description of Alternatives	Length of Coarse-grained Description of Conditions	Length of Computation of Conditional Probabilities
Classical Physics	Very Short	Short	Very Short
Astronomy	Short	Short	Short — Long
Fluid Mechanics	Short — Long	Short	Short — Long
Chemistry	Short — Long	Short — Long	Long — Very Long
Geology	Long	Long	Long
Biology	Long — Very Long	Long — Very Long	Long — Very Long
Psychology	Very Long	Very Long	Very, Very Long(?)

in discriminating among different theories of the initial condition and dynamics. It is for such reasons that it is not of pressing interest — either for other areas of science or for quantum cosmology itself — to express the predictions of such phenomena as quantum cosmological probabilities, even though it is in principle possible to do so.

Even if we were to wish to carry out a calculation of the conditional probabilities in quantum cosmology necessary for prediction in the other sciences, an examination of what it would take yields three measures which distinguish the other sciences from quantum cosmology and from one another. To yield a conditional probability, the theory requires:

- A description of the coarse-grained alternatives whose probabilities are to be predicted in terms of fundamental quantum fields.
- A description of the circumstances on which the probabilities are conditioned in terms of fundamental quantum fields.
- A computation of the conditional probabilities.

The table above shows some simplistic guesses of the lengths of these three parameters for typical problems in the various sciences. We can discuss a few of these.

By classical physics, I simply mean Newton's laws of mechanics and gravity, the laws of continuum mechanics, Maxwell's electrodynamics, the laws of thermodynamics, etc. — in short, the basic laws of physics as they were formulated in the 19th century. (I do not mean some specific application of these laws, as to the breaking of ocean waves.) Classical physics might almost be counted as a science separate from physics, for the laws of classical physics do not hold universally, but only for certain kinds of subsystems in particular circumstances. However, the table shows the reason that these laws are usually considered part of the science of physics. There is just a short list of quasiclassical variables (volume averages of fields, densities of energy, momentum, chemical composition, etc.) whose ranges of values define the coarse-grained alternatives of classical physics.[7] It is a somewhat longer business to spell out, in quantum mechanical terms, the circumstances in which classical physics applies. But the derivation of the laws of classical physics can be as short as a journal paper.[h]

As we move down the table to astronomy, we encounter more specific classes of physical systems — stars, clusters, galaxies, etc. However, the difficulty of obtaining data on such distant objects prevents us from learning much individual detail. The length of the coarse-grained descriptions of both conditions and alternatives are typically short. The computations utilizing the equations of classical physics, however, range from very short dimensional estimates to long simulations of supernovae explosions.

In fluid mechanics, we encounter a wide variety of particular phenomena arising from the differential equations of classical physics. One has only to mention laminar flow, turbulence, cavitation, percolation, convection, solitons, shock waves, detonation, superfluidity, clouds, dynamos, internal waves, ocean waves, the weather, etc., to recall something of the richness of phenomena studied in this subject. The coarse-grainings describing the alternative behaviors of fluids can sometimes be long, although the description of the conditions is usually shorter. Many of these phenomena can be simulated on computers today by solving the differential equations of classical physics. These calculations could be considered calculations in quantum cosmology were we to append to them a standard description of the alternatives and conditions, together with the computations that justify the use of these approximate equations in terms of the fundamental theory of quantum fields and the initial condition.

---

[h]For a one-journal-paper derivation from the quantum cosmological point of view, see, e.g. Ref. 7.

The description of the molecules of interest in chemistry can vary from short — as in typical chemical formulae — to long, as in the base sequence in human DNA. There is a similar range of conditions for chemical reactions ranging from a few reagents in a test tube to the interiors of cells. Quantum chemists *can* compute certain chemical properties, such as those of chemical bonds, directly from the equations of an effective low-energy theory of the elementary particles, but these computations can only be described as long.

In geology we have a science concerned with a very specific system — the earth — observed in considerable detail. A lengthy string is needed to describe the alternative configurations and compositions of the materials on the surfaces in the details that we know. A long history would have to be described to set the conditions for calculating the probabilities, and calculations of these probabilities, even assuming the laws of classical physics, would still be very long.

The reader probably needs little convincing that the description of the behavior of a complex biological organism plus its evolutionary history and its present environment in the language of quantum field theory would be a long business indeed! We should not pretend that we are anywhere close to being able to give such a description or to being able to carry out the relevant computations of conditional probabilities in quantum cosmology. Psychology and human history are yet more difficult. We may have a rough idea of how to describe the action of a bird's beak in the language of quantum field theory, but very little idea of the coarse-grainings that describe an individual's thoughts and emotions or the vicissitudes of empires.

Dear reader, please do not write to the author concerning the inadequacies of the above discussion. He is aware that the boundaries between the sciences are not precisely defined and that there is wide variation in these three parameters within each one. In astronomy, for example, the description of our nearest star — the sun — can be just as complex as that of any phenomena in fluid mechanics (and indeed it is a part of fluid mechanics). The smallest self-reproducing biological units may be simulable by conceivable computers.[14] There may be universal principles of mind which derive rather directly from the basics of physics.[15] The important point is, that at a basic level, every prediction in science may be viewed as the prediction of a conditional probability for alternatives in quantum cosmology and that the probabilities relevant to different sciences may be distinguished, in part, by their sensitivity to the theories of the initial condition and dynamics, by the length of the description of the alternatives, by the length of the description of the conditions, and by the length of the computation needed to produce them.

7. Limits to Implementation

The preceding section discussed some limitations of practice in our effort to implement the predictions of quantum cosmology for interesting specific subsystems in the universe. These limits were of the general character described in Sec. 3.2. Are there more fundamental and general limits arising from computational intractability? Quantum cosmology provides some examples.

There are physical reasons for computational intractability and mathematical ones. Landauer[16] has raised the issue of whether there are predictions whose computation would require more resources of space, material and time than are available in the universe. Quantum cosmology may also present an example of what might be regarded as an extreme example of mathematical computational intractability. There is some evidence that the wave function of the universe might be noncomputable in the technical mathematical sense. (See Chapter 1.)

One idea for a theory of the wave function of the universe is the "no-boundary" proposal.[17] To understand a little of this idea assume, for simplicity, that the universe is spatially closed and that gravity is the only quantum field. A cosmological wave function is then a function of the possible geometries of three-dimensional space. The "no-boundary" idea is that the value of the wave function of the universe Ψ at one particular spatial geometry is a sum over all locally Euclidean four-dimensional geometries which have this three-dimensional space as a boundary and *no other boundaries*. Each four-dimensional geometry $\mathcal{G}$ in the sum is weighted by $\exp(-I[\mathcal{G}])$, where $I[\mathcal{G}]$ is the classical action for the geometry. Mathematically, a geometry is a specification of a notion of distance (a metric) on a space such that any small region can be smoothly mapped into a region of flat Euclidean space (a manifold). A sum over geometries would therefore naturally include a sum over manifolds as well as a sum over metrics. By suppressing two of the four dimensions we can give a crude pictorial representation of this double sum, as shown in Fig. 3.

The mathematics of quantum gravity have not been developed to the point that we have a precise mathematical formulation of what the relation schematically represented in Fig. 3 might mean. One idea for making it precise is to approximate each term in the sum by a manifold constructed of flat four-simplices — the four-dimensional analogs of triangles in two dimensions and tetrahedra in three dimensions. The two-dimensional analog of such a simplicial manifold would be a surface made up of triangles like a geodesic dome as illustrated in Fig. 4. To calculate in four dimensions the sum crudely pictured in Fig. 3, one would proceed as follows: Choose a large number of four-simplices N. Find all possible manifolds that can be made by joining these four-simplices together. Choose *one* such assembly to represent each manifold in the sum. Integrate $\exp(-I[\mathcal{G}])$ over the edge lengths of the simplices that are compatible with the triangle and similar inequalities to approximate the sum over metrics. Sum the result over all manifolds. Take the limit as $N \to \infty$. That is one possible way the sum over geometries in the "no-boundary" proposal for the wave function of the universe might be implemented.[i]

A computer program to carry out this task would first have to try all possible ways of assembling N four-simplices together and reject those which do not give a manifold. This is already a formidable mathematical problem and it has only been recently proven that an algorithm exists to carry out this computation for four-dimensional manifolds.[19,20] The next step would be for the computer to take this list of four-manifolds and eliminate duplications. However, it is known that the issue of whether two simplicial four-manifolds are identical is undecidable.[j] More precisely, there does not exist a

[i]For more details and references to the earlier literature see, e.g. Ref. 18.
[j]For a review, see Ref. 21.

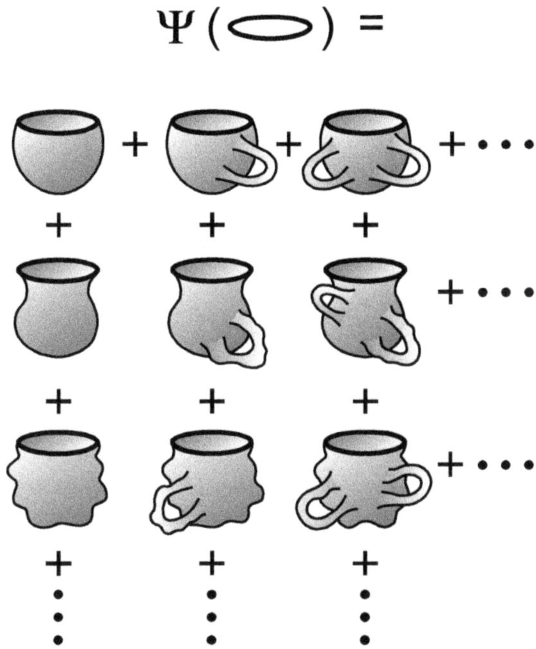

Fig. 3. The wave function of the universe as a sum over manifolds and metrics. This figure uses two-dimensional analogs to illustrate some of the ideas that enter into the construction of the "no-boundary" wave function of the universe. That wave function is a function of three-dimensional spatial geometries, one of which is represented here in two fewer dimensions by the heavy circular curve. For that given three-geometry, the "no-boundary" wave function is a sum over Euclidean four-geometries that have it as one boundary and *no other boundary.* This sum can be divided into a sum over four-manifolds and a sum over different four-metrics on those manifolds. The two-dimensional analog of this sum is shown above. The surfaces in each column represent different metrics on the *same* manifold. The manifolds in each column are the same because the surfaces can be smoothly deformed into one another by changing their shape. The metrics are different from one surface to another in a given column because the distance between two points is generally different from one shape to another. For example, the overall surface area may differ from one shape to another. The two-dimensional surfaces in different columns are different manifolds because they have different numbers of handles, and surfaces with different number of handles cannot be smoothly deformed into one another. A sum over manifolds is thus analogous to the sum over columns. A sum over metrics is analogous to the sum over different surfaces in each column.

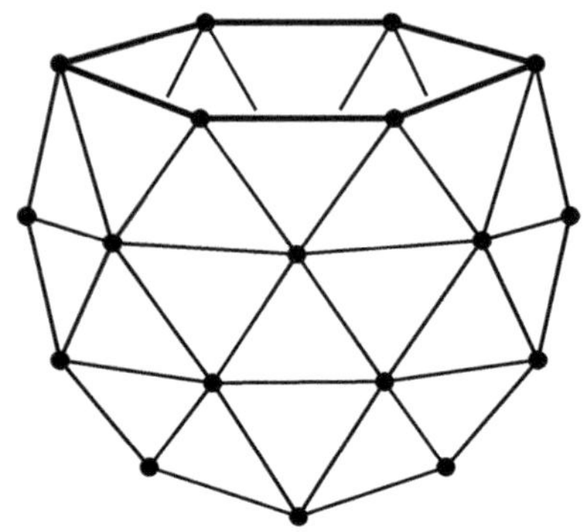

Fig. 4. A smooth two-dimensional surface may be approximated by an assembly of flat triangles like a geodesic dome. In an analogous way, a curved four-dimensional geometry may be approximated by an assembly of four-simplices — the four-dimensional analog of two-dimensional triangles or three-dimensional tetrahedra. The four-simplices must be assembled so as to make a manifold — a space such that any small region can be smoothly mapped to a region of Euclidean space. Suppose we are given N four-simplices. Assembling them in different ways can give different manifolds (the different vertical columns in Fig. 3). Different assignments of lengths to the edges give different sizes and shapes on a given manifold, that is, a different metric (the different shapes within each column in Fig. 3). A sum over geometries, which is a sum over manifolds and metrics, may therefore be approximated by choosing one assembly to represent each manifold in the sum, integrating over its possible edge-lengths, and summing over all manifolds. The resulting sum may be a non-computable number because there is no algorithm for deciding when two simplicial four-manifolds are identical.

computer program which, *for any* N, can compare two input assemblies of N four-simplices making up manifolds, and halt after having printed out "yes" if the manifolds are identical and "no" if they are not.

This suggests that the wave function of the universe defined by a sum over geometries that includes a sum over manifolds is a non-computable number.[k] However, appearances can be deceptive. Whether a number is non-computable or not, is a property of

[k]We are specifically assuming the Turing model of computability.

the number and not of the way it is represented. Merely exhibiting one non-computable representation like the series in Fig. 3 does not establish that there is no other representation in which it *is* computable. Demonstrating non-computability in such cases is likely to be a difficult mathematical problem.

This suggested non-computability of sums over topologies has been taken as motivation for modifying the theory of the initial condition so that it is clearly computable.[22,23] But suppose that the wave function of the universe were non-computable. What would be the implications for science? Bob Geroch and I analyzed the implications of non-computability for physics in 1986.[24] Our conclusion was that the prediction of non-computable numbers would not be a disaster for physics. That is because at any one time, one needs theoretical predictions only to an accuracy consistent with experimental possibilities. Suppose, for example, it was sufficient for comparison with present observations to know the wave function of the universe to an accuracy of 10%. Suppose further it could be shown that to achieve this accuracy, only simplicial manifolds with less than 100 four-simplices need be included in the series defining the wave function. The theorem concerning the non-existence of an algorithm for deciding the identity of simplicial four-manifolds refers to an algorithm that would work for *any* N. It does not rule out establishing the identity of two four-manifolds with less than 100 four-simplices. Indeed, being a problem that involves a finite number of specific cases, one imagines it could be solved with sufficient work on those cases. What the theorem ensures is that, if observations improve, and the wave function is later needed to an accuracy of 1%, requiring manifolds with a larger number of four-simplices (say 10,000), a new intellectual effort will be required to compute it. The algorithms that worked for manifolds assembled from less than 100 four-simplices are unlikely to work for manifolds assembled from less than 10,000 four-simplices.

Thus, the prediction of non-computable numbers would not mean the end of comparison between theory and observation. It would mean that the process of computing the predictions could be as conceptually challenging a problem, as posing the theory itself.

8. Limits to Verification

Quantum mechanics predicts the probabilities of alternative histories of the universe. We cannot interpret these probabilities as predictions of frequencies which are accessible to test, for we have access to but a single universe and but a single history of it. Our ability to test the theory or to infer the theory from empirical data is therefore limited.

An example of current interest is "cosmic variance" in the predictions of temperature fluctuations in the cosmic background radiation. The observed pattern of temperature determines the correlation function $C(\theta)$ between the temperature fluctuations δT at two different directions $\vec{n}_1$ and $\vec{n}_2$ in the sky separated by an angle θ:

$$C(\theta) = \left\langle \frac{\delta T(\vec{n}_1)}{T} \; \frac{\delta T(\vec{n}_2)}{T} \right\rangle, \tag{14a}$$

where $\langle \cdot \rangle$ denotes an average over all directions $\vec{n}_1$ and $\vec{n}_2$ such that $\vec{n}_1 \cdot \vec{n}_2 = \cos\theta$. This correlation function can be conveniently expanded in spherical harmonics $P_\ell(\cos\theta)$:

$$C(\theta) = \sum_{\ell=0}^{\infty} \frac{2\ell+1}{4\pi} \, C_\ell P_\ell(\cos\theta). \tag{14b}$$

The coefficients C_ℓ so defined, are the way the data from observations are usually quoted and are the objects of theoretical prediction.[1]

The probabilities of temperature fluctuations in the cosmic background radiation are predicted from a spectrum of fluctuations implied by the initial quantum state. The probabilities of these fluctuations are thus a detailed prediction of quantum cosmology that stem directly from the initial condition. They are not conditional probabilities requiring other information. The theory does not predict high probabilities for particular fluctuations in temperature at particular locations in the sky. Rather, it predicts distributed probabilities for these fluctuations (see e.g. Ref. 26), or, equivalently, the probabilities for various values of the C_ℓ. The expected value and the standard deviation of this distribution is shown in Fig. 5. The width of the distribution is "cosmic variance".

---

[1]See e.g. Ref. 25 for a detailed review.

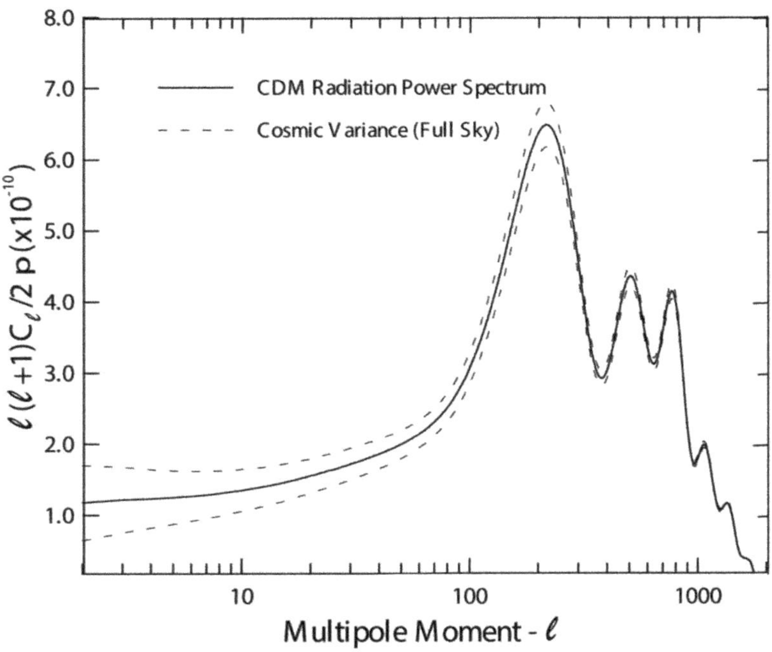

Fig. 5. Cosmic variance. The heavy line on this figure shows the expected value of the multipole moments of the two-point correlation function defined by Eq. (14) for temperature fluctuations in the cosmic background radiation as predicted from the probabilities of these fluctuations arising from a simple theory of the universe's initial condition. The dotted lines show the standard deviation of the predicted distribution called the "cosmic variance". Observations of our single universe yield the correlation function and one particular distribution of observed multipole moments. These observations will not distinguish two theories of the initial condition whose "cosmic variance" surrounds the observed distribution. [Graph by J. Gundersen.]

We cannot test these probabilistic predictions for the cosmic background temperature fluctuations by measuring these fluctuations in a large number of identical cases. We have only one universe and only one set of observed temperature fluctuations! An observed distribution of C_ℓ's inside this "cosmic variance" would be confirmation of the theory of the initial condition. An observed distribution outside it would be evidence against it. However, observations will not distinguish two theories of the initial condition whose "cosmic variance" surrounds the observed distribution. Thus, for such probabilistic predictions we are inevitably limited in our ability to test a theory of the initial condition.

More generally, as mentioned above, a theory of the initial condition can be tested only through predictions whose probabilities are so near certain that we would reject the theory if they were not observed. The sets of histories which lead to near certain predictions are just a small set of those for which probabilities are predicted.

There are limits, therefore, to the process of inferring the initial state of the universe from observation. If $C_{\rm obs}$ is the operator describing the entirety of our collective observations, then strictly speaking all we can conclude about an initial state $|\Psi\rangle$ is that it is not such that

$$C_{\rm obs}|\Psi\rangle = 0. \tag{15}$$

That is not much of a restriction. For example, suppose that the projection $P_{\rm pres.\ data}$ represents all our present data, including our records of the past history. It is not possible on the basis of either present or future observations to distinguish an initial state $|\Psi\rangle$ from that defined by

$$|\Psi'\rangle = \frac{P_{\rm pres.\ data}|\Psi\rangle}{\|P_{\rm pres.\ data}|\Psi\rangle\|}. \tag{16}$$

Retrodictions of the past from the present data and $|\Psi\rangle$ could differ greatly from those from the same data and $|\Psi'\rangle$.[m] But retrodictions are not accessible to experimental check and therefore do not distinguish the two candidate initial states. The two initial conditions $|\Psi\rangle$ and $|\Psi'\rangle$ could differ greatly in complexity if the description of $|\Psi\rangle$ is short but that of $P_{\rm pres.\ data}$ is long, and we may choose between these physically equivalent possibilities on the basis of simplicity. The search for a theory of the initial condition must therefore rely

[m]Unlike classical physics, where the past can be retrodicted from sufficiently precise present data alone, retrodiction in quantum theory requires present data *and* the initial condition of the system in question. For further discussions see, e.g. Ref. 5, Sec. II.3.1.

on the principles of simplicity and connection with the fundamental dynamical theory in an essential way.

Why is it that the fundamental dynamical theory — the Hamiltonian of the elementary particle system — seems so much more accessible to experimental test and so much easier to infer from observational data than the theory of the initial condition? Strictly speaking, it is not. Were the Hamiltonian of the elementary particle system to vary on cosmological scales — to be a function of spacetime position of the form $H(x)$ — then inferring H would be just as difficult a process as inferring the initial $|\Psi\rangle$. However, we *assume* the principle that the elementary particle interactions are local in space and time. With that assumption, the Hamiltonian describing these interactions becomes accessible to many local tests on all sorts of scales ranging from those accessible in particle accelerators to the expansion of the universe itself. The problem of inferring the initial $|\Psi\rangle$, therefore, is not so very different from that of inferring H in making use of the theoretical assumptions. It is just that the assumption of locality is so well-adapted to the quasiclassical realm of familiar experience that many more tests can be devised on small scales of a theory of H than we are ever likely to find of a theory of $|\Psi\rangle$ on cosmological scales.

9. Conclusions: The Necessity of Limits to Scientific Knowledge

If the world is complex and the laws of nature are simple, then there are inevitable limits to science. Not everything that is observed can be predicted; only certain regularities of those observations can be predicted. Even given a theory, computational intractability or observational difficulty may limit our ability to predict. In a world of finite observations, there are inevitable limits to our ability to discriminate between different theories by the process of induction and test.

Quantum cosmology — the most general context for prediction in science — exhibits examples of all three kinds of limits to scientific knowledge. There are only very few predictions of useful probabilities that are conditioned solely on simple theories of dynamics and the universe's initial condition. There is a far richer variety of useful probabilities conditioned on further empirical data that are the basis for most of the predictions in science. There are some indications that the "no-boundary" initial wave function is non-computable in the technical sense of yielding non-computable numbers. That does not limit our ability to extract predictions from the theory in principle, but may be an indication that predictions sensitive to the topological structure of spacetime on small scales, could be conceptually challenging to compute. Finally, it is possible to exhibit different theories of the initial condition with identical present and future predictions which can only be discriminated apart by an appeal to principles of simplicity and harmony with fundamental dynamical laws.

We should not conclude a discussion of limits in science without mentioning that science is useful *because* of its limits. Complex adaptive systems are successful in evolution and individual behavior because they identify and exploit the regularities that the universe exhibits. Scientific theories predict what these regularities are and explain their origin. Theories can be used to estimate how tractable these predictions are to compute or practical to measure. By comparing different theories induced from the same data, an idea can be gained of the reliability of our predictions. The existence of limits of the kind we have discussed, therefore, does not represent a failure of the scientific enterprise. Limits are inherent in the nature of that enterprise, and their demarcation is an important scientific question.

Acknowledgments

The author is grateful to M. Gell-Mann for discussions of a number of the issues discussed in Sec. 8 over many years. Thanks are also due to H. Morowitz and R. Shepard for instruction in certain aspects of biology and psychology respectively, and to J. Gundersen for supplying several of the figures. This work was supported in part by NSF grants PHY90-08502, PHY95-075065 and PHY94-07194.

References

1. N. Bohr, *Atomic Physics and the Description of Nature* (Cambridge University Press, Cambridge, UK, 1934).
2. M. Gell-Mann, *The Quark and the Jaguar* (W. Freeman, San Francisco, 1994).
3. J. Halliwell, in *Quantum Cosmology and Baby Universes: Proc. 1989 Jerusalem Winter School for Theoretical Physics*, eds. S. Coleman, J. B. Hartle, T. Piran and S. Weinberg (World Scientific, Singapore, 1991), pp. 65–157.
4. C. H. Woo, *Phys. Rev. D* **39**, 3174 (1989); *Found. Phys.* **19**, 57 (1989); and in *Complexity, Entropy, and the Physics of Information*, ed. by W. Zurek (Addison Wesley, Reading, MA, 1990).
5. J. B. Hartle, in *Quantum Cosmology and Baby Universes: Proc. 1989 Jerusalem Winter School for Theoretical Physics*, eds. S. Coleman, J. B. Hartle, T. Piran and S. Weinberg (World Scientific, Singapore, 1991), pp. 65–157.
6. J. B. Hartle, in *Directions in General Relativity, Volume 1: A Symposium and Collection of Essays in Honor of Professor Charles W. Misner's 60th Birthday*, eds. B.-L. Hu, M. P. Ryan and C. V. Vishveshwara (Cambridge University Press, Cambridge, 1993), gr-qc/9210006.
7. M. Gell-Mann and J. B. Hartle, *Phys. Rev. D* **47**, 3345 (1993), gr-qc/9210010.
8. R. Griffiths, *J. Stat. Phys.* **36**, 219 (1984).
9. R. Omnès, *J. Stat. Phys.* **53**, 893 (1988); *ibid.* **53**, 933 (1988); *ibid.* **53**, 957 (1988); *ibid.* **57**, 357 (1989); *Rev. Mod. Phys.* **64**, 339 (1992); *Interpretation of Quantum Mechanics* (Princeton University Press, Princeton, 1994).
10. M. Gell-Mann and J. B. Hartle, in *Complexity, Entropy, and the Physics of Information*, SFI Studies in the Sciences of Complexity, Vol. VIII, ed. W. Zurek (Addison Wesley, Reading, MA) or in *Proc. 3rd Int. Symp. Foundations of Quantum Mechanics in the Light of New Technology*, eds. S. Kobayashi, H. Ezawa, Y. Murayama and S. Nomura (Physical Society of Japan, Tokyo, 1990).
11. E. Joos and H. D. Zeh, *Zeit. Phys. B* **59**, 223 (1985).

12. W. Zurek, *Phys. Rev. D* **24**, 1516 (1981); *ibid.* **26**, 1862 (1982).
13. J. B. Hartle, in *Proc. Cornelius Lanczos International Centenary Conf.*, North Carolina State University, December 1992, eds. J. D. Brown, M. T. Chu, D. C. Ellison and R. J. Plemmons (SIAM, Philadelphia, 1994), gr-qc/9404017; also J. B. Hartle, in the Proceedings of the SFI conference on *Fundamental Sources of Unpredictability* to be held on March 28–30, 1996.
14. C. M. Fraser *et al.*, *Science* **270**, 397 (1995).
15. R. N. Shepard, *Psychon. Bull. Rev.* **1**, 2 (1994), *World and Mind* (to be published).
16. R. Landauer, *IEEE Spectrum* **4**, 105 (1967); in *Proc. 3rd Int. Symp. Foundations of Quantum Mechanics in the Light of New Technology*, eds. S. Kobayashi, H. Ezawa, Y. Murayama and S. Nomura (Physical Society of Japan, Tokyo, 1990); and *Phys. Today* **44**, 23 (1990).
17. J. B. Hartle and S. W. Hawking, *Phys. Rev. D* **28**, 2960 (1983).
18. J. B. Hartle, *J. Math. Phys.* **26**, 804 (1985a).
19. H. Rubenstein, *The Solution to the Recognition Problem for* S^3, unpublished lectures, Haifa, Israel, 1992.
20. A. Thompson, *Math. Res. Lett.* **1**, 613 (1994).
21. W. Haken, in *Word Problems*, eds. W. W. Boone, F. B. Cannonito and R. C. Lyndon (North Holland, Amsterdam, 1973).
22. J. B. Hartle, *Class. Quant. Grav.* **2**, 707 (1985).
23. K. Schleich and D. Witt, *Nucl. Phys.* **402**, 411 (1993); *ibid.* **402**, 469 (1993).
24. R. Geroch and J. B. Hartle, *Found. Phys.* **16**, 533 (1986).
25. J. R. Bond, in *Cosmology and Large Scale Structure, Proc. Les Houches School, Session LX*, August 1993, ed. R. Schaefer (Elsevier Science Publishers, Amsterdam, 1995).
26. J. Halliwell and S. W. Hawking, *Phys. Rev. D* **31**, 1777 (1985).

CHAPTER 6

The Physics of 'Now'

The world is four-dimensional according to fundamental physics, governed by basic laws that operate in a spacetime that has no unique division into space and time. Yet our subjective experience of this world is divided into present, past and future. What is the origin of this division? What is its four-dimensional description? Is this the only way experience can be organized consistently with the basic laws of physics? This paper reviews such questions through simple models of information gathering and utilizing systems (IGUSes), such as ourselves.

Past, present and future are not properties of four-dimensional spacetime but notions describing how individual IGUSes process information. Their origin is to be found in how these IGUSes evolved or were constructed. The past, present and future of an IGUS is consistent with the four-dimensional laws of physics and can be described in four-dimensional terms. The present, for instance, is not a moment of time in the sense of a spacelike surface in spacetime. Rather, there is a localized notion of present at each point along an IGUS's world line. The common present of many localized IGUSes is an approximate notion appropriate when they are sufficiently close to each other and have relative velocities much less than that of light.

(continued overleaf)

http://arxiv.org/abs/gr-qc/0403001v2

Some features of the present, past and future organization are closely related to basic physical laws. For example, the retarded nature of electromagnetic radiation and the second law of thermodynamics are the likely reasons we cannot remember the future. But modes of organization that are different from present, past and future can be imagined that are also consistent with the basic laws of physics. We speculate why the present, past and future organization might be favored by evolution and suggest that therefore it may be a cognitive universal.

1. Introduction

A lesson from physics of the last century is that, up here, on length scales much greater than the Planck length, the world is four-dimensional with a classical spacetime geometry. There is neither a unique notion of space nor a unique notion of time. Rather, at each point in spacetime there are a family of timelike directions and three times as many spacelike directions. Yet, in this four-dimensional world, we divide our subjective experience up into past, present and future. These seem very different. We experience the present, remember the past, and predict the future. How is our experience organized in this way? Why is it so organized? What is the four-dimensional description of our past, present and future? Is this the only way experience can be organized? This paper is concerned with such questions.

The general laws of physics by themselves provide no answers. Past, present and future are not properties of four-dimensional spacetime. Rather, they are properties of a specific class of subsystems of the universe that can usefully be called *information gathering and utilizing systems* (IGUSes).[1] The term is broad enough to include both single representatives of biological species that have evolved naturally and mechanical robots that were constructed artificially. It includes human beings both individually and collectively as members of groups, cultures and civilizations.

To understand past, present and future it is necessary to understand how an IGUS employs such notions in the processing of information. To understand why is it organized in this way, it is necessary to understand how it is constructed and ultimately how it evolved. Questions about past, present and future therefore are most naturally the province of psychology, artificial intelligence, evolutionary biology and philosophy.[a]

However, questions concerning past, present and future cannot be completely divorced from physics. For instance, the notions must be describable in four-dimensional terms just to be consistent with the fundamental picture of spacetime. Further, as we will review, the distinctions between the past, present and future of an IGUS depend upon some of the arrows of time that our universe exhibits such as that summarized by the second law of thermodynamics. In the tradition of theoretical physics, we illustrate these connections with simple models of an IGUS — achieving clarity at the risk of irrelevance. Our considerations are entirely based on classical physics.[b] One such model IGUS — a robot — is described in Sec. 2. It is simple enough to be easily analyzed, but complex enough to suggest how realistic IGUSes distinguish between past, present and future. The four-dimensional description of this robot is discussed in Sec. 3. There we will see that the robot's present is not a moment in spacetime. Rather, there is a present at each instant along the robot's world line consisting of its most recently acquired data about its external environment. The approximate common notion of "now" that could be utilized by a collection of nearby robots moving slowly with respect to one another is also described.

Section 4 describes the connection of present, past and future with the thermodynamic arrow of time and the radiation arrow of time. It seeks to answer the question, "Could we construct a robot that would remember the future?" Section 5 describes alternative organizations of a robot's experience that are different from past, present and future. These are consistent with the four-dimensional laws of physics. But the possibility of these alternative organizations shows that past, present and future are not

[a]See, e.g. Ref. 2 for a collection of philosophical papers on time. This paper does not aim to discuss or resolve any of the philosophical debates on the nature of time.

[b]Arrows of time in the context of quantum mechanics, as well as the quantum mechanical arrow of time, are discussed in Ref. 3 in the framework of a time-neutral generalized formulation of quantum theory. The author knows of no obstacle of principle to extending the present classical discussion to quantum mechanics in that framework. For the special features of history in quantum mechanics see, e.g. Ref. 4.

consequences of these laws. We speculate, however, that, as a consequence of the laws of physics, the past, present and future organization may offer an evolutionary advantage over the other modes of organization. This supports a conjecture that past, present and future may be a cognitive universal[5] for sufficiently localized IGUSes.

2. A Model IGUS

Imagine constructing a robot which gathers and utilizes information in the following manner (Fig. 1):

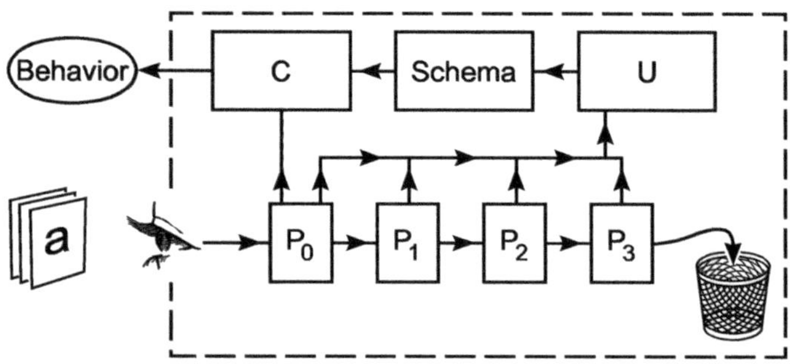

Fig. 1. Information processing in a model robot. The information flow in the robot described in the text is represented schematically in this diagram. The internal workings of the robot are within the dotted box; its external environment is without. At every proper time interval τ_* the robot captures an image of its external environment. In the example illustrated, this is of a stack of cards labeled a, b, c, etc., whose top member changes from time to time. The captured image is stored in register P_0 which constitutes the robot's present. Just before the next capture the image in P_3 is erased and images in P_0, P_1 and P_2 are shifted to the right, making room for the new image in P_0. The registers P_1, P_2 and P_3 therefore constitute the robot's memory of the past. At each capture, the robot forgets the image in register P_3. The robot uses the images in P_0, P_1, P_2 and P_3 in two processes of computation: C ("conscious") and U ("unconscious"). The process U uses the data in all registers to update a simplified model or schema of the external environment. That is used by the robot in C together with the most recently acquired data in P_0 to make predictions about its environment to the future of the data in P_0, make decisions, and direct behaviors accordingly. The robot may therefore be said to experience (through C) the present in P_0, predict the future, and remember the past in P_1, P_2 and P_3.

Information Gathering: The robot has $n+1$ memory locations P_0, $P_1, \ldots, P_n$, which we call "registers" for short. These contain a time series of images of its external environment assembled as follows: At times separated by intervals τ_*, the image in register P_n is erased and replaced by the image in P_{n-1}. Then the image in P_{n-1} is erased and replaced by the image in P_{n-2}, and so on. For the last step, the robot captures a new image of its external environment and stores it in register P_0. Thus, at any one time, the robot possesses a coarse-grained image history of its environment extending over a time $(n+1)\tau_*$. The most recent image is in P_0; the oldest is in P_n.

Information Utilization: The robot employs the information in the registers $P_0, P_1, \ldots, P_n$ to compute predictions about its environment at times to the future of the data in P_0 and direct its behavior based upon these predictions. It does this in two steps employing two different processes of computation:

- *Schema:* The robot's memory stores a simplified model of its environment containing, not all the information in $P_0, P_1, \ldots, P_n$, but only those parts important for the robot's functioning. This model is called a *schema.*[1] At each time interval τ_*, the robot updates its schema, making use of the information in P_0 by a process of computation we denote by U.

  The schema might contain the locations and trajectories of food, predators, obstacles to locomotion, fellow robots, etc. It might contain hard-wired rules for success (e.g. get food — yes, be food — no) and perhaps even crude approximations to the rules of geometry and the laws of physics (e.g. objects generally fall down). It might contain summaries of regularities of the environment abstracted from the information gathered long before the period covered by registers $P_0, \ldots, P_n$ or explained to it by other robots, etc.[c]

[c]A history book is a familiar part of the schema of the collective IGUS linked by human culture. It is a summary and analysis of records gathered at diverse times and places. That is true whether the history is of human actions or the scientific history of the universe. The schema resulting from the reconstruction of present records simplifies the process of future prediction. (For more on utility of history, see Ref. 4.)

- *Decisions and Behavior:* At each time interval τ_*, the robot uses its schema and the fresh image in P_0 to assess its situation, predict the future, and make decisions on what behavior to exhibit next by a process of computation that we denote by C. This is distinct from U. The important point for this paper is that the robot directly employs only the most recently acquired image in register P_0 in this process of computation C. The information in $P_1, \ldots, P_n$ is employed only through the schema.

It seems possible that such a robot could be constructed. As a model of sophisticated IGUSes such as ourselves, it is grossly oversimplified. Yet, it possesses a number of features that are similar to those in sophisticated IGUSes that are relevant for understanding past, present and future: The robot has a coarse-grained memory of its external environment contained in registers $P_0, P_1, \ldots, P_n$. The robot has two processes of computation, C and U. Without entering into the treacherous issue of whether the robot is conscious, the two processes have a number of similarities with our own processes of conscious and unconscious computation:

- U computation provides input to decision-making C computation.
- There is direct input to C computation only from the most recently acquired image in the register P_0. The images in $P_1, \ldots, P_n$ affect C only through the schema computed in U.
- Information flows into and out from the register P_0 directly used by C.

Equally evident are some significant differences between the robot and ourselves. Our information about the external environment is not exclusively visual, it is not stored in a linear array of registers, nor is it transferred from one to the other in the simple manner described. Input and records are not separated by sharp time divisions. We can consciously access memories of other than the most recent data, although often imperfectly and after modification by unconscious computation. This list of differences can easily be extended, but that should not obscure the similarities discussed above.

The analogies between the robot and ourselves can be emphasized by employing everyday subjective terminology to describe the robot. For example, we will call C and U computation "conscious" and "unconscious", secure in the confidence that such terms can be eliminated in favor of the mechanical description we have employed up to this point if necessary for clarity. Proceeding in this way, we can say that the robot "observes" its environment. The register P_0 contains a record of the "present", and the registers $P_1, \ldots, P_n$ are records of the "past".[d] When the register P_n is erased, the robot has "forgotten" its contents. The present extends[e] over a finite interval[f] τ_*.

The robot has conscious focus on the present, but only access to the past through the records that are inputs to the unconscious computation of its schema. The robot can thus be said to "experience" the present and "remember" the past. The "flow of time" is the movement of information into the register of conscious focus and out again. Prediction requires computation — either conscious or unconscious — from memories of the present and past acquired by observation and is thus distinct from remembering.

The subjective past, present and future, the flow of time, and the distinction between predicting and remembering are represented concretely and physically in the structure and function of the model robot. We now proceed to describe this structure and function in four-dimensional terms.

3. The Present is Not a Moment in Time

This section gives the four-dimensional, spacetime description of the robot specified in the previous section.[g] For simplicity, we consider the flat spacetime of special relativity (Minkowski space). But with little change it could be a curved spacetime of general relativity.

---

[d]For the moment, we take the records $P_1, \ldots, P_n$ to *define* the past. Section 4 will connect this notion of past with other physical notions of "past", that defined by the time direction toward the big bang, for instance.

[e]As James[6] put it more eloquently: "...the practically cognized present is no knife-edge, but a saddleback, with a certain breadth of its own, on which we sit perched, and from which we look in two directions in time."

[f]For human IGUSes, the time τ_* can be taken to be of order of the 0.1 s separation time needed to discriminate between two visual signals.[7]

[g]An abbreviated version of this discussion was given in Ref. 4.

3.1. *Some Features of Minkowski Space*

We begin by recalling[h] a few important features of four-dimensional spacetime that are illustrated in Fig. 2. Events occur at points in spacetime. At each point Q there is a *light cone* consisting of two parts. The future light cone is the three-dimensional surface generated by the light rays emerging from Q. The past light cone is similarly defined by the light rays converging on Q. (The labels "future" and "past" are conventions at this point in the discussion. In Sec. 4 we will define them in the cosmological context.) Points inside the light cone of Q are timelike separated from it; points outside the light cone are spacelike separated. Points inside the future light cone of Q are in its future; points inside the past light cone are in its past. Points outside the light cone are neither.

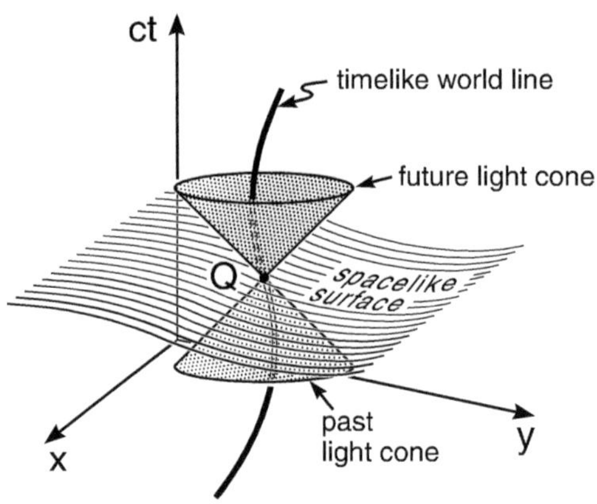

Fig. 2. Spacetime concepts. This spacetime diagram represents a three-dimensional slice of a four-dimensional flat spacetime defined by three axes ct, x and y of the four specifying an inertial (Lorentz) frame. The ideas of event, light cone, world line, and spacelike surface are illustrated. An event is a point in spacetime like Q. Each point has a future and past light cone. A spacelike surface like the one illustrated defines an instant in time. Each such surface divides spacetime into two regions conventionally called the future of this surface and its past. There are an infinity of such families and thus infinitely many different ways of defining instants in time and their futures and pasts. In the context of cosmology, the past of a spacelike surface is defined to be the region closest to the big bang and the future is the region furthest away.

[h]For a classic text on special relativity from a spacetime point of view, see Ref. 8.

The center of mass of a localized IGUS, such as the robot of the previous section, describes a timelike world line in spacetime. At each point along the world line, any tangent to it lies inside the light cone so that the IGUS is moving at less than the speed of light in any inertial frame.

A moment in time is a three-dimensional spacelike surface in spacetime — one in which any two nearby points are spacelike separated. Each spacelike surface divides spacetime into two regions — one to its future and one to its past. A family of spacelike surfaces such that each point in spacetime lies on one and only one member of the family specifies a division of spacetime into space and time. The family of surfaces defined by constant values of the time of a particular inertial frame is an example (e.g. the surfaces of constant ct in Figs. 2–4). Different families of spacelike surfaces define different notions of space and different notions of time, none of which is preferred over the other.

3.2. *The Past, Present and Future of the Robot*

Figure 3 shows the world line of the robot introduced in Sec. 2 together with the world line of an object in its environment that appears in the robot's stored images. The robot illustrated has a short memory with only four registers P_0, P_1, P_2, P_3 whose contents in each interval τ_* are indicated by the contents of the boxes to the right of the world line. These contents change at proper time intervals τ_* as described in Sec. 2.

The contents of the register P_0 defining the robot's present do not define a spacelike surface representing a moment in time. They do not even define an instant along the robot's world line since the contents of P_0 are constant over proper time intervals τ_*. Rather, there is content defining the present for *every* instant along the world line. For each point along the world line the most recently acquired image defines the present. That is the four-dimensional description of the present. In a similar way, the data acquired earlier and stored in registers P_1, P_2, P_3 define the robot's past for each point along the world line.

Thus, there is no conflict between the four-dimensional reality of physics and the subjective past, present and future of an IGUS. Indeed, as defined above, the subjective past, present and future *are* four-dimensional notions. They are not properties of spacetime but of the history of a particular IGUS. In Sec. 5, we will see that IGUSes constructed differently from

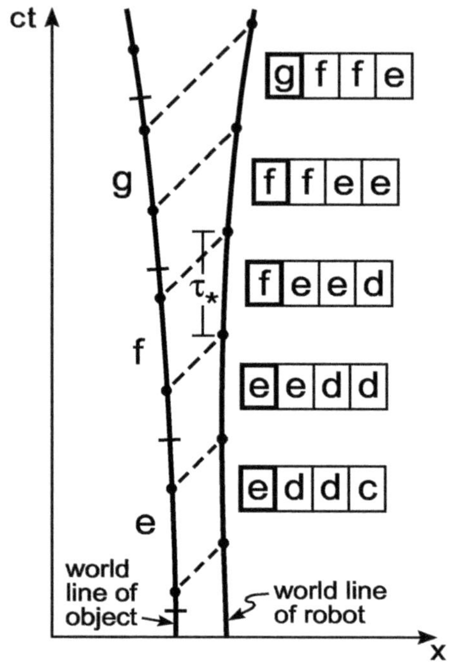

Fig. 3. A spacetime description of the present and past of the robot whose information processing is illustrated in Fig. 1. In addition to the world line of the robot, the figure shows the world line of an external object that is the source of its images, such as the stack of cards in Fig. 1. This source changes its shape at discrete instants of time demarcated by ticks, running through configurations $\ldots b, c, d, e, f, g, \ldots$. The configuration in each time interval of the object's world line is labeled to its left. At discrete instants separated by proper time τ_*, the robot captures an image of the object. The light rays conveying the image from object to robot are indicated by dotted lines. The images are stored in the registers P_0, P_1, P_2 and P_3 described in Fig. 1. The contents of these registers in between image captures are displayed in the boxes with P_0 on the left and P_3 on the right. The history of the contents of the register P_0 constitute the four-dimensional notion of "now" for this robot (the heavily outlined boxes in the figure). The present is not one instant along the robot's world line, much less a spacelike surface in spacetime. Rather, there is a "now" for *each* instant along the robot's world line extending over proper time τ_*. The evolution of the contents P_0 can be described four-dimensionally and is fully consistent with special relativity. In a similar way the contents of P_1, P_2 and P_3 constitute a four-dimensional notion of "past".

our robot could have different notions of past, present and future. All of these are fully consistent with a four-dimensional physical reality.

However, there *is* a conflict between ordinary language and the four-dimensional, IGUS-specific notions of past, present and future. To speak of the "present moment" of an IGUS, for instance, risks confusion because it could be construed to refer to a spacelike surface in spacetime stretching over the whole universe. No such surface is defined by physics.[i] In fact, the "moment" in the context of this section refers to the most recently acquired data of an IGUS. This is not a notion restricted to one point on the IGUS's world line which somehow moves along it. Rather, it is a notion present at *every* point along the world line.

3.3. *The Common Present*

The previous discussion has concerned the present, past and future of individual localized IGUSes. We now turn to the notion of a common present that may be held by collections of IGUSes separated in space.

When someone asked Yogi Berra[j] what time it was, he is reported to have replied "Do you mean now?" The laughter usually evoked by this anecdote shows how strongly we hold a common notion of the present. More precisely, different IGUSes agree on "what is happening now". This section is concerned with the limitations on the accuracy of that agreement arising both from the construction of the IGUSes and the limitations of defining simultaneity in special relativity. We continue to use robot model IGUSes to make the discussion concrete.

Figure 4 shows the world lines of two robots in spacetime together with the intervals on their world lines that define their individual notions of "now". There are at least two reasons that there is no unambiguous notion of a common "now" that can be shared by the two robots. The first

---

[i]There is no evidence for preferred frames in spacetime and modern versions of the Michelson–Morley experiment and other tests of special relativity set stringent limits on their existence. The fractional accuracy of these experiments range down to 10^{-21}, making Lorentz invariance at accessible energy scales one of the most accurately tested principles in physics. See e.g. Ref. 9.

[j]Lawrence E. Berra, a catcher for the New York Yankees baseball team in the 1950's. See Ref. 10.

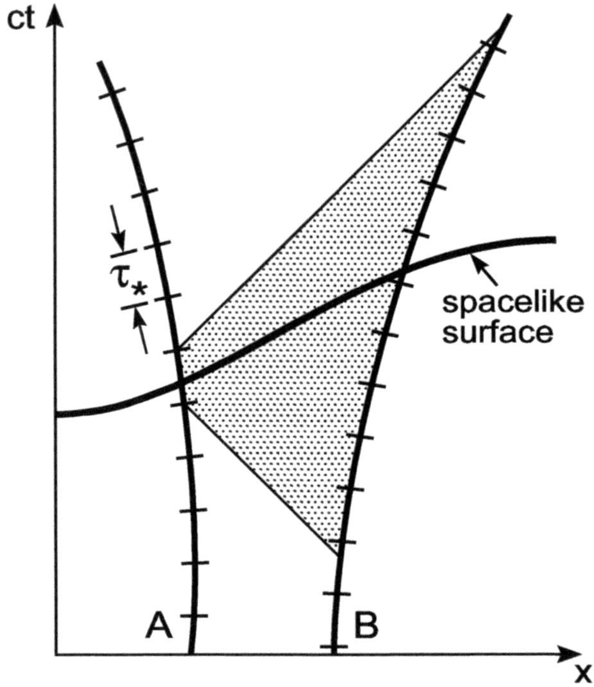

Fig. 4. The ambiguous common present. This spacetime diagram shows the world lines of two similarly constructed robots A and B. The intervals of proper time of length τ_* over which the contents of the registers defining their individual presents being constant are demarcated by ticks. A common present would be defined by an identification of each interval on one world line with that on the other "at the same time" (to the accuracy τ_*). But special relativity allows many different such identifications. Using the constant t surfaces of the inertial frame illustrated is one way to define a common present; but any other spacelike surface such as the one shown would do equally well. The range of ambiguity for intervals on B that could be said to be "at the same time" as one interval on A is shown as a shaded region. The figure shows two robots separated by a distance (e.g. as defined in the rest frame of one) over which the light travel time is longer than τ_*. In this situation the ambiguity in the definition of a common present is much larger than τ_*. However, if the distance between the robots is much smaller than $c\tau_*$, and if their relative velocity is much less than c, and this continues to be the case, then the ambiguity is much smaller than τ_*. An *approximate* common present can then be defined.

is the elementary observation[k] that the present for each individual robot is not defined to an accuracy better than τ_*. The second reason arises from special relativity.

A precise common "now" would specify a correspondence between events on the two world lines. Such a correspondence would specify a notion of simultaneity between events on the world lines. But there is no unique notion of simultaneity provided by special relativity. Rather there are many different notions corresponding to the different possible spacelike surfaces that can intersect the two world lines. Figure 4 illustrates the range of ambiguity.

To illustrate the ambiguity of the present more dramatically, imagine you are a newscaster on the capital planet Trantor of a galaxy-wide empire some hundreds of thousands of years in the future. News of events all over the galaxy pour in constantly via electromagnetic signals. You want to broadcast a program called "The Galaxy Today", reviewing important events in the last 24 hours (galactic standard). But what time do you assign to the latest news from the planet Terminus at the edge of the galaxy 60,000 light years away? There is an inertial frame[l] in which those events happened within the last 24 hours. But in the approximate rest frame of the galaxy, they happened 60,000 years ago. There is thus no unambiguous notion of "present" for the collective IGUS consisting of the citizens of the galactic empire because the ambiguities in defining simultaneity are large compared to the time scales on which human events happen. You could, of course, fix on the time of the galactic rest frame as a standard for simultaneity. But in that case the only comprehensive program you could broadcast would be "The Galaxy 60,000 Years Ago". One imagines the audience for this on Terminus would not be large since the program would be seen 120,000 years after the events there had happened.

---

[k]The ambiguity in the common present arising from the finite time of IGUS operation has been discussed in Ref. 11, which includes a review of current neurophysiological data bearing on this question.

[l]For example, there is the inertial frame moving with speed V with respect to the galactic rest frame such that

$$\Delta t' = \gamma[\Delta t - V(\Delta x/c)]$$

where $\gamma = (1-V^2)^{-1/2}$, $\Delta t' = 24$ hrs, $\Delta t = 6\times 10^4$ yrs, and $\Delta x = 6\times 10^4$ light years. The required velocity is within a few parts in 10^7 of the velocity of light.

The satellites comprising the Global Positioning System are an example of a collective IGUS closer to home that faces a similar problem. The special relativistic ambiguity in defining simultaneity for two satellites is of order of the light travel time between them in the approximate inertial frame in which the Earth is at rest. That is much larger than the light travel time across the few meters accuracy to which the system aims to locate events. Precise agreement on a definition of simultaneity is therefore needed. Each satellite clock is corrected so that it broadcasts the time of a clock on the Earth's geoid (approximately the ocean surface).[12]

No such agreement on a definition of simultaneity appears to be a prerequisite for the everyday notion of "now" employed by human IGUSes. Rather, we seem to be employing an approximate, imprecise notion of the common present appropriate in everyday situations and characterized by the following contingencies:

1. The time scale of perception τ_* is short compared to the time scales on which interesting features of the environment vary.
2. Individual IGUSes are moving relatively to one another at velocities which are small compared to c.
3. The light travel time between IGUSes in an inertial frame in which they are nearly at rest, is small compared to the time scales τ_*.

Contingency (1) means that the ambiguity in the "now" of each IGUS is negligible in the construction of a common present. Contingency (3), based on (2), means that the ambiguity arising from the definition of simultaneity is negligible.[m]

Collections of robots satisfying contingencies 1–3 can agree on what is happening "now". Consider just two robots — Alice and Bob. Alice can send Bob a description of the essential features of the image currently in her register P_0. Bob can check whether these essential features are the same as those of the image in his register P_0 at the time of receipt. He can then signal back agreement or disagreement. As long as the light travel time is much shorter than τ_* (contingency 3), and as long as the essential features vary on much longer time scales (contingency 1), Alice and Bob will agree.

---

[m]These kinds of contingencies and the synchronization protocols necessary when they are violated have been discussed in Ref. 13.

Contingency 2 ensures that this agreement will persist over an interesting time scale. Thus Alice and Bob can construct a common present, but it is a present that is local, inherently approximate, and contingent upon their relation to each other and their environment. This approximate common "now" is not a surface in spacetime.

No modification of the laws of physics is needed to understand the common now of a group of IGUSes, as has sometimes been suggested.[14] The common nows of IGUSes meeting the above contingencies will coincide approximately with constant time surfaces in any inertial frame in which they are approximately at rest. But these frames are not singled out by the laws of physics. Indeed, the experimental evidence against preferred frames in special relativity is extraordinarily good.[9] Rather, the frames are singled out by the particular situations of the IGUSes themselves

4. Why Don't We Remember the Future?

The fundamental dynamical laws of physics are invariant under time reversal to an accuracy adequate for organizing everyday experience.[n] They are *time-neutral.* The Einstein equation of general relativity and Maxwell's equations for electrodynamics are examples. But the boundary conditions specifying solutions to these equations which describe our universe are not time symmetric. The universe has a smooth (near homogeneous and isotropic) big bang at one end of time and a very different condition at the other end. This might be the unending expansion driven by a cosmological constant of the simplest cosmological models favored by observation.[15] Or, with different assumptions on the matter, it could be a highly irregular big crunch. In any event, one end is different from the other.[o]

By convention, this paper refers throughout to the times closer to the big bang as the "past" and times further away as the "future". Asymmetry between past and future boundary conditions is the origin of the various

[n]The effective theory of the weak interactions applicable well below the Planck scale is not time-reversal invariant. This is important, for instance, for the synthesis of baryons in the early universe, but negligible for the functioning of our robot. See, e.g. Ref. 3 for further discussion.

[o]We thus exclude, mainly for simplicity, the kind of cosmological model where initial and final conditions are related by time symmetry that have sometimes been discussed (e.g. Refs. 3, 19–21).

time asymmetries — "arrows of time" — exhibited by our universe. The arrow of time associated with the second law of thermodynamics is an example.[p]

The operation of the robot in Sec. 2 is not time-neutral in at least two respects. First the robot receives information about external events in its past (closer to the big bang) and not the future.[q] Second, its processing of the received information is not time neutral. The flow of information from recording to erasure defines a direction in time. As mentioned in Sec. 2, that direction gives a concrete model for the subjective feeling of inexorable forward progression in time commonly called the "psychological arrow of time". In natural IGUSes, such as ourselves, information flows from past to future. More specifically, the records in registers $P_1,\ldots,P_n$ are of external events to the *past* of those in P_0 in decreasing order of time from the big bang. This is the reason that the robot in Sec. 2 could be said to experience the present, remember the past, and predict the future.[r]

Could a robot be constructed that receives information from the future? Could one be constructed whose psychological arrow of time runs from future to past, with the consequence that it would remember the future? Both of these possibilities are consistent with time neutral dynamical laws. But two familiar time asymmetries of our universe prohibit such constructions as a practical matter. These are the radiation arrow of time and the arrow of time associated with the second law of thermodynamics.

[p]For reviews of the physics of time asymmetry from various perspectives, see Refs. 3, 16–19. There are also some discussions in the Appendix. These are only a few of the references where these issues are treated.

[q]More precisely, registered signals originate from events within the past light cone of their reception event. We use "future" and "past" in the present discussion, understanding that in each case these are defined by an appropriate light cone as described in Sec. 3.1.

[r]In Sec. 2, we *defined* the robot's past to be the records in the registers $P_1,\ldots,P_n$. If information flows from past to future in the robot, that notion coincides with the physical past defined as the direction in time towards the big bang. From now on we assume this congruence except where discussing its possible violation, as in this section.

Records of the future *are* possible as in a table of future lunar eclipses.[s] Indeed, records of the future are the outcomes of any useful process of prediction. But, our robot's records of the future are obtained by computation, whereas its records of the past are created by simple, automatic, sensory mechanisms. These are very different processes both physically and from the point of view of information processing by the robot.[t] By a robot that remembers the future, we would mean one constructed[u] as in Sec. 2, but with the records in registers $P_1, \ldots, P_n$ being of events to the future of P_0.

First consider the question of whether information from the future could be recorded by the robot. In our universe, electromagnetic radiation is retarded — propagating its emission event to the future. That time asymmetry is the radiation arrow of time. The electromagnetic signals recorded by the robot are propagated to it along the *past* light cone of the reception event. The images received by the robot, whether of the cosmic background radiation, distant stars and galaxies, or the happenings in its immediate environment, are therefore all from past events. As far as we know, all other carriers, neutrinos for example,[22] are similarly retarded. One reason the robot does not remember the future is that it receives no information about it.

Irrespective of the time at which its input originates, could a robot like that in Sec. 2 be constructed whose psychological arrow of time is reversed, so that internally information flows from future to past? In such a robot the events recorded in registers $P_1, \ldots, P_n$ would lie to the future of that in P_0 — further from the big bang. The robot would thus remember the future. Such a construction would run counter to the arrow of time specified by the second law of thermodynamics, as we now review.[v]

---

[s]We use the term "record" in a time-neutral sense of an alternative at one time correlated with high probability to an alternative at another time — future or past. Thus, there can be records of the future.

[t]Realistic IGUSes, such as human beings, also create records of the past by computation as in differing interpretations of past experience, and as in the collective construction of human history, and the history of the universe. However, we did not explicitly endow our robot with these functions.

[u]Later in Sec. 4, we will describe a different construction of a robot which could be said to remember the future.

[v]Many authors have connected the "psychological" arrow of time with that of the second law. See, e.g. Refs. 23, 17, 24.

All isolated subsystems of the universe evolve toward equilibrium. That is statistics. But the preponderance of isolated systems in our universe are evolving toward equilibrium from past to future, defining an arrow of time. That is the second law of thermodynamics that is expressed quantitatively by the inexorable increase of an appropriately defined total entropy.

If the robot processes information irreversibly, then its psychological arrow of time must generally be congruent with the thermodynamic arrow of time. The formation of records are crucial steps. An increase in total entropy accompanies the formation of many realistic records. An impact crater in the moon, an ancient fission track in mica, a darkened photographic grain, or the absorption of a photon by the retina are all examples.

But an increase in total entropy is not a necessary consequence of forming a record. Entropy increase is necessary only on the erasure of a record.[25] For the model robot discussed in Sec. 2, the only part of its operation in which entropy must necessarily increase is in the erasure of the contents of the register P_n at each step.[w] However, that is enough. To see that, imagine the process of erasure run backwards from future to past. It would be like bits of smashed shell reassembling to form an egg.

To construct a robot with a reversed psychological arrow of time, it would be necessary to reverse the thermodynamic arrow, not only of the robot, but also of the local environment it is observing. That is possible in principle. However, since we have a system of matter coupled to electromagnetic radiation, it would be necessary to deal with every molecule and photon within a radius of 2×10^{10} km to reverse the system for a day. More advanced civilizations may find this amusing. We can have the same fun more cheaply by running the film through the projector in reverse.

The origin of both the thermodynamic and radiation arrows of time are the time-asymmetric boundary conditions that single out our universe from the many allowed by time-reversible dynamical laws. These boundary conditions connect the two arrows. A brief sketch of the relevant physics is given in the Appendix, although it is not necessary for understanding the main argument of this paper. But it is interesting to think that our

[w]Possibly an isolated robot could be constructed on the principles of reversible computation[26] that would not have an erasure step. However, it seems unlikely that the whole system of a robot plus a realistic observed environment could be reversible.

subjective distinction between future and past can ultimately be traced to the cosmological boundary conditions that distinguish the future and past of the universe.

5. Alternatives to Past, Present and Future

The preceding discussion suggests that the laws of physics do not define unambiguous notions of past, present and future by themselves. Rather, these are features of how specific IGUSes gather and utilize information. What, then, is the origin of the past, present and future organization of information in familiar, naturally-occurring IGUSes? Is it the only organization compatible with the laws of physics? If not, does it arise uniquely from evolutionary imperatives, or is it a frozen accident that took place in the course of three billion years of biological evolution? This section discusses such questions.

Certainly some features of the laws of physics are essential prerequisites to the functioning of the robot in Sec. 2. There would be no past, present and future at all if spacetime did not have timelike directions. The fact that IGUSes move on timelike rather than spacelike world lines is the main reason they can have an approximate common "now" rather than an approximate common "here". An IGUS functioning in a spacetime where it moved along a closed timelike curve could not maintain a consistent notion of past and future. Likewise, a local distinction between past and future would be difficult to maintain in the absence of the arrows of time discussed in Sec. 2. But the features of the physical laws of dynamics and the initial condition of the universe that are necessary for a past, present, future organization of temporal information are consistent with other organizations of this information, as we now show.

5.1. *Different Organizations of Temporal Information*

Perhaps the easiest way of convincing oneself that the notions of past, present and future do not follow from the laws of physics is to imagine constructing robots that process information differently from the one described in Sec. 2. We consider just three examples:

The Split Screen (SS) Robot. This robot has input to C computation from both the most recently acquired data in P_0 and from that in a different register P_J that was acquired at a proper time $\tau_s \equiv J\tau_*$ earlier along its world line. There is thus input to conscious computation from *two* times.

The Always Behind (AB) Robot. This robot has input to C computation only from a particular register $P_K, K > 0$ and the schema. That input is thus always a proper time $\tau_d \equiv K\tau_*$ *behind* the most recently acquired data.

The No Schema (NS) Robot. This robot has input to C computation from all the registers $P_0, \ldots, P_n$ equally. It employs no unconscious computation and constructs no schema, but rather makes decisions by conscious computation from all the data it has.

There seems to be no obstacle to constructing robots wired up in these ways, and they process information differently from the present, past and future organization that we are familiar with.[x]

An *SS* robot would have a tripartite division of recorded information. Its present experience — its "now" — would consist of two times (P_0, P_J) — equally vivid and immediate. It would "remember" the intermediate times ($P_1, \ldots, P_{J-1}$), and the past ($P_{J+1}, \ldots, P_n$) through the U process of computation and its influence on the schema.

The *AB* robot would also have a tripartite division of recorded information. Its present experience would be the contents of the register P_K. It would remember the past stored in registers $P_{K+1}, \ldots, P_n$. But also, it would remember its future stored $P_0, \ldots, P_{K-1}$ a time τ_d ahead of its present experience.[y] (Or perhaps we should say that it would have premonitions of the future.)

What would discussions with an *AB* robot be like, assuming that our information processing is similar to the robot discussed in Sec. 2? Assume for simplicity that we and the *AB* robot are both nearly at rest in one inertial

[x]Some idea of what the notion of "present" would be like for some of these robots could be had by serving time in a virtual reality suit in which the data displayed was delayed as in the *AB* case, or in which there was an actual split screen as in the *SS* case. An alternative realization of the *SS* robot's experience might be produced by electrical stimulation of the cortex that evokes memories of the past which are comparably immediate to the present.[27]

[y]That is not in conflict with the discussion in Sec. 4 because each record is still of events in the past to the proper time it was recorded.

frame and that contingencies (1) through (3) of Sec. 3.3 are satisfied. The *AB* robot would seem a little slow — responding in a time τ_d or longer to questions. Its answers to queries about "What's happening now?" would seem out of date. It would be always behind.

The *NS* robot would just have one category of recorded information. Conversations with an *NS* robot would be impressive since it would recall every detail that it has recorded about the past as immediately and vividly as the present.[z]

The laws of physics supply no obstacle of principle to the construction of robots with exotic organizations of information processing such as the *SS*, *AB* and *NS* robots. But are such organizations a likely outcome of biological evolution? Can we expect to find such IGUSes in nature in this or other planets? We speculate that we will not. It is adaptive for an IGUS of everyday size to focus mainly on the most recently acquired data as input to making decisions. The effective low-energy laws of physics in our universe are *local* in spacetime and the nearest data in space and time is usually the most relevant for what happens next. A frog predicting the future position of a fly needs the present position and velocity of the fly, not its location 10 s ago. An *AB* frog would be at a great competitive disadvantage in not focusing on the most current information. An *SS* frog would be wasting precious conscious focus on data from the past that is less relevant for immediate prediction than current data.

An *NS* robot is making inefficient use of computing resources in giving equal focus to present data and data from the past whose details may not have effect on relevant future prediction. Employing a schema to process the data is plausibly adaptive because it is a more efficient and faster way of processing data with limited computing resources. The collective IGUS linked by human culture certainly evolved to make use of schema rather than focus on the individual records that went into them. For instance, prediction of the future of the universe is much simpler from a Friedman–Robertson–Walker model characterized by a few cosmological parameters than directly from the records of the measurements that determined them.

[z]Perhaps not unlike conversations with Ireneo Funes in the Borges story, *Funes the Memorious*.[28]

5.2. *Different Laws, Different Scales*

Something like the *SS* organization of temporal information might be favored by evolution if the laws of physics were not local in time. Suppose, for example, that the position of objects to the future of a time t depended,[aa] not just on the force acting and their position and velocity at that time, but rather on their position at time t and at earlier times $t-h$ and $t-2h$ for some fundamental fixed time interval h. Then an organization such as the *SS* robot with conscious focus on both the most recently acquired data and that acquired at times h and $2h$ ago might be favored by evolution.

Similarly, different organizations might evolve if the IGUS is not smaller than the scale over which light travels on the characteristic times of relevant change in its environment. The present, past, future organization is unlikely to serve such an IGUS well because these notions are not well defined in these situations, as discussed in Sec. 3.3. As mentioned there, the galactic empires beloved of science fiction would be examples of such IGUSes. Faster-than-light travel inconsistent with special relativity is often posited by authors whose stories feature these empires to make their narratives accessible to IGUSes like ourselves that do employ — however, approximately — a present, past and future organization of information.

6. Conclusion

A subjective past, present, and future are not the only conceivable way an IGUS can organize temporal data in a four-dimensional physical world consistently with the known laws of physics. But it is a way that may be adaptive for localized IGUSes governed by local physical laws. We can conjecture that a subjective past, present and future is a cognitive universal[5] of such localized IGUSes. That is a statement accessible to observational test, at least in principle.

[aa]It is not difficult to write down dynamical difference equations with this property, for instance, in a one-dimensional model we could take

$$F(t) = m\frac{[x(t)-2x(t-h)+x(t-2h)]}{2h^2}$$

where $x(t)$ is the body's position, m its mass, and $F(t)$ the force. However, such equations are not consistent with special relativity and the author is not suggesting a serious investigation of alternatives to Newtonian mechanics.

Acknowledgments

The author thanks Murray Gell-Mann for discussions of complex adaptive systems over a long period of time. He is grateful to Terry Sejnowski and Roger Shepard for information about the literature in psychology that bears on the subject of this paper. Special thanks are due to Roger Shepard for an extended correspondence on these issues. Communications with Jeremy Butterfield, Craig Callender and Matt Davidson have been similarly helpful with parts of the philosophical literature. However, the responsibility rests with the author for any deficiencies in providing relevant references to the vast literatures on these subjects of which he is largely ignorant. Research for this work was supported in part by the National Science Foundation under grant PHY02-44764.

Appendix A

The Cosmological Origin of Time's Arrows

The origin of our universe's time asymmetries is not to be found in the fundamental dynamical laws which are essentially time-reversible. Rather, both the radiation and the thermodynamic arrows of time arise from special properties of the initial condition of our universe.[bb] This appendix gives a simplified discussion of these special features starting, not at the very beginning, but at the time the hot initial plasma had become cool enough to be transparent to electromagnetic radiation. This is the time of "decoupling" in cosmological parlance — about 400,000 years after the big bang or a little over 13 billion years ago.

As Boltzmann put it over a century ago: "The second law of thermodynamics can be proved from the [time-reversible] mechanical theory if one assumes that the present state of the universe... started to evolve from an improbable [i.e. special] state".[30] The entropy of matter and radiation usually defined in physics and chemistry is about 10^{80} in the region visible from today at the time of decoupling (in units of Boltzmann's constant). This seems high, but it is in fact vastly smaller than the maximal value of about 10^{120} if all that matter was dumped into a black hole.[17] The entropy of the matter early in the universe is high because most constituents are in

[bb]For example, Hawking's "no boundary" wave function of the universe.[29]

approximate thermal equilibrium. However, the gravitational contribution of the smooth early universe to the entropy is near minimal, and entropy can grow by the clumping of the matter arising from gravitational attraction leading to the galaxies, stars and other inhomogeneities in the universe we see today.

Amplifying on Boltzmann's statement above, the explanation of why the entropies of isolated subsystems are mostly increasing in the same direction of time is this: The progenitors of these isolated systems were all further out of equilibrium at times closer to the big bang (the past) than they are today. Earlier, the total entropy was low compared to what it could have been. It has therefore tended to increase since.

The radiation arrow of time can also be understood as arising from time-asymmetric cosmological boundary conditions applied to time-reversible dynamical laws. These are Maxwell's equations for the electromagnetic field in the presence of charged sources. Their time-reversal invariance implies that *any* solution for specified sources at a moment of time can be written in *either* of two ways: (R) a sum of a free field (no sources) coming from the past plus *retarded* fields whose sources are charges in the past, or (A) a sum of a free field coming from the future plus *advanced* fields whose sources are charges in the future. More quantitatively, the four-vector potential $A_\mu(x)$ at a point x in spacetime can be expressed in the presence of four-current sources $j_\mu(x)$ in Lorentz gauge as either

$$A_\mu(x) = A_\mu^{\text{in}}(x) + \int d^4x' D_{\text{ret}}(x-x')\, j_\mu(x') \qquad (R)$$

or

$$A_\mu(x) = A_\mu^{\text{out}}(x) + \int d^4x' D_{\text{adv}}(x-x')\, j_\mu(x') \qquad (A).$$

Here, D_{ret} and D_{adv} are the retarded and advanced Green's functions for the wave equation and $A_\mu^{\text{in}}(x)$ and $A_\mu^{\text{out}}(x)$ are free fields defined by these decompositions.

Suppose there were no free electromagnetic fields in the distant past so that $A_\mu^{\text{in}}(x) \approx 0$. Using the R description above, this time asymmetric boundary condition would imply that present fields can be entirely ascribed to sources in the past. This is retardation and that is the electromagnetic arrow of time.

This explanation needs to be refined for our universe because, at least if we start at decoupling, there *is* a significant amount of free electromagnetic radiation in the early universe constituting the cosmic background radiation (CMB). Indeed, at the time of decoupling the energy density in this radiation was approximately equal to that of matter. Even today, approximately 13 billion years later, after being cooled and diluted by the expansion of the universe, the CMB is still the largest contributor to the electromagnetic energy density in the universe by far.

The CMB's spectrum is very well fit by a black body law.[31] That strongly suggests that the radiation is *disordered* with maximal entropy for its energy density. There is no evidence for the kind of correlations (sometimes called "conspiracies") that would tend to cancel $A_\mu^{\rm out}(x)$ in the far future and give rise to advanced rather than retarded effects.[cc]

The expansion of the universe has red-shifted the peak luminosity of the CMB at decoupling to microwave wavelengths today. There is thus a negligible amount of energy left over from the big bang in the wavelengths we use for vision, for instance. The radiation used by realistic IGUSes is therefore retarded. A contemporary robot functioning at wavelengths where the CMB is absent will therefore be receiving information about charges in the past. This selection of wavelengths is plausibly not accidental, but adaptive. A contemporary robot seeking to function with input from microwave wavelengths would find little emission of interest, and what there was would have been overwhelmed by the all-pervasive CMB, nearly equally bright in all directions, and carrying no information.

References

1. M. Gell-Mann, *The Quark and the Jaguar* (W. Freeman, San Francisco, 1994).
2. J. Butterfield, *The Arguments of Time* (British Academy and Oxford University Press, Oxford, 1993).
3. M. Gell-Mann and J. B. Hartle, Time symmetry and asymmetry in quantum mechanics and quantum cosmology, in *Proc. NATO Workshop on the Physical Origins of Time Asymmetry*, Mazagón, Spain, September 30–October 4, 1991, eds. J. Halliwell, J. Pérez-Mercader and W. Zurek (Cambridge University Press, Cambridge, 1993), gr-qc/9309012.

[cc]For an experiment that checked on advanced effects, see Ref. 32.

4. J. B. Hartle, Quantum pasts and the utility of history, *Phys. Scripta* **T76**, 67–77 (1998), gr-qc/9712001.
5. R. N. Shepard, Perceptual-cognitive universals as reflections of the world, *Psychon. Bull. Rev.* **1**, 2 (1994); How a cognitive scientist came to seek universal laws, *ibid.* **11**, 1–23 (2004).
6. Wm. James, *Principles of Psychology* (Henry Holt, New York, 1890).
7. See, *e.g.* G. Westheimer, Discrimination of short time intervals by the human observer, *Exp. Brain Res.* **129**, 121–126 (2000); G. Westheimer, Visual signals used in time interval discrimination, *Vis. Neurosci.* **17**, 551–556 (2000).
8. E. F. Taylor and J. A. Wheeler, *Spacetime Physics* (W. Freeman, San Francisco, 1963).
9. For a review, see M. P. Haugan and C. M. Will, Modern tests of special relativity, *Phys. Today* **40**, 69–76 (1987) or C. M. Will, The confrontation between general relativity and experiment, *Living Rev. Relativ.* **17**, 4 (2001), (http://www.livingreviews.org/lrr-2001-4). For a modern version of the Michelson–Morley experiment see, in particular, A. Brillet and J. L. Hall, Improved laser test of the isotropy of space, *Phys. Rev. Lett.* **42**, 549 (1979).
10. Y. Berra, *The Yogi Book* (Workman Publishing, New York, 1998).
11. C. Callender, *The Subjectivity of the Present*, to be published.
12. N. Ashby, Relativity and the Global Positioning System, *Phys. Today* **55**, 41 (2002).
13. J. Butterfield, Seeing the present, *Mind* **93**, 161–176 (1984). [Reprinted in *Questions of Time and Tense*, ed. R. Le Poidevin (Oxford University Press, Oxford, 1998).]
14. See, *e.g.* the preface to F. Hoyle and G. Hoyle, *The Fifth Planet* (Harper and Row, New York, 1963).
15. *e.g.* D. N. Spergel *et al.*, First year Wilkinson microwave anisotropy probe (WMAP) observations: Determination of cosmological parameters, *Ap. J. Suppl.* **148**, 175 (2003); J. L. Tonry *et al.*, Cosmological results from high-z supernovae, *Ap. J.* **594**, 1–24 (2003).
16. P. C. W. Davies, *The Physics of Time Asymmetry* (University of California Press, Berkeley, 1976).

17. R. Penrose, *Singularities and Time Asymmetry* in *General Relativity: An Einstein Centenary Survey*, eds. S. W. Hawking and W. Israel (Cambridge University Press, Cambridge, 1979).
18. H. D. Zeh, *The Physical Basis of the Direction of Time* (Springer, Berlin, 1989).
19. H. Price, *Time's Arrow and Archimedes' Point* (Oxford University Press, Oxford, 1996).
20. R. Laflamme, Time-symmetric quantum cosmology and our universe, *Class. Quant. Grav.* **10**, L79 (1993), gr-qc/9301005.
21. D. Craig, Observation of the final condition: Extragalactic background radiation and the time symmetry of the universe, *Ann. Phys. (NY)* **251**, 384 (1995), gr-qc/9704031.
22. *e.g.* K. Hirata *et al.*, Observations of a neutrino burst from the supernova SN1987A, *Phys. Rev. Lett.* **58**, 1490 (1987).
23. H. Reichenbach, *The Direction of Time* (University of California Press, Berkeley, 1956).
24. S. W. Hawking, The direction of time, *New Scient.* **115**, 46 (1987).
25. R. Landauer, Irreversibility and heat generation in the computing process, *IBM J. Res. Dev.* **5**, 183–191 (1961).
26. *e.g.* C. Bennett, Logical reversibility of computation, *IBM J. Res. Dev.* **17**, 525–532 (1973); E. Fredkin and T. Toffoli, Conservative logic, *Int. J. Theor. Phys.* **21**, 219–253 (1982).
27. W. Penfield and L. Roberts, *Speech and Brain Mechanisms* (Princeton University Press, Princeton, 1959).
28. J. L. Borges, *Ficciones* (Grove Press, New York, 1962).
29. S. W. Hawking, The quantum state of the universe, *Nucl. Phys. B* **239**, 257 (1984).
30. L. Boltzmann, Zu Hrn. Zermelo's Abhandlung Über die mechanische Erklärung irreversibler Vorgange, *Ann. Physik* **60**, 392 (1897); translated in S. G. Brush, *Kinetic Theory* (Pergamon Press, New York, 1965).
31. D. J. Fixsen *et al.*, The cosmic microwave background spectrum from the full COBE FIRAS data set, *Ap. J.* **473**, 576 (1996).
32. R. B. Partridge, Absorber theory of radiation and the future of the universe, *Nature* **244**, 263–265 (1973).

CHAPTER 7

The Quantum Mechanics of Closed Systems*

A pedagogical introduction is given to the quantum mechanics of closed systems, most generally the universe as a whole. Quantum mechanics aims at predicting the probabilities of alternative coarse-grained time histories of a closed system. Not every set of alternative coarse-grained histories that can be described may be consistently assigned probabilities because of quantum mechanical interference between individual histories of the set. In "Copenhagen" quantum mechanics, probabilities can be assigned to histories of a subsystem that have been "measured". In the quantum mechanics of closed systems, containing both observer and observed, probabilities are assigned to those sets of alternative histories for which there is negligible interference between individual histories as a consequence of the system's initial condition and dynamics. Such sets of histories are said to decohere. We define decoherence for closed systems in the simplified case when quantum gravity can be neglected and the initial state is pure. Typical

(*continued overleaf*)

*A slightly shortened version of an article that appeared in the festschrift for C. W. Misner, eds. B. L. Hu, M. P. Ryan and C. V. Vishveshwara (Cambridge University Press, Cambridge, 1993).
http://arxiv.org/abs/gr-qc/9210006v2

mechanisms of decoherence that are widespread in our universe are illustrated.

Copenhagen quantum mechanics is an approximation to the more general quantum framework of closed subsystems. It is appropriate when there is an approximately isolated subsystem that is a participant in a measurement situation in which (among other things) the decoherence of alternative registrations of the apparatus can be idealized as exact.

Since the quantum mechanics of closed systems does not posit the existence of the quasiclassical realm of everyday experience, the domain of the approximate applicability of classical physics must be explained. We describe how a quasiclassical realm described by averages of densities of approximately conserved quantities could be an emergent feature of an initial condition of the universe that implies the approximate classical behavior of spacetime on accessible scales.

1. Introduction

It is an inescapable inference from the physics of the last sixty years that we live in a quantum mechanical universe — a world in which the basic laws of physics conform to that framework for prediction we call quantum mechanics. We perhaps have little evidence of peculiarly quantum-mechanical phenomena on large and even familiar scales, but there is no evidence that the phenomena that we do see cannot be described in quantum mechanical terms and explained by quantum mechanical laws. If this inference is correct, then there must be a description of the universe as a whole and everything in it, in quantum mechanical terms. The nature of this description and its observable consequences are the subject of quantum cosmology

Our observations of the present universe on the largest scales are crude and a classical description of them is entirely adequate. Providing a quantum mechanical description of these observations alone might be an interesting intellectual challenge, but it would be unlikely to yield testable predictions differing from those of classical physics. Today, however, we have a more ambitious aim. We aim, in quantum cosmology, to provide a theory of the initial condition of the universe which will predict testable correlations among observations today. There are no realistic predictions of any kind that do not depend on this initial condition, if only very weakly. Predictions of certain observations may be testably sensitive to its details. These include the large scale homogeneity and isotropy of the universe, its approximate spatial flatness, the spectrum of density fluctuations that produced the galaxies, the homogeneity of the thermodynamic arrow of time, and the existence of classical spacetime. Recently, there has been speculation that even the coupling constants of the effective interactions of the elementary particles at accessible energy scales may be probabilistically distributed with a distribution which may depend, in part, on the initial condition of the universe.[4–6] It is for such reasons that the search for a theory of the initial condition of the universe is just as necessary and just as fundamental as the search for a theory of the dynamics of the elementary particles. They may even be the same searches.

The physics of the very early universe is likely to be quantum mechanical in an essential way. The singularity theorems of classical general relativity suggest that an early era preceded ours in which even the geometry of spacetime exhibited significant quantum fluctuations. It is for a theory of the initial condition that describes this era, and all later ones, that we need to spell out how to apply quantum mechanics to cosmology. Recent years have seen much promising progress in the search for a theory of the quantum initial condition. However, it is not my purpose to review these developments here.[a] Rather, I shall argue that this somewhat obscure branch of astrophysics may have implications for the formulation and interpretation of quantum mechanics on day-to-day scales. My thesis will be that by looking at the universe as a whole, one is led to an understanding of quantum mechanics which clarifies many of the long standing interpretative difficulties of the subject.

The Copenhagen frameworks for quantum mechanics, as they were formulated in the '30s and '40s and as they exist in most textbooks today, are inadequate for quantum cosmology. Characteristically these formulations assumed, as *external* to the framework of wave function and Schrödinger equation, the quasiclassical realm we

[a]For a recent review of quantum cosmology, see Ref. 7.

see all about us. Bohr[8] spoke of phenomena which could be alternatively described in classical language. In their classic text, Landau and Lifschitz[9] formulated quantum mechanics in terms of a separate classical physics. Heisenberg and others stressed the central role of an external, essentially classical, observer.[b] Characteristically, these formulations assumed a possible division of the world into "observer" and "observed", assumed that "measurements" are the primary focus of scientific statements and, in effect, posited the existence of an external "quasiclassical realm". However, in a theory of the whole thing, there can be no fundamental division into observer and observed. Measurements and observers cannot be fundamental notions in a theory that seeks to describe the early universe when neither existed. In a basic formulation of quantum mechanics there is no reason in general for there to be any variables that exhibit classical behavior in all circumstances. Copenhagen quantum mechanics thus needs to be generalized to provide a quantum framework for cosmology.

In a generalization of quantum mechanics which does not *posit* the existence of a quasiclassical realm, the domain of applicability of classical physics must be *explained.* For a quantum mechanical system to exhibit classical behavior there must be some restriction on its state and some coarseness in how it is described. This is clearly illustrated in the quantum mechanics of a single particle. Ehrenfest's theorem shows that generally

$$M\frac{d^2\langle x\rangle}{dt^2} = \left\langle -\frac{\partial V}{\partial x}\right\rangle. \tag{1}$$

However, only for special states, typically narrow wave packets, will become an equation of motion for $\langle x\rangle$ of the form

$$M\frac{d^2\langle x\rangle}{dt^2} = -\frac{\partial V(\langle x\rangle)}{\partial x}. \tag{2}$$

For such special states, successive observations of position in time will exhibit the classical correlations predicted by the equation of motion (2) *provided* that these observations are coarse enough so that the

[b]For a clear statement of this point of view, see Ref. 10.

properties of the state which allow (2) to replace the general relation (1) are not affected by these observations. An *exact* determination of position, for example, would yield a completely delocalized wave packet an instant later and (2) would no longer be a good approximation to (1). Thus, even for large systems, and in particular for the universe as a whole, we can expect classical behavior only for certain initial states and then only when a sufficiently coarse-grained description is used.

If classical behavior is *in general* a consequence only of a certain class of states in quantum mechanics, then, as a particular case, we can expect to have classical spacetime only for certain states in quantum gravity. The classical spacetime geometry we see all about us in the late universe is not the property of every state in a theory where geometry fluctuates quantum mechanically. Rather, it is traceable fundamentally to restrictions on the initial condition. Such restrictions are likely to be generous in that, as in the single particle case, many different states will exhibit classical features. The existence of classical spacetime and the applicability of classical physics are thus not likely to be very restrictive conditions on constructing a theory of the initial condition.

It was Everett who, in 1957, first suggested how to generalize the Copenhagen frameworks so as to apply quantum mechanics to cosmology.[c] Everett's idea was to take quantum mechanics seriously and apply it to the universe as a whole. He showed how an observer could be considered part of this system and how its activities — measuring, recording, calculating probabilities, etc. — could be described within quantum mechanics. Yet the Everett analysis was not complete. It did not adequately describe within quantum mechanics the origin of the "quasiclassical realm" of familiar experience,[d] nor, in an observer-independent way, the meaning of the "branching" that replaced the notion of measurement. It did not distinguish from among

---

[c]The original reference is Ref. 11. For a useful collection of reprints see Ref. 12.

[d]In our earlier work, e.g. Ref. 2, this was called a "quasiclassical domain", but this risked confusion with the use of the word "domain" in condensed matter physics.

the vast number of choices of quantum mechanical observables that are in principle available to an observer, the particular choices that, in fact, describe the quasiclassical realm.

In this essay, I will describe the joint work with Murray Gell-Mann[2,3] which aims at a coherent formulation of quantum mechanics for the universe as a whole that is a framework to explain rather than posit the quasiclassical realm of everyday experience. It is an attempt at an extension, clarification, and completion of the Everett interpretation. It builds on many aspects of the, so-called post-Everett development, especially the work of Zeh,[13] Zurek,[14,15] and Joos and Zeh.[16] At important points it coincides with the, independent, earlier works of Bob Griffiths[17] and Roland Omnès (e.g. as reviewed in Ref. 18).

Our work is not complete, but I hope to sketch how it might become so. It is by now a very long story, but I will try to describe the important parts in simplified terms.

2. Probabilities in General and Probabilities in Quantum Mechanics

Even apart from quantum mechanics, there is no certainty in this world and therefore physics deals in probabilities. It deals most generally with the probabilities for alternative time histories of the universe. From these, conditional probabilities can be constructed that are appropriate when some features about our specific history are known and further ones are to be predicted.

To understand what probabilities mean for a single closed system, it is best to understand how they are used. We deal, first of all, with probabilities for *single* events of the *single* system. When these probabilities become sufficiently close to zero or one, there is a definite prediction on which we may act. How sufficiently close to zero or one the probabilities must be depends on the circumstances in which they are applied. There is no certainty that the sun will come up tomorrow at the time printed in our daily newspapers. The sun may be destroyed by a neutron star now racing across the galaxy at near light speed. The earth's rotation rate could undergo a quantum fluctuation. An error could have been made in the computer that extrapolates the motion of the earth. The printer could have made a mistake in setting the type. Our eyes may deceive us in reading the time. Yet, we watch the sunrise at the appointed time because we compute, however imperfectly, that the probability of these things happening is sufficiently low.

Various strategies can be employed to identify situations where probabilities are near zero or one. Acquiring information and considering the conditional probabilities based on it is one such strategy. Current theories of the initial condition of the universe predict almost no probabilities near zero or one without further conditions. The "no-boundary" wave function of the universe, for example, does not predict the present position of the sun on the sky. However, it will predict that the conditional probability for the sun to be at the position predicted by classical celestial mechanics given a few previous positions is a number very near unity.

Another strategy to isolate probabilities near zero or one is to consider ensembles of repeated observations of identical subsystems in the closed system. There are no genuinely infinite ensembles in the world, so we are necessarily concerned with the probabilities of deviations of the behavior of a finite ensemble from the expected behavior of an infinite one. These are probabilities for a single feature (the deviation) of a single system (the whole ensemble).

The existence of large ensembles of repeated observations in identical circumstances and their ubiquity in laboratory science should not, therefore, obscure the fact that in the last analysis, physics must predict probabilities for the single system that is the ensemble as a whole. Whether it is the probability of a successful marriage, the probability of the present galaxy–galaxy correlation function, or the probability of the fluctuations in an ensemble of repeated observations, we must deal with the probabilities of single events in single systems. In geology, astronomy, history and cosmology, most predictions of interest have this character. The goal of physical theory is, therefore, most generally to predict the probabilities of histories of single events of a single system.

Probabilities need be assigned to histories by physical theory only up to the accuracy they are used. Two theories that predict probabilities for the sun not rising tomorrow at its classically calculated time that are both well beneath the standard on which we act are equivalent for all practical purposes as far as this prediction is concerned. It is often convenient, therefore, to deal with approximate probabilities which satisfy the rules of probability theory up to the standard they are used.

The characteristic feature of a quantum mechanical theory is that not every set of alternative histories that may be described can be assigned probabilities. Nowhere is this more clearly illustrated than in the two-slit experiment illustrated in Fig. 1. In the usual "Copenhagen" discussion, if we have not measured which of the two slits the electron passed through on its way to being detected at the screen, then we are not permitted to assign probabilities to these alternative histories. It would be inconsistent to do so since the correct probability sum rule would not be satisfied. Because of interference, the probability to arrive at point y on the screen is not the sum of the probabilities to arrive at y going through the upper or lower slit:

$$p(y) \neq p_U(y) + p_L(y) \tag{3}$$

because in quantum theory, probabilities are squares of amplitudes and

$$|\psi_L(y) + \psi_U(y)|^2 \neq |\psi_L(y)|^2 + |\psi_U(y)|^2 \,. \tag{4}$$

If we *have* measured which slit the electron went through, then the interference is destroyed, the sum rule obeyed, and we *can* meaningfully assign probabilities to these alternative histories.

A rule is thus needed in quantum theory to determine which sets of alternative histories may be assigned probabilities and which may not. In Copenhagen quantum mechanics, the rule is that probabilities are assigned to histories of alternatives of a subsystem that are *measured* and not in general otherwise.

3. Probabilities for a Time Sequence of Measurements

To establish some notation, let us review in more detail the usual rules for the probabilities of time sequences of ideal measurements of subsystem using the two-slit experiment of Fig. 1 as an example.

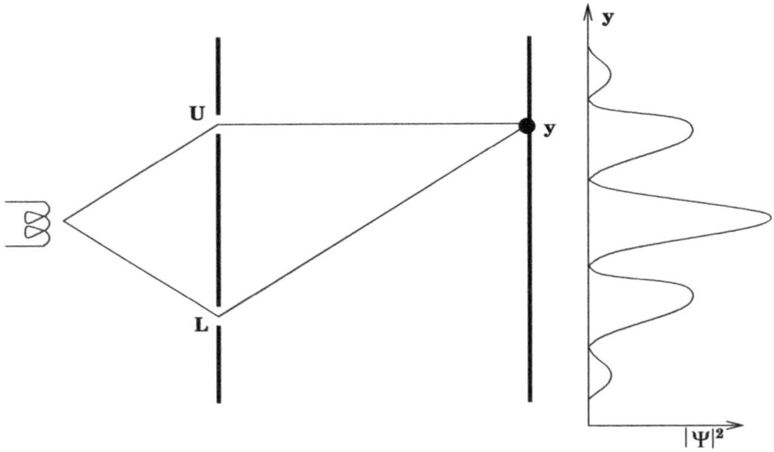

Fig. 1. The two-slit experiment. An electron gun at left emits an electron traveling towards a screen with two slits, its progress in space recapitulating its evolution in time. When precise detections are made of an ensemble of such electrons at the screen it is not possible, because of interference, to assign a probability to the alternatives of whether an individual electron went through the upper slit or the lower slit. However, if the electron interacts with apparatus that measures which slit it passed through, then these alternatives decohere and probabilities can be assigned.

Alternatives for the electron are represented by projection operators in its Hilbert space. Thus, in the two-slit experiment, the alternative that the electron passed through the lower slit is represented by the projection operator

$$P_U = \Sigma_s \int_U d^3x \, |\vec{x}, s\rangle\langle\vec{x}, s| \tag{5}$$

where $|\vec{x}, s\rangle$ is a localized state of the electron with spin component s, and the integral is over a volume around the upper slit. There is a similar projection operator P_L for the alternative that the electron goes through the lower slit. These are exclusive alternatives and they are exhaustive. These properties, as well as the requirements of being projections, are represented by the relations

$$P_L P_U = 0, \quad P_U + P_L = 1, \quad P_L^2 = P_L, \quad P_U^2 = P_U. \tag{6}$$

There is a similarly defined set of projection operators $\{P_y\}$ representing the alternative positions of arrival at the screen.

We can now state the rule for the joint probability that the electron initially in a state $|\psi(t_0)\rangle$ at $t = t_0$ is determined by an ideal measurement at time t_1 to have passed through the upper slit and measured at time t_2 to arrive at point y on the screen. If one likes, one can imagine the case in which the electron is in a narrow wave packet in the horizontal direction with a velocity defined as sharply as possible consistent with the uncertainty principle. The joint probability is negligible unless t_1 and t_2 correspond to the times of flight to the slits and to the screen respectively.

The first step in calculating the joint probability is to evolve the state of the electron to the time t_1 of the first measurement

$$\big|\psi(t_1)\big\rangle = e^{-iH(t_1-t_0)/\hbar}\big|\psi(t_0)\big\rangle. \tag{7}$$

The probability that the outcome of the measurement at time t_1 is that the electron has passed through the upper slit is:

$$(\text{Probability of } U) = \big\|P_U\big|\psi(t_1)\big\rangle\big\|^2 \tag{8}$$

where $\|\cdot\|$ denotes the norm of a vector in the electron's Hilbert space. If the outcome was the upper slit, and the measurement was an "ideal" one, that disturbed the electron as little as possible in making the determination, then after the measurement the state vector is reduced to

$$\frac{P_U|\psi(t_1)\rangle}{\|P_U|\psi(t_1)\rangle\|}. \tag{9}$$

This is evolved to the time of the next measurement

$$|\psi(t_2)\rangle = e^{-iH(t_2-t_1)/\hbar}\frac{P_U|\psi(t_1)\rangle}{\|P_U|\psi(t_1)\rangle\|}. \tag{10}$$

The probability of being detected at point y on the screen at time t_2 *given* that the electron passed through the upper slit is

$$(\text{Probability of } y \text{ given } U) = \|P_y|\psi(t_2)\rangle\|^2. \tag{11}$$

The *joint* probability that the electron is measured to have gone through the upper slit *and* is detected at y is the product of the conditional probability (11) with the probability (8) that the electron passed through U. The latter factor cancels the denominator in (10) so that combining all of the above equations in this section, we have

$$(\text{Probability of } y \textit{ and } U) = \left\| P_y e^{-iH(t_2-t_1)/\hbar} P_U e^{-iH(t_1-t_0)/\hbar} |\psi(t_0)\rangle \right\|^2 . \quad (12)$$

With Heisenberg picture projections this takes the even simpler form

$$(\text{Probability of } y \text{ and } U) = \left\| P_y(t_2) P_U(t_1)\ |\psi(t_0)\rangle \right\|^2 , \quad (13)$$

where, for example,

$$P_U(t) = e^{iHt/\hbar} P_U e^{-iHt/\hbar} . \quad (14)$$

The formula (13) is a compact and unified expression of the two laws of evolution that characterize the quantum mechanics of measured subsystems — unitary evolution in between measurements and reduction of the wave packet at a measurement.[e] The important thing to remember about the expression (13) is that everything in it — projections, state vectors, Hamiltonian — refer to the Hilbert space of a subsystem, in this example the Hilbert space of the electron that is measured.

In "Copenhagen" quantum mechanics, it is measurement that determines which histories of a subsystem can be assigned probabilities and formulae like (13) that determine what these probabilities are. We cannot have such rules in the quantum mechanics of closed systems. There is no fundamental division of a closed system into measured subsystem and measuring apparatus. There is no fundamental reason for the closed system to contain classically behaving measuring apparatus in all circumstances. In particular, in the early universe, none of these concepts seem relevant. We need a more observer-independent, measurement-independent, quasiclassical realm-independent rule for which histories of a closed system can be assigned probabilities and what these probabilities are. The next section describes this rule.

---

[e] As has been noted by many authors, e.g. Refs. 19 and 20 among the earliest.

4. Post-Everett Quantum Mechanics

To describe the rules of post-Everett quantum mechanics, I shall make a simplifying assumption. I shall neglect gross quantum fluctuations in the geometry of spacetime, and assume a fixed background spacetime geometry which supplies a definite meaning to the notion of time. This is an excellent approximation on accessible scales for times later than 10^{-43} s after the big bang. The familiar apparatus of Hilbert space, states, Hamiltonian, and other operators may then be applied to the process of prediction. Indeed, in this context the quantum mechanics of cosmology is in no way distinguished from the quantum mechanics of a large isolated box, perhaps expanding, but containing both the observed and its observers (if any).

A set of alternative histories for a closed system is specified by giving exhaustive sets of exclusive alternatives at a sequence of times. Consider a model closed system initially in a pure state that can be described as containing an observer and a two-slit experiment, with appropriate apparatus for producing the electrons, detecting which slit they passed through, and measuring their position of arrival on the screen (Fig. 2). Some alternatives for the whole system are:

1. Whether or not the observer decided to measure which slit the electron went through.
2. Whether the electron went through the upper or lower slit.
3. The alternative positions, $y_1, \ldots, y_N$, that the electron could have arrived at the screen.

This set of alternatives at a sequence of times defines a set of histories whose characteristic branching structure is shown in Fig. 3. An individual history in the set is specified by some particular sequence of alternatives, e.g. (measured, upper, y_9).

Many other sets of alternative histories are possible for the closed system. For example, we could have included alternatives describing the readouts of the apparatus that detects the position that the electron arrived at the screen. If the initial condition corresponded to a good experiment, there should be a high correlation between these alternatives and the position that the electron arrives at the screen. In a more refined model, we could discuss alternatives corresponding

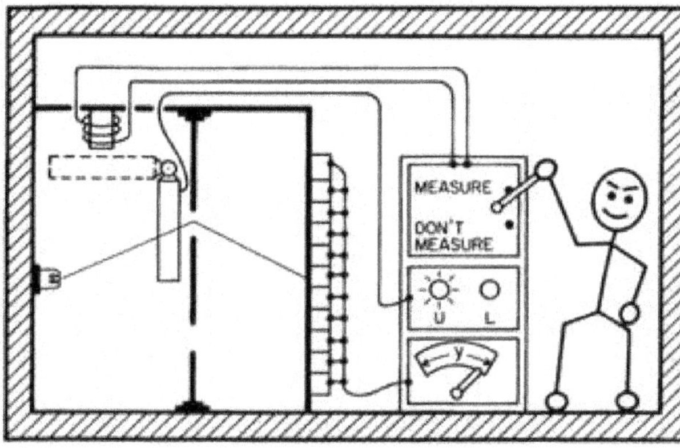

Fig. 2. A model closed quantum system containing an observer together with the necessary apparatus for carrying out a two-slit experiment. Alternatives for the system include whether the observer measured which slit the electron passed through or not, whether the electron passed through the upper or lower slit, the alternative positions of arrival of the electron at the screen, the alternative arrival positions registered by the apparatus, the registration of these in the brain of the observer, etc. Each exhaustive set of exclusive alternatives is represented by an exhaustive set of orthogonal projection operators on the Hilbert space of the closed system. Time sequences of such sets of alternatives describe sets of alternative coarse-grained histories of the closed system. Quantum theory assigns probabilities to the individual alternative histories in such a set when there is negligible quantum mechanical interference between them, that is, when the set of histories decohere. A more refined model might consider a quantity of matter in a closed box. One could then consider alternatives such as whether the box contains a two-slit experiment or not, as well as alternative configurations of the atoms in the box.

to thoughts in the observer's brain, or to the individual positions of the atoms in the apparatus, or to the possibilities that these atoms reassemble in some completely different configuration. There are a vast number of possibilities.

Characteristically, the alternatives that are of use to us as observers are very coarse-grained, distinguishing only very few of the degrees of freedom of a large closed system. This is especially true if we recall that our box with observer and two-slit experiment is only an idealized model. The most general closed system is the universe

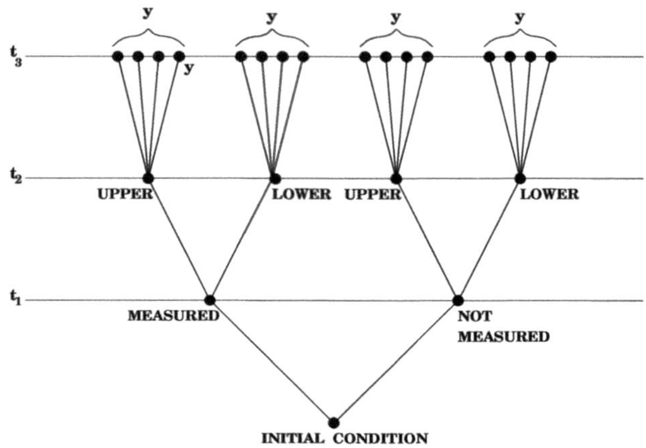

Fig. 3. Branching structure of a set of alternative histories. This figure illustrates the set of alternative histories of the closed system illustrated in Fig. 2 defined by the alternatives of whether the observer did decide to measure or did not decide to measure which slit the electron went through at time t_1, whether the electron went through the upper slit or through the lower slit at time t_2, and the alternative positions of arrival y at the screen at time t_3. The illustrated set of histories does *not* decohere because there is significant quantum mechanical interference between the branch where no measurement was carried out and the electron went through the upper slit and the similar branch where it went through the lower slit. A related set of histories that does decohere can be obtained by replacing the alternatives at time t_2 by the following set of three alternatives: (a record of the decision shows a measurement was initiated and the electron went through the upper slit); (a record of the decision shows a measurement was initiated and the electron went through the lower slit); (a record of the decision shows that the measurement was not initiated). The vanishing of the interference between the alternative values of the record and the alternative configurations of apparatus ensures the decoherence of this set of alternative histories.

itself, and, as I hope to show, the only realistic closed systems are of cosmological dimensions. Certainly, as observers of the universe we utilize only very, very coarse-grained descriptions of the universe as a whole.

I would now like to state the rules that determine which coarse-grained sets of histories may be assigned probabilities and what those probabilities are. The essence of the rules I shall describe can be found in the work of Bob Griffiths.[17] The general framework was extended by Roland Omnès[18] and was independently, but later, arrived at by Murray Gell-Mann and myself.[2] The idea is simple: The failure of probability sum rules due to quantum interference is the obstacle to assigning probabilities. *Probabilities can be assigned to just those sets of alternative histories of a closed system for which there is negligible interference between the individual histories in the set as a consequence of the Hamiltonian and particular initial state the closed system has. All probability sum rules are satisfied as a result of the absence of interference.* Let us now give this idea a precise expression.

Sets of alternatives at one moment of time are represented by sets of orthogonal projection operators. Employing the Heisenberg picture these can be denoted $\{P^k_{\alpha_k}(t_k)\}$. The superscript k denotes the set of alternatives being considered at time t_k (for example, the set of alternative position intervals $\{y_1, \ldots, y_N\}$ at which the electron might arrive at the screen at time t_3), α_k denotes the particular alternative in the set (for example, y_9) and t_k is the time. The set of P's satisfy

$$\sum\nolimits_{\alpha_k} P^k_{\alpha_k}(t_k) = 1, \quad P^k_{\alpha_k}(t_k) P^k_{\alpha'_k}(t_k) = \delta_{\alpha_k \alpha'_k} P^k_{\alpha_k}(t_k) \qquad (15)$$

showing that they represent an exhaustive set of exclusive alternatives.

Sets of alternative histories are defined by giving sequences of sets of alternatives at definite moments of time, e.g. $\{P^1_{\alpha_1}(t_1)\}$, $\{P^2_{\alpha_2}(t_2)\}, \ldots, \{P^n_{\alpha_n}(t_n)\}$. Different choices for $\{P^1_{\alpha_1}(t_1)\}$, $\{P^2_{\alpha_2}(t_2)\}$, etc. describe different sets of alternative histories of the closed system. An individual history in a given set corresponds to a particular sequence $(\alpha_1, \ldots, \alpha_n) \equiv \alpha$ and, for each history, there is a corresponding chain of projection operators

$$C_\alpha \equiv P^n_{\alpha_n}(t_n) \cdots P^1_{\alpha_1}(t_1) \,. \qquad (16)$$

For example, in the two-slit experiment in a box illustrated in Fig. 2, the history in which the observer decided at time t_1 to measure which slit the electron goes through, in which the electron goes through the upper slit at time t_2, and arrives at the screen in position interval y_9 at time t_3, would be represented by the chain

$$P^3_{y_9}(t_3)P^2_U(t_2)P^1_{\text{meas}}(t_1) \tag{17}$$

in an obvious notation. The only difference between this situation and that of the "Copenhagen" quantum mechanics of measured subsystems is the following: The sets of operators $\{P^k_{\alpha_k}(t_k)\}$ defining alternatives for the closed system act on the Hilbert space of the closed system that includes the variables describing any apparatus, observers, and anything else. The operators defining alternatives in Copenhagen quantum mechanics act only on the Hilbert space of the measured subsystem.

When the initial state is pure, it can be resolved into *branches* corresponding to the individual members of any set of alternative histories. The generalization to an impure initial density matrix is not difficult,[2] but for simplicity we shall assume a pure initial state throughout this article. Denote the initial state by $|\Psi\rangle$ in the Heisenberg picture. Then

$$|\Psi\rangle = \sum\nolimits_\alpha C_\alpha|\Psi\rangle = \sum_{\alpha_1,\ldots,\alpha_n} P^n_{\alpha_n}(t_n)\cdots P^1_{\alpha_1}(t_1)|\Psi\rangle. \tag{18}$$

This identity follows by applying the first of (15) to all the sums over α_k in turn. The vector

$$C_\alpha|\Psi\rangle \tag{19}$$

is the *branch*[f] corresponding to the individual history α and (18) is the resolution of the initial state into branches.

When the branches corresponding to a set of alternative histories are sufficiently orthogonal, the set of histories is said to *decohere*. More precisely, a set of histories decoheres when

$$\langle\Psi|C^\dagger_{\alpha'}C_\alpha|\Psi\rangle \approx 0, \quad \text{for any} \quad \alpha'_k \neq \alpha_k. \tag{20}$$

We shall return to the standard with which decoherence should be enforced, but first let us examine its meaning and consequences.

---

[f]More specifically a *branch state vector*.

Decoherence means the absence of quantum mechanical interference between the individual histories of a coarse-grained set.[g] Probabilities can be assigned to the individual histories in a decoherent set of alternative histories because decoherence implies the probability sum rules necessary for a consistent assignment. The probability of an individual history α is

$$p(\alpha) = \|C_\alpha|\Psi\rangle\|^2 . \tag{21}$$

To see how decoherence implies the probability sum rules, let us consider an example in which there are just three sets of alternatives at times t_1, t_2 and t_3. A typical sum rule might be

$$\sum\nolimits_{\alpha_2} p\left(\alpha_3, \alpha_2, \alpha_1\right) = p\left(\alpha_3, \alpha_1\right). \tag{22}$$

---

[g]The term "decoherence" is used in several different ways in the literature. Therefore, for those familiar with other works, a comment is in order to specify how we are employing the term in this simplified presentation. We have followed our previous work,[2,3] in using the term "decoherence" to refer to a property of a set of alternative time *histories* of a closed system. A decoherent set of histories is one for which the quantum mechanical interference between individual histories is small enough to guarantee an appropriate set of probability sum rules. Different notions of decoherence can be defined by utilizing different measures of interference. The weakest notion is just the consistency of the probability sum rules that was called "consistency" by Griffiths[17] and Omnès[18] and that term is used by some to refer to all measures of interference. Vanishing of the real part of (20) is a sufficient condition for the consistency of the probability sum rules called the "weak decoherence condition". We are using the stronger condition (20) because it characterizes widespread and typical mechanisms of decoherence. Equation (20) has been called the "medium decoherence condition". "Decoherence" in the context of this paper, thus, means the medium decoherence of sets of histories. In the literature the term "decoherence" has also been used to refer to the decay in time of the off-diagonal elements of a reduced density matrix defined by tracing the full density matrix over a given set of variables.[21] The two notions of "decoherence of reduced density matrices" and "decoherence of histories" are not generally equivalent but also not unconnected in the sense that, particularly, models with certain physical processes can ensure both. (See, e.g. the remarks in Sec. II.6.4 of Ref. 22.)

We show that (20) and (21) imply (22). To do that, write out the left-hand side of (22) using (21) and suppress the time labels for compactness.

$$\sum_{\alpha_2} p(\alpha_3, \alpha_2, \alpha_1) = \sum_{\alpha_2} \langle \Psi | P^1_{\alpha_1} P^2_{\alpha_2} P^3_{\alpha_3} P^3_{\alpha_3} P^2_{\alpha_2} P^1_{\alpha_1} | \Psi \rangle. \tag{23}$$

Decoherence means that the sum on the right-hand side of (23) can be written with negligible error as

$$\sum_{\alpha_2} p(\alpha_3, \alpha_2, \alpha_1) \approx \sum_{\alpha_2' \alpha_2} \left\langle \Psi | P^1_{\alpha_1} P^2_{\alpha_2'} P^3_{\alpha_3} P^3_{\alpha_3} P^2_{\alpha_2} P^1_{\alpha_1} | \Psi \right\rangle, \tag{24}$$

the extra terms in the sum being vanishingly small. But now, applying the first of (15) we see

$$\sum_{\alpha_2} p(\alpha_3, \alpha_2, \alpha_1) \approx \langle \Psi | P^1_{\alpha_1} P^3_{\alpha_3} P^3_{\alpha_3} P^1_{\alpha_1} | \Psi \rangle = p(\alpha_3, \alpha_1) \tag{25}$$

so that the sum rule (22) is satisfied.

Given an initial state $|\Psi\rangle$ and a Hamiltonian H, one could, in principle, identify all possible sets of decohering histories. Among these will be the exactly decohering sets where the orthogonality of the branches is exact. Indeed, trivial examples can be supplied by resolving $|\Psi\rangle$ into a sum of orthogonal vectors at time t_1, resolving those vectors into sums of further vectors such that the whole set is orthogonal at time t_2, and so on. However, such sets of exactly decohering histories will not, in general, have a simple description in terms of fundamental fields nor any connection, for example, with the quasiclassical realm of familiar experience. For this reason, sets of histories that approximately decohere are of interest. As we will argue in the next two sections, realistic mechanisms lead to the decoherence of histories constituting a quasiclassical realm to an excellent approximation. When the decoherence condition (20) is approximately enforced, the probability sum rules such as (22) will only be approximately obeyed. However, as discussed earlier, these probabilities for single systems are meaningful up to the standard they are used. Approximate probabilities for which the sum rules are satisfied to a comparable standard may therefore also be employed in the process of prediction. When we speak of approximate decoherence and approximate probabilities, we mean decoherence achieved and probability sum rules satisfied beyond any standard that might be conceivably contemplated for the accuracy of prediction and the comparison of theory with experiment.

Decoherent sets of histories of the universe are what we may utilize in the process of prediction in quantum mechanics, for they may be assigned probabilities. Decoherence thus generalizes and replaces the notion of "measurement", which served this role in the Copenhagen interpretations. Decoherence is a more precise, more objective, more observer-independent idea and gives a definite meaning to Everett's branches. For example, if their associated histories decohere, we may assign probabilities to various values of reasonable scale density fluctuations in the early universe whether or not anything like a "measurement" was carried out on them and certainly whether or not there was an "observer" to do it.

5. The Origins of Decoherence in Our Universe

What are the features of coarse-grained sets of histories that decohere in our universe? In seeking to answer this question it is important to keep in mind the basic aspects of the theoretical framework on which decoherence depends. Decoherence of a set of alternative histories is not a property of their operators *alone*. It depends on the relations of those operators to the initial state $|\Psi\rangle$, the Hamiltonian H, and the fundamental fields. Given these, we could, in principle, *compute* which sets of alternative histories decohere.

We are not likely to carry out a computation of all decohering sets of alternative histories for the universe, described in terms of the fundamental fields, anytime in the near future, if ever. It is therefore important to investigate specific mechanisms by which decoherence occurs. Let us begin with a very simple model, due in its essential features to Joos and Zeh.[16] We consider the two-slit example again, but this time suppose that in the neighborhood of the slits there is a gas of photons or other light particles colliding with the electrons (Fig. 4). Physically, it is easy to see what happens, the random uncorrelated collisions can carry away delicate phase correlations between the beams even if the trajectories of the electrons are not affected much. The interference pattern will then be destroyed and it will be possible to assign probabilities as to whether the electron went through the upper slit or the lower slit.

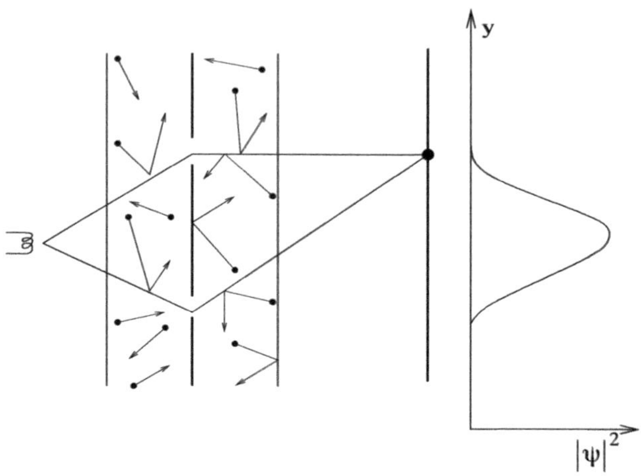

Fig. 4. The two-slit experiment with an interacting gas. Near the slits, light particles of a gas collide with the electrons. Even if the collisions do not affect the trajectories of the electrons very much, they can still carry away the phase correlations between the histories in which the electron arrived at point y on the screen by passing through the upper slit and that in which it arrived at the same point by passing through the lower slit. A coarse-graining that is described by only these two alternative histories of the electron would then approximately decohere as a consequence of the interactions with the gas given adequate density, cross-section, etc. Interference is destroyed and probabilities can be assigned to these alternative histories of the electron in a way that they could not be if the gas were not present (cf. Fig. 1). The lost phase information is still available in correlations between states of the gas and states of the electron. The alternative histories of the electron would not decohere in a coarse-graining that included both the histories of the electron *and* operators that were sensitive to the correlations between the electrons and the gas.

This model illustrates a widely occurring mechanism by which certain types of coarse-grained sets of alternative histories decohere in our universe.

Let us see how this picture in words is given precise meaning in mathematics. Initially, suppose the state of the entire system is a state of the electron $|\psi>$ and N distinguishable "photons" in states $|\varphi_1\rangle$, $|\varphi_2\rangle$, etc., viz.

$$|\Psi\rangle = |\psi\rangle|\varphi_1\rangle|\varphi_2 > \cdots |\varphi_N\rangle. \tag{26}$$

Suppose further that $|\psi\rangle$ is a coherent superposition of a state in which the electron passes through the upper slit $|U\rangle$ and the lower slit $|L\rangle$. Explicitly:

$$|\psi\rangle = \alpha|U\rangle + \beta|L\rangle. \tag{27}$$

Both states are wave packets in x, so that position in x recapitulates history in time. We now ask whether the history where the electron passes through the upper slit and arrives at a detector at point y on the screen, decoheres from that in which it passes through the lower slit and arrives at point y as a consequence of the initial condition of this "universe". That is, as in Sec. 4, we ask whether the two branches

$$P_y(t_2)P_U(t_1)|\Psi\rangle, \qquad P_y(t_2)P_L(t_1)|\Psi\rangle \tag{28}$$

are nearly orthogonal, the times of the projections being those for the nearly classical motion in x. We work this out in the Schrödinger picture where the initial state evolves, and the projections on the electron's position are applied to it at the appropriate times.

Collisions occur, but the states $|U\rangle$ and $|L\rangle$ are left more or less undisturbed. The states of the "photons", of course, are significantly affected. If the photons are dilute enough to be scattered once by the electron in its time to traverse the gas, the two branches (28) will be approximately

$$\alpha P_y|U\rangle S_U|\varphi_1\rangle S_U|\varphi_2\rangle \cdots S_U|\varphi_N\rangle, \tag{29a}$$

and

$$\beta P_y|L\rangle S_L|\varphi_1\rangle S_L|\varphi_2\rangle \cdots S_L|\varphi_N\rangle. \tag{29b}$$

Here, S_U and S_L are the scattering matrices from an electron in the vicinity of the upper slit and the lower slit respectively. The two branches in (29) decohere because the states of the "photons" are nearly orthogonal. The overlap of the branches is proportional to

$$\langle\varphi_1|S_U^\dagger S_L|\varphi_1\rangle\langle\varphi_2|S_U^\dagger S_L|\varphi_2\rangle \cdots \langle\varphi_N|S_U^\dagger S_L|\varphi_N\rangle. \tag{30}$$

Now, the S-matrices for scattering off the upper position or the lower position can be connected to that of an electron at the origin by a translation

$$S_U = \exp(-i\mathbf{k}\cdot\mathbf{x}_U)S\ \exp(+i\mathbf{k}\cdot\mathbf{x}_U), \tag{31a}$$

$$S_L = \exp(-i\mathbf{k}\cdot\mathbf{x}_L)S\ \exp(+i\mathbf{k}\cdot\mathbf{x}_L). \tag{31b}$$

Here, $\hbar\mathbf{k}$ is the momentum of a photon, $\mathbf{x}_U$ and $\mathbf{x}_L$ are the positions of the slits and S is the scattering matrix from an electron at the origin.

$$\langle\mathbf{k}'|S|\mathbf{k}\rangle = \delta^{(3)}(\mathbf{k}-\mathbf{k}') + \frac{i}{2\pi\omega_{\mathbf{k}}} f(\mathbf{k},\mathbf{k}')\delta(\omega_k - \omega_k'), \tag{32}$$

where f is the scattering amplitude and $\omega_k = |\vec{k}|$.

Consider the case where initially all the photons are in plane wave states in an interaction volume V, all having the same energy $\hbar\omega$, but with random orientations for their momenta. Suppose further that the energy is low so that the electron is not much disturbed by a scattering and low enough so the wavelength is much longer than the separation between the slits, $k|\mathbf{x}_U - \mathbf{x}_L| \ll 1$. It is then possible to work out the overlap. The answer according to Joos and Zeh[16] is

$$\left(1 - \frac{(k|\mathbf{x}_U - \mathbf{x}_L|)^2}{8\pi^2 V^{2/3}}\sigma\right)^N \tag{33}$$

where σ is the effective scattering cross-section and the individual terms have been averaged over incoming directions. Even if σ is small, as N becomes large, this tends to zero. In this way, decoherence becomes a quantitative phenomenon.

What such models convincingly show is that decoherence is frequent and widespread in the universe for histories of certain kinds of variables. Joos and Zeh calculated that a superposition of two positions of a grain of dust, 1 mm apart, is decohered simply by the scattering of the cosmic background radiation on the time scale of a nanosecond. The existence of such mechanisms means that the only realistic isolated systems are of cosmological dimensions. So widespread is this kind of phenomena with the initial condition and dynamics of our universe that we may meaningfully speak of habitually decohering variables such as the center of mass positions of massive bodies.

6. The Copenhagen Approximation

What is the relation of the familiar Copenhagen quantum mechanics described in Sec. 3 to the more general "post-Everett" quantum mechanics of closed systems described in Secs. 4 and 5? Copenhagen quantum mechanics predicts the probabilities of the histories of measured subsystems. Measurement situations may be described in a closed system that contains both measured subsystem and measuring apparatus. In a typical measurement situation, the values of a variable not normally decohering become correlated with alternatives of the apparatus that decohere because of *its* interactions with the rest of the closed system. The correlation means that the measured alternatives decohere because the alternatives of the apparatus with which they are correlated decohere.

The recovery of the Copenhagen rule for when probabilities may be assigned is immediate. Measured quantities are correlated with decohering histories. Decohering histories can be assigned probabilities. Thus in the two-slit experiment (Fig. 1), when the electron interacts with an apparatus that determines which slit it passed through, it is the decoherence of the alternative configurations of the apparatus that enables probabilities to be assigned for the electron.

There is nothing incorrect about Copenhagen quantum mechanics. Neither is it, in any sense, opposed to the post-Everett formulation of the quantum mechanics of closed systems. It is an *approximation* to the more general framework appropriate in the special cases of measurement situations and when the decoherence of alternative configurations of the apparatus may be idealized as exact and instantaneous. However, while measurement situations imply decoherence, they are only special cases of decohering histories. Probabilities may be assigned to alternative positions of the moon and to alternative values of density fluctuations near the big bang in a universe in which these alternatives decohere, whether or not they were participants in a measurement situation and certainly whether or not there was an observer registering their values.

7. Quasiclassical Realms

As observers of the universe, we deal with coarse-grained histories that reflect our own limited sensory perceptions, extended by instruments, communication, and records, but, in the end, characterized by a large amount of ignorance. Yet, we have the impression that the universe exhibits a much finer-grained set of histories, independent of us, defining an always decohering "quasiclassical realm", to which our senses are adapted, but deal with only a small part of it. If we are preparing for a journey into a yet unseen part of the universe, we do not believe that we need to equip ourselves with spacesuits having detectors sensitive, say, to coherent superpositions of position or other unfamiliar quantum variables. We expect that the familiar quasiclassical variables will decohere and be approximately correlated in time by classical deterministic laws in any new part of the universe we may visit, just as they are here and now.

Since the post-Everett quantum mechanics of closed systems does not posit a quasiclassical realm, it must provide an explanation of this manifest fact of everyday experience. No such explanation can be provided from the dynamics of quantum theory alone. Rather, like decoherence, the existence of a quasiclassical realm in the universe must be a consequence of both the initial condition of the universe and the Hamiltonain describing evolution.

Roughly speaking, a quasiclassical realm should be a set of alternative histories that decoheres according to a realistic principle of decoherence, that is maximally refined consistent with that notion of decoherence, and whose individual histories are described largely by alternative values of a limited set of quasiclassical variables at different moments of time, that exhibit, as much as possible, patterns of classical correlation in time. To make the question of the existence of one or more quasiclassical realms into a *calculable* question in quantum cosmology, we need measures of how close a set of histories comes to constituting a "quasiclassical realm". A quasiclassical realm cannot be a *completely* fine-grained description, for then it would not decohere. It cannot consist *entirely* of a few "quasiclassical variables" repeated over and over, because sometimes we may measure something highly quantum mechanical. Quasiclassical variables cannot be *always* correlated in time by classical laws, because sometimes quantum mechanical phenomena cause deviations from classical physics. We need measures for maximality and classicality.[2]

It is possible to give crude arguments for the type of habitually decohering operators we would expect to occur over and over again in a set of histories defining a quasiclassical realm.[2] Such habitually decohering operators are called "quasiclassical operators". In the earliest instants of the universe, the operators defining spacetime on scales well above the Planck scale emerge from the quantum fog as quasiclassical. Any theory of the initial condition that does not imply this is simply inconsistent with observation in a manifest way. A background spacetime is thus defined and conservation laws arising from its symmetries have meaning. Then, where there are suitable conditions of low temperature, density, etc., various sorts of hydrodynamic variables may emerge as quasiclassical operators. These are integrals over suitably small volumes of densities of conserved or nearly conserved quantities. Examples are densities of energy, momentum, baryon number, and, in later epochs, nuclei, and even chemical species. The sizes of the volumes are limited above by maximality and are limited below by classicality because they require sufficient "inertia" resulting from their approximate conservation, to enable them to resist deviations from predictability caused by their interactions with one another, by quantum spreading, and by the quantum and statistical fluctuations resulting from interactions with the rest of the universe that accomplish decoherence.[23] Suitable integrals of densities of approximately conserved quantities are thus candidates for habitually decohering quasiclassical operators. These "hydrodynamic variables" *are* among the principal variables of classical physics.

It would be in such ways that the quasiclassical realm of familiar experience would be an emergent property of the fundamental description of the universe, not generally in quantum mechanics, but as a consequence of our specific initial condition and the Hamiltonian describing evolution. Whether the universe exhibits a quasiclassical realm, and, indeed, whether it exhibits more than one essentially inequivalent realm, thus become calculable questions in the quantum mechanics of closed systems.

8. Conclusion

Quantum mechanics is best and most fundamentally understood in the context of quantum mechanics of closed systems, most generally the universe as a whole. The founders of quantum mechanics were right in pointing out that something external to the framework of wave function and the Schrödinger equation *is* needed to interpret the theory. But it is not a postulated classical realm to which quantum mechanics does not apply. Rather it is the initial condition of the universe that, together with the action function of the elementary particles and the throws of the quantum dice since the beginning, is the likely origin of quasiclassical realm(s) within quantum theory itself.

Acknowledgments

The formulation of quantum mechanics described in this paper is a result of joint work with Murray Gell-Mann. It is a pleasure to thank him for the many conversations over the years and for permission to summarize aspects of our work here. Preparation of the report was supported in part by the National Science Foundation under grant PHY90-08502.

References

1. In the spirit of providing a simplified introduction to the quantum mechanics of closed systems rather than a comprehensive review, no attempt has been made to provide anything more than the references that are directly relevant to the points raised in the text as they were in 1992.
2. M. Gell-Mann and J. B. Hartle, in *Complexity, Entropy, and the Physics of Information*, SFI Studies in the Sciences of Complexity, Vol. VIII, ed. W. Zurek (Addison Wesley, Reading); or in *Proc. 3rd Int. Symp. Foundations of Quantum Mechanics in the*

Light of New Technology, eds. S. Kobayashi, H. Ezawa, Y. Murayama and S. Nomura (Physical Society of Japan, Tokyo, 1990).
3. M. Gell-Mann and J. B. Hartle, in the *Proc. 25th Int. Conf. High Energy Physics*, Singapore, August, 2–8, 1990, eds. K. K. Phua and Y. Yamaguchi (South East Asia Theoretical Physics Association and Physical Society of Japan), distributed by (World Scientific, Singapore, 1990).
4. S. W. Hawking, *Phys. Lett. B* **196**, 337 (1983).
5. S. Coleman, *Nucl. Phys. B* **310**, 643 (1988).
6. S. Giddings and A. Strominger, *Nucl. Phys. B* **307**, 854 (1988).
7. J. Halliwell, in *Quantum Cosmology and Baby Universes: Proc. 1989 Jerusalem Winter School for Theoretical Physics*, eds. S. Coleman, J. B. Hartle, T. Piran and S. Weinberg (World Scientific, Singapore, 1991), pp. 65–157.
8. N. Bohr, *Atomic Physics and Human Knowledge* (Science Editions, New York, 1958).
9. L. Landau and E. Lifshitz, *Quantum Mechanics* (Pergamon, London, 1958).
10. F. London and E. Bauer, *La Théorie de L'observation en Mécanique Quantique* (Hermann, Paris, 1939).
11. H. Everett, *Rev. Mod. Phys.* **29**, 454 (1957).
12. B. DeWitt and R. N. Graham (eds.), *The Many Worlds Interpretation of Quantum Mechanics* (Princeton University Press, Princeton, 1973).
13. H. D. Zeh, *Found. Phys.* **1**, 69 (1971).
14. W. Zurek, *Phys. Rev. D* **24**, 1516 (1981).
15. W. Zurek, *Phys. Rev. D* **26**, 1862 (1982).
16. E. Joos and H. D. Zeh, *Zeit. Phys. B* **59**, 223 (1985).
17. R. Griffiths, *J. Stat. Phys.* **36**, 219 (1984).
18. R. Omnès, *Rev. Mod. Phys.* **64**, 339 (1992).
19. H. J. Groenewold, *Proc. Kon. Akad. Wetensch. Ned. Ser. B* **55**, 219 (1952).
20. E. Wigner, *Am. J. Phys.* **31**, 6 (1963).
21. W. Zurek, *Phys. Today* **44**, 36 (1991).

22. J. B. Hartle, The quantum mechanics of cosmology, in *Quantum Cosmology and Baby Universes: Proc. 1989 Jerusalem Winter School for Theoretical Physics*, eds. S. Coleman, J. B. Hartle, T. Piran and S. Weinberg (World Scientific, Singapore, 1991), pp. 65–157.
23. M. Gell-Mann and J. B. Hartle, *Phys. Rev. D* **47**, 3345 (1993).

CHAPTER 8

The Spacetime Approach to Quantum Mechanics*

Feynman's sum-over-histories formulation of quantum mechanics is reviewed as an independent statement of quantum theory in spacetime form. It is different from the usual Schrödinger–Heisenberg formulation that utilizes states on spacelike surfaces because it assigns probabilities to different sets of alternatives. In a sum-over-histories formulation, alternatives at definite moments of time are more restricted than in usual quantum mechanics because they refer only to the coordinates in terms of which the histories are defined. However, in the context of the quantum mechanics of closed systems, sum-over-histories quantum mechanics can be generalized to deal with spacetime alternatives that are not "at definite moments of time". An example in field theory is the set of alternative

(*continued overleaf*)

*This essay is dedicated to Louis Witten on the occasion of his retirement from the University of Cinncinati. It appeared in the proceedings of the symposium held in his honor on April 3–4, 1992, *Topics on Quantum Gravity and Beyond*, eds. F. Mansouri and J. J. Scanio, in *Proc. Int. Symp. Quantum Physics and the Universe*, Waseda University, Tokyo, Japan, August 23–27, 1992, and *Vistas Astron.* **37**, 569 (1993).
arXiv:gr-qc/9210004v4

ranges of values of a field averaged over a spacetime region. An example in particle mechanics is the set of the alternatives defined by whether a particle never crosses a fixed spacetime region or crosses it at least once. The general notion of a set of spacetime alternatives is a partition (coarse-graining) of the histories into an exhaustive set of exclusive classes. With this generalization, the sum-over-histories formulation can be said to be in fully spacetime form with dynamics represented by path integrals over spacetime histories and alternatives defined as spacetime partitions of these histories. When restricted to alternatives at definite moments of times, this generalization is equivalent to Schrödinger–Heisenberg quantum mechanics. However, the quantum mechanics of more general spacetime alternatives does not have an equivalent Schrödinger–Heisenberg formulation. We suggest that, in the quantum theory of gravity, the general notion of "observable" is supplied by diffeomorphism invariant partitions of spacetime metrics and matter field configurations. By generalizing the usual alternatives so as to put quantum theory in fully spacetime form, we may be led to a covariant generalized quantum mechanics of spacetime free from the problem of time.

1. Introduction

In 1948, Feynman,[1] building on the work of Dirac,[2] introduced his sum-over-histories formulation of quantum mechanics and with it the path integral that has proved a powerful tool in many branches of physics. It is possible to see the sum-over-histories as merely a technical tool — a useful device for computing certain amplitudes within the usual Schrödinger–Heisenberg formulation of quantum mechanics in terms of states on spacelike surfaces. That is not, however, how I think Feynman saw it. Rather, the sum-over-histories formulation of quantum mechanics can be regarded as an independent formulation of quantum theory. Its equivalence with the usual formulations is not automatic, but rather a question, the answer for which may be different in different theories and for different physical systems. In this talk, I shall review the current status of the sum-over-histories formulation as an independent statement of quantum theory. I shall argue that, viewed most fundamentally, it is *different* from the Schrödinger–Heisenberg formulations because the totality of alternatives to which it potentially assigns probabilities is different from that of Schrödinger–Heisenberg quantum mechanics. Sum-over-histories alternatives at definite moments of time are more restricted than in usual quantum mechanics because they refer only to the coordinates of the configuration space in which the paths are defined. By contrast, at a moment of time, Schrödinger–Heisenberg quantum mechanics has all the alternatives provided by transformation theory. However, even in non-relativistic quantum mechanics, the sum-over-histories formulation allows a generalization of the alternatives at definite moments of time to genuine spacetime alternatives that are not considered in usual Schrödinger–Heisenberg formulations. This generalization allows a more realistic description of everyday experiments. But, more importantly, it may be central for the construction of a quantum theory of gravity in which there is no well-defined notion of time. There, alternatives "at a moment of time" may be difficult to find, while spacetime alternatives may be the natural "observables" for which the theory makes predictions.

2. The Spacetime Approach to Quantum Mechanics

In the Schrödinger–Heisenberg quantum mechanics of particles or fields moving in a fixed background spacetime, the quantum dynamics of measured subsystems is formulated in terms of state vectors defined on spacelike surfaces that evolve unitarily in between measurements and are reduced at measurements. Unitary evolution is represented by

$$\left|\psi(t)\right\rangle = e^{-iHt/\hbar}\left|\psi\right\rangle \tag{1}$$

where H is the Hamiltonian and $|\psi\rangle$ is the state at $t = 0$. From (1) we could calculate the transition amplitude between coordinates q' at time t' to coordinates q'' at time t'', that is an equivalent summary of unitary evolution, viz.

$$\left\langle q''t''\middle|q't'\right\rangle = \left\langle q''\middle|e^{-iH(t''-t')/\hbar}\middle|q'\right\rangle. \tag{2}$$

(Coordinate indices, which may refer to either particles or fields, are often omitted to keep the notation compact.) In a sum-over-histories formulation of quantum mechanics, such amplitudes, are specified *directly* as path integrals

$$\langle q''t''|q't'\rangle = \int_{[q',q'']} \delta q \ \exp(iS[q(\tau)]/\hbar). \tag{3}$$

Here, $S[q(\tau)]$ is the action functional corresponding to the Hamiltonian H and the sum is over paths $q(t)$ that starts at q' at time t', ends at q'' at time t'', and are single-valued functions of time. In an abbreviated notation, we may write

$$\int \delta q' \ \exp\bigl(iS[q(\tau)]/\hbar\bigr)|\psi\rangle \tag{4}$$

for the state vector $|\psi(t)\rangle$ that is evolved by the propagator (3). More explicitly, (4) stands for the state vector $|\psi(t)\rangle$ whose representative wave function $\psi(q,t) = \langle q|\psi(t)\rangle$ is

$$\psi(q,t) = \int dq' \left(\int_{[q',q]} \delta q \ \exp\bigl(iS[q(\tau)]/\hbar\bigr)\right) \psi(q',t'). \tag{5}$$

In this way, quantum dynamics corresponding to unitary evolution is cast into manifestly spacetime form involving spacetime histories directly. This is an important advantage in dealing with spacetime symmetries such as Lorentz invariance.

Unitary evolution, however, is not the only law by which the state vector evolves in quantum mechanics. In the usual discussion, at an ideal measurement that disturbs the measured system as little as possible, the state vector is instantaneously "reduced" according to the "second law of evolution"

$$|\psi(t)\rangle \longrightarrow \frac{P_\alpha|\psi(t)\rangle}{\|P_\alpha|\psi(t)\rangle\|} \tag{6}$$

where P_α is the projection operator on the subspace corresponding to the outcome of the measurement and $\|\cdot\|$ denotes the norm of a vector in Hilbert space. This "second law of evolution" may not be needed to calculate the transition probabilities in scattering experiments, but it is essential for calculating the probabilities of the *sequences* of observations that define the histories of everyday life, such as the orbit of the earth around the sun. It is every bit as essential for the prediction of realistic probabilities as is unitary evolution.

Using the two laws of evolution, the joint probability for a sequence of measured outcomes $\alpha_1, \ldots, \alpha_n$ at times $t_1, \ldots, t_n$ may be calculated. It finds its most compact expression in the Heisenberg picture:

$$\left\| P^n_{\alpha_n}(t_n) P^{n-1}_{\alpha_{n-1}}(t_{n-1}) \cdots P^1_{\alpha_1}(t_1) |\psi\rangle \right\|^2. \tag{7}$$

Here, $\{P^k_{\alpha_k}(t_k)\}$ is an exhaustive set of orthogonal Heisenberg picture projections representing the alternatives α_k in the set k at time t_k. For example, the set of alternatives might be an exhaustive set of alternative regions $\Delta^k_{\alpha_k}$ of the coordinates q at time t_k.

The second law of evolution can be expressed simply in sum-over-histories form if attention is restricted to alternatives defined by sequences of configuration space regions $\{\Delta^1_{\alpha_1}\}, \{\Delta^2_{\alpha_2}\}, \ldots$ at times $t_1, \ldots, t_n$.[4,5] The joint probability that the system passes through the particular sequence of regions $\alpha = (\alpha_1, \ldots, \alpha_n)$ is

$$\left\| \int_{c_\alpha} \delta q \; \exp\bigl(iS[q(\tau)]/\hbar\bigr) |\psi\rangle \right\|^2 \tag{8}$$

where the path integral is over the class of paths c_α that pass through the region $\Delta^1_{\alpha_1}$ at time t_1, $\Delta^2_{\alpha_2}$ at time t_2, etc. Equation (8) is a unified expression for the two laws of evolution in quantum mechanics.

Equation (8) is equivalent to (7) for alternatives defined by exhaustive sets of exclusive regions of configuration space at definite moments of time. In this case, the Schrödinger–Heisenberg formulation of quantum mechanics and the sum-over-histories formulation coincide. But they are not *fully* equivalent because the effect of projections onto ranges of momentum, for example, cannot be represented as restrictions on a configuration space path integral as in (8). Probabilities for momentum measurements *can* be predicted using path integrals, but only approximately, by modeling a time of flight determination of velocity in configuration space terms.[21] The sum-over-histories formulation of quantum mechanics therefore deals directly and exactly with a more restricted class of alternatives at sequences of moments of time than is available from the transformation theory of the Schrödinger–Heisenberg formulation.

3. Spacetime Alternatives

As we have presented it so far, the sum-over-histories formulation of quantum mechanics is not in fully spacetime form. The use of path integrals over spacetime histories has put the dynamics corresponding to unitary evolution into spacetime form. But the alternatives to which the theory assigns probabilities are not general spacetime alternatives. They have been restricted to sequences of alternative regions of configuration space at definite moments of time. More general spacetime alternatives are easy to imagine. We mention two: Consider the quantum mechanics of a particle in which the histories are paths in spacetime. Fix a spacetime region R with extent both in space and time (Fig. 1). A given particle path may never cross R or, alternatively, it may cross R sometime, perhaps more than once. These are an exhaustive set of spacetime alternatives for the particle that are not "at a moment of time". A second example is provided by field averages over spacetime regions with extent both in space and time, such as were considered by Bohr and Rosenfeld[6] in their discussion of the measurability of the electromagnetic field. An exhaustive set of ranges of such average values of a field is an example of a set of spacetime alternatives in field theory. Such spacetime alternatives are not directly assigned probabilities in Schrödinger–Heisenberg quantum mechanics because they are not "at a moment in time".

In his 1948 paper, Feynmann discussed alternatives defined with respect to spacetime regions such as we described above. In particular, he offered a sum-over-histories definition of the probability that "if an ideal measurement is performed to determine whether a particle has a path lying in a region of spacetime ... the result will be affirmative". However, that discussion, as well as more recent ones[4,7–9,29,38] were incomplete because they did not specify clearly what such an ideal measurement consisted of or what was to replace the "second law of evolution", in Eq. (6), following its completion. "I have not been able to find a precise definition," Feynman said.[1] Only recently has it become clear how to generalize sum-over-histories quantum mechanics to predict probabilities for such spacetime alternatives within the quantum mechanics of closed systems in which

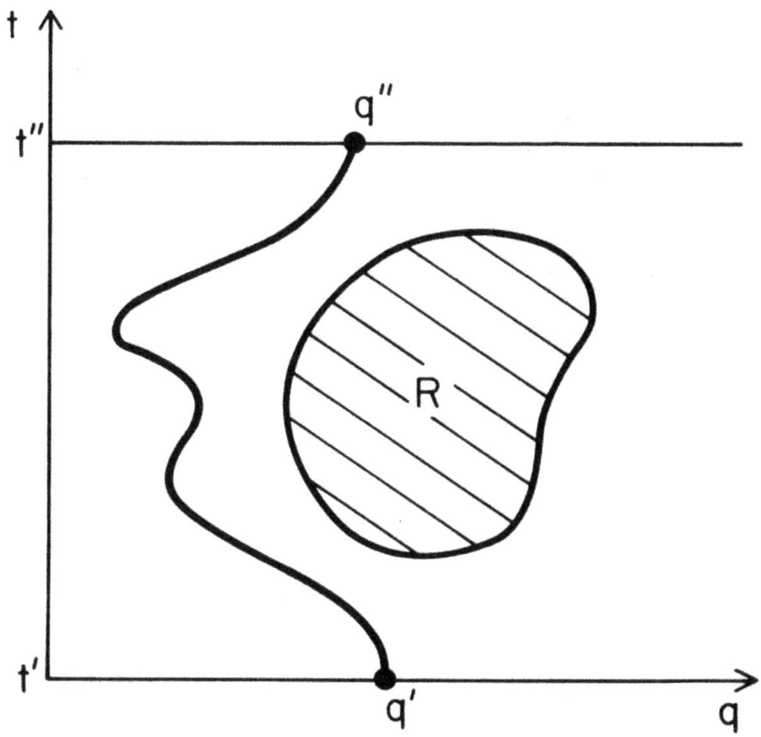

Fig. 1. Spacetime alternatives defined by a spacetime region. The figure shows a spacetime region R with extent in both space and time. The paths of a non-relativistic particle between q' at time t' and q'' at time t'' may be divided into two classes: First, the class of paths that never cross the region R (illustrated). Second, the class of paths that cross R sometime. These two classes constitute an exhaustive set of alternatives for the particle that is not "at a moment of time".

the notion of "measurement" does not play a fundamental role.[3,10–12] I shall discuss this generalization, but first we must review a bit of the quantum mechanics of closed systems.

4. The Quantum Mechanics of Closed Systems

The most general objective of quantum theory is to predict the probabilities of individual histories in an exhaustive set of alternative, coarse-grained histories of a closed system. A characteristic feature of a quantum-mechanical theory is that not every set of histories that may be described can be assigned probabilities because of quantum-mechanical interference between the individual histories in the set. Nowhere is this more clearly illustrated than in the two-slit experiment (Fig. 2). In the usual discussion, if we have not measured which slit the electron went through on its way to being detected at the screen, then we are not permitted to assign probabilities to the alternative histories in which it passed through the upper or lower slit. It would be inconsistent to do so since the correct probability sum rules would not be satisfied. Because of interference, the probability to arrive at a point y on the screen is not the sum of the probabilities to arrive at y going through the upper and the lower slits:

$$\left|\psi_L(y) + \psi_U(y)\right|^2 \neq \left|\psi_L(y)\right|^2 + \left|\psi_U(y)\right|^2. \tag{9}$$

In quantum mechanics, a rule is needed to determine which sets of histories can be assigned probabilities and then what those probabilities are.

In the "Copenhagen" quantum mechanics of measured subsystems, probabilities can be assigned to alternative histories that have been *measured.* In the two-slit experiment, for example, if we *have* measured which slit the electron went through, then interference is destroyed, the sum rule obeyed, and we *can* consistently assign probabilities to the alternative histories in which the electron passed through the upper or lower slit.

In the quantum mechanics of closed systems, containing both observer and observed, measuring apparatus and measured subsystem, the above rule is but a special case of a more general one of much wider applicability.[11,13–15] Probabilities can be assigned to the individual members of a set of alternative coarse-grained histories of a closed system when there is negligible quantum-mechanical interference between these histories as a consequence of the system's initial condition and dynamics. Such sets of histories are said to *decohere.*

To describe more precisely what is meant by decoherence, consider for simplicity, a pure initial state $|\psi\rangle$ and a set of histories defined

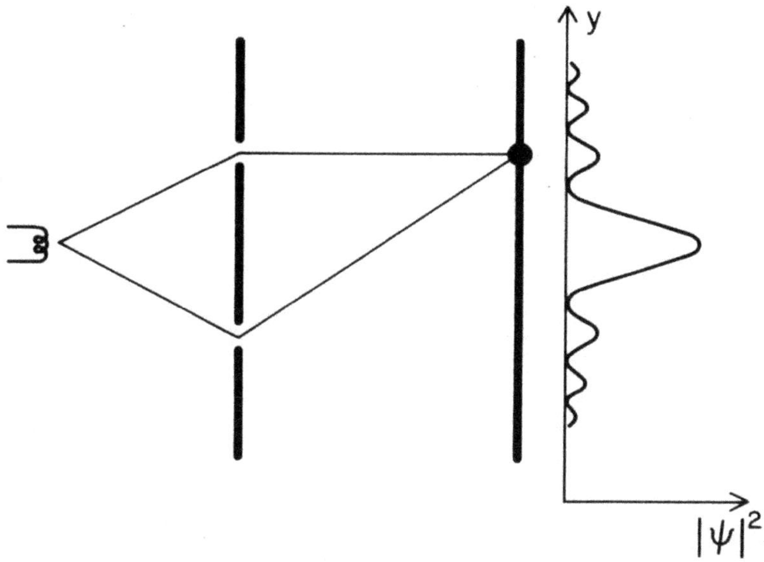

Fig. 2. The two-slit experiment. An electron gun at left emits an electron traveling towards a screen with two slits, its progress in space recapitulating its evolution in time. When precise detections are made of an ensemble of such electrons at the screen it is not possible, because of interference, to assign a probability to the alternatives of whether an individual electron went through the upper slit or the lower slit. However, if the electron interacts with an apparatus that measures which slit it passed through, then these alternatives decohere and probabilities can be assigned.

by sets of alternatives at definite moments of time $t_1, \ldots, t_n$. The alternatives at the moment of time t_k are described by an exhaustive set of exclusive Schrödinger picture projection operators $\{P^k_{\alpha_k}\}$ satisfying

$$\sum\nolimits_{\alpha_k} P^k_{\alpha_k} = 1, \quad P^k_{\alpha_k} P^k_{\beta_k} = \delta_{\alpha_k \beta_k} P^k_{\alpha_k}. \tag{10}$$

In this notation, k denotes the set of alternatives at time t_k (e.g. a set of position ranges or a set of momentum ranges, etc.) and α_k the particular alternative. The alternatives are *fine-grained* if the P's project onto the one-dimensional subspaces defined by a complete set of states, and are otherwise *coarse-grained*.

The sequences of alternatives at definite moments of time, $0 < t_1 < t_2 < \cdots < t_n < T$, define a set of coarse-grained alternative histories on a time interval $[0, T]$. The individual histories correspond to particular sequences $\alpha = (\alpha_1, \ldots, \alpha_n)$ which are represented by the corresponding chains of projection operators interrupted by unitary evolution

$$C_\alpha = e^{-iH(T-t_n)/\hbar} P^n_{\alpha_n} e^{-iH(t_n-t_{n-1})/\hbar} P^{n-1}_{\alpha_{n-1}} \cdots e^{-iH(t_2-t_1)/\hbar} P^1_{\alpha_1} e^{-iHt_1/\hbar}. \tag{11}$$

These may be written somewhat more compactly using Heisenberg picture projections as

$$C_\alpha = e^{-iHT/\hbar} P^n_{\alpha_n}(t_n) \cdots P^1_{\alpha_1}(t_1), \tag{12}$$

where

$$P^k_{\alpha_k}(t) = e^{iHt/\hbar} P^k_{\alpha_k} e^{-iHt/\hbar}. \tag{13}$$

Since $\sum_\alpha C_\alpha = e^{-iHT/\hbar}$ as a consequence of (10) and (11), the evolution of the initial state $|\psi\rangle$ may be resolved into branches, $|\psi_\alpha\rangle$, corresponding to the individual histories

$$e^{-iHT/\hbar}|\psi\rangle = \sum_\alpha C_\alpha|\psi\rangle \equiv \sum_\alpha |\psi_\alpha\rangle. \tag{14}$$

A set of histories decoheres[a] when the individual branches are essentially orthogonal

$$\langle \psi_{\alpha'} | \psi_\alpha \rangle \approx 0, \; \alpha' \neq \alpha. \tag{15}$$

---

[a]The term "decoherence" is used in several different ways in the literature. We have followed our earlier work[15] in using the term to refer to a property of a set of coarse-grained *histories* of a closed system, namely the absence of interference between individual histories in the set at a level to ensure the consistency of the probability sum rules. There are several different measures of decoherence.[30] The condition (15) is called the "medium decoherence condition". In this simplified presentation, "decoherence" therefore means the medium decoherence of histories. Similar conditions are called "consistency conditions"[13,14] or "no-interference conditions".[11] The term decoherence has also been used to refer to the approach to diagonality of a reduced density matrix in a particular basis. The decoherence of density matrices is not the same as the decoherence of histories in general, but the two ideas can be related in special models.

The probabilities of the individual histories in such a decoherent set are the square of the norms of the corresponding branches

$$p(\alpha) = \| \, |\psi_\alpha\rangle \, \|^2. \tag{16}$$

Equation (16) is a consistent assignment of probabilities to a decoherent set of histories because decoherence implies that the probability sum rules are satisfied in their most general form. To give a simple example, consider a set of histories defined by alternatives at just two moments of time t_1 and t_2 and the probability sum rule

$$\textstyle\sum_{\alpha_1} p(\alpha_2, \alpha_1) = p(\alpha_2). \tag{17}$$

For the left-hand side of (17) we may write

$$\begin{aligned}\textstyle\sum_{\alpha_1} p(\alpha_2, \alpha_1) &= \textstyle\sum_{\alpha_1} \langle\psi|P^1_{\alpha_1}(t_1)P^2_{\alpha_2}(t_2) \cdot P^2_{\alpha_2}(t_2)P^1_{\alpha_1}(t_1)|\psi\rangle \\ &= \textstyle\sum_{\alpha_1'\alpha_1} \langle\psi|P^1_{\alpha_1'}(t_1)P^2_{\alpha_2}(t_2) \cdot P^2_{\alpha_2}(t_2)P^1_{\alpha_1}(t_1)|\psi\rangle \\ &= \langle\psi|P^2_{\alpha_2}(t_2) \cdot P^2_{\alpha_2}(t_2)|\psi\rangle = p(\alpha_2). \end{aligned} \tag{18}$$

The first and last equalities are as in the definition (16). The second equality is true because of the decoherence condition (15), and the third because of (10). Thus, decoherence implies the probability sum rules needed for a consistent assignment of probabilities.

Measured alternatives decohere, but an alternative need not be a participant in a measurement situation in order to decohere. In cosmology, quantum theory predicts the probabilities of alternative sizes of density fluctuations one minute after the big bang, in a universe in which these alternatives decohere, whether or not anything like a measurement was carried out on them and certainly whether or not there was an observer around to do it. Decoherence is thus a more precise, more general, and more observer-independent notion than measurement, and replaces it in the quantum mechanics of closed systems as the criterion determining which sets of histories can be assigned probabilities.

5. A Generalized Sum-Over-Histories Quantum Mechanics for Spacetime Alternatives

Building on the sum-over-histories ideas, the quantum mechanics of alternatives at definite moments of time that was described in the previous section may be generalized to deal with the spacetime alternatives described in Sec. 3. The result is a quantum framework for prediction with dynamics and alternatives fully in spacetime form. We shall describe this generalization for the case of a configuration space spanned by coordinates, q^i, and assume a pure initial state. There is nothing essential about these restrictions. The generalization to an initial density matrix requires only a modest expansion of the formalism, and the coordinates q^i could be the values of fields $\phi(\vec{x})$ at each point in space. The important assumption is that there is a fixed background spacetime that supplies a well-defined notion of time. We follow the discussion in Ref. 3.

The most refined possible description of a closed system are its fine-grained histories. These are the paths $q^i(t)$ that are single-valued functions of the time. Partitions of these fine-grained histories into exhaustive sets of exclusive classes c_α yield sets of coarse-grained histories. An individual coarse-grained history c_α is thus a class of fine-grained histories. Exhaustive sets of alternative coarse-grained histories $\{c_\alpha\}$ are the general notion of "observable" for which quantum theory predicts probabilities when the set of coarse-grained histories is decoherent.

Sequences of alternative ranges of coordinates $\{\Delta^1_{\alpha_1}\}$, $\{\Delta^2_{\alpha_2}\}, \ldots, \{\Delta^n_{\alpha_n}\}$ at, say, times $t_1, \ldots, t_n$ define one kind of partition of the paths $q^i(t)$. An individual coarse-grained history corresponding to the sequence of alternatives $\alpha = (\alpha_1, \ldots, \alpha_n)$ is the class of paths c_α that thread the region $\Delta^1_{\alpha_1}$ at time t_1, $\Delta^2_{\alpha_2}$ at time t_2, etc. (Fig. 3). These correspond to sequences of the familiar "observables" at definite moments of time. However, much more general partitions are possible. For example, following the discussion of Sec. 3, paths may be partitioned into two classes according to their behavior with respect to a spacetime region R. One class, c_0, consists of all paths that never intersect R and the other, c_1, consists of all paths that intersect R at least once. These classes are exclusive, and together

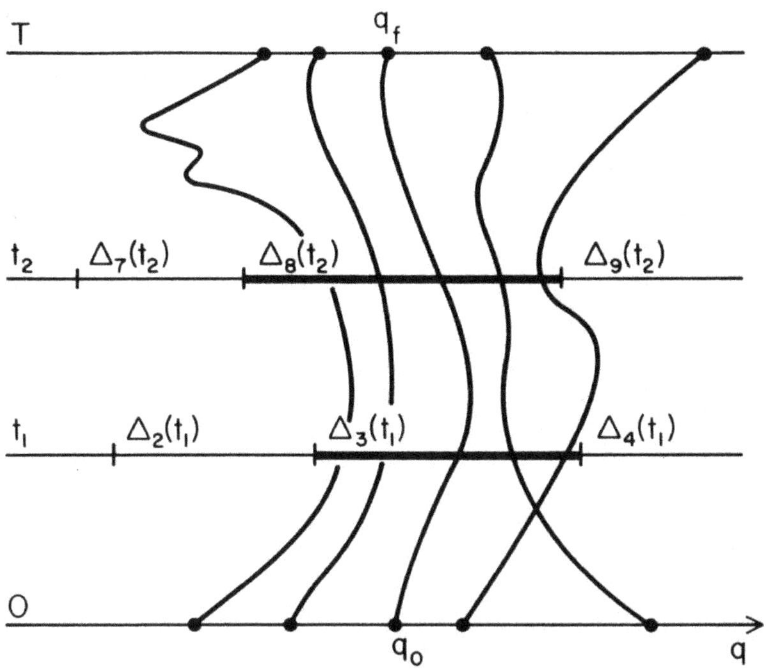

Fig. 3. Coarse-graining by regions of configuration space at successive moments of time. The figure shows a spacetime that is a product of a one-dimensional configuration space (q) and the time interval $[0, T]$. At times t_1 and t_2 the configuration space is divided into exhaustive sets of non-overlapping intervals: $\{\Delta_{\alpha_1}(t_1)\}$ at time t_1, and $\{\Delta_{\alpha_2}(t_2)\}$ at time t_2 (written in the text as $\{\Delta^1_{\alpha_1}\}$, $\{\Delta^2_{\alpha_2}\}$, etc). Some of these intervals are illustrated. The fine-grained histories are the paths which pass between $t = 0$ and $t = T$. Because the paths are assumed to be single-valued in time, the set of fine-grained histories may be partitioned into two classes according to which intervals they pass through at times t_1 and t_2. The figure illustrates a few representative paths in the class which pass through region $\Delta_3(t_1)$ at time t_1 and region $\Delta_8(t_2)$ at time t_2.

they are exhaustive. They constitute a coarse-grained set of histories.

The most general notion of coarse-grained histories is a partition by values of functionals of histories.[4,37] Consider, for example, just a single functional $F[q(\tau)]$ and a set of ranges $\{\Delta_\alpha\}$ of the real line. The class c_α consists of all paths $q^i(t)$ such that $F[q(\tau)]$ lies in the range Δ_α. We could partition the paths, for example, by the value of some particular coordinate, q^k, averaged over a time interval, that is, by the functional

$$F[q(\tau)] = \frac{1}{T}\int_0^T dt\ q^k(t). \tag{19}$$

In this way we could deal with the average values of fields over space-time regions whose importance were stressed by Bohr and Rosenfeld.[6] In a field theory with a spinor field $\psi(x)$, we could partition the field histories by the values of currents, e.g. $\psi^\dagger(\vec{x},t)\psi(\vec{x},t)$. In this way the theory can incorporate observables associated with spin. In the theory of a nonrelativistic particle, we could partition the paths by the value of the position *difference* between two times separated by a time interval T. In the limit that T becomes large, but still short compared to dynamical time scales of any interaction, these partitions define momentum alternatives determined by the time of flight.[21] In this way, probabilities for momenta can be predicted by the theory.

Sums over the fine-grained histories contained in a coarse-grained history define the branch of the initial state corresponding to that history. To make this explicit, consider histories on the time interval $[0,T]$ and suppose that the fine-grained histories on this interval are partitioned into an exhaustive set of exclusive classes c_α. We define branches of the initial state and class operators C_α corresponding to each class by

$$|\psi_\alpha\rangle \equiv C_\alpha|\psi\rangle = \int_{c_\alpha} \delta q\ \exp\bigl(iS[q(\tau)]/\hbar\bigr)|\psi\rangle \tag{20a}$$

where the integral is over all paths in the class c_α. We are using the same abbreviated notation as in (4). More explicitly, (20a) stands for

$$\psi_\alpha(q,T) = \langle q|C_\alpha|\psi\rangle = \int dq' \int_{[q'c_\alpha q]} \delta q\ \exp\Bigl(iS\bigl[q(\tau)\bigr]/\hbar\Bigr)\psi(q',0). \tag{20b}$$

Evidently, since the sum over all paths just gives unitary evolution as in (4), we have

$$\sum\nolimits_\alpha C_\alpha = e^{-iHT/\hbar}. \tag{21}$$

Decoherence and probabilities are defined as before. The set of coarse-grained histories $\{c_\alpha\}$ decoheres if

$$\langle\psi_{\alpha'}|\psi_\alpha\rangle \approx 0, \quad \alpha' \neq \alpha \tag{22}$$

and the probability of an individual history c_α in a decoherent coarse-grained set is

$$p(c_\alpha) = \| \, |\psi_\alpha\rangle \|^2. \tag{23}$$

The important point about this construction is that the probability sum rules, which are the obstacle to consistently assigning probabilities in quantum mechanics, are satisfied for a decoherent set of coarse-grained histories as a consequence of (22). To see this consider a partition of the $\{c_\alpha\}$ into an exhaustive set of exclusive classes yielding a coarser-grained partition of the fine-grained histories $\{\bar{c}_\beta\}$. The most general form of the probability sum rules is

$$p(\bar{c}_\beta) = \sum_{\alpha\epsilon\beta} p(c_\alpha) \tag{24}$$

where the sum is over α such that c_α is included in $\bar{c}_\beta$. This is easily seen to be satisfied as a consequence of (20) and (22). From the linearity of (20) that reflects the principle of superposition it follows that

$$\bar{C}_\beta = \sum_{\alpha\epsilon\beta} C_\alpha. \tag{25}$$

But then a repetition of the argument that led to (17) shows the validity of the more general sum rule (24). In this way, sum-over-histories ideas can be used to formulate a generalized quantum mechanics of closed systems that is fully in spacetime form with dynamics summarized by path integrals over alternatives that are general partitions of spacetime histories.

6. Comparison with Schrödinger–Heisenberg Quantum Mechanics

We mentioned that, for alternatives at moments of time, sum-over-histories quantum mechanics deals directly only with configuration space alternatives, in contrast to Schrödinger–Heisenberg quantum mechanics which utilizes all the possibilities of transformation theory. The sum-over-histories formulation is thus more restricted in its alternatives at moments of time. Configuration space variables have a preferred place in the formalism.

However, in the preceding section we have shown how sum-over-histories ideas can be used to formulate a quantum mechanics that deals with spacetime alternatives that are much more general than those "at moments of time" with which the Schrödinger–Heisenberg formulation is concerned. This generalized spacetime quantum mechanics cannot be reformulated in terms of the two laws of evolution of the Schrödinger–Heisenberg formulation. If it could be so formulated, the class operators C_α defined by (20) for each history in a spacetime coarse-graining could be represented as a chain of projections of the form (12), or perhaps a continuous product of such projections, one for each time. Then one could describe the evolution completely in terms of unitary evolution interrupted (perhaps continuously in time) by reductions of the state vectors.

However, the class operators of a spacetime coarse-graining cannot, in general, be represented as products of projections, even continuous ones. Consider, by way of example, the coarse-graining of the paths of a particle by their behavior with respect to a spacetime region R that was discussed earlier. The class operator C_0 for the class of paths that never cross R *can* be represented as a continuous chain of projections on the region of q outside of R at each time. However, the class operator C_1 for the class of paths that crosses R sometime, cannot be so represented because at each time the particle could be either inside R or outside it. The class operator C_1 can be represented formally as a *sum* of continuous chains of projections

$$C_\alpha = \sum_{\alpha(t)\epsilon c_\alpha} \prod_t P^{k(t)}_{\alpha(t)}(t). \tag{26}$$

A quantum mechanics of alternatives represented by class operators that are *sums* of chains of projections already constitutes a generalization of familiar quantum mechanics[10,32] although a very natural one.

The probabilities of spacetime coarse-grained histories thus cannot be expressed in terms of a unitarily evolving state vector that is "reduced" at various moments of time. The reason, it should be stressed, does not lie in the use of path integrals versus operators. Indeed, as (20) shows, there is a correspondence between path integrals and operators and operator techniques provide the most convenient way of defining the path integrals on the right-hand side of (20).[3] Rather, it is the space*time* nature of the alternatives that does not allow meaningful notions of the state of the system at a moment of time.

This generalized sum-over-histories quantum mechanics is thus not equivalent to the Schrödinger–Heisenberg formulation because the two formulations deal with different alternatives. Sums of continuous products like (26) could be taken as the starting point of a yet more generalized quantum mechanics that would contain all the alternatives of both. Such a generalization presents interesting mathematical problems. In the meantime, to the extent that our experience can be expressed in terms of spacetime alternatives, nothing seems to be lost by a restriction to these, and much would seem to be gained from a more realistic description of alternatives extended over time. In the next section I shall argue that the spacetime approach to quantum mechanics has definite advantages in the quantum theory of spacetime where there is no well-defined notion of "at a moment of time".

7. A Generalized Quantum Mechanics of Spacetime

The nature of the "observables" to which a quantum theory of spacetime geometry assigns probabilities has always been something of a puzzle in quantum gravity. We cannot straightforwardly and covariantly define alternatives at a moment of time because there is no fixed notion of time. In a theory where the geometry of spacetime fluctuates quantum-mechanically, there is no fixed interval between spacetime points, and not even a fixed notion of whether that interval is timelike or spacelike. Put differently, there is no covariant choice of geometrical variable to play the role of "t" in the predictive framework summarized by (12), (15) and (16). In essence, there is a conflict between the diffeomorphism invariance of spacetime theories of gravity and the requirements of Schrödinger–Heisenberg quantum mechanics. This is called the "problem of time" in quantum gravity.

For recent critical surveys of various approaches to its resolution, see Refs. 17–19 and 36.

One thread of thought is that alternatives in quantum gravity should be represented by operators that commute with the constraints implied by diffeomorphism invariance. However, time reparametrizations are among the diffeomorphisms and for such theories the constraints generate the dynamics. Restricting the observables to operators that commute with the constraints corresponds classically to observables that are constants of the motion. This is a very restricted class of observables![b]

The spacetime approach to quantum mechanics provides a different resolution to the problem of time. In a quantum theory of spacetime, the fine-grained histories are the possible four-dimensional metrics and matter field configurations on a fixed manifold M (in the simplest case). Quantum gravitational dynamics can be expressed in spacetime form through sums of $\exp(iS[g,\phi]/\hbar)$ over metrics and matter fields on M, where S is the action for spacetime and matter. The use of sums over histories to formulate a covariant quantum gravitational dynamics has been extensively investigated.[c] However, the spacetime approach to quantum mechanics can also be used to specify the alternatives to which a quantum theory of spacetime assigns probabilities in a covariant way.

A generalized sum-over-histories quantum mechanics for spacetime can be constructed in which the fine-grained histories are the four-dimensional metrics and matter field configurations as discussed above.[10,37] Allowed coarse-grainings are partitions of these fine-grained histories into exhaustive sets of *diffeomorphism invariant* classes $\{c_\alpha\}$. These diffeomorphism invariant classes are the analogs of the spacetime coarse-grainings we have discussed for quantum mechanics in fixed background geometries. They supply a very broad

[b]However, argued to be sufficient by some, e.g. Rovelli.[33]

[c]It has been investigated in a literature far too large to be cited here. However, some representative and important early papers are those of Misner,[22] Leutwyler,[34] DeWitt,[23] Fradkin and Vilkovisky,[35] Faddeev and Popov,[24] Hawking,[25] Teitelboim,[26] and Polyakov.[27] There are many other important ones.

class of generally covariant alternatives to which, when decoherent, a quantum theory of spacetime will assign probabilities.

Many examples of diffeomorphism invariant coarse-grainings could be given, but in the limited space available I will confine myself to just two. First, we consider the probability that a closed cosmology reaches a maximum spatial volume greater than, say, V_0. This question can be given a precise meaning by partitioning the class of all cosmological four-metrics into the class that has at least one spacelike three-surface with a total volume greater than V_0, and the class of three-metrics that have no such three-surface. This is clearly an exhaustive partition into two exclusive classes that are each diffeomorphism invariant. Equally clearly, the specification of these alternatives involves no preferred notion of time.

As a second example, we consider the natural lattice formulation of general relativity — the Regge calculus.[28,31] A two-dimensional surface can be constructed from triangles (Fig. 4). The topology of the surface is specified by how the triangles are joined together. The geometry of the surface is specified by an assignment of the squared edge-lengths of the triangles and a flat geometry to their interior. Similarly, four-dimensional Lorentzian geometries can be constructed out of four-simplices with squared edge-lengths that may be positive or negative. The space of geometries is parametrized by the n_1 squared edge-lengths $s^1, \ldots, s^{n_1}$ consistent with the analogs of triangle inequalities. A point in this space of squared edge-lengths is a fine-grained history. A partition of this space into regions that are invariant under the symmetries of the simplicial net provides a very general class of coarse-grainings for these lattice geometries that does not require a preferred notion of time for its specification.

Beyond these examples, however, diffeomorphism invariant partitions of metrics and field configurations supply the *most general* notion of an alternative in quantum theory that is describable in spacetime terms. Every property of the universe, the probability of which we may seek to calculate, corresponds to some such coarse-graining, namely the partition of all four-metrics and matter field configurations into the class in which the property is true and the class in which it is false. If we cannot distinguish which fine-grained

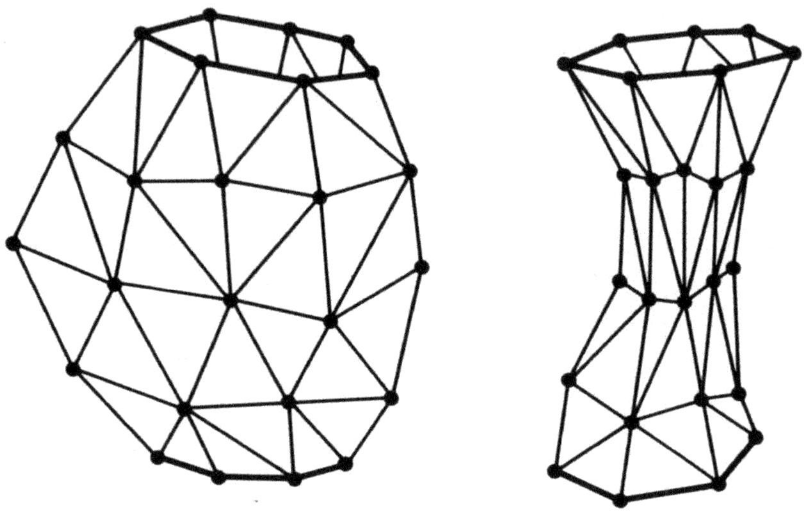

Fig. 4. Simplicial geometries. Two-dimensional surfaces can be made up by joining together flat triangles to form a simplicial manifold. A geometry of the surface is specified by an assignment of squared edge-lengths to the triangles. The figure shows two different geometries obtained by a different assignment of squared edge-lengths to the same simplicial manifold. The generalization of these ideas to four dimensions and Lorentz signature gives the natural lattice version of general relativity — the Regge calculus. In a sum-over-histories quantum theory of simplicial spacetimes, sums over geometries are represented by integrals over the squared edge-lengths. Diffeomorphism invariant alternatives can be defined by partitioning the space of allowed squared edge-lengths into exhaustive sets of exclusive regions. For example, one could partition closed cosmological geometries into the class that has no simplicial spacelike three-surface greater than a certain volume and the class that has at least one such surface. In a given simplicial manifold it is possible to enumerate all three surfaces and identify the regions in the space of squared edge-lengths to which each class corresponds.

histories have the property and which do not, then the property is not well defined.

To complete the construction of a generalized quantum mechanics, a notion of decoherence must be specified for these coarse-grained histories. I shall only indicate how to do this schematically. More details can be found in Ref. 37. Branches, corresponding to individual coarse-grained histories in a set $\{c_\alpha\}$, can be represented as wave functions $\Psi_\alpha[h,\chi]$ on the superspace of spatial three-metrics, $h_{ij}(\mathbf{x})$, and spatial matter field configurations, $\chi(\mathbf{x})$. These are defined in analogy to (20) or, more explicitly, to (5)

$$\Psi_\alpha[h,\chi]=\int \delta h'\delta\chi' \int_{[(h',\chi'),c_\alpha,(h,\chi)]} \delta g\, \delta\phi \exp\Big(iS[g,\phi]/\hbar\Big) \circ \Psi\left[h',\chi'\right]. \tag{27}$$

In this expression, $\Psi[h,\chi]$ is the wave function representing the initial condition of the universe, say, the "no-boundary" wave function.[16] The integral is over metrics and fields in the class c_α on a manifold M with two boundaries. On one boundary, metrics and fields match the arguments (h,χ) of the branch wave function. On the other boundary they match the arguments (h',χ') of the initial condition. Much remains to be spelled out to make such a construction concrete, not least the details of the measure and the product $\circ$ with which the initial condition is attached to the functional integral, more details are in Ref. 37. The important point is that with these branches, one can define a decoherence condition for coarse-grained histories analogous to (22) and probabilities for the individual histories in decoherent sets by expressions analogous to (23). The quantum mechanics of spacetime is thus cast into a generally covariant, fully spacetime form free from the problem of time.

How is the usual formulation of quantum mechanics with its preferred time variables connected with this generalization that requires no such preferred time? The answer is to be found by examining the origin of the classical spacetime of everyday experience. Classical spacetime is not a general feature of every state in a quantum theory of gravity. We expect classically behaving spacetime only for particular states and then only for coarse-grainings that define geometry well above the Planck scale. For such states and coarse-grainings, the integral over metrics in (27) may be carried out by steepest descents. Suppose, in the simplest case, that only a single classical geometry $\hat{g}$ dominates the sum. The remaining integrals over matter fields are equivalent to those of a field theory in the fixed background geometry of $\hat{g}$. The quantum mechanics of matter fields thus inherits its notions of time from the timelike directions of the classical background $\hat{g}$. It could be in this way that the familiar Hamiltonian formulation of quantum mechanics, with its preferred time(s), emerges as an approximation appropriate to the existence of an approximately quasiclassical spacetime in a more general, covariant, spacetime, formulation of quantum theory that is free from the problem of time.

Acknowledgments

The author has benefited from many conversations with many physicists on these subjects, but especially with M. Gell-Mann, on the quantum mechanics of closed systems, with K. Kuchař on the problem of time in quantum gravity, and with R. Sorkin on sum-over-histories formulations of quantum mechanics. Preparation of this report as well as the work it describes was supported, in part, by the National Science Foundation under grant PHY90-08502.

References

1. R. P. Feynman, *Rev. Mod. Phys.* **20**, 267 (1948).
2. P. A. M. Dirac, *Phys. Zeit. Sowjetunion* **3**, 64 (1933).
3. J. B. Hartle, *Phys. Rev. D* **10**, 3173 (1991).
4. C. Caves, *Phys. Rev. D* **33**, 1643 (1986); *ibid.* **35**, 1815 (1987).
5. J. Stachel, in *From Quarks to Quasars*, ed. R. G. Colodny (University of Pittsburg Press, Pittsburgh, 1986), p. 331ff.
6. N. Bohr and L. Rosenfeld, *Det Kgl. Danske Vidensk. Selskab Mat.-Fys. Medd.* **12**, nr. 8 (1933).
7. J. B. Hartle, *Phys. Rev. D* **37**, 2818 (1988).
8. R. Sorkin, in *History of Modern Gauge Theories*, eds. M. Dresden and A. Rosenblum (Plenum Press, New York, 1989); in *Conceptual Problems of Quantum Gravity*, eds. A. Ashtekar and J. Stachel (Birkhauser, Boston, 1991); S. Sinha and R. Sorkin, preprint.
9. N. Yamada and S. Takagi, *Prog. Theor. Phys.* **85**, 985 (1991).
10. J. B. Hartle, in *Quantum Cosmology and Baby Universes: Proc. 1989 Jerusalem Winter School for Theoretical Physics*, eds. S. Coleman, J. Hartle, T. Piran and S. Weinberg (World Scientific, Singapore, 1991).

11. N. Yamada and S. Takagi, *Prog. Theor. Phys.* **86**, 599 (1991).
12. N. Yamada and S. Takagi, *Prog. Theor. Phys.* **87**, 77 (1992).
13. R. Griffiths, *J. Stat. Phys.* **36**, 219 (1984).
14. R. Omnès, *J. Stat. Phys.* **53**, 893 (1988); *ibid.* **53**, 933 (1988); *ibid.* **53**, 957 (1988); *ibid.* **57**, 357 (1989); *Rev. Mod. Phys.* **64**, 339 (1992).
15. M. Gell-Mann and J. B. Hartle, in *Complexity, Entropy, and the Physics of Information*, SFI Studies in the Sciences of Complexity, Vol. VIII, ed. W. Zurek (Addison Wesley, Reading) or in *Proc. 3rd Int. Symp. Foundations of Quantum Mechanics in the Light of New Technology*, eds. S. Kobayashi, H. Ezawa, Y. Murayama and S. Nomura (Physical Society of Japan, Tokyo, 1990).
16. J. B. Hartle and S. W. Hawking, *Phys. Rev. D* **28**, 2960 (1983).
17. K. Kuchař, in *Quantum Gravity 2*, eds. C. Isham, R. Penrose and D. Sciama (Clarendon Press, Oxford, 1981).
18. C. Isham, in *Recent Aspects of Quantum Fields*, eds. H. Mitter and H. Gausterer (Springer-Verlag, Berlin, 1992); in Lectures presented at the NATO Advanced Summer Institute, *Recent Problems in Mathematical Physics*, Salamanca, June 15–17, 1992 (to be published).
19. K. Kuchař, in *Proc. 4th Canadian Conference on General Relativity and Relativistic Astrophysics*, eds. G. Kunstatter, D. Vincent and J. Williams (World Scientific, Singapore, 1992).
20. J. B. Hartle, Spacetime quantum mechanics and the quantum mechanics of spacetime, in Proc. 1992 Les Houches Summer School, *Gravitation and Quantizations* (to be published).
21. R. P. Feynman and A. Hibbs, *Quantum Mechanics and Path Integrals* (McGraw-Hill, New York, 1965).
22. C. W. Misner, *Rev. Mod. Phys.* **29**, 497 (1957).
23. B. DeWitt, in *General Relativity: An Einstein Centenary Survey*, eds. S. W. Hawking and W. Israel (Cambridge University Press, Cambridge, 1979).
24. L. Faddeev and V. Popov, *Usp. Fiz. Nauk.* **111**, 427 (1973); *Sov. Phys. Usp.* **16**, 777 (1974).
25. S. W. Hawking, in *General Relativity: An Einstein Centenary Survey*, eds. S. W. Hawking and W. Israel (Cambridge University Press, Cambridge, 1979).

26. C. Teitelboim, *Phys. Rev. D* **25**, 3159 (1983); *ibid.* **28**, 297 (1983); *ibid.* **28**, 310 (1983); *Phys. Rev. Lett.* **50**, 795 (1983).
27. A. M. Polyakov, *Phys. Lett. B* **103**, 211 (1981).
28. T. Regge, *Nuovo Cimento* **19**, 558 (1961).
29. M. B. Mensky, *Phys. Rev. D* **20**, 384 (1979); *Theor. Math. Phys.* **75**, 357 (1988).
30. M. Gell-Mann and J. B. Hartle, in *Proc. 25th Int. Conf. High Energy Physics*, Singapore, August 2–8, 1990, eds. K. K. Phua and Y. Yamaguchi (South East Asia Theoretical Physics Association and Physical Society of Japan), distributed by (World Scientific, Singapore, 1990).
31. R. Williams, *Class. Quant. Grav.* **9**, 1409 (1992).
32. M. Gell-Mann and J. B. Hartle, *Classical Equations for Quantum Systems*; preprint #UCSBTH-91-15.
33. C. Rovelli, *Phys. Rev. D* **42**, 2638 (1990); *ibid.* **43**, 442 (1991).
34. H. Leutwyler, *Phys. Rev.* **134**, B1155 (1964).
35. E. Fradkin and G. Vilkovisky, *Phys. Rev. D* **8**, 4241 (1973).
36. W. Unruh, in *Gravitation: A Banff Summer Institute*, eds. R. Mann and P. Wesson (World Scientific, Singapore, 1991).
37. J. B. Hartle (to be published)
38. A. Schmid, *Ann. Phys.* **173**, 103 (1987).

CHAPTER 9

Quantum Physics and Human Language*

Human languages employ constructions that tacitly assume specific properties of the limited range of phenomena they have evolved to describe. These assumed properties are true features of that limited context, but may not be general or precise properties of all the physical situations allowed by fundamental physics. In brief, human languages contain "excess baggage" that must be qualified, discarded, or otherwise reformed to give a clear account, in the context of fundamental physics, of even the everyday phenomena that languages have evolved to describe. The surest route to clarity is to express the constructions of human languages in the language of fundamental physical theory, not the other way around. These ideas are illustrated by an analysis of the verb "to happen" and the word "reality" in special relativity and the modern quantum mechanics of closed systems.

*Dedicated to GianCarlo Ghirardi on his 70th birthday.
arXiv:quant-ph/0610131

"Mind and world... have evolved together and in consequence are something of a mutual fit."

— Wm. James, 1893[1]

"We are deceived at every level by our introspection."

— F. H. C. Crick, 1979[2]

1. Introduction

Human languages are features of particular kinds of information gathering and utilizing systems (IGUSes),[a] living late in the universe, dwelling on a minor planet, circling a garden-variety star, that is but one of about ten billion stars in a galaxy, that is but one of about a hundred billion other galaxies within the visible universe. Human languages are thus features of our universe, but very special ones. Languages evolved over the history of our species through a combination of frozen accidents and selection by both physical and cultural evolutionary pressures. Human languages are adapted to provide certain highly coarse-grained descriptions of the quasiclassical realm of everyday experience — the "world" to which James presumably referred in the above quote. Yet, human languages are not restricted to such quasiclassical descriptions. Suitably extended, they also permit the discussion of regimes under exploration in contemporary fundamental physics that are characterized by concepts that can be quite far from those native to the quasiclassical realm. This essay explores some aspects of the tension between the domains in which human languages have evolved and those to which they can be applied. In particular, we have in mind applications to the modern quantum mechanics of closed systems, most generally the universe as a whole. This deals not only with quasiclassical realms but many mutually incompatible ones as well.

[a]The term IGUS is broad enough to include both single representatives of biological species that have evolved naturally and certain kinds of mechanical devices. It includes human beings, both individually and collectively as members of groups, cultures, and civilizations. It also includes intelligent beings that we might meet in the future. For more discussion, see Ref. 3.

The theses of this essay are these: Human languages employ constructions that tacitly assume properties of the limited range of phenomena they have evolved to describe. These assumed properties are true features of that limited context, but may not be general properties of all the physical situations allowed by fundamental physics. In brief, human languages contain "excess baggage" that must be qualified, discarded, or otherwise reformed to give a clear account, in the context of fundamental physics, of even the everyday phenomena that languages have evolved to describe.

It is no more surprising to find that human language contains tacit assumptions than it is to discover that we possess a useless appendix. Indeed, these limitations of language are evidence of its evolution, just as our appendix is evidence of human evolution. Rather, the important circumstance is that the flexible, open-ended nature of human language allows it to be employed in the discussion of concepts very far from those it has evolved to describe, with only a few precautions and modifications. This essay is concerned with those precautions and modifications.

The surest route to clarity is to express the constructions of human languages in the language of fundamental physical theory, not the other way around. Alternatively, the constructions can be qualified so that tacit assumptions are made explicit. By "human language" here we mean roughly the language of everyday discourse. By "language of physics" we mean roughly the language, terms, concepts, and mathematics found in physics textbooks. Making a precise general distinction will not be necessary, because we will do that explicitly in specific examples.

There are many different physical theories that could be used to illustrate these theses, many different linguistic constructions that could be considered, and certainly many different human languages. For simplicity and clarity we shall focus on limited examples of each of these possibilities.

We will consider just one human language — English — and hope that our analysis of the connection between English and physics extends to other human languages in an essentially equivalent way. Assuming this equivalence, we shall refer just to "human language" — singular rather than plural.

We shall consider just two linguistic constructions. The first is the use of the verb "to happen" in its various forms: "happened", "is happening", and "will happen". Other existential verbs ("to be", "to exist", etc.) may be treated similarly. The other construction is the word "reality".

We shall consider just two theories: the quantum theory of closed systems[3–5] and special relativity. Clarifying the quantum mechanics of closed systems is the objective of this essay. But special relativity provides a simple, uncontroversial example.[b]

When we refer to "quantum mechanics" or "quantum theory" in this essay, we mean the decoherent (or consistent) histories quantum mechanics of closed systems, most generally the universe.[3–5] This is a modern synthesis of the ideas of many that extend the work begun by Everett. Decoherent histories quantum theory is logically consistent, consistent with experiment as far as is known, applicable to cosmology, consistent with the rest of modern physics such as special relativity, and generalizable to include quantum gravity (e.g. as in Ref. 7). It incorporates Copenhagen quantum theory as an approximation appropriate for measurement situations. It may not be the only theory with these properties, but it is the most promising of those presently available, in the author's opinion.

Briefly, the discussion will be as follows: A conflict arises between special relativity and the use of "happened", "is happening", and "will happen" in human language. These constructions tacitly assume that there is a division of spacetime into past, present, and future. But special relativity does not provide such a unique division, but rather many of them. The resulting conflict can be resolved by abandoning "to happen" and using the language of special relativity. Alternatively, spacetime can be conventionally divided into past, present, and future, and the use of "to happen" qualified to refer to that convention.

In quantum mechanics, "happened", "is happening", and "will happen" are probabilistic statements to be qualified with the

[b]It should be noted that none of these considerations is necessary to discuss the work of G. Ghirardi, our dedicatee, as, for example, in Ref. 6.

relevant probability, e.g. "happened with probability p". These probabilities are constructed from those that quantum mechanics assigns to a set of alternative histories of the universe. Human language tacitly assumes that there is a unique such set, mostly the histories of the quasiclassical realm. But quantum theory exhibits many alternative sets of histories to describe past, present, and future which may be incompatible roughly in the sense that position and momentum are incompatible. The resulting conflict can be resolved by replacing statements which involve "to happen" with statements about quantum-mechanical probabilities. Alternatively, "to happen" can be qualified to refer to the particular set of histories which define those probabilities.

This essay is based on the author's experience in explaining and teaching the quantum mechanics of closed systems along with some modest reflection on the subject. It is intended for *physicists* as a practical guide to a clearer understanding of this area by dealing with some of the linguistic tangles that naturally arise. No new physics is considered, only issues concerning the discussion of physics. The ideas expressed are merely the author's opinions on routes to clarity. Others may find that different routes work better. This essay is therefore not a review of all that has been written on these questions. In particular, there is a long history of discussion of related questions in philosophy and linguistics, but this essay does not pretend to address any of the deep issues that may arise there (see, e.g., Ref. 8).

The paper is structured as follows: Sec. 2 discusses the use of "to happen" in the context of special relativity. Section 3 discusses the use of "to happen" in the quantum-mechanical context, where there are many mutually incompatible decoherent sets of coarse-grained alternative histories of the universe. Section 4 ventures to discuss the implications of this for our notions of reality. Section 5 discusses dispensible words and what to do about them. Section 6 concludes with practical advice on achieving clarity.

2. Spacetime

Up here, on length scales much greater than the Planck length, the world is four-dimensional with a classical spacetime geometry. There is neither a unique notion of space nor a unique notion of time. Rather, from each point in spacetime, there is a family of timelike directions and three times as many spacelike directions. Spacetime geometry is curved by mass-energy according to the laws of general relativity, but in sufficiently small patches, every geometry is well-approximated by the flat spacetime of special relativity. To simplify our discussion we begin by restricting attention to a patch of spacetime where this approximation holds.

First, recall a few basic facts about flat spacetime. Events occur at points. At each point Q there is a light cone consisting of two parts: The future light cone is the three-dimensional surface generated by light rays emerging from Q. The past light cone is similarly defined by light rays converging on Q. (The labels "past" and "future" are conventional, but may be conveniently specified by the directions toward the big bang and away from it.) Points inside the light cone of Q are timelike separated from it; points outside it are spacelike separated.

The center of mass of a localized IGUS such as ourselves describes a timelike world line whose points can be labeled by the timelike distance along the curve, also known as proper time.

Consider an event A on your world line and another event B elsewhere. Suppose you are located along your world line at event A. Could you answer the questions: "Will B happen in the future?"; "Is B happening now?" and "Did B happen in the past?" You cannot, because these questions are meaningless in special relativity without further qualifications. The questions presume that spacetime can be divided into past, present (now), and future just on the basis of the location of A in spacetime. There is no such division.

More specifically, imagine there are astronauts on Saturn's moon Titan. Questions like "What are the astronauts doing now?" or "Did they begin breakfast half an hour ago?" are meaningless without further qualifications. The light travel time between Titan and Earth can be over an hour, implying a comparable special relativistic ambiguity in the notion of simultaneity. There is thus a conflict between familiar constructions of human language and the facts of special relativity.

The simplest resolution of the conflict is to discard the verbs like "to happen", and to formulate physically meaningful questions in the language of physics. Questions like "Is B in my future light cone at A?", "Is B spacelike separated from me at A?", and "Is B in my past light cone at A?" do make sense and have unambiguous answers.[c]

Another route to resolving the conflict is to retain the verb "to happen" but to qualify its usage in some conventional way. Specifically, we could divide spacetime up into a spacelike surface S containing A, the future of S, and the past of S. Then the questions "Will B happen in the future of S?", "Is B happening on S?", and "Did B happen in the past of S?" do make sense and have definite answers depending on where B is in spacetime. They are made meaningful by the qualifications referring to the spacelike surface S. However, there are an arbitrarily large number of spacelike surfaces containing A, and the answers to the questions can be different for another spacelike surface S' when B is spacelike separated from A. Unqualified constructions involving "to happen" are meaningless, but when qualified they have many different meanings.[d]

Introducing a coordinate system in our patch of spacetime is another arbitrary way of fixing a convention to give "to happen" a qualified meaning. Riemann normal coordinates based at the point A are a simple example, if they cover the whole patch. If the geometry of the patch happens to be exactly flat, these reduce to a choice of a particular Lorentz frame. Past, present, and future are now defined by the timelike coordinate (call it t) being greater, equal, or less

[c] "Is" here is understood in a tenseless, four-dimensional sense referring to the properties of spacetime. In a similar way "happen" is sometimes used (mostly by physicists) in a four-dimensional sense of occurring in spacetime. That usage won't be discussed in this essay but can be a further source of confusion.

[d] Using the language of physics and employing appropriate qualifications are not the only ways to resolve the conflict between special relativity and human language. Raphael Sorkin tells the author that there have been various proposals for *reforming* human language. These include redefining "now" to mean "spacelike" and using adverbs rather than verb forms to indicate tense, as in "It happen spacelike." The author guesses that the two resolutions discussed in the text have better chances of success.

than its value at A. As a bonus, a coordinate system provides a convention for answering further questions of the form "Does B occur before C or afterward?", which would be ambiguous if unqualified when B and C are spacelike separated.

The satellites comprising the Global Positioning System (GPS) are an example of a collective IGUS that effectively employs a coordinate system to define temporal relations. Both the special relativistic effects arising from the velocities of the satellites, and the general relativistic effects arising from the slight curvature of spacetime in the vicinity of the Earth are important for GPS operation.[9] The system would fail in about an hour if these were not both accounted for. Precise agreement among the satellites on a notion of simultaneity is needed. To define that, the GPS uses a version of a standard set of coordinates for the weak field metric of general relativity, centered on Earth, with spatial axes pointing towards fixed stars, and the time coordinate normalized so that on the Earth's geoid (approximately the ocean surface) it coincides with the time of clocks co-rotating with the Earth there. It is unlikely that the satellites are employing the verb "to happen" in any communication. But, if they did, they could define it by the time of their effective coordinate system.

Human IGUSes were using constructions like "will happen", "is happening", and "did happen" long before they had accurate clocks, and certainly before there was a precise notion of coordinate systems. How would the notion of "present" or "now" that is implicit in these constructions be described in the language of physics? The author has discussed the physics of "now" in Ref. 10. The next few paragraphs summarize some of these ideas.

Human IGUSes have an individual notion of "now" which can be modeled as a feature of their conscious focus on their most recently acquired information. This is already an approximate idea only defined up to the time scale of human perception[e] which we denote by τ_*. Individual IGUSes can agree on "what is happening now" by reporting their current observations and checking the reports they receive against their individual notions of "now". The result is an

[e]This is of order 0.1 s, e.g. Ref. 11.

approximate, imprecise, common present useful in everyday circumstances.

Agreement on a common present can be reached by a group of IGUSes in a patch where spacetime is approximately flat if the following contingencies are met:

1. The relative velocities of the IGUSes are small compared to the velocity of light.
2. The light travel time between IGUSes in a Lorentz frame in which they are nearly at rest is small compared to the time scale τ_*, characterizing perception.
3. The time scale for perception τ_* is short compared to the time scales on which interesting features of the IGUSes' environment vary.

Contingency (2), based on (1), means special relativistic ambiguities in the meaning of simultaneity are negligible in the construction of a common present. Contingency (3) means that the ambiguity in the "now" of each IGUS is negligible in this construction. Contingency (1) means that agreement can be reached over an interesting length of time.

Under these contingencies, a collection of IGUSes can construct a common present, but it is a present that is local, approximate and contingent on their relation to each other and to their environment. All the contingencies are easily satisfied for human IGUSes on Earth. However, astronauts on Titan will not be able to participate in this common "now"; contingency (2) will be violated. Further conventions will be needed to use the verb "to happen" in talking to them.

The purpose of this section was not to analyze human linguistic constructions in any detail. Rather, it was to emphasize that certain of these constructions referring to time are ambiguous even in the context of our understanding of the physics of spacetime through special and general relativity. The existence of such ambiguities is not surprising given that human language has evolved to describe limited and specific circumstances in which conditions (1)–(3) hold. The human language implicitly assumes certain features of these limited circumstances summarized roughly by the three contingencies above. The ambiguities can be resolved either by replacing parts of the human language with the language of physics, or by using that language to specify conventions that resolve the ambiguity.

As our understanding of spacetime progresses to ever more general contexts, the ambiguities in the use of human language can be expected to become larger. In a quantum theory of gravity, there is no fixed spacetime geometry. Rather, geometry is a quantum variable, fluctuating and without definite value. Then, even statements like "This happened in my past light cone" become meaningless without further qualification. Beyond that, some explorers expect that spacetime will only be a coarse-grained phenomenon at some deeper level of description.[12] How will "to happen" be unambiguously defined then?

3. The Quantum Mechanics of Closed Systems

3.1. *A Model Universe*

To keep the discussion manageable, we consider a closed quantum system, most generally, the universe, in the approximation that gross quantum fluctuations in the geometry of spacetime can be neglected. The closed system can then be thought of as a large (say $\gtrsim 20,000$ Mpc), perhaps expanding, box of quantum fields moving in a fixed background spacetime. Everything is contained within the box, in particular galaxies, planets, observers, and observed (if any), measured subsystems, and the apparatus that measures them. This is a model of the most general physical context for prediction.

The fixed background spacetime means that the notions of time are fixed and that the usual apparatus of Hilbert space, states, and operators can be employed in a quantum description of the box. The essential theoretical inputs to the process of prediction are the Hamiltonian H governing evolution and the initial quantum state $|\Psi\rangle$. These are assumed to be fixed and given by fundamental theory. We assume that H, $|\Psi\rangle$, and the operators representing alternatives can be described in terms of a set of fundamental fields and their conjugate momenta. For definiteness, we work in a fixed Lorentz frame whose time is t.

All the special relativistic concerns regarding the use of "to happen" discussed in Sec. 2 arise in this context, but we resolve these by fixing the Lorentz frame in order to concentrate on quantum mechanical issues.

For the reader not familiar with it, a simplified, bare-bones account of the quantum mechanics of closed systems is given in Appendix A. However, very little of even the modest detail given there is necessary for the present discussion. The points essential for the present discussion are these:

1. The assumed inputs are theories of the quantum dynamics (H) and the quantum initial state ($|\Psi\rangle$).
2. The outputs are the probabilities of the individual members of sets of alternative coarse-grained histories of the closed system. Consistent probabilities are predicted only for sets of histories for which there is negligible quantum interference between all pairs of histories in the set as a consequence of H and $|\Psi\rangle$. Such sets of histories are said to *decohere* and are called *realms* for short.
3. There are no non-trivial completely fine-grained realms. Coarse-graining is therefore necessary for decoherence. Some realms are *compatible*, in the sense that they can be related by the operations of fine and coarse-graining. But quantum theory also exhibits *incompatible* realms. Two realms are mutually *incompatible* if there is no common finer-grained realm of which they are both coarse-grainings. Realms defined by coarse-grainings of incompatible variables such as position and momentum provide simple examples.
4. *Quasiclassical realms*[f] exhibit the regularities of classical physics, and in particular the approximate correlations in time summarized by effective classical equations of motion. At a sufficiently fine-grained level, quasiclassical realms are defined by coarse-grainings of familiar *quasiclassical variables* such as averages of energy, momentum, and number over suitable volumes. The

[f]We use the term *quasi*classical realm to emphasize that the classical behavior is probabilistic and on occasion significantly interrupted by quantum mechanics.

quasiclassical realms of everyday experience are a subset of the totality of realms provided by quantum theory for a description of the universe.

5. Quantum theory does not distinguish between its different realms, although IGUSes may distinguish between them by their utility, as for example in the almost exclusive focus of human IGUSes on quasiclassical realms.

If any of this is not immediately clear, the reader should consult Appendix A.

3.2. *Probabilities*

Probabilities are measures of ignorance in classical physics, but in quantum physics they are fundamental. This subsection is devoted to reconciling human language which incorporates assumptions of classical certainty with a probabilistic fundamental theory. To this end, we restrict the discussion of this subsection to one realm. We begin by reviewing a little notation explained more fully in Appendix A.

We consider histories that are sequence alternatives labeled by $\alpha_1, \alpha_2, \ldots, \alpha_n$ at a series of times $t_1, t_2, \ldots, t_n$. The probabilities of these are given by

$$p(\alpha_n, \ldots, \alpha_1) = \|P^n_{\alpha_n}(t_n) \cdots P^1_{\alpha_1}(t_1)\,|\Psi\rangle\|^2 \tag{1}$$

where the $P^k_{\alpha_k}(t_k)$ are Heisenberg picture projections onto these alternatives. The label k denotes the particular exhaustive *set* of exclusive alternatives, α_k labels the particular alternative within the set, and t_k is the time. To conserve on notation, we denote individual histories by $\alpha \equiv (\alpha_1, \alpha_2, \ldots, \alpha_n)$ and denote the corresponding chains of projections by C_α so that

$$p(\alpha) = \|C_\alpha|\Psi\rangle\|^2 \tag{2}$$

is a shorthand for (1).

Conditional probabilities for alternatives α given another alternative β are constructed in the usual way

$$p(\alpha|\beta) \equiv p(\alpha, \beta)/p(\beta). \tag{3}$$

Most useful probabilities are conditional. Suppose, for instance, we know certain data d about the universe at a present time t_0 represented by a projection $P_d(t_0)$. The predictions for future histories of the universe given this data are specified by the conditional probabilities

$$p(\alpha_{\rm fut}|d) = \frac{\|C_{\alpha_{\rm fut}} P_d(t_0)|\Psi\rangle\|^2}{\|P_d(t_0)|\Psi\rangle\|^2} \tag{4}$$

where the $\{C_{\alpha_{\rm fut}}\}$ represent an exhaustive set of alternative histories to the future of t_0. Similarly, the probabilities for retrodiction of the past are given by

$$p(\beta_{\rm pst}|d) = \frac{\|P_d(t_0)\, C_{\beta_{\rm pst}}|\Psi\rangle\|^2}{\|P_d(t_0)|\Psi\rangle\|^2} \tag{5}$$

where $\{C_{\beta_{\rm pst}}\}$ represent an exhaustive set of alternative histories to the past of t_0. In each case, we assume that the set of histories consisting of P_d and the C's is decoherent.

The alternatives $P_d(t_0)$ can refer to the data possessed by an IGUS at a time t_0 along its own history. Similarly, $\alpha_{\rm fut}$ and $\beta_{\rm pst}$ can refer to data possessed by that IGUS in the future or past of t_0. Thus, it is possible to provide probabilities that answer questions like "Given that I observe a tree here today, what is the probability that there was a tree here yesterday?" or "Given that I observe a tree here today, what is the probability that I will observe a tree here tomorrow?", or even "Given that I observe a tree here today, what is the probability that there is a tree here today?" (This is the probability that our observations don't deceive us.)

Like all assertions in a probabilistic theory, those involving the verb "to happen" should always be qualified by a probability. For instance, from present data that includes texts giving 55 BC as the date of the Roman invasion of Britain, we would like to infer that Caesar did invade Britain in 55 BC. But the probability for this is not unity. There is some probability that the texts are forgeries, or contain propagating mistakes, or that the ink on their pages made a quantum transition from a configuration spelling a different date. If the probability is sufficiently close to unity we say simply that the

Roman invasion of Britain *happened* in 55 BC. Similar qualifications are needed for "is happening" and "will happen".

This need for qualifying "happen" because of probabilities is a trivial observation. Such qualifications are generally needed in classical physics as well. Probabilities are inescapable as a practical matter because of ignorance of present data or inability to determine classical evolution. Quantum fluctuations add one more source of uncertainty which is often negligible in everyday circumstances, as in the probability for the date of the Roman invasion of Britain. In the next subsection we consider a more serious need for qualification.

3.3. *Incompatible Realms*

Human IGUSes focus almost exclusively on coarse-grainings of the quasiclassical realm. Our senses are adapted to perceive quasiclassical variables and our language is adapted to describe quasiclassical histories.[g]

As we have already mentioned, the quantum universe exhibits distinct realms which are incompatible, in the sense that there is no finer-grained realm of which they are coarse-grainings. Questions, answers, predictions, retrodictions, etc. are all in the context of a particular realm which must be specified to understand what they mean.

Consider, by way of example, the reconstruction of a past history of the universe from data we have gathered in the present together with the theory of the initial state.[h] As mentioned above, that is accomplished by retrodicting the probabilities of past events from the given present data using the conditional probabilities (5).

Suppose present data is given in quasiclassical variables. The most familiar and useful retrodictions from this data are made using quasiclassical past histories. With these we retrodict the date 55 BC for the first Roman invasion of Britain from present textual records.

---

[g]Explanation for our quasiclassical focus can be sought in the physical structure of human IGUSes and the origin of that structure.

[h]For more on the reconstruction of the past in quantum mechanics, see Ref. 13.

We use present observations of the planets to reconstruct their past trajectories. We use fossil records to estimate that there is a high probability that dinosaurs roamed the Earth 150 million years ago. We infer that matter and radiation were in thermal equilibrium at the beginning of the universe from the present values of the Hubble constant, spatial curvature, and the temperature of the cosmic background radiation. These are all past histories that are members of quasiclassical realms based on coarse-grainings of quasiclassical variables.

But in quantum theory there is no unique past conditioned on given present data. Incompatible past realms can provide different stories of what had happened. A striking, if artificial, example of this is provided by the three-box model introduced by Aharonov and Vaidman for a different purpose.[14]

Consider a particle that can be in one of three boxes, A, B, or C, in corresponding orthogonal states $|A\rangle$, $|B\rangle$, and $|C\rangle$. For simplicity, take the Hamiltonian to be zero, and suppose the system to initially be in the state

$$|\Psi\rangle \equiv \frac{1}{\sqrt{3}}\left(|A\rangle + |B\rangle + |C\rangle\right). \tag{6}$$

Suppose for present data at t_0 the particle is in the state

$$|\Phi\rangle \equiv \frac{1}{\sqrt{3}}\left(|A\rangle + |B\rangle - |C\rangle\right). \tag{7}$$

Denote the projection operators on $|\Phi\rangle$, $|A\rangle$, $|B\rangle$, and $|C\rangle$ by P_Φ, P_A, P_B, and P_C respectively. Denote by $\bar{A}$ the negation of A ("not in box A") represented by the projection $P_{\bar{A}} = I - P_A$. The negations $\bar{\Phi}$, $\bar{B}$, and $\bar{C}$ and their projections $P_{\bar{\Phi}}$, $P_{\bar{B}}$, and $P_{\bar{C}}$ are similarly defined.

From the present data $|\Phi\rangle$ and the initial condition $|\Psi\rangle$ let us ask whether the particle was in the box A at a time earlier than t_0. (The exact values of the times are unimportant since $H = 0$. Only the order matters.) The relevant past realm consists of the histories

$$P_\Phi P_A,\ P_\Phi P_{\bar{A}},\ P_{\bar{\Phi}} P_A,\ P_{\bar{\Phi}} P_{\bar{A}}, \tag{8}$$

and is easily checked to decohere exactly. The conditional probabilities for A and $\bar{A}$ given Φ can be calculated from (5) with $P_d = P_\Phi$

and $\{C_\alpha\} = \{P_A, P_{\bar{A}}\}$. The result is

$$p(A|\Phi) = 1, \quad p(\bar{A}|\Phi) = 0. \tag{9}$$

Thus, we can say *in this past realm* that the event that the particle was in box A happened.

But an examination of (6) and (7) shows that both the initial condition and present data are symmetric under interchange of A and B. Therefore, using the decoherent set of histories

$$P_\Phi P_B, \; P_\Phi P_{\bar{B}}, \; P_{\bar{\Phi}} P_B, \; P_{\bar{\Phi}} P_{\bar{B}} \tag{10}$$

we can compute

$$p(B|\Phi) = 1, \; p(\bar{B}|\Phi) = 0. \tag{11}$$

Thus, we can say *in this past realm* that the event that the particle was in box B happened.

There is no contradiction because the sets of histories (8) and (10) are incompatible realms. The finer-grained set of histories describing both A and B is

$$P_\Phi P_A P_B, \; P_\Phi P_A P_{\bar{B}}, \; P_\Phi P_{\bar{A}} P_B, \ldots, \text{ etc.} \tag{12}$$

But this set does not decohere. The inference "if in A then not in B" cannot be drawn since there are no consistent probabilities for it.

Equations (8) and (10) do not exhaust the possible realms defining possible pasts for the present data P_Φ. For example, we could consider

$$P_\Phi P_\Psi, \; P_\Phi P_{\bar{\Psi}}, \; P_{\bar{\Phi}} P_\Psi, \; P_{\bar{\Phi}} P_{\bar{\Psi}}. \tag{13}$$

This is trivially decoherent with easily anticipated probabilities, but also clearly distinct from (8) and (10). In this past realm we could say that P_Ψ happened rather than anything about the above alternatives.

The usual use of "happened" assumes that there is only one realm. In a theory that permits incompatible pasts, its use must be reformed. As with the other conflicts between human language and fundamental physics discussed in this paper, there are two routes to improving precision and clarity. One is to use the language of physics and speak of the past in terms of the conditional probabilities in different past realms. The other route is to qualify "happen" so that it refers to a particular realm. For instance, in the three-box example we could say that "the event that the particle was in box A happened in the realm that referred to A" (or whatever other characterization of (8) one prefers) and similarly for B.

If "happened" means high probability for an event to have occurred in the past conditioned on certain present data, the above examples show that different events can have happened in different incompatible pasts, even seemingly contradictory events. If someone asks you "What happened yesterday?", you should, strictly speaking, respond with the question "In what realm?".

It should be stressed, however, that the same event cannot have happened in one realm and not have happened in another. The probability for a past history is given uniquely by (5) in all the realms of which it is a member. If it is high in one realm, it is high in all the others. In this sense, "happened" is non-contextual.

Needless to say, similar considerations apply to "happening" and "will happen". We next turn to the implications of all this for the word "reality".

4. Reality

The words "real" and "reality" are used in many different ways in human language. In the following we attempt to draw crude distinctions between a few of these without suggesting that other distinctions are not possible. The general point is that notions of reality reside in the models (schemata) that IGUSes construct of the world around them, both individually and collectively. Different models have different notions of reality. Therefore, when using the words "real" or "reality" it is important, for clarity, to specify which model is being referred to.

4.1. *Everyday Physical Reality*

Everyday notions of physical reality arise from the agreement among human IGUSes, both individually and collectively, on their observations and on the models of the world (schemata) that they infer from these observations. These models are formed from the gathered data by processes of selection, communication, and schematization, consistent with built-in biases. The models are constantly updated as the IGUSes acquire new information, integrate it with previous experience, infer new useful regularities, and check the model against other schemata. The everyday notions of physical reality reside in these models. This is the reality of tables and chairs, stars and galaxies, biological species, fellow humans, and the records of experiments revealing quantum phenomena, among many other things. These are the notions of reality that human language has evolved to describe and assist in constructing. Explaining the regularities found in such models is an important objective of science. The limits of physical reality are illustrated by the lack of agreement on mirages and illusions, and by the delusions of schizophrenics who are said to be "out of touch with reality". In human language, this everyday physical reality is often what is meant when the word "reality" is used without qualification.

How do we understand the agreement among human IGUSes on the facts of their physical reality in a quantum universe characterized fundamentally by the distributed probabilities of the alternative histories of a vast number of incompatible realms? The simplest explanation is that human IGUSes are all making observations utilizing coarse-grainings of the quasiclassical realms in order to exploit the quasiclassical regularities that these realms exhibit. They thus are adapted to develop schemata in more or less the same way. Occasionally they slip up, as when they are subject to delusions, or see canals on Mars, or find ghosts under the bed. But they agree, by and large. Indeed, we could not function in social units without this agreement; the ability to construct a common physical reality must have been a highly adaptive trait. Plausibly, many other IGUSes on Earth, such as dolphins and ants, make use of similar coarse-grainings.

If we find intelligent life on other planets, will they have the same notion of physical reality that we do? It's plausible that many kinds of IGUSes will have evolved to exploit the regularities of quasiclassical realms, as we have. In that case, we can expect to reach agreement with them. But could there be IGUSes focused on coarse-grainings of a distinct, incompatible realm with a correspondingly different notion of physical reality? The statistics of the schemata of extra-terrestrial IGUSes will constitute a test of the conjecture of the adaptive utility of the quasiclassical realms.

Closer to home, we imagine we could construct mechanical IGUSes (robots) that have notions of reality differing from the human kind. Even restricting to input data streams that are coarse-grainings of a quasiclassical realm, we imagine that we could construct robots that create different schemata from that data. For instance, the built-in biases for selecting what to schematize and the rules for how to schematize could both be varied. A thermostat is a very simple example of an IGUS with a restricted schema. It should be possible to construct IGUSes (robots) with schemata that do not employ our past, present, and future organization of temporal information[i].[10] Science fiction abounds with robots that construct different schemata. It is also not beyond possibility that we could construct robots utilizing non-quasiclassical input data streams. Such robots would consequently have qualitatively different schemata and qualitatively different notions of everyday reality.

4.2. *The Realities of Physical Theories*

As Bohr said, "the task of science is both to extend the range of our experience and reduce it to order".[15] Elementary notions of physical reality are extended by fundamental physical theories. These are realities that are agreed on by physicists and reside in physics literature. Indeed, there is very little, if any, distinction between the model itself and the notion of reality which follows from it.[j]

[i]Our conscious focus on the most recently acquired data is plausibly the reason that the present is sometimes characterized as more "real" than the imperfectly recorded past and the unknown future. What is meant, presumably, is that present data could also figure more prominently and accurately in a robot's schema. However, the "NS" robot of Ref. 10 would treat the past and present equally, and the "AB" robot would have some premonitions of the future.

[j]That is possibly the reason that "reality" is so little discussed in physics textbooks — it is already implicit in the model under discussion.

We hope that the reality of our fundamental theoretical frameworks is objective because they summarize the universal regularities of the universe independently of any selection by us. In particular, we hope that they would be agreed on by other IGUSes we might meet that are interested in physics, whether they share our notion of everyday physical reality or not.

The everyday physical reality described in Sec. 4.1 is an approximate, particular feature of these fundamental models arising from particular coarse-grainings and particular initial conditions. Characterizing the emergence of physical reality already raises interesting questions in classical physics and more profound ones in quantum mechanics.

A fundamental classical model for the universe would consist of the fine-grained histories of particles and fields evolving from an initial condition according to deterministic laws. These fine-grained histories can be specified by giving the coordinates and momenta of both particles and fields as a function of time.[k] An initial condition could be specified by giving a distribution function for these variables at an initial time. The important point for present considerations is that the reality of classical theory consists of a unique family of fine-grained histories.

The constituents of the everyday notion of physical reality are not atoms, molecules, and electromagnetic fields, but rather tables and chairs, stars and galaxies, etc., and more generally the forms, velocities, and locations of individual objects. Indeed such notions were used by our species long before atoms, molecules, and fields were discovered. Rather, everyday physical reality arises from the fundamental classical model by appropriate coarse-graining. For instance, coarse-graining by quasiclassical variables such as the averages over suitable volumes of densities of approximately conserved quantities such as energy, momentum, and number leads to phenomenological equations of motion such as the Navier–Stokes equation for a wide

---

[k]More precisely, they are specified by giving these values on a foliating family of spacelike surfaces in a curved spacetime obeying the Einstein equation when general relativity is taken into account.

class of initial conditions (see, e.g. Ref. 16). Further coarse-graining is needed to define individual physical objects, such as particular trees. The important point is that everyday physical reality is an approximate feature of classical physics contingent on particular choices of coarse-graining.

The story of the emergence of everyday physical reality from the quasiclassical realms in quantum theory is similar to that in classical physics. The coarse-grainings defining the relevant sets of histories are the same. The two new features are the following: First, the initial quantum state must be such that the set of coarse-grained histories decoheres. Second, the correlations in time that define classical determinism are now only approximate — occurring with high probability in particular initial states.[l]

What is very different from classical physics is the reality of the fundamental quantum theory. There are no non-trivial fine-grained decoherent sets of histories for quantum theory as there are for classical physics.[21] Coarse-graining is necessary for decoherence, and there are many different coarse-grainings, not all of which are compatible. As we have already discussed, quantum theory therefore provides many mutually incompatible realms, of which the quasiclassical ones are a small subset. The theory does not distinguish between these, although IGUSes may do so. In the three-box example in Sec. 3.3, the past realm that refers to box A is no more or less "real" than the incompatible past realm that refers to box B. The reality of quantum theory may rather be said to consist of all the different possible realms.[m]

To use the word "reality" in the context of quantum theory without qualification is to risk confusing everyday physical reality, which is constructed from quasiclassical realms, with the reality of quantum theory, which consists of all realms.

[l]For some further discussion and models, see, e.g. Refs. 17–20.

[m]The far-reaching consequences of this generalization of the reality of classical theory have been stressed in Ref. 22.

4.3. *Other Realities*

Human IGUSes exhibit a wide range of individual notions of reality that go beyond physical reality by incorporating in their schemata the supernatural, prejudice, revelation, hearsay, wishful thinking, etc., along with scientific evidence. Today, one can find human IGUSes of different persuasions on the reality of UFOs, paranormal phenomena, and biological evolution. Beyond these individual notions, there are many other kinds of agreed-upon models constructed by human IGUSes, each defining a notion of reality. For instance, a notion of mathematical reality arises from the agreement of mathematicians on axioms which, as Gödel put it,[23] "force themselves on us as being true".[n]

4.4. *Usage*

The above discussion probably does not exhaust the uses of the word "reality" in human language. But it does illustrate that its meanings are diverse. What they seem to have in common is an agreement by at least some IGUSes on a model for some class of coarse-grained physical phenomena. This diversity of meanings can lead to confusion if the word is used without qualification. In particular, it is important to distinguish between everyday physical reality and the notions of reality provided by fundamental physical theory. That is especially the case if everyday experience is special and contingent among many other possibilities, as it is in quantum theory. The trend in fundamental physics today seems to lie in the direction of increasing disparity between everyday physical reality and the reality of the fundamental theory. If that trend continues, maintaining appropriate linguistic qualifications will become even more necessary for clarity.

---

[n]When we encounter other intelligent beings we can confidently predict that they will have the same arithmetic that we do. But will they have ZFC?

5. Dispensible Words

Dispensible words are ones that can be added to or deleted from an exposition of a physical theory without effecting the theory's experimentally verifiable predictions or its utility. Such words can be important for motivation, for evoking analogies, for building intuition, and for suggesting future research. In short, they can help understanding. But they can also be confusing, the source of false problems, and an obstacle to understanding. Not surprisingly, some dispensible words arise from natural constructions in human language. Also, not surprisingly, many of the confusing dispensible words occur in quantum theory.[o]

It is useful to know when language describing a theory is dispensible and when it is not. In particular, a question of whether or not dispensible words are appropriate need not be settled by experiment.

There is a simple test for dispensible words: Dispense with them and see if the predictions of the theory are unchanged. The author has usually found that dispensing with the dispensibles is a route to clarity. This section illustrates these ideas with two examples from quantum mechanics.

5.1. *Probabilities "to happen"*

Consider the two sentences, "The probability of rain this afternoon is 80%" and "The probability for rain to happen this afternoon is 80%." To the author these two sentences mean the same thing in any understanding of probability of which he is aware. His decision whether or not to carry an umbrella would be the same under either assertion. The words "to happen" are dispensible.

Qualifying "the probability of A" so that it becomes "the probability of A to happen" can be a source of confusion because it suggests to some that "happening" is a physical process. Indeed, some have held that quantum theory is incomplete until it explains why, in a set of alternatives for which it supplies probabilities, one of them "happens" (or is observed). Quantum theory as developed here has

[o]For one reason this might be so, see Ref. 24.

no such mechanism and is yet consistent with all experiment, as far as we know, through the probabilities which are its only output.

5.2. *Equally real histories*

Another candidate for dispensible words is the statement, "all the histories in a given realm are equally real".[p] This statement has a simple meaning if "real" is understood to refer to the theoretical model — the reality of the theory as discussed in Sec. 4.3. In that context the statement could be rephrased as "quantum mechanics does not distinguish between the histories in a given realm except by their probabilities." Indeed, using the word "real" in this sense, it would be possible to correctly say that "all the realms of quantum theory are equally real". This could be similarly rephrased as "quantum theory does not distinguish between different realms". In each case, the first statement can be replaced by the second without affecting the predictions of the theory for experiment. The second formulations are easier to understand for many physicists. But all these statements are dispensible as the exposition of quantum theory given in this article shows.

Deutsch[25] especially has stressed the naturalness and interpretative value of the equal status of histories in a given realm, for instance, for understanding the power of quantum computation. Such advantages, however great, do not alter that fact that words like "all the other histories are equally real" can be dispensed with without affecting the experimental implications of the theory.

As discussed in Sec. 4, the linguistic difficulty with using the word "real" is that it can mean different things in different circumstances, even in physics. Maintaining that "all the histories in a set are equally real" risks confusion unless the meaning of "real" is explained. The author's experience is that otherwise it can be confused with the everyday physical reality discussed in Sec. 4.1. The resulting conflict can be an obstacle for some to accept Everett's powerful and natural idea of taking the state of the universe seriously.

---

[p]This is a translation of the statement "all the other worlds are equally real", where what is meant by "world" is what is called "history" here, even if it consists only of an alternative at one time.

6. Advice

The aim of fundamental physics is to find the laws governing the regularities exhibited universally by *all* physical systems, without exception, without qualification, and without approximation. The search for these laws has been seriously underway at least since the time of Newton. Classical mechanics, Newtonian gravity, Maxwell's electrodynamics, the atomic theory of matter, special and general relativity, quantum field theory, superstring theory, the quantum theory of geometry, and quantum cosmology are just some of the milestones in the history of this search. As new regimes of experiment and observation have been explored, more general theoretical frameworks have evolved. Old theories have become effective theories applicable in limited circumstances of the extended context.

Excess theoretical baggage is typically shed in this process of generalization.[26] Ideas that were once accepted as fundamental, general, and inescapable have come to be seen as consequent, special, and dispensible. Examples from the history of physics are an Earth-centered cosmology, a single universal time, an exact second law of thermodynamics, fixed Euclidean spatial geometry, and a quantum mechanics restricted to measurements. These ideas were not true general features of the world, but only perceived to be general because of our special universe, our special place in it, and the limited range of our experience.

Over this history, the models provided by fundamental physical theory have moved far beyond our everyday notions of physical reality.[q] The evolution of physical theory beyond everyday notions of physical reality has complicated its description using a human

[q]Excess baggage in the language of physics is also typically shed in the process of extension and generalization of physical theory. Indeed, the relationship between the languages of two effective theories often presents problems similar to those between the human language and the language of physics that we have discussed. For example, applying the language of classical physics to quantum mechanical situations can lead to paradoxes which can be resolved by sticking to the language of quantum theory.[4,27] By and large, however, these physics linguistic problems have not caused much difficulty and, in any event, are not the subject of this essay.

language that is adapted to that everyday reality. That has been the subject of this essay.

By way of conclusion, the author offers a few words of advice on routes to clarity in the face of the disparity between the languages of fundamental physics and that of human IGUSes. This advice is not directed as to how to *find* a fundamental theory. Rather, it is only about how to deal with the conflicts with human language that may arise.

- *Identify Theoretical Excess Baggage.* Remember that ideas that were fundamental and obvious in one theory can become emergent and dispensible in a more general one. The idea that there is a unique past, present, and future defined by physics is one example that we have discussed. Another is the idea that there should be a unique theoretical reality specified by one decoherent set of alternative fine-grained histories, rather than many different ones that may be incompatible and yet fit into a consistent theoretical framework.
- *Identify Linguistic Excess Baggage.* Remember that human language can contain tacit assumptions that reflect the limited context in which it evolved. We have discussed examples associated with "to happen" and "reality". Linguistic excess baggage can be dispensed with in favor of the more precise language of physics. *To clearly discuss quantum theory, learn to speak the language of quantum mechanics.* Alternatively, linguistic excess baggage can be qualified so that tacit assumptions become explicit.
- *Identify Dispensible Language.* Remember that an exposition of a theory can contain language that can be dispensed with without affecting its experimental predictions. Examples are the use of "to happen" to qualify probabilities and the use of "all equally real" to qualify sets of histories. The author often finds it useful to drop dispensible language in introducing quantum theory. That is especially the case if the audience carries its own linguistic baggage which conflicts with the dispensible language, making it an obstacle to understanding and acceptance. However, don't forget that adding interpretative but dispensible language can be an important route to insight, motivation,

understanding, and generalization. Either way, it is important to recognize when language is dispensible and when it is not. In arguments concerning the interpretation of the theory, one then understands which statements are experimentally verifiable and which are not.

- *Beware of Introspection.* Remember that human beings are physical systems in the universe that have a long, specific evolutionary history. That evolution is consistent with the universal laws of physics. But the present nature of these IGUSes has far more to do with the frozen accidents of their evolution than with those fundamental laws. As a consequence, introspection is unlikely to be a good guide to the character of those laws. That is especially the case if introspection is seen to require precise representations in the fundamental theory for constructions of the human language. To take an extreme example, some have concluded from the strong impression of "now" held by human IGUSes that the fundamental theory *must* incorporate it despite the overwhelming body of experimental evidence against preferred frames with figures of merit approaching 10^{-21},[28] and despite the possibility of constructing IGUSes which do not prefer the present.[10] Rather, "now" can be seen as a feature of how certain IGUSes process temporal information as described in Sec. 2.
- *Beware of Agendas.* Remember that the basic criteria for physical theory are logical consistency and consistency with experiment. Agendas for physical theory motivate research, as in the quest for unification or a selection principle that would distinguish one realm from all others.[22,29] But if a theory that is logically consistent and consistent with experiment disagrees with your agenda, that should not be called a problem for the theory, it is more likely to be a problem with your agenda.

It seems likely that there are limits to quantum theory and the validity of its principle of superposition, as there have been for every other candidate for a fundamental theory to date. The remarkable fact about the history of this most successful of all physical theories is that, despite the limited range over which it has been experimentally verified, there are no alternative theories that are consistent with these experiments, consistent with the rest of modern physics, but which differ in their predictions in domains not yet tested. As Steve Weinberg puts it: "It is striking that it has not so far been possible to find a logically consistent theory that is close to quantum mechanics other than quantum mechanics itself".[30]

Alternatives to quantum theory would be of great interest, if only to guide experiment. Given the trend in the development of fundamental theory, it is very possible that the disparity between human language and the language of fundamental physics will increase as quantum theory is replaced or extended. If that is the case, careful analyses of the relationship between human language and the language of physics of the kind sketched all too superficially in this essay will be increasingly important for clarity of understanding.

Appendix A

The Quantum Mechanics of a Closed System

This appendix gives a simplified, bare-bones account of some essential elements of the modern synthesis of ideas constituting the quantum mechanics of closed systems[3–5] using the closed box model described in Sec. 3.1.

A.1. *Realms*

The most general objective of a quantum theory of a closed system is the prediction of probabilities for the individual members of exhaustive sets of coarse-grained alternative time histories of the system. For instance, we might be interested in alternative histories of the center of mass of the Earth in its progress around the Sun, or in histories of the correlation between the registrations of a measuring apparatus and a measured subsystem. Alternatives at one moment of time can always be reduced to a set of yes/no questions. For example, alternative positions of the Earth's center of mass can be reduced to asking, "Is it in this region — yes or no?", "Is it in that region — yes or no?", etc. An exhaustive set of yes/no alternatives is represented in the Heisenberg picture by an exhaustive set of orthogonal projection operators $\{P_\alpha(t)\}$, $\alpha = 1, 2, 3 \ldots$. These satisfy

$$\sum_\alpha P_\alpha(t) = I, \text{ and } P_\alpha(t)\, P_\beta(t) = \delta_{\alpha\beta} P_\alpha(t) \tag{A.1}$$

showing that they represent an exhaustive set of exclusive alternatives. In the Heisenberg picture, the operators $P_\alpha(t)$ evolve with time according to

$$P_\alpha(t) = e^{+iHt/\hbar} P_\alpha(0)\, e^{-iHt/\hbar}. \tag{A.2}$$

The state $|\Psi\rangle$ is unchanging in time.

An important kind of set of histories is specified by a series of sets of alternatives $\{P^1_{\alpha_1}(t_1)\}$, $\{P^2_{\alpha_2}(t_2)\}, \ldots, \{P^n_{\alpha_n}(t_n)\}$ at a sequence of times $t_1 < t_2 < \cdots < t_n$. The sets at distinct times can differ, and are distinguished by the superscript on the Ps. For instance, projections on ranges of position might be followed by projections on ranges of momentum, etc.[r] An individual history c_α in such a set is a particular sequence of alternatives $(\alpha_1, \alpha_2, \ldots, \alpha_n) \equiv \alpha$ and is represented by the corresponding chain of projections called a *class operator*

$$C_\alpha \equiv P^n_{\alpha_n}(t_n) \cdots P_{\alpha_1}(t_1). \tag{A.3}$$

Such a set of histories is generally *coarse-grained* because alternatives are specified at some times and not at every time, and because the alternatives at a given time are projections onto subspaces with dimension greater than one and not projections onto a complete set of states. *Fine-grained* sets of histories consist of one-dimensional projections at each and every time.

Operations of fine- and coarse-graining may be defined on sets of histories. A set of histories $\{c_\alpha\}$ may be *coarse-grained* by partitioning it into exhaustive and exclusive classes $\bar{c}_{\bar{\alpha}}, \bar{\alpha} = 1, 2, \ldots$. Each class consists of some number of histories in the finer-grained set, and every finer-grained history is in some class. Suppose, for example, that the position of the Earth's center of mass is specified by dividing

---

[r]In general, realistic sets of histories will be *branch-dependent*, with sets of projections at a given time depending on the particular sequence of previous alternatives, but we ignore this in the present simplified exposition.

space into cubical regions of a certain size. A coarser-grained description of position could consist of larger regions made up of unions of the smaller ones. *Fine-graining* is the inverse operation of dividing sets of histories up into smaller classes. The class operator $\bar{C}_{\bar{\alpha}}$ for a history in a coarse-graining of a set whose class operators are $\{C_\alpha\}$, is related to those operators by summation, viz.

$$\bar{C}_{\bar{\alpha}} = \sum_{\alpha \in \bar{\alpha}} C_\alpha \tag{A.4}$$

where the sum defining $\bar{C}_{\bar{\alpha}}$ for the class $\bar{c}_{\bar{\alpha}}$ is the sum over the C_α for all finer-grained histories contained within it.

For any set of histories $\{c_\alpha\}$, there is a *branch state vector* for each history in the set defined by

$$|\Psi_\alpha\rangle = C_\alpha|\Psi\rangle. \tag{A.5}$$

When probabilities can be consistently assigned to the individual histories in a set, they are given by

$$p(\alpha) = \||\Psi_\alpha\rangle\|^2 = \|C_\alpha|\Psi\rangle\|^2. \tag{A.6}$$

However, quantum interference prevents consistent probabilities from being assigned to every set of alternative histories that may be described. The two-slit experiment provides an elementary example: An electron emitted by a source can pass through either of two slits on its way to detection at a farther screen. It is not possible to consistently assign probabilities to the two histories distinguished by which slit the electron goes through. Because of quantum interference, the probability to arrive at a point on the screen would not be the sum of the probabilities to arrive there by going through each of the slits. In quantum theory, probabilities are squares of amplitudes, and the square of a sum is not generally the sum of the squares. On the other hand, if other interactions of the electron destroy the interference between the two histories (as when a measurement determines which slit it passes through), then probabilities can be consistently assigned.

Negligible interference between the branches of a set

$$\langle\Psi_\alpha|\Psi_\beta\rangle \approx 0, \quad \alpha \neq \beta \tag{A.7}$$

is a sufficient condition for the probabilities (A.6) to be consistent with the rules of probability theory. Specifically, as a consequence of the decoherence condition (A.7), the probabilities (A.6) obey the most general form of the probability sum rules

$$p(\bar{\alpha}) \approx \sum_{\alpha \in \bar{\alpha}} p(\alpha) \tag{A.8}$$

for any coarse-graining $\{\bar{c}_{\bar{\alpha}}\}$ of the $\{c_\alpha\}$. Sets of histories obeying (A.7) are said to (medium) decohere.[s] These are sets for which quantum mechanics makes predictions. They are determined through (A.6) by the Hamiltonian H and the quantum state of the universe $|\Psi\rangle$. We use the term *realm* as a synonym for a *decoherent set of alternative coarse-grained histories.*

A coarse-graining of a decoherent set is again decoherent. A fine-graining of a decoherent set risks losing decoherence.

An important mechanism of decoherence is the dissipation of phase coherence between branches into variables not followed by the coarse-graining. Consider, by way of example, a dust grain in a superposition of two positions deep in interstellar space.[35] In our universe, about 10^{11} cosmic background photons scatter from the dust grain each second. The two positions become correlated with different, nearly orthogonal states of the photons. Coarse-grainings that follow only the position of the dust grain at a few times therefore correspond to branch state vectors that are nearly orthogonal and satisfy (A.7). The orthogonality is approximate but in realistic situations sufficient to define consistent probabilities well beyond the standard to which they can be checked or, indeed, the physical situation modeled.[36]

Measurements and observers play no fundamental role in this general formulation of quantum theory. Measurement situations can, of course, be described.[36,37] In a typical measurement situation, one subsystem of the universe (the measured subsystem) interacts with

[s]For a discussion of the linear positive, weak, medium, and strong decoherence conditions, see Refs. 31–33. However, as L. Diósi has shown,[34] medium decoherence is the weakest of this chain that is consistent with elementary notions of the independence of isolated systems.

another (the apparatus). A variable of the measured subsystem, not otherwise decohering, becomes correlated with a variable of the apparatus which decoheres because of its interaction with the rest of the universe. The correlation thus effects the decoherence of the measured variables so that probabilities can be predicted for its values. With suitable idealizations and assumptions, probabilities for the measured outcomes are given to an excellent approximation by usual textbook quantum theory. But, in a set of histories where they decohere, probabilities can be assigned to the position of the Moon when it is not receiving the attention of observers and to the values of density fluctuations in the early universe when there were neither measurements taking place nor observers to carry them out.

The probabilities of the histories of the possible *decoherent* sets of coarse-grained histories and the conditional probabilities constructed from them are the totality of predictions of the quantum mechanics of a closed system given the Hamiltonian H and initial state $|\Psi\rangle$.

A.2. *Incompatible Realms*

Coarse-graining is generally necessary for decoherence. There are only trivial decoherent sets of completely fine-grained histories.[21]

A completely fine-grained set of histories can be coarse-grained in many different ways to yield a decoherent set whose probabilities can be employed in the processes of prediction and retrodiction. Further, there are many different completely fine-grained sets to start from, corresponding to the different possible choices of one-dimensional projections at each time arising from different complete sets of commuting observables. Once coarse-grained enough to achieve decoherence, further coarse-graining preserves it. Some decoherent sets can be organized into families connected by the operations of fine- and coarse-graining. Such sets are said to be *compatible.*

Realms for which there are no common finer-grained decoherent sets are *incompatible.* We may not draw inferences by combining probabilities from incompatible realms.[t] That would implicitly assume that there are the probabilities of a finer-grained description

[t] As the work of R. Griffiths[4] especially has shown, essentially all inconsis-

which is not available. Incompatible realms provide different descriptions of the universe. All of the totality of incompatible realms are necessary to give a complete account of the universe because they are, in principle, equally available for exhibiting regularities and constructing explanations. Quantum theory does not distinguish one of these realms over another without further criteria.

Incompatibility is not inconsistency in the sense of making different predictions for the same history. The probability of a history c_α is given by (A.6) in all the realms of which it is a member.

A.3. *Quasiclassical Realms*

While quantum theory permits a great many incompatible descriptions of a closed system by different realms, we as human observers utilize mainly realms that are coarse-grainings of one family of compatible sets — the quasiclassical realms of everyday experience. These are the sets of decoherent histories whose probabilities manifest the classical regularities of the universe that are exploitable in our various pursuits, in particular, to get food, reproduce, avoid destruction, and achieve recognition. These are the sets of histories defined by quasiclassical alternatives which our perception is adapted to distinguish.

Quasiclassical realms are defined as a coarse-graining specifying ranges of values of the variables of classical physics. These include the averages over small volumes of approximately conserved quantities such as energy, momentum, and various kinds of particle species. We call such variables *quasiclassical variables.* The useful properties of quasiclassical realms follow from the approximate conservation of the variables that define them. In particular, quasiclassical realms exhibit correlations in time governed by the approximate deterministic laws of motion of classical physics. In our quantum universe, classical laws are applicable over a wide range of time, place, scale, and epoch.

More specifically, by a *quasiclassical realm* we mean an exhaustive

---

tencies alleged against consistent histories quantum mechanics arise from violating this logical prohibition.

set of mutually exclusive coarse-grained alternative histories, that obey a realistic principle of decoherence, that consist largely of related but branch-dependent projections onto ranges of quasiclassical variables at a succession of times, with individual histories exhibiting patterns of correlation implied by closed sets of effective equations of motion subject to frequent small fluctuations and occasional major branchings (as in measurement situations). By a *family of quasiclassical realms* we mean a set of compatible ones that are all coarse-grainings of a common one. Thus defined, the quasiclassical realms are a feature of our universe that arise from its quantum state and dynamics which we are adapted to exploit.[21]

Acknowledgments

The author is grateful to Fay Dowker, Simon Saunders, and Rafael Sorkin for critical readings of the manuscript. Thanks are due to David Deutsch for clarifying discussions of the many worlds' formulation of quantum theory and to Simon Saunders' help with the references. The author thanks the Aspen Center for Physics for hospitality while this work was being completed. This work was supported in part by the National Science Foundation under grants PHY02-44764 and PHY05-55669.

References

1. W. James, *Principles of Psychology* (Henry Holt, New York, 1893).
2. F. H. C. Crick, Thinking about the brain, *Sci. Am.* **241**, September, 219–232 (1979).
3. M. Gell-Mann, *The Quark and the Jaguar* (W. Freeman, San Francisco, 1994).
4. R. B. Griffiths, *Consistent Quantum Theory* (Cambridge University Press, Cambridge, UK, 2002).
5. R. Omnès, *Interpretation of Quantum Mechanics* (Princeton University Press, Princeton, 1994).
6. A. Bassi and G. Ghirardi, Dynamical reduction models, *Phys. Reps.* **379**, 257–426 (2003).

7. J. B. Hartle, Spacetime quantum mechanics and the quantum mechanics of spacetime, in *Gravitation and Quantizations*, in *Proc. 1992 Les Houches Summer School*, eds. B. Julia and J. Zinn-Justin, Les Houches Summer School Proc. Vol. LVII (North Holland, Amsterdam, 1995), gr-qc/9304006. A précis of these lectures is given in *Quantum Mechanics at the Planck Scale*, gr-qc/9508023.
8. Various sources suggest that the following may be useful introductions to these subjects in philosophy: S. Saunders, Time, quantum mechanics, and decoherence, *Synthese* **102**, 235–266 (1995); S. Saunders, Tense and indeterminateness, *Phil. Sci.* **33**, 637–661 (2000); T. Sider, *Four-Dimensionalism* (Oxford University Press, Oxford, 2001); D. Wallace, Language use in a branching universe, *Stud. Hist. Philos. Sci.* **33**, 637 (2002).
9. N. Ashby, Relativity and the Global Positioning System, *Phys. Today* **55**, no. 5, 41–47 (2002).
10. J. B. Hartle, The physics of 'now', *Am. J. Phys.* **73**, 101–109 (2005).
11. G. Westheimer, Discrimination of short time intervals by the human observer, *Exper. Brain Res.* **129**, 121–126 (2000), and G. Westheimer, *Visual signals used in time interval discrimination, Vis. Neurosci.* **17**, 551–556 (2000).
12. N. Seiberg, Emergent spacetime, to appear in the proceedings of the 23rd Solvay Conference, *The Quantum Structure of Space and Time*, hep-th/0601234.
13. J. B. Hartle, Quantum pasts and the utility of history, in *The Proc. Nobel Symposium: Modern Studies of Basic Quantum Concepts and Phenomena*, Gimo, Sweden, June 13–17, 1997, *Phys. Scripta* **T76**, 67–77 (1998), gr-qc/9712001.
14. Y. Aharonov and L. Vaidman, Complete description of a quantum system at a given time, *J. Phys. A* **24**, 2315 (1991).
15. N. Bohr, *Atomic Theory and the Description of Nature* (Cambridge University Press, Cambridge, UK, 1961).
16. D. Foster, *Hydrodynamic Fluctuations, Broken Symmetry, and Correlation Functions* (Addison-Wesley, Redwood City, 1975).

17. M. Gell-Mann and J. B. Hartle, Classical equations for quantum systems, *Phys. Rev. D* **47**, 3345 (1993), gr-qc/9210010.
18. T. Brun and J. B. Hartle, Classical dynamics of the quantum harmonic chain, *Phys. Rev. D* **60**, 123503 (1999), quant-ph/9905079.
19. J. Halliwell, Decoherent histories and hydrodynamic equations, *Phys. Rev. D* **58**, 105015 (1998), quant-ph/9805062.
20. J. Halliwell, Decoherence of histories and hydrodynamic equations for linear oscillator chain, *Phys. Rev. D* **68**, 025018 (2003), quant-ph/0305084.
21. M. Gell-Mann and J. B. Hartle, Quasiclassical coarse graining and thermodynamic entropy, quant-ph/0609190.
22. H. F. Dowker and A. Kent, On the consistent histories approach to quantum mechanics, *J. Stat. Phys.* **82**, 1574 (1996), gr-qc/9412067.
23. K. Gödel, What is Cantor's continuum problem? in *Philosophy of Mathematics: Selected Readings*, eds. P. Benacerraf and H. Putnam (Cambridge University Press, Cambridge, UK, 1983).
24. J. B. Hartle, What connects different interpretations of quantum mechanics? in *Quo Vadis Quantum Mechanics*, eds. A. Elitzur, S. Dolev, and N. Kolenda (Springer-Verlag, Heidelberg, 2005), pp. 73–82, quant-ph/0305089.
25. D. Deutsch, *The Fabric of Reality* (Penguin Books, London, 1997).
26. J. B. Hartle, *Excess baggage*, in *Elementary Particles and the Universe: Essays in Honor of Murray Gell-Mann*, ed. J. H. Schwarz (Cambridge University Press, Cambridge, UK, 1991), pp. 1–16, gr-qc/0508001.
27. Y. Aharonov and D. Rohrlich, *Quantum Paradoxes: Quantum Theory for the Perplexed* (John Wiley, New York, 2005).
28. For reviews, see D. Mattingly, Modern tests of Lorentz invariance, *Living Rev. Rel.* **8**, 5 (2005), (http://www.livingreviews.org/lrr-2005-5), gr-qc/0502097; M. P. Haugan and C. M. Will, Modern tests of special relativity, *Phys. Today* **40**, 69–76 (1987); C. M. Will, The confrontation between general relativity and experiment, *Living Rev. Rel.* **17**, 4 (2006), (http://www.livingreviews.org/lrr-2006-3).

29. A. Kent, Quantum histories and their implications, in *Relativistic Quantum Measurement and Decoherence*, eds. H.-P. Breuer and F. Petruccione, Lecture Notes in Physics, Vol. 559, pp. 93–115 (Springer-Verlag, Berlin, 2000).
30. S. Weinberg, *Dreams of a Final Theory* (Pantheon Books, New York, 1992).
31. M. Gell-Mann and J. B. Hartle, Alternative decohering histories in quantum mechanics, in the *Proc. 25th Int. Conf. High Energy Physics*, Singapore, August 2–8, 1990, eds. K. K. Phua and Y. Yamaguchi (South East Asia Theoretical Physics Association and Physical Society of Japan) (distributed by World Scientific, Singapore, 1990).
32. M. Gell-Mann and J. B. Hartle, Strong decoherence, in the *Proc. 4th Drexel Symp. Quantum Non-Integrability — The Quantum-Classical Correspondence*, eds. D.-H. Feng and B.-L. Hu (International Press, Boston/Hong-Kong, 1995), gr-qc/9509054.
33. J. B. Hartle, Linear positivity and virtual probability, *Phys. Rev. A* **70**, 022104 (2004), quant-ph/0401108.
34. L. Diósi, Anomalies of weakened decoherence criteria for quantum histories, *Phys. Rev. Lett.* **92**, 170401 (2004).
35. E. Joos and H. D. Zeh, The emergence of classical properties through interaction with the environment, *Zeit. Phys. B* **59**, 223 (1985).
36. J. B. Hartle, The quantum mechanics of cosmology, in *Quantum Cosmology and Baby Universes: Proc. 1989 Jerusalem Winter School for Theoretical Physics*, eds. S. Coleman, J. B. Hartle, T. Piran, and S. Weinberg (World Scientific, Singapore, 1991), pp. 65–157.
37. M. Gell-Mann and J. B. Hartle, Quantum mechanics in the light of quantum cosmology, in *Complexity, Entropy, and the Physics of Information*, SFI Studies in the Sciences of Complexity, Vol. VIII, ed. W. Zurek (Addison Wesley, Reading, MA, 1990).

CHAPTER 10

Quantum Pasts and the Utility of History*

From data in the present we can predict the future and retrodict the past. These predictions and retrodictions are for histories — most simply time sequences of events. Quantum mechanics gives probabilities for individual histories in a decoherent set of alternative histories. This paper discusses several issues connected with the distinction between prediction and retrodiction in quantum cosmology: the difference between classical and quantum retrodiction, the permanence of the past, why we predict the future but remember the past, the nature and utility of reconstructing the past(s), and information theoretic measures of the utility of history.

*Talk presented at *The Nobel Symposium: Modern Studies of Basic Quantum Concepts and Phenomena*, Gimo, Sweden, June 13–17, 1997.
arXiv:gr-qc/9712001

1. Introduction

Up here on length scales above the Planck length, where the notion of an approximately classical spacetime makes sense, the world is four-dimensional with three spacelike dimensions and one for time. Classically, we most accurately describe the world in terms of the four-dimensional concepts of points, world lines, and field configurations in spacetime. Yet, we can divide our subjective experience up into past, present and future. From our point of view these have very different properties: We know the present, remember the past and predict the future.

The processes of predicting the future and retrodicting the past are familiar and elementary in classical physics. However, fundamentally the world is not classical but quantum mechanical. The process of prediction — the prediction of a measurement outcome, for instance — is also familiar in quantum mechanics. The quantum past, however, is more of a foreign country. This essay is concerned with the past in quantum mechanics. What is the nature of quantum mechanical retrodiction? What is the origin of the differences between past and future? Most importantly, why are we interested in the past at all? — It's over. What is the utility of history?

2. Prediction and Retrodiction in Quantum Mechanics

2.1. *Classical Prediction and Retrodiction*

Classically, prediction and retrodiction are symmetrically related if we presume that the fundamental equations of motion are time-reversal invariant. From present values for the positions and velocities of the molecules in a gas, we may predict their future positions and velocities and retrodict their past ones by using the deterministic equations of motion. Effective classical equations for coarse-grained quantities such as the density of the gas may exhibit dissipation, chaos and history dependence which distinguish prediction from retrodiction, but *in principle*, if present data contained enough information, the processes of prediction and retrodiction would be symmetrically related. That is not the case in quantum mechanics.

2.2. *The Quantum Mechanics of a Closed System*

We consider the quantum processes of prediction and retrodiction from the most comprehensive perspective — the quantum mechanics of a closed system, most generally the universe as a whole. (See, e.g. Ref. 1.) We neglect quantum gravity to keep the discussion manageable. This is an excellent approximation for many useful purposes for any time later than 10^{-43} s after the big bang. In this approximation, the universe may be thought of as a collection of matter fields inside a large, and perhaps expanding, box in fixed background spacetime geometry. The two fundamental inputs that specify a closed quantum mechanical system are its Hamiltonian H and initial quantum state $|\Psi\rangle$. We now very briefly review how these lead to predictions and retrodictions.

The general objectives of quantum theory are the probabilities of individual time histories in sets of alternative coarse-grained histories of the universe. Examples are the probabilities of a set of possible orbits of the earth around the sun. Such histories are said to be coarse-grained because they do not specify the coordinates of every particle in the universe but only those of the center of mass of the Sun and Earth, and these only crudely and not at every time.

The simplest sets of alternative histories are alternatives at a moment of time. These can always be reduced to a set of yes/no alternatives. For instance, questions about the position of the earth's center of mass can be reduced to questions of the form: "Is the earth's center of mass in this region — yes or no?", "Is it in that region — yes or no?", etc. A set of yes/no alternatives at a moment of time, say $t = 0$, is represented by a set of orthogonal projection operators $\{P_\alpha\}$, $\alpha = 1, 2, \ldots$ — one projection operator for each alternative. These projection operators must satisfy

$$\sum_\alpha P_\alpha = I, \quad \text{and} \quad P_\alpha P_\beta = 0, \ \alpha \neq \beta, \tag{1}$$

showing that they represent an exhaustive set of exclusive alternatives. In the Heisenberg picture that we shall employ, the same set of alternatives at a later time is represented by a different set of operators related to the first by

$$P_\alpha(t) = e^{iHt} P_\alpha e^{-iHt}. \tag{2}$$

(Here and throughout we use units where $\hbar = 1$.)

A set of alternative histories may be specified by giving sets of alternatives at a series of times $t_1 < t_2 < \cdots < t_n$. We denote the corresponding sets of projections by $\{P^1_{\alpha_1}(t_1)\}, \ldots, \{P^n_{\alpha_n}(t_n)\}$. An individual history corresponds to a particular sequence of alternatives $(\alpha_1, \ldots, \alpha_n) \equiv \alpha$ and is represented by the corresponding chain of projections

$$C_\alpha = P^n_{\alpha_n}(t_n) \cdots P^1_{\alpha_1}(t_1). \tag{3}$$

Such a set of histories is said to be *coarse-grained* when there is not a set of alternatives at each and every time or when the projections are not one-dimensional onto complete sets of states. For a set of histories describing alternative orbits of the earth, the $\{P^k_{\alpha_k}(t_k)\}$ might be sets of projections onto an exhaustive set of ranges of positions of the earth's center of mass at different times t_k. A coarse-grained individual orbit is a sequence of these ranges at a series of times.

When the C_α act on the initial state they give branch state vectors $C_\alpha|\Psi\rangle$ for each of the alternative histories in the set. The probabilities of the individual histories are

$$p(\alpha) = \|C_\alpha|\Psi\rangle\|^2 . \tag{4}$$

Equation (4) displays an important asymmetry in time. There is $|\Psi\rangle$ on one end of a history and nothing on the other. That is the arrow of time in quantum theory. This is the only fundamental time asymmetry, and all other "arrows of time" must arise from it and particular properties of the $|\Psi\rangle$ of our universe. Similarly this is the only source for the distinction between past and future. The question is not "Where does the distinction between past and future come from?" but rather "How does the distinction between past and future arise from special properties of the initial condition of the universe in conjunction with the quantum mechanical arrow of time?"

However, it is a convention that the state $|\Psi\rangle$ is an *initial* condition at a moment earlier than all others in history. Field theory is CPT invariant, and by making use of a CPT transformation, the time order of the operators and $|\Psi\rangle$ could be inverted so that $|\Psi\rangle$ becomes a final condition. That convention, however, is in conflict with ordinary language and therefore risks confusion. Thus, in this paper, by "past" we mean the times on the side of the present which are closest to the initial condition.

Quantum mechanics can be formulated time-neutrally with initial and final density matrices ρ^i and ρ^f entering symmetrically into the expressions for probabilities.[2] So formulated, quantum mechanics contains no intrinsic direction of time, and all arrows of time arise from asymmetries between the initial and final conditions. It appears that a final condition of indifference, $\rho^f = I/Tr(I)$, and a special initial condition, possibly pure, $\rho^i = |\Psi\rangle\langle\Psi|$, are consistent with known data.[3,4] In that case, the time neutral formulation reduces to the one used here summarized by (4).

Probabilities cannot be consistently assigned to *every* set of alternative histories of a closed system because of quantum mechanical interference. In the two-slit experiment, an electron emitted by an electron gun can pass through either of two slits in a screen on its way to detection at a further screen. It would be inconsistent to assign probabilities to the two histories distinguished by which slit the electron goes through. The probability to arrive at a definite point on the detecting screen would not be the sum of the probabilities to arrive there by going through each of the two different slits because of interference. The sets of histories to which probabilities *can* be assigned, are those which have negligible interference between their individual members, that is

$$\left\langle \Psi \left| C_\alpha^\dagger C_\beta \right| \Psi \right\rangle \approx 0, \quad \alpha \neq \beta. \tag{5}$$

Such sets of histories are said to *decohere* and (5) is a decoherence condition. The probabilities (4) satisfy the additivity properties required of probabilities as a result of (5). As a consequence (4) is a consistent assignment of probabilities for decoherent sets. The sets of alternative histories which can be assigned probabilities are thus determined through (5) by the initial condition of the universe.

As an example of a mechanism of decoherence, think of a millimeter-sized dust grain in a superposition of two positions deep in intergalactic space.[5] Consider alternative histories of the position of this particle at a series of a few times. Were the grain truly isolated, this situation would be analogous to the two-slit experiment and alternative histories of its position would not decohere. However, in our universe, even deep in space, about 10^{11} cosmic blackbody photons scatter from it every second. Through these interactions, states of the seemingly isolated dust grain become strongly correlated with states of the radiation. Two states of position differing by a millimeter become correlated with nearly orthogonal states of the radiation after about a nanosecond. The branch state vectors corresponding to different suitably coarse-grained histories of position thereby become nearly orthogonal, Eq. (5) is satisfied, and decoherence is achieved. Mechanisms such as this are widespread in our universe and cause the habitual decoherence of the kinds of quasiclassical variables we so often find useful to follow. We shall then always assume that we are dealing with a decoherent set of histories.

2.3. *Prediction and Retrodiction*

From known present data we aim to predict the future and retrodict the past. The exact nature of this present data is not important for the subsequent discussion. It could be the memories of an individual observer, the records of a collection of them, or an alternative having nothing to do with observers, such as the present size of the universe. Neither does the data from which predictions and retrodictions are inferred have to be at one moment of time, and indeed typically it is not. However, to simplify the exposition, we shall assume that it is; the generalization to data at several times should be clear.

Probabilities of future alternatives conditioned on present data are relevant for prediction. Probabilities for past alternatives conditioned on the same present data are required for retrodiction. These are easily constructed from the joint probabilities (4) of histories that include present data, as we now show.

Let t_0 be the time of the present and denote the projection corresponding to present data by $P_{pd(t_0)}(t_0)$. The awkward subscript $pd(t)$ is necessary because present data will change in time as new data are acquired and old data are lost. The operator $P_{pd(t_0)}(t)$, evolved from t_0 according to (4), would represent that same alternative as $P_{pd(t_0)}(t_0)$ at time t, but that is not present data at time t.

However, since we only use the operator $P_{pd(t_0)}(t_0)$, we shall often abbreviate it as $P_{pd(t_0)}$ or even P_{pd}.

The conditional probability for a chain of alternatives $(\alpha_n, \ldots, \alpha_1)$ at times $t_n > \cdots > t_1$ all to the future of t_0 is

$$p\,(\alpha_n t_n, \ldots, \alpha_1 t_1 | pd\, t_0) = \frac{p\,(\alpha_n t_n, \ldots, \alpha_1 t_1,\, pd\, t_0)}{p(pd\, t_0)}. \tag{6}$$

Using (4), and assuming the set of histories is decoherent, the joint probabilities in the numerator and denominator may be evaluated from the relevant chains of projections and the state $|\Psi\rangle$. The results are conditional probabilities for prediction.

It is similar for retrodiction. The conditional probabilities for a history of alternatives $(\alpha_1, \ldots, \alpha_n)$ at times $\tau_1 > \cdots > \tau_n$ all *prior* to t_0 are

$$p\,(\alpha_1 \tau_1, \ldots, \alpha_n \tau_n | pd\, t_0) = \frac{p\,(\alpha_1 \tau_1, \ldots, \alpha_n \tau_n, pd\, t_0)}{p(pd\, t_0)}, \tag{7}$$

where the chain relevant for computing the numerator is

$$P_{pd(t_0)}(t_0) P^1_{\alpha_1}(\tau_1) \cdots P^n_{\alpha_n}(\tau_n). \tag{8}$$

Equation (7) gives conditional probabilities for retrodiction.

For example, to predict or retrodict the orbit of the earth about the sun, $P_{pd(t_0)}$ might be the projection on a present record of previous observations of the earth's position to certain accuracies, and $P^k_{\alpha_k}(t_k), k = 1, \ldots, n$ might be projections on a series of future or past positions to given accuracies. For typical accuracies to which the position of the earth's center of mass is determined, we expect there to be a conditional probability near unity for classical orbits consistent with present data.

2.4. *Retrodiction in the Approximate Quantum Mechanics of Measured Subsystems*

The Copenhagen formulation of quantum mechanics is an approximation to the more general quantum mechanics of closed systems. It is appropriate when dealing with measurement situations in which a subsystem is approximately isolated from the rest of the universe and evolves independently except for occasional interactions with a measuring apparatus.[a] In ideal situations, the decoherence of the alternative histories of the registrations by the apparatus may be approximated as exact, and the probabilities for the outcomes of measurements calculated from a formula similar to (4) but *involving the Hilbert space of the measured subsystem alone*. The change in focus from the Hilbert space of the 10^{80} particles in the visible universe to, say, the Hilbert space of the few particles involved in a typical experiment is an approximate simplification of overwhelming practical advantage.

There is an obstacle to using the Copenhagen probabilities to retrodict the past of a subsystem. The output of the Copenhagen approximation are the probabilities of the outcomes of measurements. But, there is not enough information in the present state of a subsystem to fix what past measurements were performed on it and when they were performed. The influence of these measurements on the history of the subsystem's state is generally non-negligible. To retrodict, the information about the nature and times of measurements must either be supplied, or perhaps retrodicted from a separate classical physics governing the measuring apparatus.

In this practical sense it is not possible to retrodict the past from the present data in the Copenhagen formulation of quantum mechanics, except for those situations where evolution is negligibly perturbed by external measurements as can be arranged in classical physics. A similar obstacle exists for prediction. This appears less serious because, as observers, we can control the nature of time of *future* measurements, whereas those in the past seem beyond our reach. True quantum mechanical retrodiction is only possible for a closed system.

3. Quantum Pasts

The reconstruction of the past is most honestly viewed in the context of retrodiction from present records. We retrodict the date 55 BC for the first Roman invasion of Britain on the basis of present textual records. We use present observations of the sun and moon to

[a]See e.g. Ref. 6, Sec. II.10.

reconstruct their past trajectories. We use fossil records to estimate that the probability is high that dinosaurs roamed the earth 250 million years ago. We infer that matter and radiation were in thermal equilibrium at the beginning of the universe from the present values of the Hubble constant, mean mass density, and the temperature of the cosmic background radiation.

We have seen how past history can be reconstructed in quantum mechanics through retrospective probabilities conditioned on present data. We now describe a number of non-classical features of quantum retrodiction.

3.1. *The Present State is not Enough to Retrodict the Past*

The conditional probabilities for *predictions* can be obtained from a present state. Define

$$|\Psi_{\rm present}\rangle = \frac{P_{pd(t_0)}|\Psi\rangle}{\|P_{pd(t_0)}|\Psi\rangle\|}. \tag{9}$$

The probabilities (6) for future alternatives conditioned on present data are easily seen to be expressible as

$$p\,(\alpha_n t_n,\ldots,\alpha_1 t_1|pd\,t_0) = \left\|P^n_{\alpha_n}(t_n)\cdots P^1_{\alpha_1}(t_1)|\Psi_{\rm present}\rangle\right\|^2. \tag{10}$$

The vector (9) is the usual notion of present state of the system. In this Heisenberg picture, it is constant from the time of the initial condition up to the present and then "reduced" by the action of the projection representing present data. This present state contains all the information necessary for predicting the future as (10) shows.

By contrast, no such summary is possible for retrodiction. The probabilities (7) for alternatives conditioned on present data cannot be expressed in terms of the projections and $|\Psi_{\rm present}\rangle$ alone. They require the initial condition $|\Psi\rangle$ in addition to present data $P_{pd(t_0)}$. Put differently, correct probabilities for the past cannot generally be constructed simply by running the Schrödinger equation backwards in time from the present state. A theory of the initial state is required as well. In this respect, the quantum mechanical notion of state at a moment of time is very different from the corresponding classical notion. Almost all retrodictions therefore depend, at least to some degree,[b] on assumptions about the initial state $|\Psi\rangle$. We cannot ignore this dependence. Complete ignorance of the initial state would be represented by an initial density matrix $\rho = I/Tr(I)$ where I is the unit operator. But that is an initial condition of infinite temperature — patently inconsistent with observations today.

3.2. *The Past is not Unique*

We naturally think of the past as defined by alternative prior histories of the usual quasiclassical variables which we are adapted to observe — histories of alternative values of densities of energy, momentum, chemical species, averaged over small volumes, etc. However, in quantum mechanics there are many other mutually incompatible possible pasts.

A simple mathematical example may help to illustrate this point in a trivial way. Consider a set of projections $\{Q_\alpha\}$ on a basis of orthogonal states that include $|\Psi\rangle$. The set of histories

$$P_{pd(t_0)}(t_0)Q_{\alpha_1}(\tau_1)Q_{\alpha_2}(\tau_2)\cdots Q_{\alpha_n}(\tau_n) \tag{11}$$

is decoherent for any assignment of times $\tau_1 > \tau_2 > \cdots > \tau_n$ (all earlier than t_0). The conditional probabilities

$$p\,(\alpha_1\tau_1, \ldots, \alpha_n\tau_n|pd\,t_0) \tag{12}$$

retrodict a past which is nothing like the usual quasiclassical one. Indeed the only retrodiction with nonzero probability is for the history where the Q's are $|\Psi\rangle\langle\Psi|$ repeated at every time.

We can illustrate this using Schrödinger's cat. Restricting attention to the usual thought experiment, we may infer from present data showing the cat to be alive with a conditional probability near unity

---

[b]The trivial exception is the probability of one retrodiction of a history defined by repeating the *operator* P_{pd} over and over at a series of past times. That is a kind of retrodiction following from the determinism of the Schrödinger equation alone. However, even if P_{pd} represents a quasiclassical alternative at the present time, it corresponds to nothing like that at past times. Rather, operators representing quasiclassical alternatives in the past are related to quasiclassical alternatives in the present by the Heisenberg equations of motion, Eq. (2).

that the atom did not decay in the past. However, the conditional probability is also unity for the past alternative $|\phi\rangle\langle\phi|$, where $|\phi\rangle$ is the state which is a superposition of states in which the cat is alive and dead.

Indeed, an arbitrary number of possible pasts can be constructed mathematically as follows: Take the vectors $P_{pd(t_0)}|\Psi\rangle$ and $\bar{P}_{pd}(t_0)|\Psi\rangle$ where $\bar{P}_{pd(t_0)} = I - P_{pd(t_0)}$ ("not the present data"). Find a set of mutually orthogonal vectors such that half of them add to $P_{pd(t_0)}|\Psi\rangle$ and the remaining ones to $P_{pd\,(t_0)}|\Psi\rangle$. There are a great many ways of doing this. Pick a time τ_1 prior to t_0 and write these vectors as

$$P_{pd(t_0)}\, P^1_{\alpha_1}(\tau_1)\,|\Psi\rangle, \quad \bar{P}_{pd(t_0)}\, P^1_{\beta_1}(\tau_1)\,|\Psi\rangle \tag{13}$$

where α_1 and β_1 range over half the set of vectors. This can always be done because the projections $P^1_{\alpha_1}(\tau_1)$ can always be found, and, in the Heisenberg picture, a projection operator projects on a range of some quantity at any time. Repeat the process to add further projections and thereby construct a decoherent set of histories of prior alternatives that includes present data. That is a past in quantum mechanics. There are a very large number of ways of interpolating such projection operators and therefore a large number of quantum pasts.[c]

Many of these possible quantum pasts are mutually incompatible. That is, two decoherent sets of histories needed in retrodiction are not generally coarse-grainings of a finer-grained decoherent set of histories containing them both. The two pasts in the Schrödinger cat example above are instances.[d] One can draw past inferences using one set or the other set, but not both.[1,11]

[c]Other references discussing the implications of the multiplicity of quantum pasts are Refs. 7 and 8.

[d]Even two alternatives represented by orthogonal projections can both be retrodicted with probability one in different quantum pasts for certain present data and initial and final conditions.[9] Were two such alternatives to occur in the same past, they would be exclusive — if one has probability one, the other necessarily has probability zero. No logical inconsistency arises from two orthogonal alternatives having probability one in *different* pasts, because, with the relevant initial and final conditions and present data, there are no decoherent sets of histories containing both orthogonal projections from whose probabilities a contradiction could be drawn.[10]

The description of multiple quantum pasts requires care in the use of ordinary language which is adapted to a single quasiclassical past. For example, we are accustomed to say that an event "happened" in the past if its probability conditioned on present data is near unity. That is a statement which requires the history, consisting of present data and the past alternative, to be a member of *some* decoherent set of histories, but it is independent of *which* decoherent set, because the value of the probability of a history is the same in all decoherent sets of which it is a member; see Eq. (4). If we retain this meaning of "happened" in quantum theory, we must accept that mutually incompatible events can happen in different quantum pasts.[e]

What are we to make of these different quantum pasts? It is important to stress that, in general, we do not expect these to be constructed from operators representing anything like the alternatives of the usual quasiclassical realm. Further, in general, we must expect these to be pasts with distributed probabilities — there is no one past history with a probability near one on the basis of present data and the initial condition. The possible pasts can thus be expected to differ greatly in their utility. We shall return to this below.

3.3. *A Past is not Necessarily Permanent*

Moving into the future presents data changes. Individually, we make more observations and acquire more information over time. We also forget or lose access to much. Collectively, we expect information to increase, although much is erased or lost, such as Aristotle's *Comedy*.

Thus, in general, we expect $P_{pd(t)}$ to become increasingly fine-grained with t. The decoherence of any particular quantum past is thus at risk. A coarse-graining of a decoherent set is again decoherent, but a fine-graining is not necessarily decoherent. If a set of past histories no longer decoheres in the presence of new data, we lose the ability to make these retrodictions. It is not that the past alternatives become less certain; there are no probabilities at all. A past is therefore not necessarily permanent in quantum mechanics.

[e]A prescription for avoiding ambiguities of ordinary language is to turn all statements into statements about quantum mechanical probabilities which are the output of quantum mechanics.

This loss of the past occurs even in familiar laboratory situations. Consider a Stern–Gerlach thought experiment in which a beam of atoms, each initially in a superposition of spin states, is separated into two beams by passing through an inhomogeneous magnetic field, and later these beams are recombined. Suppose the atoms are in narrow wave packets so that we can meaningfully speak of the time a single atom is at a particular position along the beam. When the atom is in the region of separated beams, the set of histories defined by alternative spin states decoheres. (Or indeed at other times, since histories consisting of alternatives at a single time always decohere.) However, a history which includes, in addition to these alternatives, alternative values of the spin at a time when the atom is in the region where the beams have been recombined, will not decohere. The phases between the histories have been recovered in that present data by recombining the beams.

As we shall now describe, for quasiclassical pasts, it is genuinely or practically impossible to find present data for which alternative past histories of values of the usual quasiclassical variables fail to decohere. As long as we stick to such alternatives we may proceed into the future, secure in the knowledge that the quasiclassical past is permanent.[f]

4. The Quasiclassical Past

Classical physics is an approximation to quantum mechanics that is appropriate for particular coarse-grainings and particular initial conditions such as the one which characterizes our universe. The classical past is unique, permanent, and retrodictable from present data alone — features which are not general in quantum theory as we have discussed. We now briefly describe how these features are recovered in the classical approximation to quantum theory.

The familiar variables of classical physics include averages over suitable volumes of densities of approximately conserved quantities such as energy, momentum, chemical species, etc. When the volumes

[f]The decoherence condition (5) can be strengthened to ensure the permanence of the past. An example is the notion of *strong decoherence.*[12]

are sufficiently large, sets of alternative histories of such variables can decohere, and exhibit approximate patterns of classical, and deterministic correlations over time. The success of classical physics in describing phenomena over wide ranges of time, place and epoch, suggests that the initial condition and Hamiltonian of our universe exhibit a *quasiclassical realm* — a decohering set of alternative histories of quasiclassical variables that is maximally refined, and consistent with the requirements of decoherence and approximate classical predictability. Such a quasiclassical realm would extend over most of space and much of the history of the universe, and is a feature of the universe not of our choice. As observers in the universe, we utilize coarse-grainings of this usual quasiclassical realm to describe everyday experiences and particular experimental situations.

To restrict to past histories of quasiclassical variables is to beg the question of the uniqueness of the past. However, there is still a sense in which we may speak of a unique quasiclassical past — the past of the usual quasiclassical realm. Different observers making use of usual quasiclassical variables for their particular present data and retrodictions construct pasts which are mutually compatible if they are all employing coarse-grainings of the usual quasiclassical realm.

It is an open question whether the universe exhibits essentially distinct quasiclassical realms of high predictability characterized by variables that are different from the usual quasiclassical ones, or not.[13] If, however, the usual quasiclassical realm is essentially alone in its high level of predictability, that would be another sense in which the usual quasiclassical past is unique. In the following, to avoid cumbersome expressions, when we refer to "the quasiclassical realm" or "quasiclassical variables", we mean the usual quasiclassical realm and the usual quasiclassical variables, unless otherwise noted.

In a restricted class of cases, the equations of motion summarizing the temporal regularities of the quasiclassical realm enable coarse-grainings of the quasiclassical past to be inferred from present data alone without further knowledge of the initial condition. The history of the solar system, for instance, can be inferred from the present positions and velocities of its constituents. However, this kind of retrodiction from present data is possible only in certain circumstances. The effective deterministic equations of motion of the usual quasiclassical realm are generally dissipative, chaotic, and dependent on the past history of the system.[g] Dissipation and chaos generally mean that past evolution is unstable. Dependence on past history means that present data is not enough to retrodict. Further, thermal and quantum noise arising from coarse-graining can cause deviations from classical predictability, both for prediction and retrodiction. Only when the effects of dissipation, chaos, history dependence, and noise are sufficiently small, can we use equations of motion to retrodict the usual quasiclassical past from present data alone. Unlike classical physics where coarse-grainings and effective equations arise only out of ignorance, coarse-graining is inevitable in quantum mechanics. Coarse-graining is necessary for decoherence, and for the quasiclassical realm, coarse-graining beyond that is necessary to achieve the approximate predictability in the face of noise that typical mechanisms of decoherence produce.[14]

It is impossible or hopelessly impractical to recover the phase information in present data that would lead to a failure of decoherence of the usual quasiclassical past. We mentioned that, for an intergalactic dust grain initially in a superposition of positions, alternative histories of subsequent position decohered because of their interaction with cosmic background photons. If later data include the states of all scattered photons, then the past histories of position of the dust grain cannot be retrodicted from it — the relevant set of histories of position will not decohere. Of course, it is hopelessly impractical to recover such data. Some 10^{11} photons scatter off the dust grain every second. Indeed, if we did not prepare ahead of time to recover the information, it is too late in principle. It left earlier at the speed of light. If any of those photons went down a black hole, they are similarly irretrievably lost.

In science fiction, we might imagine that even now intelligent aliens could be poised to recover the phases between past histories that we retrodict from present data. From this data they will not be

[g]They are not dependent on future history; classical causality follows from quantum mechanical causality.[14]

able to retrodict the same past as we do. We, however, are unlikely to survive the process of collecting this data to compare notes.

In the quasiclassical past, one can say with the poet, "The Moving Finger writes; and having writ, Moves on: nor all thy Piety nor Wit Shall lure it back to cancel half a Line, Nor all Thy tears wash out a Word of it." That is not necessarily the case for all quantum mechanical pasts.

5. Why are there More Records of the Past than the Future?

In the quasiclassical realm, present data contain many more records of the past than of the future. In this paper a "record" is not presupposed to be of the past. A record is an an alternative in the present that is correlated with high probability, with another alternative *in the past or future*. That is, present data contains a record of an alternative α_0 if $p(\alpha_0, t|pd, t_0) \approx 1$ irrespective of the relation between t and t_0. A recorded alternative is thus either a near certain prediction or retrodiction from present data. For example, the arrangement of ink reading as 55 BC in many texts (and 55 BCE in others) is correlated with the Roman invasion of Britain. Certain configurations of neurons are correlated with our past experiences. We can also have records of the future. A table of the eclipses of the moon in AD2010 is just as much a record of that time as a table of eclipses in AD1000 is of its time. The present position and velocity of the earth in its orbit around the sun are a record of its future positions as well as of its past ones. However, the preponderance of records seem to be of the past. That time asymmetry, like all others, can only be explained by the particular features of the initial condition $|\Psi\rangle$. In this section, we shall attempt to identify these features. We confine our attention entirely to the usual quasiclassical realm.

The second law of thermodynamics is the reason that present data contains more records of the past than the future. The Jaynes construction may be used to associate an entropy with every decoherent set of alternatives in quantum mechanics.[h] For example, the entropy

[h]If that is not familiar, see e.g. Ref. 15. By extending the original Jaynes discussion, an entropy can be defined for sets of histories.[16]

of the coarse-grained set of histories consisting of alternatives $\{P_\alpha(t)\}$ at a single moment of time is:

$$S(\{\alpha\},t) = -\sum_\alpha p(\alpha,t)\log p(\alpha,t) + \sum_\alpha p(\alpha,t)\log Tr\,[P_\alpha(t)] \tag{14}$$

where $p(\alpha,t) = ||P_\alpha(t)|\psi\rangle||^2$ are the probabilities determined from the initial condition. Now follow this entropy with time, keeping the coarse-graining fixed. When the P's are projections on ranges of *quasiclassical variables*, this entropy is low initially for the initial condition of our universe and therefore has a general tendency to increase afterwards.[i] That is the second law of thermodynamics and the origin of the thermodynamic arrow of time. As Boltzmann has put it,[19] "The second law of thermodynamics can be proved from the [time-reversible] mechanical theory if one assumes that the present state of the universe...started to evolve from an improbable state."

A similar second law behavior is expected for the entropy of a set of future alternatives $\{P_\alpha(t)\}$ conditioned on present data. This is

---

[i]While low, the entropy is not as low as it could be. At the time of decoupling, matter and radiation were in near-thermal equilibrium, with an entropy of about $10^{80}k$ inside the region visible from today. By contrast, the geometry of the early universe is as ordered as it could be. As Penrose has stressed,[17] "gravitational entropy" is small compared to the maximal value of $10^{120}k$ it could have if all that matter were in a black hole. The second law is a consequence of this initial geometrical order and the attractive nature of gravity.

For the entropy of the initial state to be near zero with respect to quasiclassical coarse-grainings, it would have to predict a single configuration of quasiclassical variables with probability near unity [cf. Eq. (14)]. The subsequent approximate deterministic evolution of that configuration means that it must necessarily encode all the complexities of the quasiclassical realm we see about us. Such a near zero entropy initial condition would therefore be incomprehensibly complex. Rather, the initial condition of our universe seems to be as *simple* as possible in a crude, intuitive sense: Matter in near-thermal equilibrium, which is maximum entropy for it, and highly ordered geometry, which is minimal entropy for it because of the attractive nature of gravity. Quantum mechanical initial conditions which incorporate this simplicity (e.g. Ref. 18) predict distributed probabilities for histories of the quasiclassical realm, which is why there are few initial records of the future.

given analogously to (14) by

$$S(\{\alpha\}, t|pd, t_0) = -\sum_\alpha p(\alpha, t|pd, t_0) \log p(\alpha, t|pd, t_0) + \sum_\alpha p(\alpha, t|pd, t_0) \log Tr[P_\alpha(t)] \qquad (15)$$

where the conditional probabilities $p(\alpha, t|pd, t_0)$ are constructed as described in Sec. 2.3. For a generic quasiclassical coarse-graining, this conditional entropy should increase towards the future and decrease towards the past.

If present data contains a record of a particular alternative α_0, then the conditional probability for it is unity and the conditional probability is zero for all other alternatives in any decoherent set containing it. The entropy of such a set is then

$$S(\{\alpha\}, t|pd, t_0) = \log Tr[P_{\alpha_0}(t)]. \qquad (16)$$

This is *independent of time* as a consequence of the unitary evolution of the P's [(2)] and the cyclic property of the trace. That is consistent with the second law. The second law thus does not rule out in particular near certain predictions or retrodictions. For example, the entropy of a set of alternatives describing the positions of the earth and the moon to the accuracies current in celestial mechanics increases hardly at all in the near term. That is because their motion is essentially deterministic. A table of future eclipses is an example of a future record of such alternatives.[j]

The question, however, is not whether individual records are consistent with the second law, but whether present data contains more

---

[j]Note that the effort to compute the table does not enter the argument, nor should it. The present records of positions of the earth–moon system are correlated with their future positions whether or not these records are the inputs to explicit computation. From the point of view of entropy, computation is tantamount to copying and does not necessarily increase the entropy[20] unless outputs of a reversible computation are erased in accord with Landauer's principle.[21] Computation is typically required in the process of prediction to generate *useful* records of future alternatives. In the computation of a table of eclipses, an account of complex astronomical observations is converted to a more useful table of dates in the amended Gregorian calendar, both equally being records of future astronomical events.

near-certain retrodictions than near certain predictions. The second law implies that it does. Pick a set of quasiclassical alternatives $\{P_\alpha(t)\}$. Their entropy is generically greater at times after the time of present data than it is for times before. Therefore, generically the probabilities $p(\alpha, t)$ are more distributed (less concentrated on one certain alternative) when t is to the future of the time of present data than when it is to the past of it. That means a given quasiclassical coarse-graining is more likely to contain a near certain retrodiction of the past than a near certain prediction of the future.

This is in accord with the intuitive understanding that many records originate in irreversible processes. An impact crater on the moon, an ancient fission track in mica, or the distribution of ink on this page are all examples of records in which the entropy rose significantly when they were created. Records created by irreversible processes must necessarily be records of the past to be consistent with the second law.

6. The Utility of History

We reconstruct the past to help predict the future. Certain regularities in the universe that can be inferred from present data can be understood as arising from historical events. This understanding is useful for predicting future regularities and that is the utility of history. In this section we shall amplify on this theme. We shall briefly discuss the nature of regularities, the process of historical explanation, quantitative measures of explanation, the role of theoretical input to that process, why explanations are mostly in the past, and the implications of many quantum pasts for historical explanation in quantum mechanics.

6.1. *The Origin of Regularities*

Science is concerned with identifying regularities in the universe and exploiting these in the process of prediction. The laws which govern the regularities exhibited by all physical systems — without exception, without qualification, and without approximation — are the fundamental laws of physics. The present theoretical viewpoint is that there are two fundamental laws — a unified theory of the dynamical interactions, perhaps heterotic superstring theory or a generalization of it, and a theory of the initial quantum state of the universe.

Beyond the universal laws, science, including physics, aims at identifying and exploiting the regularities of specific physical systems. To give just a few examples: physics is concerned with the regularities of atoms, stars, and fluid flows; chemistry with the regularities of specific molecules; geology with the regularities of a specific planet; and the social sciences with the regularities exhibited by human beings — both collectively and individually. Specific classes of systems exhibit many more regularities than those implied by the fundamental laws of physics. What is the origin of the regularities peculiar to specific subsystems of the universe? Evidently they do not arise from the fundamental laws, as these govern only the regularities exhibited by all systems.

Every prediction in science can be regarded as a conditional probability in quantum cosmology. The regularities of specific systems can therefore only arise from the conditions entering the probabilities for these regularities. In some cases, merely specifying the system is enough. To predict the emission spectrum of an isolated atom, it is sufficient to specify that the atom is isolated in a certain excited state. For many other systems, however, the regularities arise from processes and events that have occurred over the course of the system's history[k] — some deterministic, some accidental. That is certainly the case for the regularities exhibited by the geology of the earth or by particular biological species that are the products of billions of years of chance mutation and natural selection. Even physics is concerned with regularities that arise from historical events. The large scale distribution of galaxies in the visible universe is most succinctly understood as the outcome of some 15 billion years of evolution of primordial fluctuations by the action of gravitational attraction. This historical understanding from present data can be used to predict regularities in more detailed data yet to be obtained. For

[k]See, e.g. Ref. 22 for a more detailed discussion.

example, by reconstructing the past history of the elements, we can predict the abundances of meteorites and stars yet to be observed. In the following, we describe this process of historical explanation in quantum mechanics.

6.2. *Historical Explanation*

From present data that include most texts giving 55 BC as the date of the Roman invasion of Britain, we may infer that Caesar did invade Britain in 55 BC. That event, together with inferences from present data on the validity of the texts, is a *historical explanation* of the regularity in present data that most discussions of the Roman invasion give 55 BC as its date. From this explanation, one can predict that texts yet to be discovered will also give the date 55 BC when describing the Roman invasion of Britain. (Of course, both inference and prediction are probabilistic; there is a probability that individual texts are forgeries, or contain mistakes, or that the ink on their pages made a quantum mechanical transition from a configuration spelling of a different date.) The present coherence among texts on the date of the Roman invasion is an example of a "frozen accident" in the evocative terminology of Gell-Mann[22] — "chance events of which the particular outcomes have a multiplicity of long-term consequences all related by their common ancestry".

In principle, the same prediction could be made from the present data itself — from the physical description of the texts and the configuration of ink molecules on their pages. However, it is evidently much easier to start from the event in the past. The reason is that present data contain much information that is irrelevant for this particular future prediction. It contains the specifications of the texts — their location, content, typeface language, number of pages, authors, etc. — with many details to which the prediction of 55 BC in future texts is not very sensitive. Of course, all that data is relevant for other predictions, those referring to the subsequent evolution of the texts for instance. But the reconstruction of past history permits the focus on the *essential* features of the regularity that can be inferred from present data for the prediction of the character of future texts.

The above discussion can be given both more generally and more concretely. Let $P_{pd(t_0)}$ be the projection on present data held at time t_0, and let $\{P_\alpha(t_0)\}, \alpha = 1, 2, \ldots$ be a set of alternatives in which $P_{\alpha_0}(t_0)$ is a coarser projection than $P_{pd(t_0)}(t_0)$ in describing a feature, say a regularity, in present data. Histories

$$C_{\beta_0} = P^1_{\beta_{1,0}}(\tau_1) \cdots P^n_{\beta_{n,0}}(\tau_n) \tag{17}$$

are inferences from present data if

$$p(\beta_0|pd) \approx 1. \tag{18}$$

(In this section we will suppress the time labels of alternatives for compactness.) If among the possible inferences from present data, we can find one β_0 for which

$$p(\alpha_0|\beta_0) \approx 1, \tag{19}$$

then we say that β_0 is an explanation of α_0.

In the history of the Roman invasion of Britain, *pd* contain all the details of present texts, α_0 is the coarse-grained alternative that most texts report 55 BC for the invasion, and β_0 is the chain of events in 55 BC describing the invasion. In explaining the primordial abundance of the elements, *pd* includes the details of telescopic observations of the spectra from old, metal-poor stars, α_0 is the alternative that all these show approximately 75% H and 25% He, and β_0 is the history of big bang nucleosynthesis in the early universe.

To consider probabilities such as (18) and (19) at all, the alternatives β_0 and α_0 must be members of decohering sets of histories. As throughout this section, we assume the decoherence of the relevant sets of histories. We generally do not need to mention the sets because the probabilities of specific alternatives such as β_0 and α_0 are the same in all decoherent sets, being given by (4).

There will generally be many possible explanations of a given alternative α_0. To give a trivial example, merely repeating present data or evolving it by the Schrödinger equation (2) to a different time would give a β_0 satisfying (18) and (19). However, it is not an explanation in general which is of interest, but rather an explanation which is simpler than the present data.

Quantitative measures of the simplicity of explanations depend on the context that is assumed in comparing them. The simplest measure is the information content (AIC) of the histories β_0 satisfying (18) and (19). The smaller the AIC, the simpler the explanation. The notion of AIC was defined some thirty years ago by Kolmogorov, Chaitin, and Solomonoff[23] (all working independently) and has been employed in the definition of augmented thermodynamic entropy,[24] the effective complexity of physical systems,[25] and measures of classicality.[12] For a string of bits s and a particular universal computer U, the AIC of s, written $K_U(s)$, is the length of the shortest program that will cause U to print the string and then halt. A history like β_0, (Eq. (17)), may be described by giving a description of its string of projection operators and their times, which entails describing the operator whose eigenvalues are projected upon in terms of fundamental fields, and specifying the range of these eigenvalues. If s is a description of such a history, then the AIC of the string can be regarded as the AIC of the history. There will generally be many physically equivalent descriptions of a history in terms of fields.[13] For example, physically equivalent explanations are obtained by choosing different times for the operators in (17) and describing the operators by fields at that chosen time. However, the AIC's of these equivalent histories may differ greatly. We therefore take the AIC of the history, $K_U(\gamma)$ to be the minimum over all such descriptions.[1] The AIC, $K_U(\beta_0)$, of an explanation β_0, then provides a quantitative way of distinguishing between different explanations by their relative simplicities. The β_0 with the smallest AIC is the simplest explanation. Since α_0 trivially satisfies (18) and (19), it is an explanation of itself, and the simplest explanation of α_0 will have an AIC lower than that of α_0 itself.

Not every explanation, simple or not, is useful for prediction. The probabilities $p(\gamma|pd)$ of an arbitrary set of alternatives $\{P^1_\gamma(t_1)\}$ need be neither more nor less distributed than the probabilities $p(\gamma|\beta_0)$ conditioned on an explanation of part of that data. However, if the explanation β_0 captures the essential features of present data

[1]More accurately, the AIC is defined on the physical equivalence classes of descriptions of α_0 and β_0.[13]

that are relevant for the alternatives $\{P^1_\gamma(t_1)\}$, then their prediction may be simplified. That will be the case when the probabilities of the $\{P^1_\gamma(t_1)\}$ are independent of the part of the present data not explained by β_0 in the sense that

$$p(\gamma|pd, \beta_0) = p(\gamma|\beta_0). \tag{20}$$

For such alternatives, the predictions from β_0 agree with those from present data

$$p(\gamma|pd) = p(\gamma|\beta_0), \tag{21}$$

because of

$$p(\gamma|pd) = \Sigma_\beta p(\gamma|pd, \beta)p(\beta|pd) \tag{22}$$

and Eqs. (18) and (20). When the explanation β_0 is a simpler input to the prediction of the $\{P^1_\gamma(t_1)\}$ than all of the present data, the prediction of the future is simplified by historical explanation.

The above discussion does not assume that explanations must always be in the past, and indeed they need not be. The present position and velocity of the earth in its orbit around the sun might just as well be explained by some future position and velocity towards which it is heading as by any past position and velocity from which it came. However, the second law argument of Sec. 3 implies that many explanations will be found in the past. Equation (18) shows that present data contain a record of the explanation β_0, and most, but not all, records are of the past because of the second law. Equation (19) shows that β_0 is a record of the alternative α_0 to its *future*. However, α_0 is typically only a small part of the present data — a coarse-graining of it. It is more consistent with the general tendency of entropy to increase, to have the records arranged in this way rather than the opposite time order. That is why the past is the most promising place to look for explanations.

There are no past explanations without theoretical input beyond present data. The explanation of the primordial abundance of the elements by big bang nucleosynthesis entails a number of assumptions about the nature of the early universe — its approximate homogeneity and isotropy, for instance. Those theoretical assumptions may be most properly understood as hypotheses about the universe's initial condition. Indeed, we know from the discussion in Sec. 3.1 that a theory of the initial condition is required for any retrodiction from present data in quantum mechanics.

As we have mentioned earlier, there will generally be many explanations — many β_0 satisfying Eqs. (18) and (19) — of a given part of present data. These will differ in their simplicity and their utility for future prediction. There is no requirement that these different explanations must be mutually compatible. They could belong to different quantum pasts in the sense of Sec. 3.2. With its many quantum pasts, the quantum theory of closed systems has more possibilities for historical explanation than it would have if the theory were somehow restricted to a single decoherent set of histories. Mere decoherence in the presence of present data and the initial condition are all that are required to have a set of histories defining a quantum past. But to be useful for historical explanation, the probabilities for records and explanations in (18) and (19) must be near unity. The approximate determinism of the usual quasiclassical realm makes it more likely that these conditions will be satisfied in the usual quasiclassical past for usual quasiclassical present data than in other possible pasts. That is why the usual quasiclassical past is the most promising place to look for historical explanation. But we should not ignore the possibility of explanation through many other pasts presented to us by quantum mechanics.

7. What About Us?

7.1. *Present, Past and Future*

"Observers" or "measurements" have no fundamental role in the quantum mechanics of closed systems. Of course, "observers" can be described as special physical systems (typically complex ones) within the universe, and "measurements" can be described as special interactions between subsystems of the universe. "Observers" and "measurements" are of special interest to us because that's what we are and that's how we learn about the universe. How then do information gathering and utilizing systems (IGUSes) such as ourselves distinguish the present moment, its future, and its past? The author is not equipped to speculate on the detailed operation of the human brain. However, it seems likely that mechanical systems such as computers could be constructed and programmed to draw similar distinctions between past, present and future. We proceed by discussing how such a system could operate. We restrict attention to IGUSes whose description and observations are coarse-grainings of the usual quasiclassical realm. The discussion in this section is necessarily more suppositional than that in the rest of the paper, and the reader should take note of that. Nevertheless, it concerns questions that can be seriously addressed in science, even if now only by conjecture and not calculation, as here.

An IGUS holds a set of records. As time changes (in either direction!) this set of records also changes. New records appear and others disappear or are altered. The records can therefore be approximately time ordered. How accurate the records are at any time, how long and faithfully they persist, and how accurately they can be time ordered depends on the particular IGUS. Everyday experience shows considerable variation in these qualities from one IGUS to another.

If the records of an IGUS are examined at any one time, we expect some to be correlated with external events near that time. That is the present input, and these records constitute the IGUS's notion of the present. To the extent that the IGUS's records are acquired in irreversible processes, we expect most of the remaining records of the *past* to be consequences of the second law as discussed in Sec. 3. That is why we say we remember the past and not the future.[m] However, an IGUS is not prohibited from possessing a record of the future. It might have memorized a table of future eclipses. It might have correctly conjectured the motion of a fly or the movement of tomorrow's stock market. An IGUS must predict to function, which means creating useful approximate records of the future. As the above examples illustrate, such records of the future are typically distinguishable from records of the past as the outcomes of *computation* involving other records as input, mostly records of the past.

---

[m]The author is not alone in connecting the "psychological arrow of time" to the thermodynamic arrow of time, see e.g. Refs. 17 and 26.

That is how we can say that we predict the future but remember the past.

Thus, like all other time asymmetries, the subjective distinction between past, present and future is traceable to properties of the universe's initial condition — in this case its low entropy in terms of usual quasiclassical variables. All other aspects of the phenomena loosely called the "psychological arrow of time" presumably have similar explanations. For example, we have the impression that we can control the future, but the past is over and done with. What is meant by "control" is the amplification of a pattern of weak electrical activity in the brain to large irreversible effects correlated with that pattern. The second law means that such effects are much more likely to be in the future of the pattern of activity than to its past.

7.2. *Which Past Do We Remember?*

Which of the many possible quantum pasts do we remember? As IGUSes, we are adapted to observe and record alternative values of quasiclassical operators. It is no surprise that we individually remember quasiclassical pasts if our records are almost entirely of quasiclassical alternatives. Further, these individual quasiclassical pasts are all coarse-grainings of the usual quasiclassical realm, which is why they are in approximate agreement. However, at a deeper level, we may ask: "What is the reason for our focus on quasiclassical alternatives?" Our particular properties as physical systems in the universe can only be understood in the context of our evolution within it. The usual quasiclassical realm of decoherent histories has a high level of predictability in time. Such properties make it plausible that IGUSes described in terms of usual quasiclassical variables evolved to make use of coarse-grainings of the usual quasiclassical realm because its relative predictability makes it adaptive to do so in comparison with more quantum mechanical alternatives that they might have used. Probabilities relevant for that adaptation are in principle computable from quantum mechanics, although well beyond our ability to do so at present.[n]

[n]For further discussion, see Ref. 13.

7.3. *The Marvelous Moment "Now"*

To function, an IGUS employs a self-referential, evolving model or schemata of itself and its external environment.[22] Such a model necessarily distinguishes records of present events from ones in the future and past as described above and, in the case of human beings, devotes a significant amount of conscious focus to records of the most recent input. The author knows of no physical reason why a device could not be constructed which gives equal conscious focus to records of an hour ago as well as of the present. Plausibly it has been adaptive to concentrate on the present and near past. However, our preponderance of attention to the present does not mean that "now" is somehow distinguished in physics. Neither is it evident for the more radical but problematic suggestion that physics can be formulated entirely at one marvelous moment "now".[27] To give just one reason, the moment "now" is only approximately defined to the accuracy of some processing time in the brain, and certainly not to the accuracy of Planck time, 10^{-43} s, on which spacetime can be defined.[o] "Now" is a moment in a schemata of an IGUS, not a preferred surface in the four-dimensional spacetime continuum.

8. Conclusion

The quantum mechanical past is a country of very different appearance from the past of classical physics. Present data, no matter how refined, are not enough for any retrodictions without being augmented by a theory of the initial condition. Even then, there are many mutually incompatible pasts, each a possible source of historical explanation. As new data is acquired, none of these possible quantum pasts need continue to be retrodictions — a quantum past is not necessarily permanent. The familiar properties of uniqueness, permanence, and retrodictibility from present data alone, that are characteristics of the classical past emerge in quantum theory, not generally, but as approximations appropriate to particular coarse-grainings and particular initial conditions such as that for our universe.

[o]For more reasons see, e.g. Ref. 6, Sec. V.1.2.

The multiplicity of mutually incompatible sets of alternative decoherent histories that can be employed for prediction and retrodiction in the quantum mechanics of closed systems has motivated some to search for additional principles that would restrict the available sets in some way[28,29] and thereby necessarily recover properties of the quasiclassical past that are not available generally in quantum theory. Sum-over-histories formulations,[30] strong decoherence,[13] and ordered consistency[31] can all be seen as efforts in this direction. With such restrictions, the properties of a quantum past could be different and perhaps closer to those of classical physics. It is important to stress, however, that, even without such additional principles, the quantum theory of closed systems is logically consistent with experiment as far as is known, and applicable for prediction and retrodiction to the most general physical systems.

Acknowledgments

Thanks are due to T. Brun, R. Griffiths, J. Halliwell, D. Page and W. Unruh for critical readings of the manuscript. The author has benefited from discussions with Murray Gell-Mann on many of the subjects considered in this paper over a long period of time. This work was supported in part by NSF grants PHY95-07065 and PHY94-07194.

References

1. For a survey see, R. Omnès, *Interpretation of Quantum Mechanics* (Princeton University Press, Princeton, 1994).
2. Y. Aharonov, P. Bergmann and J. Lebovitz, *Phys. Rev. B* **134**, 1410 (1964); R. B. Griffiths, *J. Stat. Phys.* **36**, 219 (1984); M. Gell-Mann and J. B. Hartle, Time symmetry and asymmetry in quantum mechanics and quantum cosmology, in *Proc. NATO Workshop on the Physical Origins of Time Asymmetry*, Mazagón, Spain, September 30–October 4, 1991, eds. J. Halliwell, J. Pérez-Mercader and W. Zurek (Cambridge University Press, Cambridge, 1993), gr-qc/9309012.
3. R. Laflamme, *Class. Quant. Grav.* **10**, L79 (1993), gr-qc/9301005.

4. D. Craig, *Ann. Phys. (NY)* **251**, 384 (1995), gr-qc/9704031.
5. E. Joos and H. D. Zeh, *Zeit. Phys.* **B59**, 223 (1985).
6. J. B. Hartle, The quantum mechanics of cosmology, in *Quantum Cosmology and Baby Universes: Proc. 1989 Jerusalem Winter School for Theoretical Physics*, eds. S. Coleman, J. B. Hartle, T. Piran and S. Weinberg (World Scientific, Singapore, 1991), pp. 65–157.
7. B. d'Espagnat, *Phys. Lett. A* **124**, 204 (1997); and the reply in R. Griffiths, *Found. Phys.* **23**, 1601 (1993).
8. H. F. Dowker and A. Kent, *J. Stat. Phys.* **82**, 1574 (1996), gr-qc/9412067.
9. A. Kent, *Phys. Rev. Lett.* **78**, 2874 (1997), gr-qc/9604012; G. Peruzzi and A. Rimini, quant-ph/9710003.
10. R. B. Griffiths and J. B. Hartle, *Comment on "Consistent Sets Yield Contrary Inferences in Quantum Theory"*, gr-qc/9710025.
11. R. B. Griffiths, *Phys. Rev. A* **54**, 2759 (1996), quant-ph/9606004.
12. M. Gell-Mann and J. B. Hartle, Strong decoherence, to be published in the *Proc. 4th Drexel Symp. Quantum Non-Integrability — The Quantum-Classical Correspondence*, Drexel University, September 8–11, 1994, eds. D.-H. Feng and B.-L. Hu (International Press, Boston/Hong-Kong), gr-qc/9509054.
13. M. Gell-Mann and J. B. Hartle, *Equivalent Sets of Histories and Multiple Quasiclassical Domains*, gr-qc/9404013.
14. M. Gell-Mann and J. B. Hartle, *Phys. Rev. D* **47**, 3345 (1993), gr-qc/9210010. An abbreviated account of this paper is given in: J. B. Hartle, Quasiclassical domains in a quantum universe, in *Proc. Cornelius Lanczos Int. Centenary Conf.*, North Carolina State University, December 1992, eds. J. D. Brown, M. T. Chu, D. C. Ellison and R. J. Plemmons (SIAM, Philadelphia, 1994), gr-qc/9404017.
15. R. D. Rosenkrantz, *Papers on Probability Statistics and Statistical Mechanics*, ed. E. T. Jaynes (D. Reidel, Dordrecht, 1983).
16. M. Gell-Mann and J. B. Hartle, Quantum mechanics in the light of quantum cosmology, in *Complexity, Entropy, and the Physics of Information*, SFI Studies in the Sciences of Complexity, Vol. VIII, ed. W. Zurek (Addison Wesley, Reading, MA) or in

Proceedings of the 3rd International Symposium on the Foundations of Quantum Mechanics in the Light of New Technology, eds. S. Kobayashi, H. Ezawa, Y. Murayama and S. Nomura (Physical Society of Japan, Tokyo, 1990).
17. R. Penrose, Singularities and time asymmetry, in *General Relativity: An Einstein Centenary Survey*, eds. S. W. Hawking and W. Israel (Cambridge University Press, Cambridge, 1979).
18. J. B. Hartle and S. W. Hawking, *Phys. Rev. D* **28**, 2960 (1983).
19. L. Boltzmann, *Ann. Physik* **60**, 392 (1897), as translated in S. G. Brush, *Kinetic Theory* (Pergamon Press, New York, 1965).
20. C. H. Bennett, *IBM J. Res. Dev.* **17**, 525 (1973); and *Int. J. Theor. Phys.* **21**, 905–940 (1982).
21. R. Landauer, *IBM J. Res. Dev.* **5**, 183 (1961); and *Nature* **355**, 779 (1988).
22. M. Gell-Mann, *The Quark and the Jaguar* (W. Freeman, San Francisco, 1994).
23. M. Li and P. Vitányi, *Introduction to Kolmogorov Complexity and Its Applications* (Springer, New York, 1993).
24. W. Zurek, *Phys. Rev. A* **40**, 4731 (1989).
25. M. Gell-Mann and S. Lloyd, *Complexity* **2**, 44 (1996).
26. S. W. Hawking, *New Scientist* **115**, 46 (1987).
27. For some recent versions of this see *e.g.* S. Coleman, *Quantum Mechanics in Your Face* (unpublished lecture) and D. Page, *Sensible Quantum Mechanics: Are Probabilities Only in the Mind?*; quant-ph/9506010.
28. See, for example, the review by A. Kent in this volume.
29. C. Anastopoulos, *On the Selection of Preferred Consistent Sets*, quant-ph/9709051.
30. See, *e.g.* J. B. Hartle, Spacetime quantum mechanics and the quantum mechanics of spacetime, in *Gravitation and Quantizations, Proc. 1992 Les Houches Summer School*, eds. B. Julia and J. Zinn-Justin, Les Houches Summer School Proceedings Vol. LVII (North Holland, Amsterdam, 1995), gr-qc/9304006; R. Sorkin, Quantum measure theory and its interpretation, eds. D. H. Feng and B.-L. Hu, *Proc. Fourth Drexel Symposium on Quantum Nonintegrability: Quantum Classical Correspondence* (International Press, 1996), gr-qc/9507057.

31. A. Kent, Quantum histories and their implications, gr-qc/9607073.

CHAPTER 11

The Quasiclassical Realms of this Quantum Universe*

The most striking observable feature of our indeterministic quantum universe is the wide range of time, place and scale on which the deterministic laws of classical physics hold to an excellent approximation. This essay describes how this domain of classical predictability of everyday experience emerges from a quantum theory of the universe's state and dynamics.

*This is an extended version of a contribution to *Many Worlds?: Everett, Quantum Theory and Reality*, eds. S. Saunders, J. Barrett, A. Kent and D. Wallace (Oxford University Press, Oxford, 2010) which is the unified proceedings of the conference *Everett at 50*, held at Oxford, July, 2007 and the conference *Many Worlds at 50*, held at the Perimeter Institute, September, 2007.
arXiv:0806.3776

1. Introduction

The most striking observable feature of our indeterministic quantum universe is the wide range of time, place and scale on which the deterministic laws of classical physics hold to an excellent approximation. What is the origin of this predictable quasiclassical realm in a quantum universe characterized by indeterminacy and distributed probabilities? This essay summarizes progress in answering this question both old and new.

The regularities that characterize the quasiclassical realm are described by the familiar classical equations for particles, bulk matter and fields, together with the Einstein equation governing the regularities of classical spacetime geometry. Our observations of the universe suggest that this quasiclassical realm extends from a moment after the big bang to the far future and over the whole of the visible volume. Were we to set out on a journey to arrive in the far future at a distant galaxy, we would count on the regularities of classical physics holding there much as they do here. The quasiclassical realm is thus a feature of the universe independent of human cognition or decision. It is not a feature that we determine, but rather, one already present that we exploit as information gathering and utilizing systems (IGUSes) acting in the universe.

So manifest is the quasiclassical realm that it is usually simply assumed in constructing effective physical theories that apply in the late universe. Classical spacetime, for instance, is the starting assumption for the standard model of the elementary particle interactions. Classical spacetime obeying the Einstein equation is assumed in cosmology to reconstruct the past history of our universe.

Even formulations of quantum mechanics assume some part of the universe's quasiclassical realm. Copenhagen quantum theory assumed a separate classical physics and quantum physics with a kind of movable boundary between them. Classical physics was the realm of observers, apparatus, measurement outcomes and spacetime geometry. Quantum physics was the realm of the particles and quantum fields that were being measured. In the Everett formulations, classical spacetime is usually assumed in order to define the branching histories which are their characteristic feature.

Classical behavior is not exhibited by every closed quantum mechanical system, only a small minority of them. For example, in the simple case of a non-relativistic particle, an initial wave function in the form of a narrow wave packet may predict a high probability for classical correlations in time between sufficiently coarse-grained determinations of position at a sequence of times. But a generic wave function will not predict high probabilities for such correlations. Classical behavior is manifested only through *certain* sets of alternative *coarse-grained* histories and then only for *particular* quantum states. In particular, we cannot expect the classical spacetime that is the central feature of our quasiclassical realm to emerge from every state in quantum gravity, although it must, from the particular quantum state of our universe.

This essay summarizes progress in understanding the origin of our quasiclassical realm from a fundamental quantum theory of the universe — a quantum cosmology.[a] There are two inputs to this theory: First, there is the specification of the quantum dynamics (the Hamiltonian in the approximation of classical spacetime). Second, there is the particular quantum state of our universe. Superstring theory is a candidate for the first input; Hawking's no-boundary wave function of the universe[6] is a candidate for the second. An explanation of the quasiclassical realm from these inputs consists roughly of exhibiting sets of suitably coarse-grained alternative histories of the universe that have high probabilities for patterns of correlations in time summarized by closed systems of deterministic classical equations of motion.

The expansion of the universe together with the properties of the strong interactions mean that nuclear densities ($\sim 10^{15}$ g/cm^3) are the largest reached by ordinary matter any time past the first

[a]This is not a review of the long history and many different approaches taken to classicality in quantum theory. Rather, it is mostly a brief summary of the author's work, much of it with Murray Gell-Mann (especially Refs. 1–3) within decoherent (or consistent) histories quantum mechanics. The references should be understood in this context. For another approach to classicality in the quantum mechanics of closed systems, see Ref. 4. For a different kind of discussion with many references, see Ref. 5.

second after the big bang. There are nearly 80 orders of magnitude separating these densities from the Planck density (10^{93} g/cm^3) characterizing quantum gravity. This large separation in scale permits the division of the explanation of the quasiclassical realm into two parts: first, the understanding of the origin of classical spacetime in quantum cosmology, and, second, the origin of the classical behavior of matter fields *assuming* classical spacetime.

This division into Planck scale physics and below corresponds to a division in contemporary theoretical uncertainty. But, more importantly, it corresponds to a division in the nature of the explanation of the parts of the quasiclassical realm. As we shall see, the classical behavior of matter follows mostly from the conservation laws implied by the local symmetries of classical spacetime, together with a few general features of the effective theory of the elementary particle interactions (e.g. locality) and the initial condition of the universe (e.g. low entropy). By contrast, the emergence of classical spacetime involves the specific theory of the universe's quantum state and a further generalization of quantum theory to deal with histories of spacetime geometry.

These differences should not obscure the point that the explanation of the quasiclassical realm is a unified problem in quantum cosmology. But because of them it is convenient to explain the origin of the quasiclassical behavior of matter first and return to the classical behavior of spacetime later.

This essay is structured as follows: In Sec. 2, we exhibit a standard textbook derivation of classical behavior largely as a foil to the kind of treatment that we aim for. Section 3 sketches the elements of decoherent histories quantum theory. In Sec. 4, we consider classicality in a familiar class of oscillator models. Section 5 sketches a general approach to classicality in terms of the approximately conserved hydrodynamic variables. In Sec. 6, we briefly discuss the origin of the second law of thermodyamics, which is necessary for the understanding of the origin of the quasiclassical realm, as well as being an important feature of it. Section 7 discusses the origin of classical spacetime that is a prerequisite for a quasiclassical realm. Section 8 asks why human observers focus on the quasiclassical realm.

Section 9 considers the Copenhagen approximation to decoherent histories quantum theory that is appropriate for measurement situations. Open questions are mentioned in Sec. 10. In Sec. 11, we return to the theme of the connection between fundamental physics and the quasiclassical realm. For readers not familiar with it, an appendix gives a bare-bones introduction to decoherent histories quantum theory in the notation that we will use.

2. Classicality from the Ehrenfest Theorem

Standard derivations of classical behavior from the laws of quantum mechanics are available in many quantum mechanics texts. One popular approach is based on Ehrenfest's theorem relating the acceleration of the expected value of a particle's position to the expected value of the force:

$$m\,\frac{d^2\langle x\rangle}{dt^2} = -\left\langle\frac{\partial V}{\partial x}\right\rangle \tag{1}$$

(written here for one-dimensional motion). Ehrenfest's theorem is true in general, but for certain states — typically narrow wave packets — we may approximately replace the expected value of the force with the force evaluated at the expected position, thereby obtaining a classical equation of motion for that expected value:

$$m\,\frac{d^2\langle x\rangle}{dt^2} = -\frac{\partial V(\langle x\rangle)}{\partial x}. \tag{2}$$

This equation shows that the center of a narrow wave packet moves on an orbit obeying Newton's laws. More precisely, if we make a succession of position and momentum measurements that are crude enough not to disturb the approximation that allows (1) to replace (2), then the expected values of the results will be correlated by Newton's deterministic law.

This kind of elementary derivation is inadequate for the type of classical behavior that we hope to discuss in quantum cosmology for the following reasons:

- *Limited to expected values:* The behavior of expected values is not enough to define classical behavior. In quantum mechanics, the statement that the Moon moves on a classical orbit is properly the statement that, among a set of alternative coarse-grained histories of its position as a function of time, the probability is high for those exhibiting the correlations in time implied by Newton's law of motion and near zero for all others. To discuss classical behavior, therefore, we should be dealing with the probabilities of sets of alternative time histories, not with expected or average values.

- *Deals only with measurements:* The Ehrenfest theorem derivation deals with the results of "measurements" on an isolated system with a few degrees of freedom. However, in quantum cosmology we are interested in classical behavior over cosmological stretches of space and time, and over a wide range of subsystems, *independently* of whether these subsystems are receiving attention from observers. Certainly our observations of the Moon's orbit, or a bit of the universe's expansion, have little to do with the classical behavior of those systems. Further, we are interested not just in classical behavior as exhibited in a few variables and at a few times of our choosing, but over the bulk of the universe in as refined a description as possible, so that classical behavior becomes a feature of the universe itself and not a choice of observers.
- *Assumes the classical equations follow from the fundamental action:* The Ehrenfest theorem derivation relies on a close connection between the equations of motion of the fundamental action and the deterministic laws that govern classical behavior. But when we speak of the classical behavior of the Moon, or of the cosmological expansion, or even of water in a pipe, we are dealing with systems with many degrees of freedom whose phenomenological classical equations of motion (e.g. the Navier–Stokes equation) may be only distantly related to the underlying fundamental theory, say, superstring theory. We need a derivation which derives the *form* of the equations as well as the probabilities that they are satisfied.
- *Posits rather than derives the variables exhibiting classical behavior:* The Ehrenfest theorem derivation posits the

variables — the position x — in which classical behavior is exhibited. But, as mentioned above, classical behavior is most properly defined in terms of the probabilities of histories. In a closed system we should be able to *derive* the variables that enter into the deterministic laws, especially because, for systems with many degrees of freedom, these may be only distantly related to the degrees of freedom entering the fundamental action.

- *Assumes classical spacetime:* The Ehrenfest derivation assumes classical spacetime, if only to define the Schrödinger equation that underlies (1). But we aim at explaining the universe's quasiclassical realms from a quantum cosmology founded on a unified theory of the fundamental interactions, including gravity. Generally, spacetime geometry will vary quantum-mechanically. Classical behavior must therefore be explained, not posited. Indeed, we do not expect to find classical spacetime geometry at the big bang where its quantum fluctuations may be large. Classical spacetime is part of a quasiclassical realm, not separate from it.

Despite these shortcomings, the elementary Ehrenfest analysis already exhibits two necessary requirements for classical behavior: Some coarseness is needed in the description of the system as well as some restriction on its initial condition. Not every initial wave function permits the replacement of (1) by (2) and therefore leads to classical behavior; only for a certain class of wave functions will this be true. Even given such a suitable initial condition, if we follow the system too closely, say, by measuring position exactly, thereby producing a completely delocalized state, we will invalidate the approximation that allows (2) to replace (1) and classical behavior will not be expected. Some coarseness in the description of histories is therefore also needed.

3. Decoherent Histories Quantum Mechanics

The conferences to which this article is a contribution, marked fifty years of Everett's formulation of quantum theory. But they were only a year away from marking twenty five years of decoherent (or consistent) histories quantum theory that can be viewed as an extension and to some extent, a completion of Everett's work (e.g. Refs. 7–9). Today, decoherent histories is the formulation of quantum theory that is logically consistent, consistent with experiment as far as is known, consistent with the rest of modern physics such as special relativity and field theory, general enough for histories, general enough for cosmology, and generalizable for quantum gravity. It may not be the only formulation with these properties, but it is the only such we have at present. Quasiclassical realms are defined through the probabilities of histories of the universe. Decoherent histories quantum theory is the framework for computing them.

The basics of decoherent histories quantum mechanics in a classical background spacetime are reviewed briefly in the appendix.[b] We recap the essential ingredients here: For simplicity we consider a model cosmology consisting of a closed system of particles and fields in a very large box. The basic theoretical inputs are a Hamiltonian H specifying quantum dynamics in the box and a density matrix ρ specifying the box's initial quantum state. Coarse-grained histories are represented by class operators C_α. In an operator formulation, these are chains of Heisenberg picture projections at a series of times formed with the aid of H [cf. (A.3)]. In a path integral formulation, they can be bundles of Feynman paths $q^i(t)$ in configuration space.

Probabilities are properties of exhaustive sets of exclusive histories $\{C_\alpha\}$, $\alpha = 1, 2, 3, \ldots$. Decoherence is a sufficient condition for their probabilities $\{p(\alpha)\}$ to satisfy the usual rules of probability theory. The central relation defining both decoherence and probability is

$$D\left(\alpha', \alpha\right) \equiv Tr\left(C_{\alpha'}\rho C_\alpha^\dagger\right) \approx \delta_{\alpha'\alpha} p(\alpha). \tag{3}$$

The first equality defines the *decoherence functional.* The second defines decoherence and the probabilities that are predicted from H and ρ. A decoherent set of alternative coarse-grained histories is called a *realm* for short.[c]

[b] Alternatively, see Ref. 10 for a tutorial in the present notation.

[c] There will generally be families of realms defined by closely related coarse-grainings that exhibit classical behavior. Realms employing slightly different intervals for defining coarse-grained position are a simple example. Thus, it would be more accurate to refer to the quasiclassical realm*s* exhibited by the universe rather than the quasiclassical realm, and we shall do so from now on.

In a path integral formulation, sets of alternative coarse-grained histories can be defined by partitioning fine-grained configuration space paths $q^i(t)$ into exhaustive sets of exclusive classes $\{c_\alpha\}$. A useful transcription of the decoherence functional (3) for such coarse-grained histories on a time interval $[0, T]$ is

$$D\left(\alpha', \alpha\right) = \int_{c_{\alpha'}} \delta q' \int_{c_\alpha} \delta q \, \delta\left(q'_f - q_f\right) e^{i(S[q'(\tau)] - S[q(\tau)])/\hbar} \rho\left(q'_0, q_0\right). \quad (4)$$

Here, the integrals are over fine-grained paths $q^i(t)$ lying in the classes $c_{\alpha'}$ and c_α, $S[q(t)]$ is the action corresponding to the Hamiltonian H, and $\rho(q'_0, q_0)$ is the configuration space representative of the initial density matrix ρ.

4. Oscillator Models

The oscillator models pioneered in Refs. 11–13 and 2 and developed by many others provide an explicitly computable setting for understanding aspects of classicality. The following assumptions define the simplest model.

- We consider a single distinguished oscillator of mass M, frequency ω_0, and coordinate x interacting with a bath of other oscillators with coordinates Q^A, $A = 1, 2, \ldots$. The coordinates q^i in (4) are then $q^i = (x, Q^A)$.
- We suppose the action to be the sum of an action for the x, an action for the Q's, and an interaction that is a linear coupling between them. That is, we assume the action has the form

$$S[q(\tau)] = S_{\text{free}}[x(\tau)] + S_0[Q(\tau)] + S_{\text{int}}[x(\tau), Q(\tau)]. \quad (5)$$

More specifically, the associated Hamiltonians are

$$H_{\text{free}} = \frac{1}{2}(M\dot{x}^2 + M\omega_0^2 x^2), \quad (6)$$

a similar form with different masses and frequencies for H_0, and

$$H_{\text{int}} = x \sum_A g_A Q^A \quad (7)$$

for some coupling constants g_A.

- We suppose the initial density matrix ρ factors into a product of one depending on the x's and another depending on the Q's which are often called the "bath" or the "environment", viz.
$$\rho\left(q_0', q_0\right) = \bar{\rho}\left(x_0', x_0\right)\, \rho_B\left(Q_0', Q_0\right). \tag{8}$$
We assume that the bath oscillators are in a thermal state ρ_B characterized by a temperature T_B.
- We restrict attention to a simple set of alternative coarse-grained histories that follow the coordinate x of the distinguished oscillator while ignoring the coordinates Q^A of the bath. The histories of the distinguished oscillator are specified by giving an exhaustive set of exclusive intervals of x at each of a series of times $t_1, t_2, \ldots, t_n$. A coarse-grained history c_α is the bundle of paths $x(t)$ passing through a particular sequence of intervals $\alpha \equiv (\alpha_1, \alpha_2, \ldots, \alpha_n)$ at the series of times $t_1, t_2, \ldots, t_n$. For simplicity, we take all the intervals to be of equal size Δ and the times to be equally separated by Δt.

Since the bath oscillators are unconstrained by the coarse-graining, the integral over the Q's in (4) can be carried out to give a decoherence functional just for coarse-grained histories of the x's of the form:

$$D\left(\alpha', \alpha\right) = \int_{c_{\alpha'}} \delta x' \int_{c_\alpha} \delta x\, \delta\left(x_f' - x_f\right)$$

$$\times \exp\left\{ i\left(S_{\text{free}}[x'(\tau)] - S_{\text{free}}[x(\tau)] + W\left[x'(\tau), x(\tau)\right]\right)/\hbar \right\} \bar{\rho}\left(x_0', x_0\right) \tag{9}$$

where $W[x'(\tau), x(\tau)]$, called the Feynman–Vernon influence phase, summarizes the results of integrations over the Q's.

In the especially simple case of a cut-off continuum of bath oscillators and high bath temperature, the imaginary part of the influence phase is given by Ref. 12:

$$\mathrm{Im}W[x'(\tau), x(\tau)] = \frac{2M\gamma k T_B}{\hbar} \int_0^T dt\, \left(x'(t) - x(t)\right)^2 \tag{10}$$

where γ is a measure of the strength of its coupling to the bath related to the g_A in (7). $\mathrm{Im}W$ becomes substantial when $x'(\tau)$ and $x(\tau)$ are very different and the time difference Δt is long enough. Then the off-diagonal elements of $D(\alpha', \alpha)$ are exponentially suppressed, meaning that the set of alternative histories approximately decoheres [cf. (3)]. Roughly, the coarse-graining time required is

$$\Delta t \gtrsim t_{\text{decoh}} \equiv \frac{\hbar^2}{2M\gamma k T_B \Delta^2}. \tag{11}$$

The time t_{decoh} is called the decoherence time.[14] This is typically very much shorter than typical dynamical time scales, for instance, $1/\gamma$.

The diagonal elements of the decoherence functional (9) are the probabilities $p(\alpha)$ of the individual histories in the set [cf. (3)]. With a little work, these can be expressed in the following form[2]:

$$p(\alpha) = \int_{c_\alpha} \delta x \, (\cdots)$$
$$\times \exp\left[-\int dt \left(\frac{M^2}{4\hbar}\right) \left(\frac{\hbar}{2M\gamma k T_B}\right) E(x(t))^2 \right] w(x_0, p_0), \tag{12}$$

the dots denoting factors irrelevant for the subsequent argument. Here $w(x_0, p_0)$ is the Wigner distribution for the density matrix of the distinguished particle $\bar{\rho}$ [cf. (8)] and E is

$$E(x(t)) \equiv \ddot{x} + \omega^2 x + 2\gamma \dot{x} \tag{13}$$

where ω is the frequency of the x-oscillator ω_0 renormalized by its interaction with the bath. Equation (12) has been organized to show that the factor in front of the imaginary part of the influence phase (10) appears inversely in the exponent of this relation.

$E = 0$ is the classical equation of motion for the distinguished oscillator. This includes a frictional force arising from the interaction of the particle with the bath. When the coefficient in front of E^2 in (12) is large, the probabilities for histories $p(\alpha)$, will peak about histories that satisfy the classical equations of motion. Thus classical behavior of the distinguished oscillator is predicted. The width of the distribution is a measure of thermal and quantum noise causing deviations from classical predictability.

In this simple case, an analysis of the requirements for classical behavior is straightforward. Equation (10) shows that high values of $M\gamma kT_B/\hbar$ are needed to achieve decoherence. Put differently, a strong coupling between the distinguished oscillator and the bath is required if interference phases are to be dissipated efficiently into the bath. However, the larger this coupling is, the smaller the coefficient in the exponent of (12) is, decreasing the size of the exponential and *increasing* deviations from classical predictability. This is reasonable: the stronger the coupling to the bath, the more noise is produced by the interactions that are carrying away the phases. To counteract that, and achieve a sharp peaking about the classical equation of motion, $M^2/4\hbar$ must be large. That is, high inertia is needed to resist the noise that arises from the interactions with the bath.

Thus, much more coarse-graining is needed to ensure classical predictability than naive arguments based on the uncertainty principle would suggest. Coarse-graining is needed to effect decoherence, and coarse-graining beyond that to achieve the inertia necessary to resist the noise that the mechanisms of decoherence produce.

This derivation of classicality deals genuinely with histories, and is not restricted to measurements. But there is still a close connection between the classical equations and the fundamental action. The variable x which behaves classically was posited, not derived, and classical spacetime was assumed. The progress in relation to the Ehrenfest derivation is summarized in the table below:

Deficiencies of the Ehrenfest Derivation

- √ *Limited to expected values, but classicality is defined through histories.*
- √ *Deals only with measurements on isolated subsystems with a few degrees of freedom.*
- ✗ *Assumes the classical equations follow directly from the fundamental action.*
- ✗ *Posits rather than derives the variables which exhibit classical behavior.*
- ✗ *Assumes classical spacetime.*

5. Quasiclassical Coarse-Grainings, Local Equilibrium and Hydrodynamic Equations

Isolated systems evolve toward equilibrium; that is a consequence of statistics. But conserved or approximately conserved quantities approach equilibrium more slowly than others. These include conserved quantities like energy and momentum that arise from the local symmetries of classical spacetime, together with conserved charges and numbers arising from the effective theory of the particle interactions. A situation of *local equilibrium* will generally be reached before complete equilibrium is established, if it ever is. This local equilibrium is characterized by the values of conserved quantities averaged over small volumes. Even for systems of modest size, time scales for small volumes to relax to local equilibrium can be very, very much shorter than the time scale for reaching complete equilibrium. Once local equilibrium is established, the subsequent evolution of the approximately conserved quantities can be described by closed sets of effective classical equations of motion, such as the Navier–Stokes equation. The local equilibrium determines the values of the phenomenological quantities, such as pressure and viscosity, that enter into these equations and the constitutive relations among them.

That in a nutshell is the explanation of the quasiclassical realms of matter given classical spacetime. It both identifies the variables in which the quasiclassical realms are defined and the mechanism by which they obey closed sets of equations of motion. To make this more concrete, we will review very briefly the standard derivation (e.g. Refs. 15 and 16) of these equations of motion in a simple model. We follow Ref. 3, where more details can be found. In Refs. 17 and 18, Jonathan Halliwell explains why sets of sufficiently coarse-grained histories of these variables decohere and lead to high probabilities for correlations in time summarized by the same equations of motion.

Consider a system of conserved particles inside a non-rotating box interacting by local short range potentials. Let the density matrix ρ — possibly pure — describe the state of the system. Divide the box up into equal volumes of size V labeled by a discrete index $\vec{y}$. Let $T^{\alpha\beta}(\vec{x}, t)$ be the stress-energy-momentum operator in the Heisenberg picture. The energy density $\epsilon(\vec{x}, t)$ and momentum density $\pi^i(x, t)$

are $T^{tt}(\vec{x},t)$ and $T^{ti}(\vec{x},t)$, respectively. Let $\nu(\vec{x},t)$ denote the number density of the conserved particles. Then define

$$\epsilon_V(\vec{y},t) \equiv \frac{1}{V}\int_{\vec{y}} d^3x\, \epsilon(\vec{x},t), \tag{14a}$$

$$\vec{\pi}_V(\vec{y},t) \equiv \frac{1}{V}\int_{\vec{y}} d^3x \vec{\pi}(\vec{x},t), \tag{14b}$$

$$\nu_V(\vec{y},t) \equiv \frac{1}{V}\int_{\vec{y}} d^3x\, \nu(\vec{x},t), \tag{14c}$$

where in each case the integral is over the volume labeled by $\vec{y}$. These are the quasiclassical variables for our model. We note that the densities in (14) are the variables for a classical hydrodynamic description of this system — for example, the variables of the Navier–Stokes equation.

Were the system in complete equilibrium, the expected values of the quasiclassical variables defined from the density matrix ρ could be accurately computed from the effective density matrix

$$\tilde{\rho}_{\rm eq} = Z^{-1}\exp[-\beta(H - \vec{U}\cdot\vec{P} - \mu N)]. \tag{15}$$

Here, H, $\vec{P}$ and N are the operators for total energy, total momentum and total conserved number inside the box — all extensive quantities. The c-number intensive quantities β, $\vec{U}$ and μ are respectively the inverse temperature (in units where Boltzmann's constant is 1), the velocity of the box, and the chemical potential. A normalizing factor Z ensures $Tr(\tilde{\rho}_{\rm eq}) = 1$. In equilibrium the expected values are, for instance,

$$\langle \epsilon_V(\vec{y},t)\rangle \equiv Tr(\epsilon_V(\vec{y},t)\rho) \approx Tr(\epsilon_V(\vec{y},t)\tilde{\rho}_{\rm eq}). \tag{16}$$

Indeed, this relation and similar ones for $\vec{\pi}_V(\vec{x},t)$ and $\nu_V(\vec{x},t)$ *define* equilibrium.

Local equilibrium is achieved when the decoherence functional for sets of histories of quasiclassical variables $(\epsilon,\vec{\pi},n)$ is given approximately by the *local* version of the equilibrium density matrix (15)

$$\tilde{\rho}_{\rm leq} = Z^{-1}\exp\Big[-\int d^3y\, \beta(\vec{y},t)\,\big(\epsilon_V(\vec{y},t) - \vec{u}(\vec{y},t)\cdot\vec{\pi}_V\,(\vec{y},t) - \mu\,(\vec{y},t)\,\nu_V\,(\vec{y},t)\big)\Big]. \tag{17}$$

(The sum over $\vec{y}$ has been approximated by an integral.) Expected values are given by (16) with $\tilde{\rho}_{\rm eq}$ replaced by $\tilde{\rho}_{\rm leq}$. The expected values of quasiclassical quantitites are thus functions of the intensive c-number quantities $\beta(\vec{y},t)$, $\vec{u}(\vec{y},t)$ and $\mu(\vec{y},t)$. These are the local inverse temperature, velocity and chemical potential, respectively. They now vary with time and place as the system evolves toward complete equilibrium.

A closed set of deterministic equations of motion for the expected values of $\epsilon(\vec{x},t)$, $\vec{\pi}(x,t)$ and $\nu(\vec{x},t)$ follows from assuming that $\tilde{\rho}_{\rm leq}$ is an effective density matrix for computing them. To see this, begin with the Heisenberg equations for the conservation of the stress-energy-momentum operator $T^{\alpha\beta}(\vec{x},t)$ and the number of the current operator $j^{\alpha}(\vec{x},t)$.

$$\frac{\partial T^{\alpha\beta}}{\partial x^{\beta}} = 0, \quad \frac{\partial j^{\alpha}}{\partial x^{\alpha}} = 0. \tag{18}$$

Noting that $\epsilon(\vec{x},t) = T^{tt}(\vec{x},t)$ and $\pi^i(\vec{x},t) = T^{ti}(\vec{x},t)$, Eqs. (18) can be written in a 3+1 form and their expected values taken. The result is the set of five equations

$$\frac{\partial\langle\pi^i\rangle}{\partial t} = -\frac{\partial\langle T^{ij}\rangle}{\partial x^j}, \tag{19a}$$

$$\frac{\partial\langle\epsilon\rangle}{\partial t} = -\vec{\nabla}\cdot\langle\vec{\pi}\rangle, \tag{19b}$$

$$\frac{\partial\langle\nu\rangle}{\partial t} = -\vec{\nabla}\cdot\langle\vec{j}\rangle. \tag{19c}$$

The expected values are all functions of $\vec{x}$ and t.

The set of equations (19) close for the following reason: Eq. (16) with $\tilde{\rho}_{\rm leq}$ could in principle be inverted to express $\beta(\vec{y},t)$, $\vec{u}(\vec{y},t)$, $\mu(\vec{y},t)$, and therefore $\tilde{\rho}_{\rm leq}$ itself, in terms of the expected values (14). Thus, the expected values on the right-hand side of (19) become functionals of the quasiclassical variables on the left-hand side and the equations close.

The process of expression and inversion sketched above could be difficult to carry out in practice. The familiar classical equations of motion arise from further approximations, in particular from assuming that the gradients of all quantities are small. For example, for a non-relativistic fluid of particles of mass m, the most general Galilean-invariant form of the stress tensor that is linear in the gradients of the fluid velocity $\vec{u}(x)$, has the approximate form[19]

$$\langle T^{ij}\rangle = p\delta^{ij} + m\nu\, u^i u^j - \eta\left[\frac{\partial u^i}{\partial x^j} + \frac{\partial u^j}{\partial x^i} - \frac{2}{3}\,\delta_{ij}\left(\vec{\nabla}\cdot\vec{u}\right)\right] - \zeta\,\delta_{ij}\left(\vec{\nabla}\cdot\vec{u}\right). \tag{20}$$

The pressure p and coefficients of viscosity η and ζ are themselves functions, say, of the expected values (14). This form of the stress tensor in (19a) leads to the Navier–Stokes equation.

What determines the volume V defining the coarse-grained variables of the quasiclassical realms? The volume V must be large enough to ensure the decoherence of histories constructed from these quasiclassical variables, and beyond that to ensure classical predictability in the face of the noise that typical mechanisms of decoherence produce. The volumes must be small enough to allow local equilibrium. Roughly speaking, the volume V should be chosen as small as possible, consistent with these requirements. That is, it should be chosen so the quasiclassical realms are maximally refined, consistent with decohence and predictability. Then they are a feature of our universe and not a matter of our choice.

We have now removed two more of the deficiencies of the Ehrenfest derivation as shown in the following table:

Deficiencies of the Ehrenfest Derivation	
✓	*Limited to expected values, but classicality is defined through histories.*
✓	*Deals only with measurements on isolated subsystems with a few degrees of freedom.*
✓	*Assumes the classical equations follow directly from the fundamental action.*
✓	*Posits rather than derives the variables which exhibit classical behavior.*
✗	*Assumes classical spacetime.*

There remains the origin of classical spacetime to which we turn after a brief discussion of the second law of thermodynamics.

6. The Second Law of Thermodynamics

The quasiclassical realms of our universe exhibit two important thermodynamic features that are not directly connected to classical determinism:

- The tendency of a total entropy of the universe to increase.
- The tendency of this entropy for nearly isolated subsystems to increase in the same direction of time. This may be called the homogeneity of the thermodynamic arrow of time.

These two features are connected. The first follows from the second, but only in the late universe when nearly isolated subsystems are actually present. In the early universe we have only the first. Together they may be called the second law of thermodynamics.

Thermodynamics, including the second law, is an essential part of classical physics, and, indeed, a prerequisite for it. In the previous section, for example, we assumed the second law when we posited the rapid approach to local equilibrium necessary to derive a closed system of deterministic equations from the conservation relations.

Entropy is generally a measure of the information missing from a coarse-grained description of a physical system. In the case of the quasiclassical variables (14), we can define it at a given time as the maximum of the information measure $-Tr(\tilde{\rho}\log\tilde{\rho})$ over density matrices $\tilde{\rho}$ that preserve the expected values of the quasiclassical variables at that time. More specifically, if ρ is the state of the system, we take

$$S(t) \equiv \max_{\tilde{\rho}}[-Tr(\tilde{\rho}\log\tilde{\rho})], \tag{21}$$

keeping fixed for each $\vec{y}$

$$\langle\epsilon_V(\vec{y},t)\rangle \equiv Tr(\epsilon_V(\vec{y},t)\rho) \approx Tr(\epsilon_V(\vec{y},t)\tilde{\rho}_{\rm eq}), \tag{22}$$

together with the similar relations for $\vec{\pi}_V(\vec{y},t)$ and $\nu_V(\vec{y},t)$. The result is the local equilibrium density matrix (17).

The entropy defined this way *is* the usual entropy of chemistry, physics and statistical mechanics. *The coarse-graining in terms of local conserved quantities that exhibits the determinism of the quasiclassical realms thus also defines the entropy for its thermodynamics.*

A special initial quantum state is needed to predict with high probability the classical spacetime whose symmetries are the origin of the conservation laws behind classical determinism. But further conditions on the state are needed for the universe to exhibit the thermodynamic features mentioned above. First, the general increase in total entropy requires that:

- The quantum state is such that the initial entropy is near the minimum it could have for the coarse-graining defining it. It then has essentially nowhere to go but up.
- The relaxation time to equilibrium is long compared to the present age of the universe, so that the general tendency of its entropy to increase will dominate its evolution.

In our simple model cosmology, we have neglected gravitation for simplicity, but to understand the origin of the second law it is necessary to consider it. That is because gravity is essential to realizing the first of the conditions above. In a self-gravitating system, gravitational clumping increases entropy. The matter in the early universe is not clumped and nearly in thermal equilibrium — already at maximal entropy. But the spacetime in the early universe is approximately homogeneous, implying that the entropy has much more room to increase through the *gravitational* growth of fluctuations. In a loose sense, as far as gravity is concerned, the entropy of the early universe is low for the coarse-graining defined by quasiclassical variables. The entropy then increases. The no-boundary quantum state in particular implies that gravitational fluctuations are small in the early universe,[20,21] giving entropy room to grow.

Coarse-graining by approximately conserved quasiclassical variables helps with the second of the two conditions above. Small volumes come to local equilibrium quickly. But the approximate conservation ensures that the whole system will approach equilibrium slowly, whether or not such equilibrium is actually attained.

The homogeneity of the thermodynamic arrow of time, which was the other aspect of the second law mentioned at the beginning of this section, cannot follow from the approximately time-reversible dynamics and statistics alone. Rather, the explanation is that the progenitors of today's nearly isolated systems were all far from equilibrium a long time ago and have been running downhill ever since. As Boltzmann put it over a century ago: "The second law of thermodynamics can be proved from the [time-reversible] mechanical theory, if one assumes that the present state of the universe... started to evolve from an improbable [i.e. special] state".[22] There is thus a stronger constraint on the initial state than merely having low total entropy. It must be locally low.

The initial quantum state of our universe must be such that it leads to the decoherence of sets of quasiclassical histories that describe coarse-grained spacetime geometry and matter fields. Our observations require this now, and the successes of the classical history of the universe suggests that there was a quasiclassical realm at a very early time. In addition, the initial state must be such that the entropy of quasiclassical coarse-graining was low in the beginning, and also be such that the entropy of presently isolated systems was also low then. Then the universe can exhibit both aspects of the second law of thermodynamics.

The quasiclassical coarse-grainings are therefore distinguished from others, not only because they exhibit predictable regularities of the universe governed by approximate deterministic equations of motion, but also because they are characterized by a sufficiently low entropy in the beginning and a slow evolution towards equilibrium — two properties which make those regularities exploitable.

7. The Origin of Classical Spacetime

The classical behavior of matter in a given background spacetime depends only weakly on the matter's fundamental quantum physics. The forms of the dynamical equations (19) follow largely from conservation laws and the conditions on the interactions necessary for local equilibrium. In a sense, the quasiclassical realms shield us from quantum physics — a happy circumstance that was of great importance historically.

By contrast, the origin of classical spacetime is strongly dependent on the physics of quantum gravity and the theory of the initial quantum state of the universe. That is both the attraction of the issue and its difficulty. It is impossible to say much about this in the space made available for this paper. That not least because the quantum theory sketched in Sec. 3 must be generalized further to deal with quantum spacetime (see, e.g. Refs. 23 and 24). The discussion in Secs. 3–5 relied on a fixed notion of time to describe histories — a notion which is not available when spacetime itself is a quantum variable. The following heuristic discussion may, however, give some sense of the issues involved.

Let us first recall one way in which quantum mechanics predicts classical behavior for the motion of a non-relativistic particle. Consider a particle of mass m moving in one dimension x in a potential $V(x)$. Wave functions $\psi(x)$ describe its states. Consider wave functions that are well approximated in the semiclassical (WKB) form

$$\psi(x) \approx A(x) \exp[iS(x)/\hbar] \tag{23}$$

where $S(x)/\hbar$ varies rapidly with x and $A(x)$ varies slowly. Such states predict classical behavior for the particle. Specifically, they imply that, in a set of alternative histories suitably coarse-grained in x at a series of times, the probabilities are high for correlations in time summarized by the classical equation of motion for the particle (e.g. Ref. 25).

A wave function satisfying (23) also predicts probabilities for *which* classical histories satisfy the equation of motion. That is, it predicts probabilities for the initial conditions to the dynamical equations. Consider histories that pass through a position x at the time the wave function is specified. Non-zero probabilities are predicted only for the history with momentum p given by

$$p \equiv m\frac{dx}{dt} = -\nabla S(x) \tag{24}$$

and the probability (density) for this history is $|A(x)|^2$. Thus, a wave function of semiclassical form (23) predicts the probabilities of an ensemble of classical histories labeled by their initial x.

An analogous discussion of the origin of classical spacetime can be given in quantum cosmology (e.g. Refs. 23 and 25). In quantum gravity, the metric on spacetime will fluctuate quantum-mechanically and generally not behave classically. Consider a simple model in which the quantum metrics are restricted to be homogeneous, isotropic and spatially closed. A model of the matter assumes a single homogeneous scalar field $\phi(t)$.

Spacetime geometry in these models is described by metrics of the form

$$ds^2 = -dt^2 + a^2(t)d\Omega_3^2 \tag{25}$$

where $d\Omega_3^2$ is the metric on the unit, round, three-sphere. The scale factor $a(t)$ determines how the size of the spatial geometry varies in time. Closed Friedmann–Robertson–Walker cosmological models describing the expansion of the universe from a big bang have metrics of this form with a scale factor $a(t)$ satisfying the Einstein equation. In quantum mechanics, $a(t)$ could have any form. Classical behavior of these minisuperspace models means high probability for coarse-grained $a(t)$'s obeying the Einstein equation.

A wave function of the universe in this model is a function $\Psi(a, \phi)$ of the scale factor and homogeneous scalar field. Suppose that the wave function in some region of (a, ϕ) space is well approximated by the semiclassical form

$$\Psi(a, \phi) \approx A(a, \phi) \exp[iS(a, \phi)/\hbar] \tag{26}$$

where $S(a, \phi)/\hbar$ is rapidly varying and $A(a, \phi)$ is slowly varying. Then, from the analogy with non-relativistic quantum mechanics, we expect[d] the wave function to predict an ensemble of classical spacetimes with initial data related by the analog of (24) and probabilities related to $|A(a, \phi)|^2$.

If our universe is a quantum mechanical system, it has a quantum state. A theory of that state is a necessary part of any "final theory" and the goal of quantum cosmology. Hawking's no-boundary wave function of the universe[6] is a leading candidate for this theory. In

[d]See, e.g. Refs. 23 and 25 for a framework in which these expectations are partially borne out.

the context of the simple model, the no-boundary wave function is specified by the following functional integral:

$$\Psi(a,\phi) = \int_{\mathcal{C}} \delta a' \delta\phi' \exp(-I[a'(\tau),\phi'(\tau)]/\hbar). \tag{27}$$

Here, the path integration is over histories $a'(\tau)$ and $\phi'(\tau)$ of the scale factor and matter field and $I[a'(\tau),\phi'(\tau)]$ is their Euclidean action. The sum is over cosmological geometries that are regular on a manifold with only one boundary at which $a'(\tau)$ and $\phi'(\tau)$ take the values a and ϕ. The integration is carried out along a suitable complex contour $\mathcal{C}$, which ensures the convergence of (27) and the reality of the result.

Does the no-boundary quantum state predict classical spacetime for the universe, and, if so, what classical spacetimes does it predict? The answer to the first part of the question is "yes". In certain regions of (a,ϕ) space, the defining path integral in (27) can be carried out by the method of steepest descents. The dominant contributions come from the complex extrema of the Euclidean action. The leading order approximation of one extremum is

$$\Psi(a,\phi) \approx \exp\{[-I_R(a,\phi) + iS(a,\phi)]/\hbar\} \tag{28}$$

where $I_R(a,\phi)$ and $-S(a,\phi)$ are the real and imaginary parts of the Euclidean action evaluated at the extremizing path.

When $S(a,\phi)/\hbar$ varies rapidly and $I_R(a,\phi)\hbar$ varies slowly, this is a wave function of the universe of semiclassical form (26). An ensemble of classical spacetimes is predicted with different probabilities. The probabilities will be different for such things as whether the universe bounces at a minimum radius or has an initial singularity, how much matter it has, and the duration of an inflationary epoch. These are important issues for cosmology (e.g. Ref. 26). But a quasiclassical realm of matter depends only on the local symmetries of a classical spacetime from the arguments of the preceding three sections. Each classical spacetime with any matter at all will therefore exhibit quasiclassical realms.

Our list of tasks now stands like this:

Deficiencies of the Ehrenfest Derivation
√ *Limited to expected values, but classicality is defined through histories.*
√ *Deals only with measurements on isolated subsystems with a few degrees of freedom.*
√ *Assumes the classical equations follow directly from the fundamental action.*
√ *Posits rather than derives the variables which exhibit classical behavior.*
√ *Assumes classical spacetime.*

8. Why We Focus on Quasiclassical Variables

A quantum universe exhibits many different decoherent sets of alternative coarse-grained histories — many different realms. Two realms are *compatible* if each one can be fine-grained to yield the same realm. But there are also mutually *incompatible realms* for which there is no finer-grained realm of which they are both coarse-grainings.[1,3] Quantum mechanics by itself does not prefer any one of these realms over the others. Why, then, do we as human IGUSes focus almost exclusively on quasiclassical realms?

Questions of the form "Why do we ...?" can be answered within quantum theory by understanding human IGUSes as particular kinds of physical systems in the universe. As human IGUSes, both individually and collectively, we are described in terms of quasiclassical variables. We are therefore not separate from the universe's quasiclassical realms but, rather, phenomena exhibited by them.[e] "Why do we ...?" questions can therefore only be formulated in terms of the probabilities of the universe's quasiclassical realms and our description within them.

---

[e]Are there realms qualitatively different from the quasiclassical ones that exhibit IGUSes? At present we lack a general enough conception of IGUS to formulate this question precisely, much less answer it.

The elementary answer to the question of why we focus on quasiclassical variables is that we are physical systems described by quasiclassical variables that possess senses that are adapted to perceive quasiclassical variables. The predictable regularities of the quasiclassical realms suggest why it is adaptive to have senses that register quasiclassical variables. But collectively we have also evolved to understand and use quantum mechanics. This has also proved adaptive at least in the short run.[f]

Could the quasiclassical realms of this universe contain quasiclassically described IGUSes elsewhere whose senses register variables substantially different from the ones we use, even non-quasiclassical ones? To answer, it would be necessary to calculate the probabilities of alternative evolutionary histories of such quasiclassically described IGUSes. It is well beyond our power at present to even formulate such a calculation precisely, much less carry it out. If we ever encounter extra-terrestrial IGUSes, this question may be settled experimentally.

Questions concerning human IGUSes in decoherent histories quantum theory are both fascinating and difficult. But we should emphasize that answers are not required to understand, utilize, or test the theory for other purposes. That is because IGUSes, including human beings, occupy no special place and play no preferred role in this formulation of quantum theory. Rather, they are but one of the many complex systems that can be described within it.

9. The Copenhagen Approximation

Copenhagen quantum mechanics can be seen as an approximation to decoherent histories quantum theory that is appropriate for situations in which a series of measurements is carried out by an apparatus on an otherwise isolated subsystem.

In Copenhagen quantum mechanics, the isolated subsystem is described quantum-mechanically. But the apparatus is described by the separate classical physics posited by the theory. The probabilities

[f]Some estimate that a large percentage of the US GDP can be attributed to our understanding of quantum theory. Our understanding of nuclear fission may prove to be less adaptive.

for the outcomes of a series of "ideal" measurements are given by unitary evolution of the subsystem's state interrupted at the time of measurements by projections onto the values of the outcomes — the infamous reduction of the wave packet.

In decoherent histories, apparatus and subsystem are separate parts of one closed system (most generally the universe). In a measurement, a variable of the subsystem, perhaps not quasiclassical and perhaps not otherwise decohering, becomes correlated with a quasiclassical variable of an apparatus. Histories of the measured variable decohere because of this correlation with the decohering histories of the quasiclassical realm.

The Copenhagen prescription for the probability of a series of measurement outcomes can be derived from the probabilities of decoherent histories quantum theory by modeling the measurement situations to which it applies (e.g. Ref. 27). Idealized measurement models have a long history in quantum theory (e.g. Ref. 28). A typical model assumes a closed system — a model universe — consisting of an apparatus, a subsystem which it measures, and perhaps other degrees of freedom. The Hilbert space is idealized as a tensor product $\mathcal{H}_s \otimes \mathcal{H}_r$ with the factor $\mathcal{H}_s$ for the subsystem and the factor $\mathcal{H}_r$ for the rest, including the apparatus. The subsystem is measured by the apparatus at a series of times $t_1, t_2, \ldots, t_n$ and it is otherwise isolated from the rest of the universe. The initial state is assumed to factor into a pure state $|\psi\rangle$ in $\mathcal{H}_s$ and a density matrix for the rest.

The measurement interaction is idealized to (i) occur at definite moments of time, (ii) create a perfect correlation between the measured alternatives of the subsystem and the registrations of the apparatus — the former represented by sets of projections $\{s_\alpha(t)\}$ in $\mathcal{H}_s$ and the latter by projections $\{R_\alpha(t)\}$ in $\mathcal{H}_r$, and (iii) disturb the subsystem as little as possible (an ideal measurement). Under these assumptions the probability of the sequence of registrations can be shown[27] to be given by

$$p(\alpha_n, \ldots, \alpha_1) = ||s^n_{\alpha_n}(t_n) \cdots s^1_{\alpha_n}(t_1)|\psi\rangle||^2. \tag{29}$$

The argument of the square in (29) can be thought of as a state of the subsystem which evolved from the initial $|\psi\rangle$ by unitary evolution (constant state in the Heisenberg picture), interrupted by the action of projections at the times of measurements (state reduction). This is the usual Copenhagen story.

Equation (29) is a huge and essential simplification when compared to the basic relation (3). Decoherence has been assumed rather than calculated. More importantly, (29) refers to a Hilbert space which may involve only a few degrees of freedom, whereas (3) involves all the degrees of freedom in the universe.

Assumptions (i) and (ii) may hold approximately for many realistic measurement situations. But assumption (iii) — the projection postulate or second law of evolution — does not hold for most.[g] But it is in this way that Copenhagen quantum mechanics is recovered from the more general decoherent histories quantum mechanics once one has a quasiclassical realm. It is not recovered generally, but only for idealized measurement situations. It is not recovered exactly, but only to an approximation calculable from the more general theory — an approximation which is truly excellent for many realistic measurement situations.[27] The separate classical physics posited by Copenhagen quantum theory is an approximation to the quasiclassical realms. *Copenhagen quantum mechanics is thus not an alternative to decoherent histories, but rather contained within it as an approximation appropriate for idealized measurement situations.*

The founders of quantum mechanics were correct that something besides the wave function and Schrödinger equation were needed to understand the theory. But it is not a posited classical world to which quantum mechanics does not apply. Rather, it is the quantum state of the universe together with the theory of quantum dynamics that explains the origin of quasiclassical realms within the more general quantum mechanics of closed systems.

[g]The idea that the two forms of evolution of the Copenhagen approximation are some kind of problem for quantum theory seems misplaced from the perspective of the quantum mechanics of closed systems which has no such division.

10. Summary and Open Questions

We now have a complete sketch of an explanation of the quasiclassical realms in our quantum universe in the context of today's fundamental physics. Our discussion has been top-down — proceeding from the classical world to the quantum — starting in today's universe and working backward to the beginning. To summarize, we recapitulate these developments from the bottom up.

- The particular quantum state of our universe implies the classical behavior of spacetime geometry coarse-grained on scales well above the Planck scale. Further, it predicts the homogeneity of this spacetime on cosmological scales that imply a low total entropy leading to the second law of thermodynamics.
- Local Lorentz symmetries of classical spacetime imply conservation of energy and momentum. The effective theory of the matter interactions implies the approximate conservation of various charges and numbers at various stages in the evolution of the universe.
- Quasiclassical variables specified by ranges of values of the averages of densities of conserved or approximately conserved quantities over small volumes are definable. Sets of alternative histories of these variables decohere and define quasiclassical realms.
- When the volumes are suitably large, the approximate conservation of the quasiclassical variables ensures that they evolve predictably despite the noise that typical mechanisms of decoherence produce.
- When the volumes are suitably small, their contents approach local equilibrium on time scales which are short compared to those on which the quasiclassical variables are changing.
- Local equilibrium implies that the evolution of the quasiclassical variables obeys a closed, deterministic set of equations of motion incorporating constitutive relations determined by local equilibrium.

The chain above gives a broad outline of how the quasiclassical realms of our universe emerge from its fundamental quantum physics and particular quantum state. However, touch this chain where you will and there are issues that remain to make it more realistic, more general, more complete, more precise, and more quantitative. The following is a short and selective list of outstanding problems:

Decoherence of Classical Spacetime: Our understanding of the *emergence of* classical spacetime from particular states in quantum gravity is more primitive than our understanding of the emergence of the classical behavior of matter *given* a fixed spacetime. Even a cursory comparison of Sec. 7 with Sec. 5 reveals this. Partly this is because we lack a complete and manageable quantum theory of gravity. But even in the low-energy effective theory of gravity based on general relativity, we do not have precise notions of the diffeomorphism invariant coarse-grainings[h] that *define* the classical behavior of geometry in everyday situations above the Planck scale. And, perforce, we have an inadequate understanding of the mechanisms effecting their decoherence.

More Realistic Models: The model universe of a static box of particles interacting by short range potentials that was discussed all too briefly in Sec. 5 is highly simplified. Models are needed which incorporate at least the following features of the realistic universe.

- *Cosmology:* The expansion of the universe, gravitational clumping, possible eternal inflation, the decay of the proton, the formation and evaporation of black holes.
- *Degrees of Freedom:* The relativistic quantum fields that are the basic variables of today's effective field theories.
- *Coarse-graining:* Branch-dependent coarse-grainings that express narratives directly in terms of realistic hydrodynamic variables.
- *Maximal Refinement:* Maximal refinement of coarse-grainings consistent with decoherence and classicality so that the quasiclassical realms are a feature of the universe and not a matter of human choice, as discussed at the end of Sec. 5.
- *Initial States:* Initial states that arise from theories of the quantum state of the universe and not from *ad hoc* assumptions about an environment, as in (8).

[h]Defining diffeomorphism invariant coarse-grainings of matter fields in quantum spacetime is itself an issue, see e.g. Ref. 29.

Comparing Different Realms: As mentioned in Sec. 8, a quantum universe can be described by many decohering sets of alternative coarse-grained histories — many realms. The quasiclassical realms are distinguished by a high level of classical predictability and a low initial entropy among other properties. Intuitively, they provide the simplest description of the general regularities of the universe that are readily exploitable by IGUSes of the kinds we know about. A genuine comparison of the quasiclassical realms with others that the universe exhibits would require quantitative measures on realms of simplicity, predictability, classicality, etc. Various approaches to such measures have been explored[30] but no complete and satisfactory result has yet emerged.

Thus, while we have gone far beyond the Ehrenfest derivation, there is still a long way to go!

11. Quasiclassical Realms and Fundamental Physics

From the present theoretical perspective, a final theory consists of two parts: (1) a dynamical theory specifying quantum evolution (the Hamiltonian in simple models), and (2) a theory of the universe's quantum state. Without both, there are no predictions of any kind. With both, probabilities for the members of every decoherent set of alternative histories of the univerese are in principle predicted.

Today the search for a final theory has taken physics further and further away from the determinism and unique reality that characterized classical physics. A final theory may incorporate quantum indeterminacy, mutually incompatible realms, and not have spacetime at a basic level. In that context, the seemingly prosaic quasiclassical realms of our universe appear remarkable.

On what features of the two parts of a final theory do the quasiclassical realms depend? The discussion in this essay suggests the following:

Requirements for Dynamics: For the most part, what is required of the dynamical part of theory, is an effective theory of the elementary interactions which has the properties necessary for local equilibrium at the matter energies well below the Planck scale that are reached in an expanding universe. Specifically, the interactions should be approximately local and dominantly short range.

However, the specific properties of the only unscreened long range interaction — gravity — are crucial for the quasiclassical realms. It is the gravitationally driven expansion of the universe that ensures the separation of the energy scales of matter from those of quantum gravity. It is the attractive and universal character of gravity which allows isolated systems to form by the growth and collapse of fluctuations. And it is the relative weakness of the gravitational interaction which allows the universe to remain out of total equilibrium on the time scale of its present age.

Requirements for the State: More is required of the initial state. It must be such as to imply that histories of cosmological geometry coarse-grained above the Planck scale behave classically. The local symmetries of this classical spacetime imply conservation laws which determine in part the variables characterizing the quasiclassical realms. Further, the quantum state must imply an initial condition of low total entropy so that the universe can exhibit the second law of thermodynamics

Almost as important as what the quantum state is required to predict, is what it is not required to predict. The beauty of quantum theory is that probabilities are basic. A simple discoverable theory of the quantum state is therefore unlikely to predict with high probability the *particular* classical history we observe with all its apparent complexity. Rather, it predicts the simple dynamical regularities common to every classical history with high probability, leaving to quantum accidents the complexity of particular configurations of matter observed. Thus, quantum mechanics allows the laws determining probabilities to be simple and still be consistent with present complexity.

It is possible to emphasize how specific these requirements for a quasiclassical realm are. Surely they will not be satisfied by every state in quantum gravity nor every conceivable theory of quantum dynamics. They are sufficiently specific that classicality could be important as a vacuum state selection principle[26] in theories like string theory that permit many.

However, it is equally striking how little is required for a final theory to exhibit a quasiclassical realm. The small number and general nature of the requirements discussed above mean that there must be many states and dynamical theories that manifest a quasiclassical realm. Indeed, historically, classical physics has shielded us from the nature of the final theory. Given classical spacetime, the form of the classical equations of motion was determined by conservation laws plus Maxwell's equations for the electromagnetic field and the Einstein equation for spacetime geometry. The equations of state, susceptibilities, etc. that entered into these equations could be determined phenomenologically. It was thus not necessary even to know about atoms, much less their quantum mechanics, to explore classical regularities. As far as quantum gravity is concerned, the expansion of the universe has shielded us from an immediate need to consider it by driving the characteristic scales of matter away from the Planck scale.

In these ways our particular universe has allowed a step by step, level by deeper level, journey of discovery of the fundamental regularities — a journey which we have not yet completed. The quasiclassical realms of everyday experience have played a central role in this journey, both as a starting point for the exploration and as the chief observational feature of our quantum universe to be explained.

Acknowledgments

Discussions with Murray Gell-Mann and Jonathan Halliwell over many years on the emergence of classical behavior in the universe are gratefully acknowledged. This work was supported in part by the National Science Foundation under grant PHY05-55669.

Appendix A

The Quantum Mechanics of Closed Systems

Largely to explain the notation, this appendix gives a bare-bones account of some essential elements of the modern synthesis of ideas constituting the decoherent histories quantum mechanics of closed systems.[7–9]

The most general objective of quantum theory is the prediction of the probabilities of individual members of sets of coarse-grained alternative histories of the closed system. For instance, we might be interested in alternative histories of the center-of-mass of the Earth in its progress around the Sun, or in histories of the correlation between the registrations of a measuring apparatus and a property of the subsystem. Alternatives at one moment of time can always be reduced to a set of yes/no questions. For example, alternative positions of the Earth's center-of-mass can be reduced to asking, "Is it in this region — yes or no?", "Is it in that region — yes or no?", etc. An exhaustive set of yes/no alternatives at one time is represented in the Heisenberg picture by an exhaustive set of orthogonal projection operators $\{P_\alpha(t)\}$, $\alpha = 1, 2, 3, \ldots$. These satisfy

$$\sum\nolimits_\alpha P_\alpha(t) = I, \text{ and } P_\alpha(t)\, P_\beta(t) = \delta_{\alpha\beta} P_\alpha(t), \tag{A.1}$$

showing that they represent an exhaustive set of exclusive alternatives. In the Heisenberg picture, the operators $P_\alpha(t)$ evolve with time according to

$$P_\alpha(t) = e^{+iHt/\hbar} P_\alpha(0)\, e^{-iHt/\hbar}. \tag{A.2}$$

The state $|\Psi\rangle$ is unchanging in time.

An important kind of set of histories is specified by a series[i] of sets of single time alternatives $\{P^1_{\alpha_1}(t_1)\}$, $\{P^2_{\alpha_2}(t_2)\}, \ldots, \{P^n_{\alpha_n}(t_n)\}$ at a sequence of times $t_1 < t_2 < \cdots < t_n$. The sets at distinct times can differ and are distinguished by the superscript on the P's. For instance, projections on ranges of position might be followed by projections on ranges of momentum, etc. An individual history α in such a set is a particular sequence of alternatives $(\alpha_1, \alpha_2, \ldots, \alpha_n) \equiv \alpha$ and is represented by the corresponding chain of projections called a *chain or class operator*

$$C_\alpha \equiv P^n_{\alpha_n}(t_n) \cdots P^1_{\alpha_1}(t_1). \tag{A.3}$$

[i]Realistically, the sets are *branch dependent* with the sets at one time depending on the particular sequence of alternatives at preceding times. However, we ignore branch dependence in this simplified exposition.

A set of histories like that specified by (A.3) is generally *coarse-grained* because alternatives are specified at some times and not at every time, and because the alternatives at a given time are typically projections on subspaces with dimension greater than one and not projections onto a complete set of states. Perfectly *fine-grained* sets of histories consist of one-dimensional projections at each and every time.

Operations of fine and coarse-graining may be defined on sets of histories. A set of histories $\{\alpha\}$ may be *fine-grained* by dividing up each class into an exhaustive set of exclusive subclasses $\{\alpha'\}$. Each subclass consists of some number of histories in a coarser-grained class, and every finer-grained subclass is in some class. *Coarse-graining* is the operation of uniting subclasses of histories into bigger classes. Suppose, for example, that the position of the Earth's center-of-mass is specified by dividing space into cubical regions of a certain size. A coarser-grained description of position could consist of larger regions made up of unions of the smaller ones. Consider a set of histories with class operators $\{C_\alpha\}$ and a coarse-graining with class operators $\{\bar{C}_{\bar{\alpha}}\}$. The operators $\{\bar{C}_{\bar{\alpha}}\}$ are then related to the operators $\{C_\alpha\}$ by summation, viz.

$$\bar{C}_{\bar{\alpha}} = \sum_{\alpha \in \bar{\alpha}} C_\alpha, \tag{A.4}$$

where the sum is over the C_α for all finer-grained histories α contained within $\bar{\alpha}$.

For any individual history α, there is a *branch state vector* defined by

$$|\Psi_\alpha\rangle = C_\alpha|\Psi\rangle. \tag{A.5}$$

When probabilities can be consistently assigned to the individual histories in a set, they are given by

$$p(\alpha) = \| |\Psi_\alpha\rangle \|^2 = \|C_\alpha|\Psi\rangle\|^2 = \|P^n_{\alpha_n}(t_n)\cdots P^1_{\alpha_1}(t_1)|\Psi\rangle\|^2. \tag{A.6}$$

However, because of quantum interference, probabilities cannot be consistently assigned to every set of alternative histories that may be described. The two-slit experiment provides an elementary example: A electron emitted by a source can pass through either of two slits on its way to detection at a farther screen. It would be inconsistent to assign probabilities to the two histories distinguished by which slit the electron goes through, if no "measurement" process determines this. Because of interference, the probability for arrival at a point on the screen would not be the sum of the probabilities to arrive there by going through each of the slits. In quantum theory, probabilities are squares of amplitudes and the square of a sum is not generally the sum of the squares.

Negligible interference between the branches of a set

$$\langle \Psi_\alpha | \Psi_\beta \rangle \approx 0, \quad \alpha \neq \beta, \tag{A.7}$$

is a sufficient condition for the probabilities (A.6) to be consistent with the rules of probability theory. The orthogonality of the branches is approximate in realistic situations. But we mean by (A.7) equality to an accuracy that defines probabilities well beyond the standard to which they can be checked or, indeed, the physical situation modeled.[27]

Specifically, as a consequence of (A.7), the probabilities (A.6) obey the most general form of the probability sum rules

$$p(\bar{\alpha}) \approx \sum_{\alpha \in \bar{\alpha}} p(\alpha) \tag{A.8}$$

for any coarse-graining $\{\bar{\alpha}\}$ of the $\{\alpha\}$. Sets of histories obeying (A.7) are said to (medium) decohere. Medium-decoherent sets are thus the ones for which quantum mechanics makes predictions of consistent probabilities through (A.6). The decoherent sets exhibited by our universe are determined through (A.7) and by the Hamiltonian H and the quantum state $|\Psi\rangle$. The term *realm* is used as a synonym for a decoherent set of coarse-grained alternative histories.

An important mechanism of decoherence is the dissipation of phase coherence between branches into variables not followed by the coarse-graining. Consider, by way of example, a dust grain in a superposition of two positions deep in interstellar space.[31] In our universe, about 10^{11} cosmic background photons scatter from the dust grain each second. The two positions of the grain become correlated with different, nearly orthogonal states of the photons.

Coarse-grainings that follow only the position of the dust grain at a few times therefore correspond to branch state vectors that are nearly orthogonal and satisfy (A.8).

Measurements and observers play no fundamental role in this general formulation of usual quantum theory. The probabilities of measured outcomes can be computed and are given to an excellent approximation by the usual story. But, in a set of histories where they decohere, probabilities can be assigned to the position of the Moon when it is not receiving the attention of observers and to the values of density fluctuations in the early universe when there were neither measurements taking place nor observers to carry them out.

References

1. M. Gell-Mann and J. B. Hartle, Quantum mechanics in the light of quantum cosmology, in *Complexity, Entropy, and the Physics of Information*, SFI Studies in the Sciences of Complexity, Vol. VIII, ed. W. Zurek (Addison Wesley, Reading, MA, 1990).
2. M. Gell-Mann and J. B. Hartle, Classical equations for quantum systems, *Phys. Rev. D* **47**, 334 (1993), arXiv: gr-qc/9210010.
3. M. Gell-Mann and J. B. Hartle, Quasiclassical coarse graining and thermodynamic entropy, *Phys. Rev. A* **76**, 022104 (2007), arXiv:quant-ph/0609190.
4. W. Zurek, Decoherence, einselection, and the quantum origins of the classical, *Rev. Mod. Phys.* **75**, 715 (2003).
5. N. P. Landsman, Between classical and quantum, in *Philosophy of Physics* (North-Holland, 2006), arXiv:quant-ph/0506082.
6. S. W. Hawking, The quantum state of the universe, *Nucl. Phys. B* **239**, 257–276 (1984).
7. R. B. Griffiths, *Consistent Quantum Theory* (Cambridge University Press, Cambridge, UK, 2002).
8. R. Omnès, *Interpretation of Quantum Mechanics* (Princeton University Press, Princeton, 1994).
9. M. Gell-Mann, *The Quark and the Jaguar* (W. H. Freeman, New York, 1994).
10. J. B. Hartle, The quantum mechanics of closed systems, in *Directions in General Relativity*, Vol. 1: *A Symposium and Collection*

of Essays in Honor of Professor Charles W. Misner's 60th Birthday, eds. B.-L. Hu, M. P. Ryan and C. V. Vishveshwara (Cambridge University Press, Cambridge, 1993), arXiv:gr-qc/9210006.
11. R. P. Feynman and J. R. Vernon, The theory of a general quantum system interacting with a linear dissipative system, *Ann. Phys. (N.Y.)* **24**, 118 (1963).
12. A. Caldeira and A. Leggett, Path integral approach to quantum Brownian motion, *Physica A* **121**, 587 (1983).
13. W. Unruh and W. Zurek, Reduction of a wave packet in quantum Brownian motion, *Phys. Rev. D* **40**, 1071 (1989).
14. W. Zurek, The reduction of the wave packet: How long does it take, in *Frontiers of Non-Equilibrium Quantum Statistical Physics*, eds. G. Moore and M. Scully (Plenum Press, New York, 1984), arXiv:quant-ph/0302044.
15. D. Forster, *Hydrodynamic Fluctuations, Broken Symmetry, and Correlation Functions* (Addison-Wesley, Redwood City, 1975).
16. D. N. Zubarev, *Nonequilibrium Statistical Thermodynamics*, eds. P. Gray and P. J. Shepherd (Consultants Bureau, New York, 1974).
17. J. Halliwell, Decoherent histories and hydrodynamic equations, *Phys. Rev. D* **58**, 1050151-12 (1998), arXiv:quant-ph/9805062.
18. J. Halliwell, Decoherent histories and the emergent classicality of local densities, *Phys. Rev. Lett.* **83**, 2481 (1999), arXiv:quant-ph/9905094.
19. L. Landau and E. Lifshitz, *Fluid Mechanics* (Pergamon, London, 1959).
20. J. Halliwell and S. W. Hawking, Origin of structure in the universe, *Phys. Rev. D* **31**, 1777 (1985).
21. S. W. Hawking, R. Laflamme and G. W. Lyons, Origin of time asymmetry, *Phys. Rev. D* **47**, 5342–5356 (1993).
22. L. Boltzmann, Zu Hrn. Zermelo's Abhandlung Über die mechanische Erklärung Irreversibler Vorgange, *Ann. Physik* **60**, 392–398 (1897).
23. J. B. Hartle, Spacetime quantum mechanics and the quantum mechanics of spacetime, in *Gravitation and quantizations, Proc. 1992 Les Houches Summer School*, eds. B. Julia and

J. Zinn-Justin, Les Houches Summer School Proceedings, Vol. LVII (North Holland, Amsterdam, 1995), gr-qc/9304006. A précis of these lectures is given in *Quantum Mechanics at the Planck Scale*, talk given at the *Workshop on Physics at the Planck Scale*, Puri, India, December 1994, arXiv:gr-qc/9508023.

24. J. B. Hartle, Generalizing quantum mechanics for quantum spacetime, in *The Quantum Structure of Space and Time: Proc. 23rd Solvay Conf. Physics*, eds. D. Gross, M. Henneaux and A. Sevrin (World Scientific, Singapore, 2007), arXiv:gr-qc/0602013.
25. J. B. Hartle and T. Hertog, *Classical behavior of quantum universes*, to appear.
26. J. B Hartle, S. W. Hawking and T. Hertog, The no-boundary measure of the universe, *Phys. Rev. Lett.* **100**, 202301 (2008), arXiv:0711:463, *The classical universes of the no-boundary quantum state*, arXiv:0803.2663.
27. J. B. Hartle, The quantum mechanics of cosmology, in *Quantum Cosmology and Baby Universes: Proc. 1989 Jerusalem Winter School for Theoretical Physics*, eds. S. Coleman, J. B. Hartle, T. Piran and S. Weinberg (World Scientific, Singapore, 1991), pp. 65–157.
28. F. London and E. Bauer, *La Théorie de L'observation en Mécanique Quantique* (Hermann, Paris, 1939).
29. S. B. Giddings, D. Marolf and J. B Hartle, Observables in effective gravity, *Phys. Rev. D* **74**, 064018 (2006), hep-th/0512200.
30. J. B. Hartle and M. Gell-Mann, Strong decoherence, in *Proc. 4th (1994) Drexel Conf. Quantum Non-Integrability: Quantum-Classical Correspondence*, eds. D.-H. Feng and B.-L. Hu (International Press of Boston/Hong Kong, 1998), arXiv:gr-qc/9509054.
31. E. Joos and H. D. Zeh, The emergence of classical properties through interaction with the environment, *Zeit. Phys. B* **59**, 223 (1985).

CHAPTER 12

Living in a Superposition*

This essay considers a model quantum universe consisting of a very large box containing a screen with two slits and an observer (us) that can pass through the slits. We apply modern quantum mechanics of closed systems to calculate the probabilities for alternative histories of how we move through the universe and what we see. After passing through the screen with the slits, the quantum state of the universe is a superposition of classically distinguishable histories. We are then living in a superposition. Some frequently asked questions about such situations are answered using this model. The model's relationship to more realistic quantum cosmologies is briefly discussed.

*A pedagogical essay.
arXiv:1511.01550

1. Introduction

The two-slit thought experiment sketched in Fig. 1 is the simplest and clearest way of illustrating the phenomena of quantum coherence and decoherence. The characteristic pattern of places of arrival at the far screen is evidence of quantum interference. Feynman in his basic physics lectures said that the two-slit experiment is "at the heart of quantum mechanics ... [containing its] only mystery".[1]

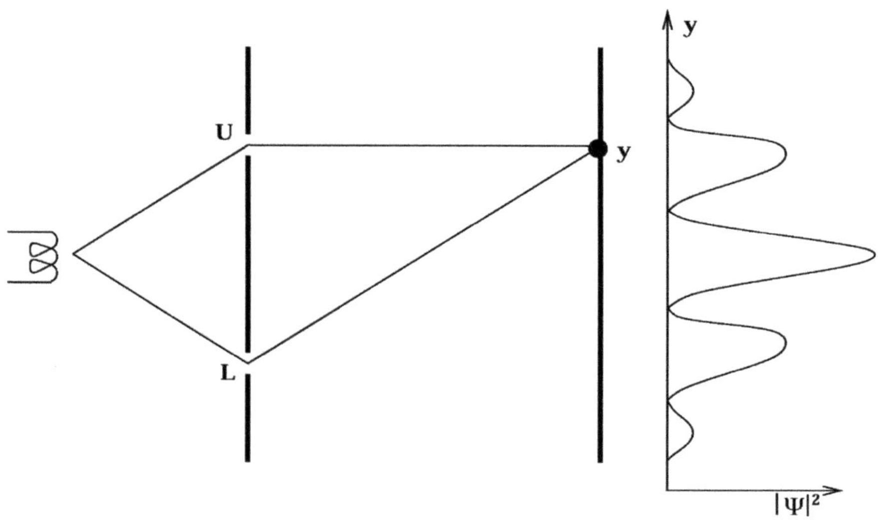

Fig. 1. The two-slit experiment. An electron gun at left emits an electron which is detected at a point y on a screen after passing through another screen with two slits. The interference pattern in arrival position that appears on the screen is a consequence of quantum interference between a history where the electron went through the upper slit and a history where it went through the lower slit.

This pedagogical essay uses a variation of the two-slit thought experiment to introduce, illustrate and clarify various aspects of the modern formulation of the quantum mechanics of closed systems, most generally the universe as a whole. This formulation is called decoherent (or consistent) histories quantum mechanics (DH) and is the work of many.[a] In applying quantum mechanics to closed systems like the universe, DH can be viewed as an extension, clarification, and to some extent, a completion of the work begun by Everett.[8]

Essential features of DH can already be understood in the context of the two-slit experiment. The most general objective of a quantum theory of a closed system is the prediction of probabilities for alternative histories of how it evolves in time — probabilities for the history of what happened in the early universe, for example. But quantum interference is an obstacle to assigning probabilities to sets of alternative histories. In the two-slit experiment, it is not possible to assign probabilities to the alternative histories in which the electron arrives at y having gone through the upper or lower slit. The probability to arrive at y should be the sum of the probabilities of the two histories. But in quantum mechanics, probabilities are squares of amplitudes, and $|\psi_L(y) + \psi_U(y)|^2 \neq |\psi_L(y)|^2 + |\psi_U(y)|^2$ because of interference. A different physical situation is illustrated in Fig. 2, where the electron interacts with the apparatus that measures which slit it passed through. Quantum interference is destroyed and the set of two histories is said to decohere. Consistent probabilities can then be assigned to these histories. In a closed system, probabilities can be consistently assigned only to sets of histories that decohere as a consequence of the system's state and Hamiltonian.

The quantum interference modeled by the two-slit experiment has been seen in many beautiful and important experiments, e.g. Refs. 9–13, using increasingly larger interfering systems over time — molecules more than a thousand au, for example.[12] Some distinguished scientists expect quantum coherence to break down for sufficiently large systems (e.g. Refs. 14 and 15), but so far there is no evidence for that, and no evidence that quantum interference phenomena could not be seen for much larger systems.

Could we, as human observers, be sent through a very large two-slit experiment, as in Fig. 3? What would we observe as we travel through the screen with the slits and where would we predict we

---

[a]For classic expositions at various levels and different emphases see, e.g. Refs. 2–6. For a tutorial in the notation used here, see Ref. 7.

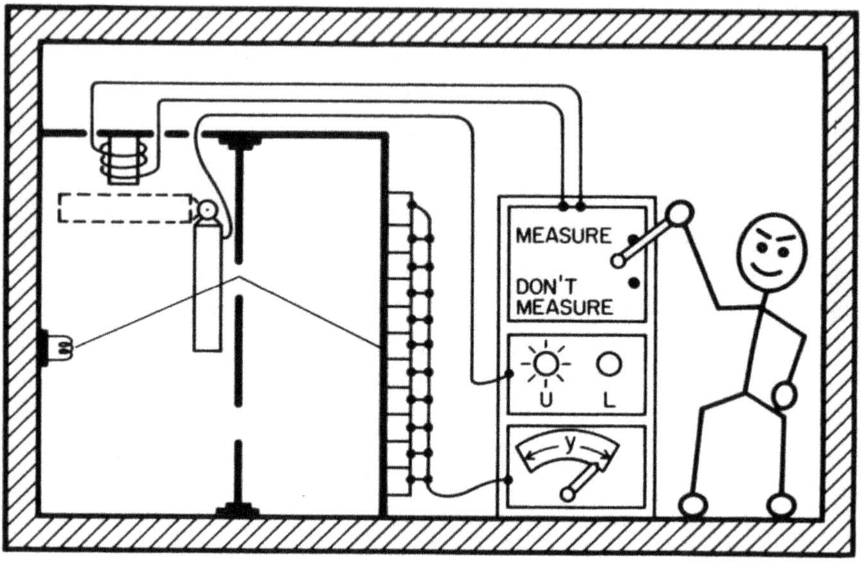

Fig. 2. A different model closed system. This picture is meant to suggest a closed system that models a laboratory measurement of a two-slit situation like those referred to in Fig. 1. A box contains both a two-slit set-up and an observer reading the results. The observer is outside the two-slit subsystem being measured but inside the box. Measuring whether the electron went through the upper slit destroys the quantum coherence between histories specifying which slit, so that probabilities can be assigned to the outcomes of the measurement. Such models can also be discussed quantum-mechanically, but in this paper we are discussing observers that are not outside a two-slit system, but rather inside one.

arrive at the farther screen? This paper addresses such questions in the context of DH utilizing a simple model closed system. We are not discussing the feasibility of carrying out such an experiment. We aim rather at a simple thought experiment to concretely illustrate, elements of the quantum mechanics of closed systems.

Beyond calculating predictions for our observations, we use this example and DH to answer a number of frequently asked questions: Are we living in a superposition? If so, why don't we see a superposition? Are we smeared out in space? Is the quantum state reduced when we make an observation? etc.

As Feynman emphasized, the essential features of the textbook quantum mechanics of measurement outcomes can be illustrated by the two-slit experiment. Here, we shall find that the essential features of the modern quantum mechanics of closed systems (DH) can be illustrated with the two-slit model universe.

The essay is structured as follows: Section 2 describes the model two-slit universe with an observer carrying a detector. The decoherent histories quantum mechanics of this system is described in Sec. 3 and used to calculate the predictions for our observations in Sec. 4. The frequently asked questions mentioned above are discussed and answered in Sec. 5. Section 6 concludes with a discussion of the relation of this example to realistic quantum cosmology.

2. A Model Two-Slit Universe

Imagine a large box containing a very large two-slit apparatus as illustrated in Fig. 3, together with a single observer (us) equipped with apparatus for survival and observation as described more fully below. We stress that everything relevant for discussion is inside the box and that nothing outside is interacting with it. We are not considering a system with an observer outside making measurements on a subsystem, as in Fig. 2. The box *is* the universe. We abbreviate "two-slit model universe" by TSMU.

We now specify the contents of the box in a little more detail. In the neighborhood of each slit, there is a gas of radiation. Near the upper slit the radiation has wavelength $\lambda_U \ll d$ where d is the distance between the slits. The radiation near the lower slit has a different wavelength $\lambda_L \ll d$. By measuring the wavelengths on the journey through the slits, an observer can tell which slit he or she is passing through.[b]

To make these measurements, we suppose that the observer is equipped with a detector that can measure which of three states the radiation is in. The three states χ_m are labeled $m = 0, 1, 2,$

[b]More fancifully we could imagine that the "**U**" and "**L**" are painted near each slit and that the observer is equipped with a flashlight (torch), and detects which slit by the pattern of reflected radiation.

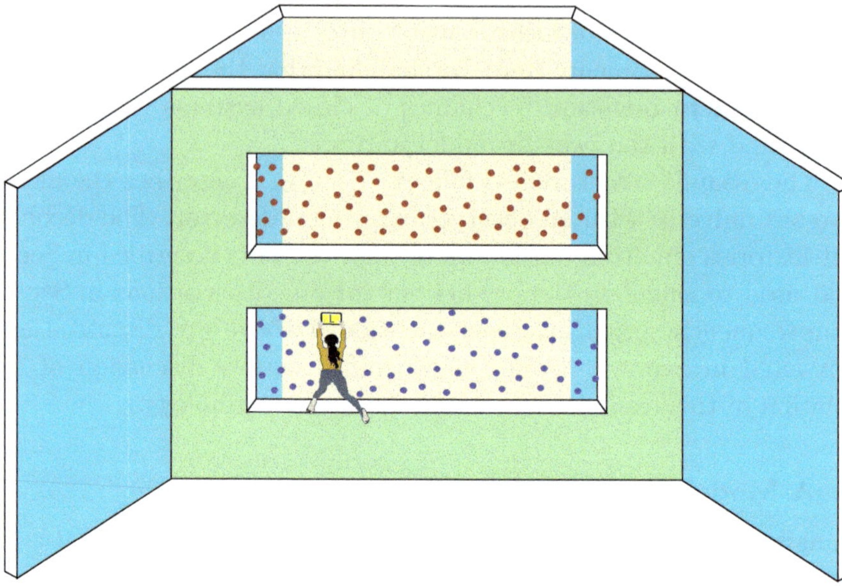

Fig. 3. A closed system (TSMU) consisting of a very large two-slit apparatus in a closed box together with an observer (us) carrying a radiation detector that goes through a slit and reaches some point on the farther screen. The top and front of the box have been omitted in the figure to aid visualization but no photons or other information are getting in or out through them. There is a gas of radiation (red) with wavelength λ_U near the upper slit, and a gas of radiation (blue) with wavelength λ_L near the lower slit. The observer's detector records the presence of radiation and its wavelength. As pictured, the observer is going through the lower slit, recording the blue radiation there with her detector.

having energy levels $0, \hbar/\lambda_U, \hbar/\lambda_L$, respectively. The detector starts its journey in the ground state 0. Levels 1 and 2 will be excited if it passes through the upper or lower slit, respectively. We assume that the observer's center of mass degree of freedom is negligibly affected by either kind of radiation.

The excitation of the detector creates a record of which slit the observer passed through. If the excitation is to level 1 with energy $\hbar/\lambda_U$, then the observer passed through the upper slit, and if to level 2 with energy $\hbar/\lambda_L$, then it was through the lower slit. If the decay time of these levels is long compared to the travel time to the farther screen, then the observer can be said to have a memory of which slit was passed through. That record constitutes a piece of data on which the observer can condition to construct probabilities for further observations, as we see in Sec. 4.

Eventually, the observer and detector reach the screen at the far end of the box. To describe her vertical position there we divide the height up into discrete intervals of length Δ. These are labeled by a discrete variable Y, where $Y = 1, 2, \ldots$.

We next discuss the quantum mechanics of this model.

3. The Quantum Mechanics of the Model

3.1. *The Wave Function of the Universe*

The only degrees of freedom that we follow in this model universe are the center of mass position of the observer and the state of the radiation detector. The box, slits, screen, radiation, etc., are all considered classically. It would be more realistic to consider them quantum-mechanically, but we aim at a simple model. For further simplification, we assume symmetry in the direction along the slits. The model is thus effectively two-dimensional, as illustrated in Fig. 4. The configuration space of the model universe is therefore spanned by the (x, y) coordinates of the observer's center of mass and the three states of the detector $m = 0, 1, 2$.

The inputs to prediction are the Hamiltonian H and the quantum state of the model universe $|\Psi(t)\rangle$. This is a function of time t in the Schrödinger picture in which we work throughout. The state can be described by a wave function on configuration space, viz.

$$\Psi = \Psi(x, y, m, t). \tag{1}$$

We move back and forth between wave functions like (1) and the representation in terms of bras and kets like $|\Psi(t)\rangle$ for convenience.

It is important to note that the observer's degrees of freedom and the registrations of the apparatus are inside the closed system, not outside it. Their degrees of freedom are arguments of the wave function just like any others.

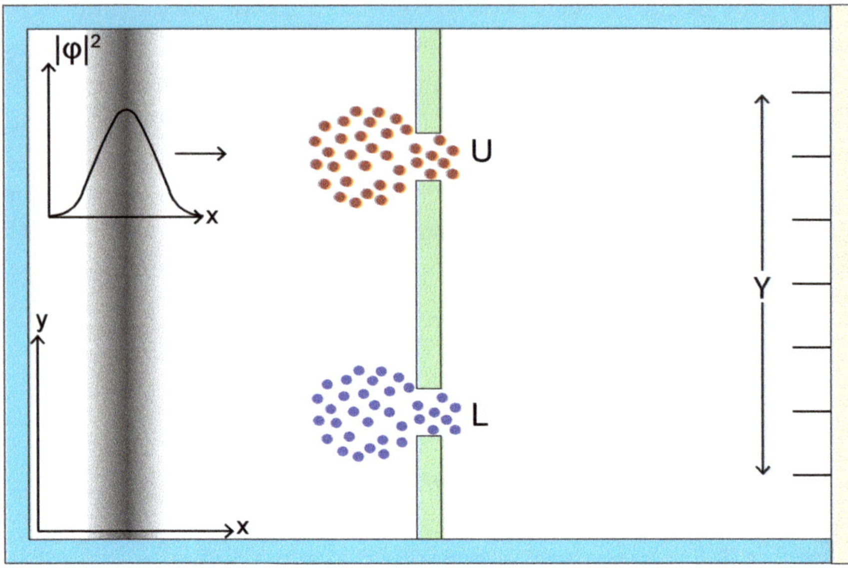

Fig. 4. A two-dimensional slice through the box in Fig. 3. This is useful because we assume symmetry in the transverse direction. The slice is shown at time t_0 when the wave function of the observer's center of mass is a narrow wave packet $\phi(x,t)$ in the x-direction (inset) [cf. (2)]. The packet is moving toward the right so as to arrive at the slits at time t_S and at the farther screen at time t_D. There is a gas of radiation of wavelength λ_U near the upper slit (red) and a cloud with a different wavelength λ_L near the lower slit (blue). We consider a set of alternative histories defined by alternatives at just two times: Whether the observer went though the upper (U) or lower (L) slit at time t_S, and at what interval Y the observer arrived at the farther screen at time t_D.

At the starting time t_0 we take the wave function of the model universe to be

$$\Psi(x,y,m,t_0) = \phi(x,t_0)\delta_{0m}, \quad t_0 < t < t_S \tag{2}$$

where $\phi(x,t_0)$ is a narrow wave packet peaked to the left of the slits but moving to the right, so as to reach the slits at time t_S and the detecting screen at t_D, as shown in Fig. 4. For simplicity, we assume that there is no y-dependence of the initial wave function over the height of the box. Thus, progress in x recapitulates evolution in time.

The wave function evolves in time by the Schrödinger equation

$$i\hbar\frac{\partial\Psi}{\partial t} = H\Psi \tag{3}$$

where the Hamiltonian H describes the evolution of the observer's center of mass position in the presence of the impenetrable walls and the interaction of the detector with the radiation. We won't solve this equation explicitly, but rather posit the plausible forms of the solutions.

After passing through the slits, the wave function of observer and detector has the form

$$\Psi(x, y, m, t) = \psi_U(y, t)\phi(x, t)\delta_{m1} + \psi_L(y, t)\phi(x, t)\delta_{m2}, \quad t_S < t < t_D. \tag{4}$$

Here, the first term $\psi_U(y, t)$ is localized near the upper slit at time t_S and spreads over a larger region of y by the time t_D such that the observer hits the detecting screen. The detector is excited to the level 1 at time t_S and remains in that state of excitation for the rest of the journey to the screen at time t_D. Similarly for the second term.

3.2. *Histories*

Suppose that the observer and detector arrive at position interval Y at the screen. What are the probabilities that they went through the upper slit (U) or lower slit (L) on the way there? These are each the probability of a history — a sequence of events at a series of times. In this case there are just two times — the time t_S when the observer and detector reach the slits, and the time t_D when they reach the screen. The individual histories in an exhaustive set of alternative histories are labeled by (Y, S) where S is U or L.

A set of histories like $\{Y, S\}$ has a branching structure illustrated in Fig. 5. A wave function called a "branch wave function" can be associated with each history. The initial wave function divides into two branch wave functions at time t_S and each of those divides into branches for different Y's at time t_D. The initial wave function is a sum of all the branch wave functions. As we will discuss in the next section, when it is consistent to assign probabilities to a set of

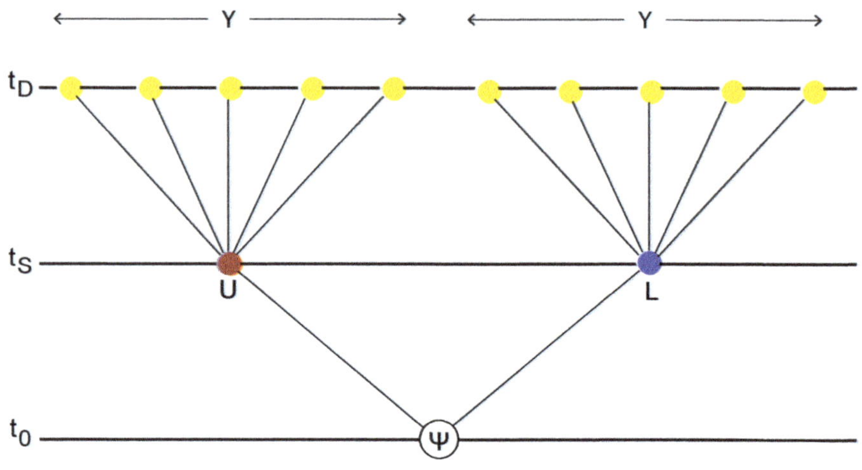

Fig. 5. A traditional representation of the set of alternative histories discussed in this section. Time runs upward. Each path from t_0 to t_D corresponds to a history in which the observer and detector went through one slit S (U or L) at time t_S and arrived at one interval Y at time t_D. The branching structure of the set is evident. The initial wave function Ψ divides into further branch wave functions at each step. When all the branch wave functions are mutually orthogonal, the set of histories decoheres.[c]

histories, their probabilities are norms of corresponding branch wave functions.

There is a general procedure for constructing branch wave functions from projection operators at representing alternatives at different times. But the set $\{Y, S\}$ is so simple that we can read the results off of (4). The branch wave functions $\Psi_{YS}(x, y, m, t)$ are the two parts of (4) restricted to the various values of Y.

The restriction to the intervals Y can be made more explicit by introducing projection operators P_Y that are one for y's inside the

[c]For the experts: In this kind of diagram, the dots correspond to projection operators representing the various alternatives at different times. The lines connecting the dots and the state represent unitary evolution. The branch state vector of a particular history can be generated from the initial by a sequence of unitary evolutions and projections with a result like (6).

interval Y and zero for y's outside it. These satisfy

$$P_Y P_{Y'} = \delta_{YY'}, \quad \sum_Y P_Y = I \tag{5}$$

showing that they are an exclusive and exhaustive set of projections.

The branch wave functions for the set $\{Y, S\}$ at the last time in the histories t_D are then, from (4),

$$\Psi_{YU}(x, y, m, t_D) = P_Y \psi_U(y, t_D)\phi(x, t_D)\delta_{m1}, \tag{6a}$$

$$\Psi_{YL}(x, y, m, t_D) = P_Y \psi_U(y, t_D)\phi(x, t_D)\delta_{m2}. \tag{6b}$$

These branch wave functions represent branch state vectors which we write as $|\Psi_{YU}(t_D)\rangle$ and $|\Psi_{YL}(t_D)\rangle$. From (5) and (4) we have

$$\Psi(x, y, m, t_D) = \sum_{Y,S} \Psi_{YS}(x, y, m, t_D). \tag{7}$$

The wave function of the model universe is the sum of all its branches. We now turn to the probabilities of these histories.

3.3. *Decoherent Histories and Probabilities*

The natural candidates for the probabilities of the history (Y, S) would be squares of the norms of its branch wave functions (6), viz.

$$p(Y, S) = \| |\Psi_{YS}(t_D)\rangle \|^2. \tag{8}$$

By "norm" here we mean the usual inner product between states represented by wave functions. For two states $|\Phi''(t)\rangle$ and $|\Phi'(t)\rangle$ with wave functions $\Phi''(x, y, m, t)$ and $\Phi'(x, y, m, t)$, we define

$$\langle \Phi''(t)|\Phi'(t)\rangle \equiv \sum_m \int dx dy \Phi''^*(x, y, m, t)\Phi'(x, y, m, t). \tag{9}$$

The norm of a state $|\Phi(t)\rangle$ is then

$$\| |\Phi(t)\rangle \|^2 \equiv \langle \Phi(t)|\Phi(t)\rangle. \tag{10}$$

The simple example in the Introduction shows that the probabilities (8) are not consistent with the rules of probability theory unless the quantum interference between the branches vanishes. More generally, the set of histories must decohere to have consistent quantum probabilities.

A set of histories decoheres if the branch state vectors are orthogonal for different histories. For the set $\{Y, S\}$ this is

$$\langle \Psi_{Y''S''}(t_D)|\Psi_{Y'S'}(t_D)\rangle \propto \delta_{Y''Y'}\delta_{S''S'} \tag{11}$$

for all values of Y and S. This orthogonality is the natural notion of the absence of quantum interference between the branches.

It is easy to see from (6) and giving the branch wave functions that the set $\{Y, S\}$ is decoherent. The branches are orthogonal in Y because $P_{Y''}P_{Y'} = 0$ for differing Y'' and Y' [cf. (5)]. The branches are orthogonal in S because the detector states χ_m are orthogonal. In the next subsection we will check the resulting consistency of the probabilities (8) explicitly.

3.4. *Consistency*

Suppose we ask just for the probability $p(Y)$ that the observer arrives at the interval Y on the far screen at time t_D. The branch state vector for this history is [cf. (6)]

$$|\Psi_Y(t_D)\rangle = P_Y|\Psi(t_D)\rangle. \tag{12}$$

Evidently, from (4) and (6) we have

$$|\Psi_Y(t_D)\rangle = |\Psi_{YU}(t_D)\rangle + |\Psi_{YL}(t_D)\rangle. \tag{13}$$

The probability to arrive at Y is

$$p(Y) = \||\Psi_Y(t_D)\rangle\|^2 = \||\Psi_{YU}(t_D)\rangle\|^2 + \|\Psi_{YL}(t_D)\rangle\|^2. \tag{14}$$

There is no interference term because the two branch state vectors are orthogonal (11). The probability distribution just for Y is then just the sum of the probabilities for these two branches

$$p(Y) = p(Y, U) + p(Y, L) \tag{15}$$

shown schematically in Fig. 6. Decoherence ensures that probabilities are consistent with the usual rules of probability theory.

3.5. *Records*

This set of histories decoheres because a record of which slit the observer had passed through is created in the state of the detector.

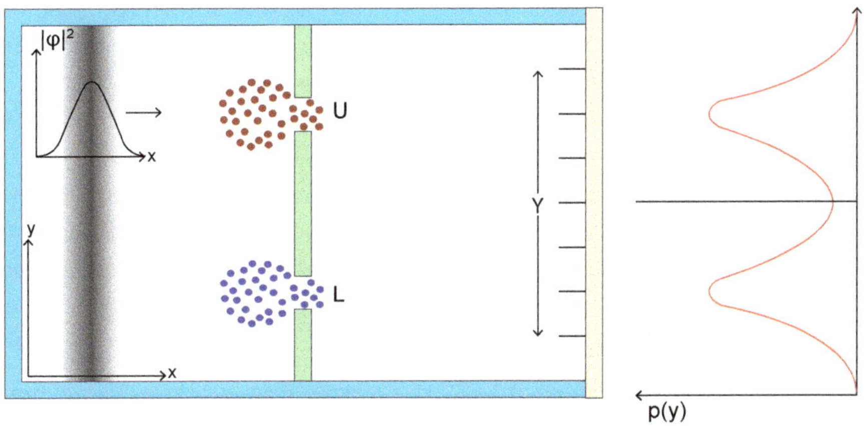

Fig. 6. A qualitative representation of the probability $p(Y)$ for the observer and detector to arrive in the position interval Y on the end screen. Because there is a record of which slit the system passed through, the set of histories decoheres, and the probability $p(Y)$ is the sum of the probabilities to go through the upper slit and arrive at Y and through the lower slit and arrive at Y. The result is not an interference pattern.

This association between records and decoherence is a very general one and a fundamental property of DH. Decoherence implies the existence of records somewhere in the closed system, although they may not be in a form that is accessible or useful to us. Conversely, the existence of records of histories implies their decoherence, as in this example.[d] Indeed, in realistic situations many records of the same history will be created (e.g. Ref. 16).

Note that the set of histories decoheres whether or not the observer looks at the state of excitation. The states χ_1 and χ_2 are still orthogonal[e] and (11) still holds.

The association of decoherence and records has been beautifully demonstrated in the experiment of Hackermüller *et al.*[11]: A heated

[d]For the experts, we are talking about medium decoherence and strong records.

[e]This is the origin of the injunction in Copenhagen quantum mechanics to sum probabilities for alternatives that "could have been measured but were not".[17]

buckyball (C_{60}) is passed through a Talbot–Lau interferometer exhibiting a quantum interference pattern. As the temperature is increased, the interference pattern disappears when a temperature is reached where the wavelength of the radiation is short enough that it would contain a record about which slit the buckyball passed through. That is the creation of a record by emission in contrast to the creation by absorption in this toy model. But both bear out the idea that histories that decohere are histories that are recorded.

3.6. *Coarse-grained Histories*

The set of histories $\{Y, S\}$ illustrated in Fig. 5 is coarse-grained. The position of the observer is followed only at two times t_S and t_D and not at all times. And these positions are not followed to arbitrary accuracy, but only to the widths of the slits and of the intervals Y. The positions in between are unspecified — a coarse-grained description of the histories. Specifying only the interval of arrival Y at the farther screen would be a coarser-grained description. Specifying more details of the path taken in between would be a finer-grained description. A coarse-graining of a decoherent set like $\{Y, S\}$ is again decoherent. But fine-graining risks losing decoherence. Some coarse-graining is essential for decoherence. A completely fine-grained description would not decohere.

4. Third Person and First Person

In the quantum mechanics of a closed system like TSMU it is useful to distinguish between two kinds of description of the system and correspondingly two kinds of probabilities.[18,19]

Third Person Descriptions and Probabilities: Descriptions of what the universe contains and how that evolves — histories of what occurs. Since observers are physical subsystems within the closed system, third person descriptions include a description of the histories of what observers see and how they behave. All of the previous discussions have been about third person histories of one observer and its detector — which slit it goes through, how the detector was excited, where it arrives at the screen, etc. The probabilities for these third person histories are called third person probabilities. They are what is supplied directly by the quantum state of the system and the dynamics. Examples are the probabilities $p(Y, S)$ in (8).

First Person Descriptions and Probabilities: Suppose that TSMU is our universe and we are the observer in it. We are interested in the first person probabilities for what *we* observe — what wavelength radiation *we* detect, which Y *we* arrive at, etc. Since we are a physical system within the universe, first person probabilities can be derived from the third person probabilities for its histories.[f] We don't observe whole four-dimensional histories, but rather limited features of the universe from an observing situation that is localized in space and time. The first person probabilities for what we observe are necessarily conditioned on the data $\mathcal{D}$ describing the observational situation, including the information about when the observation was made. The first person probability for an observable $\mathcal{O}$ is

$$p^{(1p)}(\mathcal{O}) = p(\mathcal{O}|\mathcal{D}). \tag{16}$$

Here is an example: Our detector makes a transition to level 1 at time t_S. We then know that we have just passed through the upper slit U. That is the data $\mathcal{D}$. Given that data, what do we predict for the probability that we will arrive at position Y on the screen? This is the first person probability

$$p^{(1p)}(Y) = p(Y|U) \equiv p(Y, U)/p(U). \tag{17}$$

Using (8), this is

$$p^{(1p)}(Y) = \frac{\||\Psi_{YU}(t_D)\rangle\|^2}{\||\Psi_U(t_D)\rangle\|^2}. \tag{18}$$

This distribution looks like the top part of the graph of $p(Y)$ versus Y in Fig. 6 but renormalized so that the total probability for some Y is unity.

To evaluate (18) for $p(Y|U)$, it's obviously not necessary to compute the probability of which Y would have been arrived at if we had gone through the lower slit. We only need a set of histories that

---

[f]In other papers, we have called first and third person probabilities, top down and bottom up probabilities, respectively, e.g. Ref. 20.

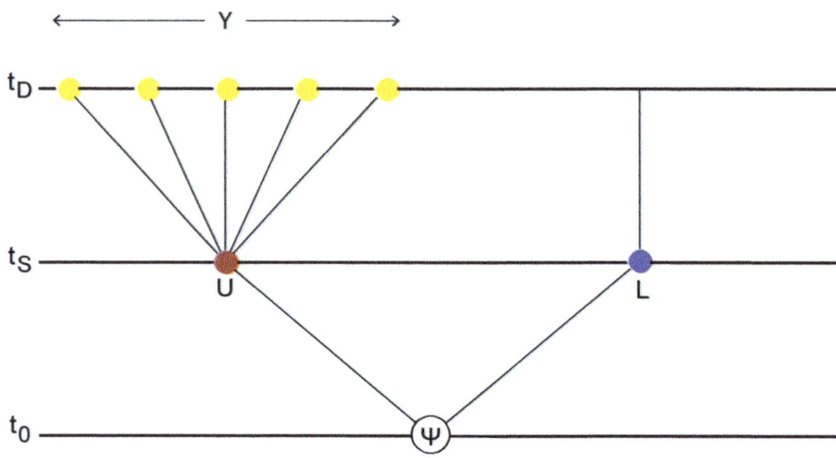

Fig. 7. There are different third person descriptions of TSMU at different levels of coarse-graining. Suppose that at time t_S our detector shows we went through the upper slit U. We want to predict where we will hit the screen at the later time t_S. We could calculate using the set in Fig. 5, but it would be inefficient because that also describes what *would* have happened if we went through the lower slit. A better set of alternative histories of TSMU is the one shown above that follows which Y is reached only if we go through the upper slit. That is a coarser-grained set than the one in Fig. 5.

follow our future observations and ignores (coarse-grains) features of TSMU that are irrelevant for that. Figure 7 shows an example.[g]

The branch wave functions of the histories in Fig. 7 are [cf. (6)]

$$\Psi_{YU}(x, y, m, t_D) = P_Y \psi_U(y, t_D)\phi(x, t_D)\delta_{m1}, \tag{19a}$$

$$\Psi_L(x, y, m, t_D) = \psi_U(y, t_D)\phi(x, t_D)\delta_{m2}. \tag{19b}$$

These are decoherent because of the orthogonality of the P_Y and the χ's.

[g]This is an example of an adaptive branch dependent set of histories.[21] Different branches at t_S branch differently in ways that are adapted to what our data are then.

5. Some Frequently Asked Questions

Imagine that we are the observer in TSMU. A number of questions arise about quantum mechanics in this context. Many of these FAQ's are not clearly defined. We mostly focus on FAQ's that can be reformulated so that they are answered by appropriate quantum probabilities.[h]

- *Are we living in a superposition?* **Yes.** Any state can be expressed as a superposition of other states, for example, by using a basis. This FAQ is therefore not precisely defined. Probably the question of interest to many is rather whether the state is a superposition of different histories that each can be "classically described" and are "macroscopically distinct". In TSMU, after the time t_S, the state of the universe has evolved to the superposition of the two branches in (4). Each of these can be described in classical terms and each could be said to be constructed from alternatives that are "macroscopically distinct". As the observer in the box, we are living in such a superposition. Schrödinger's cat also lives in a superposition.
- *If we are living in a superposition why aren't we smeared out?* The probability for the observer to be in two places at once is zero because the operators vanish that would represent this kind of situation. For example, (7) shows that the wave function of the universe is in a superposition of different arrival intervals Y. But since $P_Y P_{Y'} = 0$ for $Y \neq Y'$, see (5), it predicts zero probabilities for the observer to be in two intervals at once. It is the same for the slits.
- *If we are living in a superposition, why don't we see a superposition or feel superposed?* The detector could be part of the observer's brain. Registrations of the detector then model a physical realization of "see" and "feel". We only feel ourselves going through the upper slit or alternatively feel ourselves going through the lower slit even though the quantum state is a

[h]For FAQs that are more about the author's opinion on issues that come up in quantum mechanics like which histories are real, see e.g. Ref. 22.

superposition of the two. These are exclusive alternatives. The reason we don't see the superposition is that we are not somehow outside the universe observing whether its state is a superposition of terms. We are inside the universe participating in one of the terms of the superposition.

- *Does TSMU model a measurement?* **Yes.** There is no precise definition of "measurement" in textbook (Copenhagen) quantum mechanics or in DH. But the author would informally characterize the excitation of the detector by the radiation near the slits as "a measurement situation". A variable — the wavelength of the radiation — becomes correlated with an excited state of the detector that can be read by the observer. Certainly TSMU has many similarities with classic measurement models,[23] for instance the one in Fig. 2. But it also differs from these models in that what is measured is not fixed, but rather determined by the quantum accident of which slit the observer passed through. Specific measurement situations can be described in DH, and their outcomes predicted, but a precise general notion of measurement is not needed for DH's formulation.
- *Do the predictions of DH differ from those of textbook quantum mechanics?* **Yes and No.** Textbook quantum theory predicts the probabilities for the outcomes of measurements carried out by one subsystem of the universe on another. DH will predict the same probabilities to an excellent approximation. In TSMU, DH would yield the same probabilities for the detector's measurements of the radiation. But DH also predicts probabilities for which measurements are carried out where, and for the motion and fate of the observer — alternatives are not considered by textbook quantum theory. Textbook quantum theory is not an alternative to DH, but rather contained within it as an approximation appropriate for measurement situations.
- *Would anthropic reasoning modify the predictions of TSMU?* **No.** Anthropic reasoning is automatic in DH through the first person probabilities for observation (Sec. 4).[19] Suppose that the radiation at the upper slit were intense enough to kill the observer. The third person probabilities for the histories (Y, S) of the

motion of the observer and the registration of the detector would be unchanged. But the first person probabilities for our observations would be affected assuming that the data $\mathcal{D}$ contained information that we are alive at the farther screen. The first person probability would be unity that we passed through the lower slit. There is zero first person probability to observe the red radiation which is where we cannot exist.

- *Is the quantum state of the universe ever reduced?* **No.** In Sec. 4, we derived the first person probability $p(Y|\mathcal{D})$ to arrive at an interval Y on the far screen given data $\mathcal{D}$ about the detector registration. Suppose these data imply that we passed through the upper slit U. Equation (18) for this probability can be rewritten as

$$p(Y|U) = \|P_Y \mathsf{U}(t_D, t_S)|\Psi_U(t_S)\rangle\|^2 \tag{20a}$$

where

$$|\Psi_U(t_S)\rangle \equiv \frac{P_U|\Psi(t_S)\rangle}{\|P_U|\Psi(t_S)\rangle\|} \tag{20b}$$

and $\mathsf{U}(t_D, t_S)$ represents unitary evolution from t_S to t_D by the Schrödinger equation. Superficially, the formulae (20) are like those in textbook quantum mechanics describing the reduction of the state of a subsystem that occurs when the subsystem undergoes an "ideal" measurement by another subsystem outside it. This was von Neumann's second law of evolution.[i, 24] But this resemblance is misleading. The states and operators in these equations are not of a subsystem of the universe being measured, but rather of the whole thing. The Hilbert space includes both what is observed and the system observing it. In the quantum mechanics of the universe, there is no "other measuring system" and no "second law of evolution". Equation (20b) is not some mysterious feature of a measurement process. Rather, it is but a step in the construction of conditional probabilities. It is no different from the "reduction" that occurs in horse racing when

[i]This second law of evolution is itself problematic since almost no realistic measurements are "ideal".

a particular horse wins and the probabilities for further races which are conditioned on that event become relevant.

- *Does some physical process cause the wave function to branch?* **No.** The branching structure in Fig. 5 is a *choice* of how to describe what goes on in a TSMU. Many other descriptions are possible, leading to different branching structures. Consider, for example, the set of histories where only Y is specified and the question of which slit we go through ignored. Then there would be no branching at t_S. Or consider the set of histories in Fig. 7 where there is no branching after going through the lower slit. DH does not prefer one decoherent set of histories over any other. All are in principle available to be used by us in the process of prediction although some will be more useful to human observers than others.
- *Is coarse-graining necessary?* **Yes.** Some coarse-graining is necessary for decoherence, except in trivial cases. Realistic mechanisms of decoherence involve coarse-graining. In the classic example of Joos and Zeh,[25] the histories of the positions of a mm-sized dust grain deep in intergalactic space decohere because of the vast number of CMB photons that scatter from it every second. A decoherent set of histories follows the positions of the dust grain and ignores (coarse-grain over) the photons. More generally, "environmental decoherence" results from separating a closed system into a subsystem and an "environment" and then coarse-graining over its environment (e.g. Ref. 21). It is a remarkable fact that in the quantum mechanics of closed systems, some information must be sacrificed in order to have interesting probabilities at all.

6. Quantum Cosmology

It is an inescapable inference from the physics of the last century that we live in a quantum-mechanical universe. We perhaps have little evidence of peculiarly quantum-mechanical phenomena on large and even familiar scales, but there is no evidence that the phenomena that we do see cannot be described in quantum-mechanical terms and explained by quantum-mechanical laws. If this inference is correct,

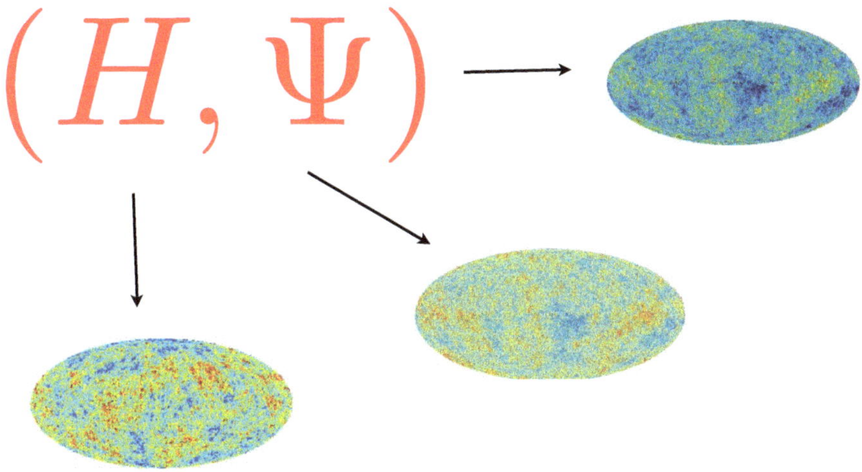

Fig. 8. There are two theoretical inputs to the process of prediction in quantum cosmology. First, there is a theory of dynamics like string theory denoted here by H. Second is the quantum state of the universe denoted here by Ψ. From these third person probabilities for the individual members, decoherent sets of alternative histories of the universe can be derived. From these and data describing our observational situation, first person probabilities predicting the results of our observations can be calculated. These include predictions for what CMB maps we will observe.

then there must be a description of the universe as a whole and everything in it in quantum-mechanical terms. The nature of this description and its predictions for observations are the subject of quantum cosmology.

The two-slit universe of this paper is a toy model to illustrate a few aspects of realistic quantum cosmology. We now describe some of the connections between the model and the ongoing program of quantum cosmology.

Theoretical Inputs: Cosmology requires a generalization of usual DH to include spacetime geometry as a quantum-mechanical variable.[26] The basic theoretical inputs are a theory of dynamics including spacetime H — say, some version of string theory and a theory of the quantum state of the universe Ψ — say, the no-boundary wave function of the universe.[27]

Histories: Our observations of the large scale universe are mostly of properties of its classical history. The expansion, the amount of inflation, the formation of the fluctuations we see in the cosmic wave background radiation (CMB), and in the large scale distribution of galaxies today, are all properties of that classical history. Quantum cosmology aims to predict probabilities for these properties by deriving probabilities for alternative classical histories from (H, Ψ), as emphasized in Fig. 8.

But classical behavior is not a given in DH. It is a matter of the quantum probabilities of decoherent sets of appropriately coarse-grained histories of geometry and field.[28] A system behaves classically when the probabilities are high for histories that exhibit correlations in time by classical deterministic laws such as the Einstein equation. Classical behavior is not built into the quantum mechanics of closed systems as it was in Copenhagen theory, but rather as an emergent feature of the probabilities supplied by (H, Ψ).

Observers and Observations: Observers and their apparatus are physical systems within the universe with only a probability to have evolved in any region of spacetime and, in a very large universe, a probability to be replicated in many regions. In TSMU, the single observer (us) and detector are physical systems within the universe assumed to exist with unit probability.

Third and First Person Probabilities: In quantum cosmology, the theory (H, Ψ) predicts third person probabilities for the history of the universe that occurs. From these, probabilities for what we will observe can be predicted. In very large universes, the histories most probable to occur may not be the histories that are most probable to be observed.[19] The branch dependent adaptive coarse-grainings discussed briefly in Sec. 4 are essential for cosmology. Our observations of the universe are limited — highly coarse-grained. The universe is vast. We can most efficiently calculate the prediction of theory for the outcomes of our observations by using sets of histories that follow what is observed and coarse-grain over features of the universe that do not affect these observations (e.g. Refs. 29 and 30).

Living in a Superposition: Just like (7) of TSMU, the quantum state of the universe is a superposition of the branch state vectors for any decoherent set of alternative classical histories. Therefore, just like the observer in the TSMU model, you and I are living in a superposition. We are all Schrödinger cats in the wave function of the universe.

Acknowledgments

The author has had the benefit of discussions with a great many scientists on the work that underlies this paper. The cited papers have those acknowledgments. However, his collaborators on the those papers over a long period of time should be thanked. They are Murray Gell-Mann, Stephen Hawking, Thomas Hertog and Mark Srednicki. Thanks are due to Mark Srednicki for critical readings of the essay and to Simon Saunders for a careful reading of the text. The work was supported in part by the US NSF grant PHY15-04541.

References

1. *The Feynman Lectures on Physics*, Vol. 3, http://www.feynmanlectures.caltech.edu.
2. R. B. Griffiths, *Consistent Quantum Theory* (Cambridge University Press, Cambridge, UK, 2002).
3. R. Omnès, *Interpretation of Quantum Mechanics* (Princeton University Press, Princeton, 1994).
4. M. Gell-Mann and J. B. Hartle, Quantum mechanics in the light of quantum cosmology, in *Complexity, Entropy, and the Physics of Information*, SFI Studies in the Sciences of Complexity, Vol. VIII, ed. W. Zurek (Addison Wesley, Reading, MA, 1990).
5. M. Gell-Mann, *The Quark and the Jaguar* (W. H. Freeman, New York, 1994).
6. P. C. Hohenberg, An introduction to consistent quantum theory, *Rev. Mod. Phys.* **82**, 2835–2844 (2010).
7. J. B. Hartle, The quantum mechanics of closed systems, in *Directions in General Relativity*, Vol. 1, eds. B.-L. Hu, M. P. Ryan and C. V. Vishveshwara (Cambridge University Press, Cambridge, 1993), arXiv: gr-qc/9210006.

8. H. Everett III, Relative state formulation of quantum mechanics, *Rev. Mod. Phys.* **29**, 454 (1957).
9. A. Tonomura, J. Endo, T. Matsuda, T. Kawasaki and H. Ezawa, Demonstration of single electron build-up of an interference pattern, *Am. J. Phys.* **57**, 117 (1989).
10. M. Arndt, L. Hackermüller, K. Hornberger and A. Zeilinger, Coherence and decoherence experiments with fullerenes, in *Decoherence, Entanglement and Information Protection in Complex Quantum Systems*, eds. W. M. Akulin *et al.* (Springer, 2005).
11. L. Hackermüller, K. Hornberger, B. Brezgus, A. Zeilinger and M. Arndt, Decoherence of matter waves by thermal emission of radiation, *Nature* **427**, 711 (2004).
12. T. Juffmann, A. Milic, M. Müllerritsch *et al.*, Real-time single-molecule imaging of quantum interference, *Nat. Nanotechnol.* **7**, 297 (2012).
13. J. R. Friedman, V. Patel, W. Chen, S. Tolpygo and J. Leukens, Quantum superposition of distinct macroscopic states, *Nature* **406**, 43 (2000).
14. A. J. Leggett and A. Garg, Quantum mechanics vs. macroscopic realism: Is the flux there when nobody looks? *Phys. Rev. Lett.* **54**, 857 (1985).
15. R. Penrose, On gravity's role in state vector reduction, *Gen. Rel. Grav.* **28**, 581 (1996).
16. C. J. Riedel, W. Zurek and M. Zwolak, *The Objective Past of a Quantum Universe, Pt. 1 Redundant Records of Consistent Histories*, arXiv:1312.0331.
17. J. B. Hartle, The reduction of the state vector and limitations on measurement in the quantum mechanics of closed systems, in *Directions in General Relativity*, Vol. 2, eds. B.-L. Hu and T. A. Jacobson (Cambridge University Press, Cambridge, 1993), arXiv:gr-qc/9301011.
18. M. Srednicki and J. B. Hartle, Science in a very large universe, *Phys. Rev. D* **81**, 123524 (2010), arXiv:0906.0042, *The Xerographic Distribution: Scientific Reasoning in a Large Universe*, arXiv:1004.3816.

19. J. B. Hartle and T. Hertog, *The Observer Strikes Back*, arXiv:1503.07205.
20. S. W. Hawking and T. Hertog, Populating the landscape: A top down approach, *Phys. Rev. D* **73**, 123527 (2006), arXiv:hep-th/0602091.
21. M. Gell-Mann and J. B. Hartle, Adaptive coarse graining, environments, strong decoherence, and quasiclassical realms, *Phys. Rev. A* **89**, 052125 (2014), arXiv:1312.7454.
22. J. B. Hartle, Quantum physics and human language, *J. Phys. A: Math. Theor.* **40**, 3101–3121 (2007), arXiv:quant-ph/0610131.
23. F. London and E. Bauer, *La Théorie de L'observation en Mécanique Quantique* (Hermann, Paris, 1939).
24. J. von Neumann, *Mathematische Grundlagen der Quantenmechanik* (Springer, Berlin, 1932); [English trans. *Mathematical Foundations of Quantum Mechanics* (Princeton University Press, Princeton, 1955)].
25. E. Joos and H. D. Zeh, The emergence of classical properties through interaction with the environment, *Zeit. Phys. B* **59**, 223 (1985).
26. J. B. Hartle, Spacetime quantum mechanics and the quantum mechanics of spacetime, in *Gravitation and Quantizations*, Proc. 1992 Les Houches Summer School, eds. B. Julia and J. Zinn-Justin (North Holland, Amsterdam, 1995), arXiv:gr-qc/9304006.
27. J. B. Hartle and S. W. Hawking, The wave function of the universe, *Phys. Rev. D* **28**, 2960–2975 (1983); S. W. Hawking, The quantum state of the universe, *Nucl. Phys. B* **239**, 257–276 (1984).
28. J. B. Hartle, The quasiclassical realms of this quantum universe, arXiv:0806.3776. A slightly shorter version is published in *Many Worlds?* eds. S. Saunders, J. Barrett, A. Kent and D. Wallace (Oxford University Press, Oxford, 2010), the longer version was published in *Found. Phys.* **41**, 982 (2011).
29. J. B. Hartle, S. W. Hawking and T. Hertog, Local observation in eternal inflation, *Phys. Rev. Lett.* **106**, 141302 (2011), arXiv:1009.252.
30. J. B. Hartle and T. Hertog, *One Bubble to Rule them All*, arXiv:1604.03580.

CHAPTER 13

The Quantum Mechanical Arrows of Time*

The familiar textbook quantum mechanics of laboratory measurements incorporates a quantum mechanical arrow of time — the direction in time in which state vector reduction operates. This arrow is usually assumed to coincide with the direction of the thermodynamic arrow of the quasiclassical realm of everyday experience. But in the more general context of cosmology we seek an explanation of all observed arrows, and the relations between them, in terms of the conditions that specify our particular universe. This paper investigates quantum mechanical and thermodynamic arrows in a time-neutral formulation of quantum mechanics for a number of model cosmologies in fixed background spacetimes. We find that a general universe may not have well defined arrows of either kind. When arrows are emergent, they need not point in the same direction over the whole of spacetime. Rather, they may be local, pointing in different directions in different spacetime regions. Local arrows can therefore be consistent with global time symmetry.

*Talk at the conference to honor Yakir Aharonov's 80th birthday, *Fundamental Aspects of Quantum Theory: A Two-Time Winner*, Chapman University, August 15–18, 2012.
arXiv:1301.2844

1. Introduction

In his 1932 book,[1] von Neumann summarized the quantum mechanics of a subsystem of the universe that is sometimes measured but otherwise isolated. Two laws of evolution for the quantum state of the subsystem were postulated. The first is the Schrödinger equation that specifies how the state evolves in time when the subsystem is isolated:

$$i\hbar\frac{d|\hat{\psi}(t)\rangle}{dt} = H|\hat{\psi}(t)\rangle \quad \text{(I)}. \tag{1}$$

The second law specifies how the state evolves when an "ideal measurement" is carried out on the subsystem at time t_m. It is

$$|\hat{\psi}(t_m)\rangle \rightarrow \frac{s|\hat{\psi}(t_m)\rangle}{\|s|\hat{\psi}(t_m)\rangle\|} \quad \text{(II)} \tag{2}$$

where s is the projection onto the measurement outcome.[a]

The Schrödinger equation (I) is time reversible — it can be run both forward and backward in time. The second law of evolution (II) is not reversible. It operates only forward in time. That defines the quantum mechanical arrow of time.

It is commonly assumed that the quantum arrow of time coincides with the thermodynamic arrow defined by the direction in which total entropy is increasing. This identification of a fundamental quantum arrow with a classical one must have seemed natural in a theory which posited separate classical and quantum worlds with a kind of movable boundary between them. A second law describing an "irreversible act of amplification" from the quantum world to the classical one in a measurement was naturally connected with the second law of thermodynamics describing more general classical irreversible processes.

[a]von Neumann called I and II, 2 and 1, respectively, but today the given convention is more used.

A thermodynamic arrow of time is not an inevitable feature of a classical world like ours, which is governed by time-neutral dynamical laws. The fact that presently isolated subsystems are mostly evolving towards equilibrium in the same direction in time cannot be a consequence of time-neutral dynamical laws. Rather, our thermodynamic arrow arises because the initial state of our universe is such that the progenitors of today's isolated subsystems were all far out of equilibrium a long time ago. As Boltzmann wrote over a century ago: "The second law of thermodynamics can be proved from the [time-reversible] mechanical theory, if one assumes that the present state of the universe...started to evolve from an improbable [i.e. special] state".[2] Our thermodynamic arrow of time is an emergent feature of the particular initial condition of our universe.

Is the quantum mechanical arrow of time a fundamental property of quantum mechanics, or can it also be seen as an emergent feature of our universe in a more general formulation of quantum mechanics free from arrows of time? Is a quantum mechanical arrow of time always codirectional with a thermodynamic arrow? Do arrows always point in one direction over the whole of spacetime or can they point in different directions in different spacetime regions? Such questions are the subject of this essay. To answer them, as Boltzmann noted, we are naturally led to cosmology.

It is almost certain that there will not be a thermodynamic arrow of time that points consistently in one direction over the whole of spacetime in the vast universes contemplated by contemporary inflationary cosmology. Rather, the thermodynamic arrow may point in different directions, in different regions of spacetime. A simple example that we will discuss in this paper is illustrated in Fig. 1.

The figure shows a curved two-dimensional slice of a four-dimensional homogeneous and isotropic cosmological spacetime embedded in a (Lorentz signatured) three-dimensional flat space. The universe begins at large radii at the bottom, contracts to smaller and smaller radii, bounces at a minimum radius and re-expands towards the top. DeSitter space is a classic example of such a bouncing spacetime.

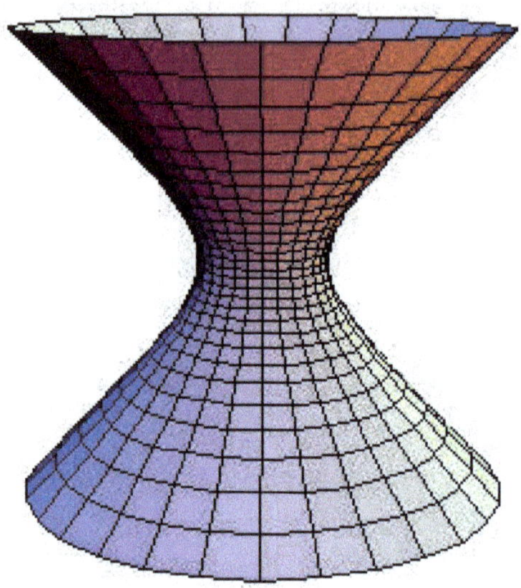

Fig. 1. A bouncing universe. The figure shows the geometry of a two-dimensional slice of a four-dimensional cosmological bouncing spacetime embedded in a Lorentz signatured three-dimensional flat space. Time is up. The universe is spatially closed. It has a large spatial volume at large negative times, collapses to smaller and smaller volumes until a minimum is reached (the bounce), and then expands to larger volumes at positive times. For simplicity, we have assumed that the contraction and expansion are time symmetric. DeSitter space is a well-known example.

Such bouncing geometries are among the classical spacetimes predicted by the no-boundary quantum state of the universe (NBWF) in simple models using a framework that includes quantum gravity.[3,4] In addition to predicting the behavior of homogeneous and isotropic classical backgrounds, the NBWF also predicts the behavior of the quantum fluctuations in matter and geometry away from these symmetries. The key result for the present discussion is the following: The NBWF predicts that fluctuations are in a state of low excitation (i.e. small) at the bounce. On either side they grow away from the bounce, become classical, collapse gravitationally, and eventually create a large scale structure of galaxies, stars, and planets. The thermodynamic arrow of time therefore points in opposite directions on opposite sides of the bounce. The universe has two spacetime regions with opposing thermodynamic arrows. We can then ask: What are the laws of quantum mechanics for the quantum fluctuations? Do they have a quantum arrow of time, and which way does it point in the two regions?

We will answer these questions in the context of a family of generalizations of textbook quantum theory that are time-neutral — not preferring one time direction over another and without any built-in quantum arrows of time. In a pioneering paper, Aharonov, Bergmann, and Lebovitz[5] showed how to use initial and final conditions to construct such a time-neutral quantum mechanics of measured subsystems. We will answer these questions, not in that context, but in the more general time-neutral formulation of quantum mechanics of closed systems that is suitable for cosmology.[6]

In a time-neutral formulation, no arrows of any kind are built in. Arrows of time of a particular universe emerge from the conditions that specify it. As we have discussed, the thermodynamic arrow of time in our universe emerges from a special initial condition (and a final condition of indifference). We will show that quantum mechanical arrows of time can emerge in a similar way.

For the simple example illustrated in Fig. 1, we will find that in a suitable generalization of quantum mechanics there is both a quantum and thermodynamic arrow of time pointing away from the bounce on each side. The overall situation is time symmetric.

More generally, we will conclude that quantum mechanical arrows of time are not an inevitable feature of quantum mechanics. Quantum mechanics can be formulated without them. The arrows of time that characterize the approximate quantum mechanics of measured subsystems obeying laws I and II in our universe arise in particular spacetime regions from the conditions that specify the universe and the region.

The paper is organized as follows: In Sec. 2, we review the work of Aharonov, Bergmann, and Lebovitz.[5] Section 3 introduces a time-neutral quantum mechanics of closed systems with initial and final conditions. All arrows of time arise from asymmetries between these two. Section 4 introduces the class of generalized quantum theories of which the one in Sec. 3 is but one example. Section 5 constructs a time-neutral generalized quantum theory for the quantum fluctuations in a bouncing universe illustrated in Fig. 1. Section 6 draws some brief conclusions.

2. Time-Neutrality in the Quantum Mechanics of Measured Subsystems

In a seminal paper, Aharonov, Bergmann, and Lebovitz[5] showed how the quantum mechanics of measured subsystems could be formulated without an intrinsic arrow of time by allowing for final states as well as initial ones. We summarize the essence of their argument here in a notation that is analogous to that which we will use for cosmology in subsequent sections.

Consider a subsystem of the universe whose states are vectors in a Hilbert space $\mathcal{H}_{\rm sub}$. Alternatives at a moment of time can be reduced to a set of yes/no questions. For instance, asking for the position of a particle is equivalent to taking an exhaustive set of position intervals and asking whether the particle is in the first interval (yes or no), then the second interval (yes or no), etc. In the Heisenberg picture, such yes/no alternatives at a time t are represented by an exhaustive set of exclusive projection operators $\{s_\alpha(t)\}$. For instance, these might be projections onto an exhaustive set of ranges of positions as discussed above. These operators satisfy

$$s_\alpha(t)s_{\alpha'}(t) = \delta_{\alpha\alpha'}s_\alpha(t), \quad \sum_\alpha s_\alpha(t) = I, \tag{3}$$

showing that they are projections, that they are exclusive, and that they are exhaustive. The projection operators representing the same alternative at two different times t_1 and t_2 are connected by the Heisenberg equations of motion

$$s_\alpha(t_2) = e^{+ih(t_2-t_1)/\hbar}s_\alpha(t_1)e^{-ih(t_2-t_1)]\hbar}, \tag{4}$$

where h is the Hamiltonian of the subsystem in isolation.

Suppose a sequence of ideal measurements[b] is carried out on the subsystem by another subsystem at a sequence of times $t_1, t_2, \ldots, t_n$. The measurements are described by a sequence of *sets* of projections $\{s^k_{\alpha_k}(t_k)\}$, $k = 1, 2, \ldots, n$. The upper index allows for the contingency that the measurements might be of different quantities at different times — a measurement of position at time t_1, of momentum at time t_2, etc.

Suppose that the subsystem is in a (Heisenberg picture) state $|\psi_i\rangle$ in $\mathcal{H}$. Then the joint probability for a history of outcomes $\alpha_1, \alpha_2, \ldots, \alpha_n$ is[7]

$$p(\alpha_n, \ldots, \alpha_1) = \|s^n_{\alpha_n}(t_n) \ldots s^1_{\alpha_1}(t_1)|\psi_i\rangle\|^2. \tag{5}$$

It is easy to work out that this compact formula for the joint probability for a sequence of ideal measurement outcomes follows from the two laws of evolution (1) and (2) — evolve, reduce, evolve, reduce, evolve .... The formula can be made more compact by defining $\alpha \equiv (\alpha_1, \alpha_2, \ldots, \alpha_n)$ and

$$c_\alpha \equiv s^n_{\alpha_n}(t_n) \ldots s^1_{\alpha_1}(t_1). \tag{6}$$

Then

$$p(\alpha) = \|c_\alpha|\psi_i\rangle\|^2 = \|s^n_{\alpha_n}(t_n) \ldots s^1_{\alpha_1}(t_1)|\psi_i\rangle\|^2. \tag{7}$$

A quantum mechanical arrow of time is manifest in (5) and (7). On one end of the chain of projections there is the state, and on the other end there is nothing.[c] Aharonov, Bergmann, and Lebovitz noticed that if one added a final state ψ_f corresponding to post-selection then the formula for the probabilities becomes

$$p(\alpha) = N|\langle\psi_f|c_\alpha|\psi_i\rangle|^2, \quad N \equiv |\langle\psi_f|\psi_i\rangle|^{-2}. \tag{8}$$

---

[b] An ideal measurement, sometimes called a projective measurement, is one that disturbs the subsystem as little as possible so that after the measurement its (Schrödinger picture) state is given by (2).

[c] The quantum mechanical arrow of time does not arise from the time-ordering of the projections. That could be reversed by a CPT transformation since field theory is invariant under CPT. But there would still be the state on one end and nothing on the other.

These formulae are symmetric in the initial and final states, in particular, one can write (8) as

$$p(\alpha) = N|\langle\psi_f|c_\alpha|\psi_i\rangle|^2 = N|\langle\psi_i|c_\alpha^\dagger|\psi_f\rangle|^2. \tag{9}$$

That is, the probabilities are the same when the order of the projections is reversed and the notion of initial and final interchanged.

From this perspective, the quantum mechanical arrow of time arises from not specifying a final state. As Aharonov and Rohrlich say,[8] "By imposing an initial and not a final condition we have already sent the arrow of time flying."

3. A Time-Neutral Formulation of the Quantum Mechanics of Closed Systems

3.1. *A Model Quantum Universe*

Cosmology provides not only the most general context for a discussion of arrows of time, but also the most relevant one. That is because the observed arrows operate on cosmological scales and can be explained by cosmological conditions. For instance, as far as we know, the thermodynamic arrow of time extends over the whole of the visible universe and holds from the time of the big bang to the distant future. The evidence of the observations is that the universe was more ordered earlier than now and that disorder has been increasing ever since.[9,10] That is the thermodynamic arrow of time. Similarly, the electromagnetic arrow — the retardation of electromagnetic radiation — arises because the early universe had very little free electromagnetic radiation that today would be at readily accessible wavelengths.[11] The psychological arrow of time can be seen to follow from the other two.[11]

To keep the discussion manageable, we consider a closed quantum system in the approximation that gross quantum fluctuations in the geometry of spacetime can be neglected.[d] A closed system can

[d]For the further generalizations that are needed for quantum spacetime, see e.g. Refs. 12 and 13. For discussions of the arrows of time in contexts that include quantum spacetime, see Ref. 4 and the earlier references therein, especially Ref. 14.

Fig. 2. A simple model of a closed quantum system is a universe of quantum matter fields inside a large closed box (say, 20,000 Mpc on a side) with fixed, flat, spacetime inside. Everything is a physical system inside the box — galaxies, stars, planets, human beings, observers and observed, measured and measuring. The most general objectives for prediction are the probabilities of the individual members of decoherent sets of alternative coarse-grained histories that describe what goes on in the box. That includes histories describing any measurements that take place there. There is no observation or other intervention from outside.

then be thought of as a large (say $\gtrsim$ 20,000 Mpc), perhaps expanding, box of particles and fields in a fixed, flat, background spacetime (Fig. 2). Everything is contained within the box, in particular galaxies, planets, observers and observed, measured subsystems, and any apparatus that measures them. This is the most general physical context for prediction.

3.2. *Time-Neutral Decoherent Histories Quantum Theory*

The quantum mechanics of this model universe is formulated in a Hilbert space $\mathcal{H}$ that is vastly larger than the Hilbert space of any isolated subsystem it contains. However, the kinematics of the prediction of probabilities for histories bears many similarities to the quantum mechanics of measured subsystems as presented in the previous section.

The most general objective of a quantum mechanics of the universe is the prediction of the probabilities for sets of alternative coarse-grained time histories of its contents. Alternatives at one moment of time are described by an exhaustive set of exclusive projection operators $\{P_\alpha(t)\}$ acting in $\mathcal{H}$. These satisfy [cf. (3)]

$$P_\alpha(t)P_{\alpha'}(t) = \delta_{\alpha\alpha'}P_\alpha(t), \quad \sum_\alpha P_\alpha(t) = I. \tag{10}$$

A set of alternative coarse-grained histories is specified by a sequence of such sets at a series of times $t_1, t_2, \ldots, t_n$. An individual history corresponds to a particular sequence of events $\alpha \equiv (\alpha_1, \alpha_2, \ldots, \alpha_n)$ and is represented by the corresponding chain of projections:

$$C_\alpha \equiv P^n_{\alpha_n}(t_n) \ldots P^1_{\alpha_1}(t_1). \tag{11}$$

An immediate consequence of this and (10) is that

$$\sum_\alpha C_\alpha = I, \tag{12}$$

showing that the set of histories is exhaustive.

This description of histories is analogous to those in the quantum mechanics of measured subsystems [cf. (6)]. However, there are at least two crucial differences. First, there is no posited separate classical world. It is all quantum. Second, the alternatives represented by the P's are not restricted to measurement outcomes. They might, for example, refer to the orbit of the Moon when no one is looking at it, or to the magnitude of density fluctuations in the early universe when there were neither observers nor apparatus to measure them. Laboratory measurements can be described in terms of correlations between two particular kinds of subsystems of the universe — one being measured and the other doing the measuring. But laboratory measurements play no central role in formulating the theory, and are just a small part of what it can predict.[e]

A time-neutral decoherent histories quantum mechanics of our model universe with both initial and final conditions was formulated by Gell-Mann and the author in Ref. 6. The formula for the probabilities for histories is

$$p(\alpha) = NTr(\rho_f C_\alpha \rho_i C_\alpha^\dagger), \quad N^{-1} \equiv Tr(\rho_f \rho_i). \tag{13}$$

Here, ρ_i and ρ_f are density matrices representing initial and final conditions. There is a clear analogy with (8). This expression is time-neutral because initial and final density matrices can be interchanged and the order of times in the C_α's reversed using the cyclic property of the trace.

However, (13) does not supply probabilities for all sets of alternative histories. The resulting probabilities might not be consistent with the usual sum rules of probability theory. Generally, probabilities cannot be assigned to interfering alternatives in quantum theory. The two-slit experiment described in Fig. 3 is a simple example.

In decoherent (or consistent) histories quantum theory, probabilities are assigned only a set of alternative histories if the quantum interference between members of the set is negligible as a consequence of the initial and final conditions and the dynamics. The measure of quantum interference is provided by the decoherence functional:

$$D(\alpha, \alpha') = NTr(\rho_f C_\alpha \rho_i C_{\alpha'}{}^\dagger) \tag{14}$$

where $N^{-1} = Tr(\rho_f \rho_i)$. A set decoheres when the off-diagonal elements of D are negligible. The diagonal elements then give probabilities (13) that are consistent with all the rules of probability theory.[f] Like (13), the decoherence functional (14) is time-neutral.

---

[e]Indeed, the quantum mechanics of measured subsystems in Sec. 2 is an approximation appropriate for measurement situations to the more general quantum mechanics of closed systems. See, e.g. Ref. 15, Sec. II.10.

[f]For more complete expositions of decoherent histories quantum theory than the brief synopsis given here, the reader can consult the classic expositions,[16–18] a short tutorial in Ref. 19, or a review in Ref. 20.

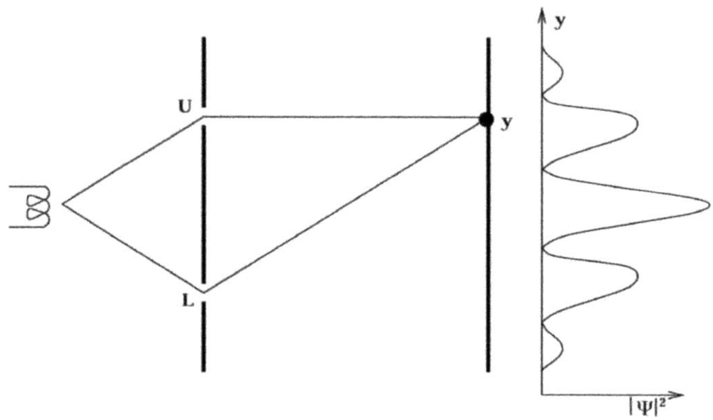

Fig. 3. The two-slit experiment. An electron gun at left emits an electron which is detected at a point y on a screen after passing through another screen with two slits. Because of quantum interference, it is not possible to assign probabilities to the alternative histories in which the electron arrives at y having gone through the upper or lower slit. The probability to arrive at y should be the sum of the probabilities of the two histories. But in quantum mechanics, probabilities are squares of amplitudes and $|\psi_L(y) + \psi_U(y)|^2 \neq |\psi_L(y)|^2 + |\psi_U(y)|^2$. In a different physical situation where the electron interacts with apparatus that measures which slit it passed through, then quantum interference is destroyed and consistent probabilities can be assigned to these histories.

In the quantum mechanics of closed systems, decoherence replaces "measured" as the criterion for when a set of histories can be consistently assigned probabilities. Measured histories decohere, but histories do not have to be of measurement outcomes in order to decohere. Decoherence is a more precise, more general, and more objective criterion than "measured" and certainly more useful in cosmology.

3.3. *Emergent Arrows of Time*

As already mentioned, the expressions both for probabilities (13) and interference (14) are time-neutral. There is thus no distinction between "initial" and "final" that is not conventional. This formulation of quantum theory for our model universe is therefore free from any built-in arrow of time.

If there is no arrow of time in the basic formulation of quantum theory, then the arrows of time observed in our particular universe can only arise from differences between the ρ_f and ρ_i that characterize it. We will then say that arrows of time *emerge* for our particular universe from ρ_f and ρ_i. We will discuss only quantum arrows and thermodynamic arrows, since, as already mentioned, other arrows are connected to these.

Our observations of the universe from laboratory to cosmological scales are consistent (so far) with one special condition that might be a pure state $\rho_i = |\Psi\rangle\langle\Psi|$ and a second condition of indifference[g] $\rho_f \propto I$. It is conventional to call the special condition "initial", as we have done here, the second one "final", and define the direction of increasing time from initial to final.

With these initial and final conditions the formula for the decoherence functional defining quantum mechanics in the box becomes

$$D(\alpha, \alpha') = Tr(C_\alpha \rho_i C^\dagger_{\alpha'}). \tag{15}$$

In particular, the probabilities for the histories in a decoherent set are:

$$p(\alpha_n, \ldots, \alpha_1) = \|C_\alpha|\Psi\rangle\|^2 = \|P^n_{\alpha_n}(t_n) \ldots P^1_{\alpha_1}(t_1)|\Psi\rangle\|^2. \tag{16}$$

This has a state on one end of the chain and nothing on the other. Thus, for $\rho_i = |\Psi\rangle\langle\Psi|$ and $\rho_f \propto I$, a quantum mechanical arrow of time emerges [cf. (7)]. It is not an arrow that is associated just with histories of measurement situations but more generally, with any set of alternative histories of the universe.

With further assumptions on ρ_i we also recover the thermodynamic arrow. Suppose the usual entropy of chemistry and physics[h] is low for ρ_i. It will be maximal for $\rho_f \propto I$. It will therefore tend to

---

[g]For some discussions of the observable information about the final condition, see, e.g. Refs. 6, 21–23.

[h]Entropy depends on coarse-graining. The usual entropy is defined in terms of a coarse-graining expressed in the variables that occur in the deterministic equations of classical physics like the Navier–Stokes equation. For more on its construction and its relation to the quasiclassical realms that are features of our universe see, e.g. Ref. 24.

increase from the time of the initial condition to that of the final one. That is the simplest characterization of the thermodynamic arrow.

However, if both ρ_i and ρ_f are non-trivial, then there is generally no clear definition of either a global thermodynamic or quantum mechanical arrow. For instance, when ρ_i and ρ_f have comparable low entropies, classical analyses[6,25] suggest that the entropy could first rise, and then after a time decrease, leading to a thermodynamic arrow that is local in time, first pointing one way and then another.[i]

There is no clear meaning to a local quantum mechanical arrow, but also no physical need for one. With non-trivial initial and final ρ's there is no notion of a single state at a moment of time from which either the future or the past could be predicted.[6] The theory is fully four-dimensional.[j]

4. Generalized Quantum Theory

The time-neutral formulation of quantum mechanics in the previous section is as notable for its simplicity as it is for its freedom from a built-in quantum arrow of time. Formulating quantum theory has been reduced to just two specifications: (1) The sets of possible alternative coarse-grained histories $\{C_\alpha\}$, and (2) a decoherence functional (14) that measures the quantum interference between histories and specifies their probabilities when the set decoheres.

The decoherence functional of time-neutral quantum mechanics (14) is a generalization of that for usual quantum mechanics (15). But it is not the only generalization. The essential features of quantum mechanics are captured by any complex valued decoherence

[i]There will also generally not be a notion of state at a moment of time. However, there might be a way of expressing the probabilities in terms of two state vectors similarly to Ref. 26.

[j]Advanced civilizations with large laboratories and enough money could in principle reverse the thermodynamic arrow of time over a region of spacetime in a universe like ours by pre- and post-selection of quantum states. If they selected initial states at one time indifferently, and states at a later time distributed according to a low entropy set of probabilities, they would have effectively reversed both the quantum mechanical and thermodynamic arrows (e.g. Ref. 15, Fig. 8).

functional $D(\alpha, \alpha')$ that satisfies the following conditions[12,27]:

(i.) Hermiticity: $D(\alpha, \alpha') = D^*(\alpha', \alpha)$,
(ii.) Normalization: $\sum_{\alpha\alpha'} D(\alpha, \alpha') = 1$,
(iii.) Positivity: $D(\alpha, \alpha) \geq 0$,

and, most importantly, consistency with the principle of superposition. This means the following: Partitioning a set of histories $\{C_\alpha\}$ into bigger sets $\{C_{\bar{\alpha}}\}$ is an operation of coarse-graining. Every history C_α is in one and only one of the sets $C_{\bar{\alpha}}$, a fact that we indicate schematically by $\alpha \in \bar{\alpha}$. Then consistency with the principle of superposition means:

(iv.) Principle of superposition:
$D(\bar{\alpha}, \bar{\alpha}') = \sum_{\alpha \in \bar{\alpha}} \sum_{\alpha' \in \bar{\alpha}'} D(\alpha, \alpha')$.

This is just the usual superposition of amplitudes applied to a quantity D that is bilinear in amplitudes.

Given a decoherence functional satisfying (i)–(iv), the central formula of quantum mechanics which specifies both sets of histories $\{C_\alpha\}$ decohere and their probability $p(\alpha)$ is:

$$D(\alpha, \alpha') \approx \delta_{\alpha\alpha'} p(\alpha). \tag{17}$$

Interference between histories vanishes when the decoherence functional is diagonal and the diagonal elements are the probabilities of the histories in a decoherent set. These probabilities satisfy all the usual rules of probability theory as a consequence of (i)–(iv).

The decoherence functional of usual quantum mechanics (15) satisfies (i)–(iv). All the other ways of satisfying these conditions give generalizations of usual quantum theory — generalized quantum mechanics. The decoherence functional (14) of the time-neutral formulation is one example. We will see another in the next section.

5. Bouncing Universe Models

The universe of quantum matter fields in a closed box that has been at the center of our discussion so far, is a much oversimplified model for cosmology. Its chief deficiency is that it ignores gravity. A better, still manageable, kind of model describes quantum matter fields moving in a fixed, classical, cosmological background such as the bouncing universe shown in Fig. 1. What kind of decoherence functional should we assume for such a model to study classical and quantum arrows of time? A model problem in quantum cosmology suggests the answer.

In Ref. 3, both spacetime geometry and matter fields were treated quantum-mechanically. Probabilities for different homogeneous and isotropic classical background spacetimes and the behavior of quantum matter fields in them were predicted from the no-boundary quantum state of the universe[28] in a simple minisuperspace model. There were two key results for the present discussion. (1) Some bouncing classical background spacetimes like that in Fig. 1 were predicted with non-zero probability.[k] (2) The predicted matter fields were small and simple at the bounce where the spatial volume is the least, and not at one or the other of the infinite volume ends of the spacetime.[4] That suggests that the quantum mechanics of matter fields in such spacetimes should not have initial and final density matrices but rather one density matrix ρ_0 at the bounce. We now produce a generalized quantum theory with this property.

Before starting on quantum mechanics, it is worthwhile to consider the thermodynamic arrow of time in this model.[l] As discussed above, the matter field fluctuations are small near the bounce. They will therefore grow in the two time directions away from the bounce. Eventually fluctuations may grow large enough to collapse and dissipate giving rise to a large scale structure of galaxies, stars, planets, biota, IGUSes, civilizations, etc., on both sides of the bounce.[m] The thermodynamic arrow of time is thus bidirectional in this model — pointing away from the bounce on both sides.

---

[k]Backgrounds that are not time symmetric were also predicted but for simplicity we are focusing on time symmetric ones.

[l]See Ref. 4 for a more detailed discussion within quantum cosmology and also Refs. 29 and 30 for related ones outside of quantum cosmology.

[m]This large scale structure will generally not be the *same* on both sides of the bounce. Individual histories do not have to be time symmetric. It is the ensemble of possible histories predicted by quantum mechanics that is time symmetric.[4,14]

Generalized quantum mechanics for quantum fields in a bouncing universe can be constructed by specifying first the histories and then a decoherence functional obeying properties (i)–(iv) in the previous section. There will be many ways of doing this, like simply generalizing the time-neutral formulation of Sec. 3.2 with initial and final density matrices at the large ends of the expansion. But motivated by the quantum cosmology model described above, we are looking for a decoherence functional with a density matrix at the bounce.[n]

To specify the histories, we arbitrarily label the two sides of the bounce as A and B, as in Fig. 4. A given history will generally have a part on the A side and a part on the B side — generally different. On each side, the parts of histories can be represented by chains of projections — $\{C^A_\alpha\}$ and $\{C^B_\beta\}$. We make the convention that the projections in the chains are *time-ordered away from the bounce*. We have separately [cf. (12)]

$$\sum_\alpha C^A_\alpha = I, \quad \sum_\beta C^B_\beta = I. \tag{18}$$

The following decoherence functional then suggests itself

$$D(\beta, \alpha; \beta', \alpha') = Tr(C^B_\beta \sqrt{\rho_0} C^{A\dagger}_\alpha C^A_{\alpha'} \sqrt{\rho_0} C^{B\dagger}_{\beta'}). \tag{19}$$

It is not difficult to verify that this satisfies requirements (i)–(iv).

The generalized quantum theory defined by (19) is time-neutral. The decoherence functional D is symmetric under interchanging A and B. It is perhaps the simplest generalized quantum theory with this property.[o]

Familiar results emerge if we consider histories just on one side, say B. The appropriate decoherence functional D^B results from coarse-graining (summing) over alternatives on the A side. Using

---

[n]We should stress that we are not deriving this decoherence functional from the more general quantum cosmological model that includes quantum spacetime, but using that as a motivation to posit a particular kind of model.

[o]It is not, however, the only one. For instance, initial and final conditions represented by density matrices ρ_i and ρ_f at the large ends could have been incorporated in addition to ρ_0 in analogy with (14).

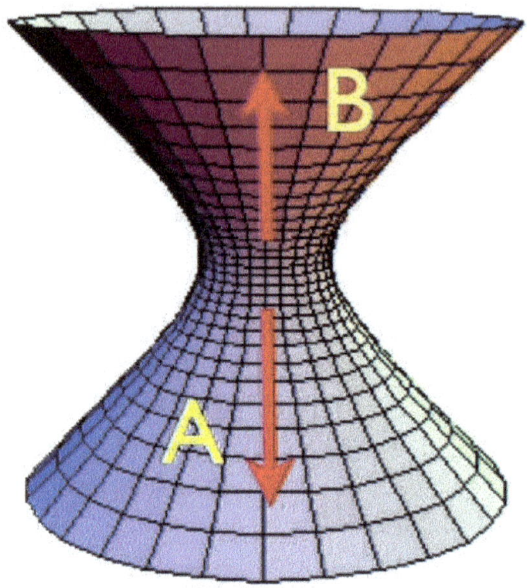

Fig. 4. A bouncing universe like that described in Fig. 1 divided into two sides A and B at the minimum volume three surface (the bounce). When the quantum mechanics of matter fields is described by a decoherence functional (19), each side has a coincident quantum and thermodynamic arrow. But the arrows point in opposite directions on opposite sides of the bounce. This is an example whose quantum mechanics is globally time-neutral but with local arrows of time.

(18) we find

$$D^B(\beta, \beta') = Tr(C_\beta \rho_0 C^\dagger_{\beta'}). \tag{20}$$

But this is just the expression (15). There will thus be a quantum mechanical arrow of time on side B and a coincident thermodynamic arrow. Similar results are obtained by following histories on side A and ignoring those on side B.

Thus, the generalized quantum theory defined by the decoherence function (19) exhibits *local* thermodynamic and quantum mechanical arrows that are codirectional on either side but point in opposite directions on opposite sides of the bounce. There are no global arrows — either quantum mechanical or thermodynamic — pointing consistently in one direction over the whole of the spacetime.

If we live in a bouncing universe, questions naturally arise as to how much present events on our side are influenced by what occurred before the bounce and what we can infer about events on the far side from observations on our side. The answers to such questions are contained in the joint probabilities $p(\beta, \alpha)$ for correlations between histories on the near side of the bounce and histories on the far side.

We can anticipate that it will be difficult to find causal correlations between the two sides because the thermodynamic arrow points in opposite directions on opposite sides of the bounce.[4] The two sides are in each other's pasts, as determined by the thermodynamic arrow. There is as much chance of events on the far side of the bounce influencing us, as we have of influencing events in our past by actions taken now.

In the simple case where the density matrix ρ_0 in (19) is pure, mutual influence is impossible. To see this write

$$\rho_0 = |\Psi\rangle\langle\Psi| = \sqrt{\rho_0}. \tag{21}$$

Then

$$D(\beta, \alpha; \beta', \alpha') = \langle\Psi|C^{B\dagger}_{\beta'} C^{B}_{\beta}|\Psi\rangle\langle\Psi|C^{A\dagger}_{\alpha'} C^{A}_{\alpha}|\Psi\rangle. \tag{22}$$

The immediate consequence is that the joint probabilities of a decoherent set of histories factors

$$p(\beta, \alpha) = p^B(\beta)p^A(\alpha), \tag{23}$$

and there is no correlation between events on one side and the events on the other. The far side might as well not exist.

6. Conclusions

Cosmology is the natural context for understanding the origin of the arrows of time of our universe. Arrows operate over cosmological distances and can be explained by cosmological conditions.

No arrows of time need be built into a fundamental formulation of quantum mechanics of closed systems like the universe. Rather, quantum mechanics can be formulated time-neutrally. The observed arrows of time are then emergent features of the asymmetries between conditions that specify our particular universe among the possibilities that the time-neutral theory allows. This general perspective allows a discussion of the different ways arrows of time can be exhibited by different universes specified by different conditions. In particular, it allows a discussion of the connections between arrows that follow from those conditions.

This essay exhibited a number of cosmological models with different possibilities for the quantum mechanical and thermodynamic arrows in the framework of time-neutral generalized quantum theory in fixed background spacetimes. From these examples we can conclude that a given universe may not exhibit well-defined arrows of either kind. Further, when arrows do emerge, they need not consistently point in one direction over the whole of spacetime. Rather, they may point in different directions in different regions of spacetime as the bouncing universe model cleanly illustrates. Local arrows can be consistent with global time-symmetry.[4,30]

In some examples there was a local thermodynamic arrow defined by the direction of local entropy increase but no obvious quantum mechanical arrow. In all examples, where both arrows were available locally, they coincided in direction. (Of course, a few examples do not make a general result.)

From this perspective, other features of quantum mechanics such as states on spacelike surfaces, their unitary evolution, and their reduction may also emerge only locally in a more general framework for quantum theory that is fully four-dimensional and time-neutral.[12]

Acknowledgments

The author thanks Thomas Hertog for many discussions of the arrows of time in quantum cosmology and Murray Gell-Mann for discussions on the quantum mechanics of the universe over many decades. He thanks the Santa Fe Institute for supporting many productive visits there. This work was supported in part by the National Science Foundation under grant PHY12-05500.

References

1. J. von Neumann, *Mathematische Grundlagen der Quantenmechanik* (Springer, Berlin, 1932). [English trans. R. T. Beyer, *Mathematical Foundations of Quantum Mechanics* (Princeton University Press, Princeton, 1955)].
2. L. Boltzmann, Zu Hrn. Zermelo's Abhandlung Über die mechanische Erklärung Irreversibler Vorgange, *Ann. Physik* **60**, 392–398 (1897).
3. J. B. Hartle, S. W. Hawking and T. Hertog, Classical universes of the no-boundary quantum state, *Phys. Rev. D* **77**, 123537 (2008), arXiv:0803.1663; The no-boundary measure of the universe, *Phys. Rev. Lett.* **100**, 202301 (2008), arXiv:0711.4630.
4. J. B. Hartle and T. Hertog, Arrows of time in the bouncing universes of the no-boundary quantum state, *Phys. Rev. D* **85**, 103524 (2012), arXiv:1104.1733.
5. Y. Aharonov, P. Bergmann and J. Lebovitz, Time asymmetry in the quantum process of measurement, *Phys. Rev.* **134**, 1410–1416 (1964).
6. M. Gell-Mann and J. B. Hartle, Time symmetry and asymmetry in quantum mechanics and quantum cosmology, in *The Physical Origins of Time Asymmetry*, eds. J. Halliwell, J. Pérez-Mercader and W. Zurek (Cambridge University Press, Cambridge, 1994), gr-qc/9309012.
7. H. J. Groenewold, Information in quantum measurement, *Proc. Akad. van Wetenschappen, Amsterdam, Ser. B* **55**, 219 (1952); E. Wigner, The problem of measurement, *Am. J. Phys.* **31**, 6 (1963).
8. Y. Aharonov and D. Rohrlich, *Quantum Paradoxes* (Wiley-VCH, Weinheim, 2005).
9. B. Basu and D. Lynden-Bell, A survey of entropy in the universe, *QJRAS* **31**, 359 (1990).
10. X. Zhao, Y. Li, Q. Zhu and D. Fox, *Co-evolution of cosmic entropy and structures in the universe*, arXiv:1211.1677.
11. J. B Hartle, The physics of now, *Am. J. Phys.* **73**, 101–109 (2005), gr-qc/0403001.

12. J. B. Hartle, Spacetime quantum mechanics and the quantum mechanics of spacetime, in *Gravitation and Quantizations*, in the *Proc. 1992 Les Houches Summer School*, eds. B. Julia and J. Zinn-Justin, Les Houches Summer School Proceedings, Vol. LVII (North Holland, Amsterdam, 1995), gr-qc/9304006.
13. J. B. Hartle, Generalizing quantum mechanics for quantum spacetime, in *The Quantum Structure of Space and Time: Proc. 23rd Solvay Conf. Physics*, eds. D. Gross, M. Henneaux and A. Sevrin (World Scientific, Singapore, 2007), gr-qc/0602013.
14. S. W. Hawking, R. Laflamme and G. W. Lyons, Origin of time asymmetry, *Phys. Rev. D* **47**, 5342–5356 (1993).
15. J. B. Hartle, The quantum mechanics of cosmology, in *Quantum Cosmology and Baby Universes: Proc. 1989 Jerusalem Winter School for Theoretical Physics*, eds. S. Coleman, J. B. Hartle, T. Piran and S. Weinberg (World Scientific, Singapore, 1991), pp. 65–157.
16. R. B. Griffiths, *Consistent Quantum Theory* (Cambridge University Press, Cambridge, UK, 2002).
17. R. Omnès, *Interpretation of Quantum Mechanics* (Princeton University Press, Princeton, 1994).
18. M. Gell-Mann, *The Quark and the Jaguar* (W. H. Freeman, New York, 1994).
19. J. B. Hartle, The quantum mechanics of closed systems, in *Directions in General Relativity, Vol. 1: A Symposium and Collection of Essays in Honor of Professor Charles W. Misner's 60th Birthday*, eds. B.-L. Hu, M. P. Ryan and C. V. Vishveshwara (Cambridge University Press, Cambridge, 1993), arXiv: gr-qc/9210006.
20. P. C. Hohenberg, An introduction to consistent quantum theory, *Rev. Mod. Phys.* **82**, 2835–2844 (2010).
21. R. Laflamme, Time-symmetric quantum cosmology and our universe, *Class. Quant. Grav.* **10**, L79 (1993), arXiv:gr-qc/9301005.
22. D. Craig, Observation of the final condition: extragalactic background radiation and the time symmetry of the universe, *Ann. Phys. (N.Y.)* **251**, 384 (1995), gr-qc/9704031.

23. H. Price, *Time's Arrow and Archimedes' Point* (Oxford University Press, New York, 1996).
24. M. Gell-Mann and J. B Hartle, Quasiclassical coarse graining and thermodynamic entropy, *Phys. Rev. A* **76**, 022104 (2007), arXiv:quant-ph/0609190.
25. W. J. Cocke, Statistical time symmetry and two-time boundary conditions in physics and cosmology, *Phys. Rev.* **160**, 1165 (1967).
26. Y. Aharonov and L. Vaidman, The two-state vector formalism: An updated review, in *Time in Quantum Mechanics*, Lecture Notes in Physics, Vol. 734 (Springer, Berlin, 2008), pp. 399–447.
27. C. J. Isham, Quantum logic and the histories approach to quantum theory, *J. Math. Phys.* **35**, 2157 (1994), arXiv:gr-qc/9308006.
28. J. B. Hartle and S. W. Hawking, The wave function of the universe, *Phys. Rev. D* **28**, 2960–2975 (1983); S. W. Hawking, The quantum state of the universe, *Nucl. Phys. B* **239**, 257–276 (1984).
29. F. Hoyle and J. V. Narlikar, Symmetric electrodynamics and the arrow of time in cosmology, *Proc. Roy. Soc. Lond. A* **277**, 1–23 (1964).
30. S. Carroll and J. Chen, *Spontaneous Inflation and the Origin of the Arrow of Time*, arXiv:hep-th/0410270.

CHAPTER 14

Why Our Universe Is Comprehensible*

Einstein wrote memorably that "The eternally incomprehensible thing about the world is its comprehensibility." This paper argues that the universe must be comprehensible at some level for information gathering and utilizing subsystems such as human observers to evolve and function.

*Based in part on a talk given at the FQXI meeting in Banff, Canada, August, 2016.
arXiv:1612.01952

1. Introduction

Albert Einstein wrote famously[1]:

> *The eternally incomprehensible thing about the world is its comprehensibility*

The thesis of this essay is that the world must be comprehensible in order for information gathering and utilizing systems (IGUSes) including human beings like us to exist.[a]

The argument in a nutshell is that IGUSes exploit the regularities of the world in order to satisfy evolutionary imperatives like "get food yes", "be food no", "make more yes", etc. A frog catching a fly is making use of a crude, approximate, (and probably hard-wired), form of Newton's laws of motion summarizing regularities of the quasiclassical realm of our quantum universe. To contain IGUSes, the universe must have regularities that can be exploited by IGUSes.

We comprehend our universe by discovering and understanding its regularities. A universe with IGUSes must therefore be comprehensible at least at the level necessary for IGUSes to function.[b]

This essay will amplify on these thoughts in the context of the decoherent (or consistent) histories quantum theory of closed systems (DH) — most generally the universe.[c] The purpose of this essay is not to understand exactly what Einstein meant by his aphorism. The author is not a historian and would not presume to comment on that. But Einstein's statement is a natural starting point for a discussion of aspects of DH in which we give more definite meanings to terms like "world" and "comprehensible". The following development is offered in that spirit.

---

[a]A more general discussion of IGUSes will be given in Sec. 4. For more discussion in the context of complex adaptive systems, see Ref. 2.

[b]It is clear from the text in Ref. 1 immediately before the quote above that, for Einstein, the comprehensibility at issue includes everyday human experience, and not merely regimes that are foreign to us, such as the interior of the Sun, or the very early universe.

[c]For classic expositions, some by founders of the subject, see Ref. 3. For a tutorial see Ref. 4.

The comprehensibility of the universe would indeed be a mystery if, as observers of the universe, we were somehow outside it. That would be the case in the Copenhagen formulations of quantum theory. Why should a physical system that we observe from the outside have any regularities at all? But we are not outside the universe. We are physical systems within the universe, subject to the laws of quantum mechanics but playing no preferred role in its formulation (cf. Ref. 5). We are just special kinds of small fluctuations that have evolved over a very long time in a very large universe. The universe must exhibit regularities for these physical systems to exist, function and proliferate.

DH offers not just one way of comprehending the universe, but many, through different decoherent sets of alternative coarse-grained histories — realms for short. The sets will differ in the variables that are followed by the histories. There can be descriptions that are complementary, somewhat like the position and momentum of a particle. Generally they will be at different levels of coarse-graining. The universe is thus not simply comprehensible or incomprehensible. It is, rather, comprehensible in different ways at different levels of precision. In the rest of the paper we will begin to address the question of which realms exhibit IGUSes.

The paper is organized as follows. Section 2 introduces a model quantum universe in a box to focus subsequent discussions on. Section 3 discusses the quasiclassical realms of everyday experience. Familiar IGUSes are described by histories of a quasiclassical realm. Section 4 discusses IGUSes and their requirements. Section 5 raises, but does not answer, the question of whether the universe might exhibit realms incompatible with our quasiclassical ones with essentially different notions of IGUSes. Section 6 contains summary conclusions.

2. A Model Quantum Universe

To keep the discussion manageable, we consider a model closed quantum system in the approximation that gross quantum fluctuations in the geometry of spacetime can be neglected. The closed system can then be thought of as a large (say $\gtrsim$ 20,000 Mpc), perhaps expanding,

Fig. 1. A simple model of a closed quantum system is a universe of quantum matter fields inside a large closed box (say, 20,000 Mpc on a side) with fixed, flat spacetime inside. Everything is a physical system inside the box — galaxies, stars, planets, human beings, observers and observed, measured and measuring. The most general objectives for prediction are the probabilities of the individual members of decoherent sets of alternative coarse-grained histories that describe what goes on in the box. That includes histories describing any measurements that take place there. There is no observation or other intervention from outside.

box of particles and fields in a fixed, flat, background spacetime (Fig. 1). There is thus a well-defined notion of time in any particular Lorentz frame. The familiar apparatus of textbook quantum mechanics then applies — a Hilbert space $\mathcal{H}$, operators, states, and their unitary evolution in time. We assume a quantum field theory in the flat space for dynamics. Assuming a fixed spacetime means neglecting quantum gravity.

The important thing is that everything is contained within the box — galaxies, planets, observers and observed, measured subsystems, and any apparatus that measures them. This is a model cosmology and the most general physical context for prediction. In this model, Einstein's "world" is this box.

2.1. *Decohering Alternative Histories of the Universe*

The basic theoretical inputs for predicting what goes on in the box are the Hamiltonian H governing evolution and the quantum state $|\Psi\rangle$, written here in the Heisenberg picture for convenience, and assumed pure for simplicity. Input theory is then $(H, |\Psi\rangle)$.

The basic outputs of the theory are the probabilities of the individual members of sets of alternative coarse-grained histories of the contents of the box. A history is a sequence of events, for example the history of the formation of the solar system, or the events in human history. A history is *coarse-grained* when it follows some variables and ignores the rest. For example, a coarse-grained history of the Moon's orbit might follow only the center of mass position of the Moon and ignore the positions of other particles in the Moon, and follow the center of mass position not at all times, but only at a sequence of times, and these positions only to an accuracy sufficient for comparing with observation. A history that also describes the Moon's rotation would be finer-grained.

There must be negligible quantum interference between the individual coarse-grained histories in a set for its members to be assigned probabilities consistent with the usual rules of probability theory. Such sets are said to *decohere*. A decoherent set of alternative coarse-grained histories of what goes on in the box is called a *realm*.

Some coarse-graining is necessary for decoherence. Ignoring some particles in the box so that they constitute an environment for the rest of the particles is a widely used example (e.g. Ref. 6). A completely fine-grained set of histories of all the particles in the box would not decohere. In quantum mechanics, some information must be ignored to have any information at all.

More explicit and mathematical detail on how this all works is provided in the Appendix.

3. Quasiclassical Realms and their Regularities

The most striking observable feature of our indeterministic quantum universe is the wide range of time, place and scale on which the deterministic laws of classical physics hold to an excellent approximation. Regularities that characterize classical physics are described by the familiar classical equations for particles, bulk matter and fields, together with the Einstein equation governing the regularities of classical spacetime geometry. Our observations of the universe suggest that a domain of classical predictability extends from a short time after the big bang to the far future and over the whole of the visible spatial volume. Were we to set out on a journey to arrive in the far future at a distant galaxy, we would count on the regularities of classical physics to be holding there just as they do here.

A quasiclassical realm is a decoherent set of alternative histories of the box that is coarse-grained by familiar quasiclassical variables. To use our model box universe as an example, first imagine dividing the space inside the box into small volumes. For each volume, give the averages over it of energy, momentum and the number of relevant conserved particles. These are the usual hydrodynamic variables that occur, for example, in the Navier–Stokes equation. Specifying such spatial configurations at a series of times defines a set of alternative quasiclassical histories. Larger volumes define coarser-grained quasiclassical histories, small volumes, a finer-grained description. The volumes must be large enough for the set of histories to decohere. Then the set is a quasiclassical realm.[d]

Useful quasiclassical realms span an enormous range of coarse-grainings. Quasiclassical realms describing everyday experience may be so highly coarse-grained as to refer merely to the features of a local environment. At the other end of the range are the quasiclassical realms that are as fine-grained as possible given decoherence and quasiclassicality. Those realms extend over the wide range of time, place, scale and epoch mentioned above.

By predicting probabilities for the histories of a quasiclassical realm, the theory $(H, |\Psi\rangle)$ predicts probabilities for the evolution of localized systems such as galaxies, stars, planets, biota, humans, tables, chairs, bacteria, etc., and, most importantly for this paper — IGUSes.

---

[d]For a review of quasiclassical realms with more references, see, e.g. Ref. 7.

3.1. *Regularities of the Quasiclassical Realm*

A quasiclassical realm exhibits a regularity when the theory $(H, |\Psi\rangle)$ predicts high quantum probabilities for histories of quasiclassical alternatives that exhibit reliable correlations in time or space or both. The dynamical laws of classical physics define regularities in time. Newton's laws of motion, Maxwell's equations, the Navier–Stokes equation, the Einstein equation, and the laws of thermodynamics are all familiar examples. These classical equations are approximate consequences of the underlying quantum theory $(H, |\Psi\rangle)$ (e.g. Ref. 8). They hold only in limited regimes of the universe and can be interrupted by quantum events.

Regularities in space can arise from symmetries in $(H, |\Psi\rangle)$. The approximate homogeneity and isotropy of the present universe on scales above several hundred megaparsecs is an example. Regularities like the second law of thermodynamics arise from a quantum state in which the entropy associated with a coarse-graining by quasiclassical variables was low all across the universe at an early time (e.g. Ref. 8).

IGUSes use approximations to the regularities of the laws of classical physics, as in the example of a frog catching a fly. But the frog is also using regularities of flies, plausibly something like: all flies are food, big flies are better than small flies, flies tend to be near ponds, etc. Such regularities typically arise from frozen accidents.

A frozen accident is a chance event whose consequences proliferated. For instance, the chance events of mutation, recombination and genetic drift produce the evolution of new species that proliferate across planets. The similarities between fly species arise from their common evolutionary origin; their differences arise from the different frozen accidents of biological evolution.

The regularities following directly from $(H, |\Psi\rangle)$ can be expressed mathematically. The regularities of specific systems like flies are generally not easily mathematically expressed. The frog may be said to be using some approximation to Newton's laws of motion, but the regularities of flies are of a different character. They do not follow with high probability from $(H, |\Psi\rangle)$. They arise from chance accidents. Individual flies are, of course, subject to the general laws of classical physics as parts of a quasiclassical realm.

IGUSes exploit the regularities that the universe presents, to evolve and proliferate. Were there no regularities to exploit there would be no IGUSes.

3.2. *Coarse-graining and Comprehensibility*

We comprehend the universe by understanding its regularities in the context of specific realms like the quasiclassical ones. This is how we interpret "comprehensible" in Einstein's aphorism. As mentioned in the Introduction, there are different kinds of comprehension in different variables at different levels of coarse-graining. Following finer-grained histories generally gives more comprehension. We comprehend the orbit of the Moon by following its center of mass. Comprehending the libration of the Moon requires a finer-grained set of histories that also follows the Earth.

Comprehension is limited by decoherence. For instance, we can never have a completely fine-grained understanding of the universe because the necessary set of histories would not decohere (e.g. Ref. 9).

4. IGUSes

4.1. *What is an IGUS?*

As human observers of the universe, both individually and collectively, we are examples of information gathering and utilizing systems (IGUSes). IGUSes are approximately localized subsystems of the universe characterized by the following three properties:

- IGUSes acquire information about their environment.
- IGUSes use the regularities in the acquired information to update a model or "schema" of their environment and beyond.
- IGUSes act on the predictions of this schema, typically acquiring new information in the process.

There are many examples of IGUSes in our quasiclassical realm. Individual humans are IGUSes, and so are societies of human beings. The humans pursuing science today constitute the human scientific IGUS. Cockroaches and bacteria qualify as IGUSes, as would any system that would be considered "living". A self-driving car is an IGUS, so is an airplane on autopilot, and a thermostat, etc.

All these examples of IGUSes can be seen as features of our quasiclassical realm. That is, they and their actions can be described by particular histories coarse-grained by quasiclassical variables as described above.[e]

4.2. *Regularity, Randomness and Complexity*

Why do quasiclassical realms exhibit IGUSes? One answer lies with what an IGUS needs to function. An IGUS requires reliable regularities to build its schema and function effectively. But it also requires some randomness to evolve. On Earth, IGUSes might have started when random fluctuation in circumstances produced the right chemical environment for life. The collective IGUS that is a species also needs the randomness of mutation, genetic drift or recombination to evolve. Beyond regularity and randomness, the varied world of IGUSes that we see today requires something like a fitness landscape that provides various niches for IGUSes to evolve into. That is, they require some level of complexity. Quasiclassical realms can provide the right mixture of regularity to exploit, randomness to evolve, and complexity for a fitness landscape that leads to the presence of IGUSes. In Sec. 5, we raise the question as to whether there are other kinds of realms with IGUSes.

4.3. *IGUSes in the Expanding Universe*

To better understand the requirements for IGUSes, let us think a little outside our model box and consider IGUSes in the expanding universe, modeled by letting the box of Sec. 2 expand.

The evidence of the observations is that our universe expanded from a hot big bang. In a simplified discussion we can usefully

[e]The observers in Copenhagen quantum mechanics might be seen as examples of IGUSes. But those observers were features of a classical world separate from the quantum one. In DH, IGUSes are subject to the quantum-mechanical laws and describable in quantum-mechanical terms.

identify four subsequent epochs at different times t after the beginning at $t = 0$.

- *The Epoch of Quantum Gravity:* A short era $t \lesssim 10^{-43}$ s in which the geometry of spacetime and matter fields exhibited large quantum fluctuations. The histories constituting a quasiclassical realm are based on classical spacetime and therefore start only after this epoch.[7]
- *The Early Universe:* The early universe consisted of a hot plasma of nucleons, electrons and photons. As revealed by the observations of the cosmic background radiation (CMB) in this epoch the universe was homogeneous, isotropic and featureless to an excellent approximation (deviations of 1 pt in 10^5).
- *The Middle Universe:* As the universe expands, the temperature drops. When it has dropped sufficiently, electrons, protons and nucleons recombine and nuclei are synthesized. The initial fluctuations from the quantum gravity era grow under the action of gravitational attraction. Eventually they produce the universe of galaxies, stars, planets, biota and IGUSes that we find today.
- *The Late Universe:* The cosmological constant causes the universe to expand and cool exponentially quickly. Stars exhaust their thermonuclear fuel and die out. Black holes evaporate. The density of matter and the temperature approach zero. The universe becomes cold, dark and inhospitable.

When in this history would one expect to find IGUSes? Not in the quantum gravity epoch, where there isn't even classical spacetime to define a notion of localized subsystem. Not in the early universe. That period has regularity in the homogeneity and isotropy, but it lacks the complexity that would give an evolutionary advantage to an IGUS. Not in the late universe for the same reason. Only in the middle universe do we have both regularities for IGUSes to exploit and enough complexity to make exploiting those regularities fruitful. And that is where we do find them.

4.4. *IGUSes in the Laboratory?*

Could these ideas for the evolution of IGUSes be testable in the laboratory? Would it be possible to vary the mix of regularity, randomness and complexity to see whether and how simple IGUSes evolve? Current experiments in the evolution of bacteria[10,11] give some hope for this

5. Beyond the Quasiclassical Realms

The quasiclassical realms are not the only decoherent sets of coarse-grained alternative histories predicted by $(H, |\Psi\rangle)$. There are other realms coarse-grained by variables other than quasiclassical ones (e.g. those in Ref. 12 in a simple example). Some of these could be incompatible with our quasiclassical one in the sense that there is no finer-grained realm of which both are coarse-grainings. The incompatible realms give *complementary* accounts of the universe, somewhat like histories' position and histories of momentum for a single particle. Both are necessary for a complete description. Quantum mechanics does not favor one over the other.

Human IGUSes focus almost exclusively on the histories of a quasiclassical realm. When we carry out certain experiments, such as measuring the spin of an electron, we are using histories of spin that are not quasiclassical. But even there, the apparatus is described and the results recorded in quasiclassical alternatives. Why this almost exclusive focus?

This seeming question is not really a question for the same reason that asking "Why is the universe comprehensible?" is not a question. It implicitly assumes that we are somehow outside the box universe (Fig. 1) and might choose different realms to focus on or even measure.[f] But we are not outside. We are physical subsystems inside the universe, described in quasiclassical terms, adapted to an environment describable in these terms, whose actions and evolution are described by a history of a quasiclassical realm. We are a feature

[f]The author made a related mistake in some early papers where he argued that we evolved to make use of the regularities that quasiclassical realms present. But there is no historical evidence for such selection.[13]

of a certain quasiclassical realm of our universe, not independent of it.

Can realms incompatible with quasiclassical ones exhibit IGUSes? This intriguing question cannot be answered at the present because we lack a sufficiently general characterization of an IGUS. We might conjecture that realms with a high level of predictability like the quasiclassical ones would exhibit IGUSes because these present enough regularity over time to permit the generation of models (schemata) with significant predictive power. Measures of classicality may help check this conjecture (see e.g. Ref. 14).

6. Conclusion

It was Everett's insight[5] that observers are physical systems within the universe, not somehow on the outside observing it. Observers are describable in quantum-mechanical terms as IGUSes subject to quantum-mechanical laws (e.g. $(H, |\Psi\rangle)$) but play no preferred role in the formulation of quantum mechanics. This change from an "outside" view of observers to an "inside" one has significant consequences for physics. We mention a few examples:
Familiar IGUSes are small subsystems of the universe. Their presence or absence therefore makes only a little change to the probabilities supplied by $(H, |\Psi\rangle)$ for the large scale history of the universe. But IGUSes are central for probabilities that predict for the results of *our* observations of the universe. That is because probabilities for observations are probabilities supplied by $(H, |\Psi\rangle)$ *conditioned* on data D that describe our observational situation, including a description of ourselves.
Observations of the temperature of the CMB provide a simple example. What does the theory $(H, |\Psi\rangle)$ predict for the probabilities of our observations of the CMB temperature? This seeming question is meaningless. The theory $(H, |\Psi\rangle)$ predicts probabilities for a four-dimensional time history of the temperature. To predict what we observe, these probabilities must be conditioned on data that specify the time from the big bang that we make these observations.

The "inside" view of observers has significant consequences for the prediction of observations, as we have described elsewhere.[15] We mention just two examples: (1) What is called "anthropic selection" is automatic in quantum cosmology — not a principle, not an option, and not subjective. (2) The most probable histories to occur are not necessarily the most probable histories to be observed. In models, these two consequences significantly affect the probabilities in predicting observations of the amount of inflation, the value of the cosmological constant, and the spectrum of the CMB, e.g. Refs. 16–20.

What emerges from the discussion in this paper is that not only are the answers to certain questions affected by the inside view, but also the questions themselves. Questions that would seem to be unanswerable when the observer is outside the system becomes answerable when the observer is seen as an IGUS inside.

We provided two examples: First there is Einstein's question about why the universe is comprehensible. We interpreted this as a question asking why the histories of the universe exhibit any understandable regularities as all. Our answer is that there must be some regularities through which the universe can be comprehended, otherwise there would be no IGUSes. Second, we considered the question of why we focus on the quasiclassical realm. The answer becomes a triviality if we understand that we are IGUSes that are a feature of a quasiclassical realm and not outside or separate from it.

We have probably only begun to appreciate the consequences of Everett's insight.

Acknowledgments

The author thanks Murray Gell-Mann, Thomas Hertog and Mark Srednicki for discussions on the quantum mechanics of the universe over many years. He thanks David Krakauer for discussions on experimental evolution. He thanks the Santa Fe Institute for supporting many productive visits there. He thanks FQXI for the invitation to speak at its meeting. This work was supported in part by the National Science Foundation under grant PHY15-04541.

Appendix A

Histories and Realms

This appendix presents a bare-bones description of how the theory $(H, |\Psi\rangle)$ predicts probabilities for which a decoherent set of alternative coarse-grained histories happens in the universe. Many more details and specific models can be found in Refs. 3 and 4.

The simplest notion of a set of histories is described by giving a sequence of yes–no alternatives at a series of times. For example: Is a particle in this region R of the box at this time — yes or no? This alternative is *coarse-grained* because it does not follow the particle's position exactly but only whether it is in R or not. The alternative "yes" is represented by the projection operator P_R on the region R and "no" by $I - P_R$. More generally coarse-grained yes–no alternatives at one moment of time are described by an exhaustive set of exclusive Heisenberg picture projection operators $\{P_\alpha(t)\}$, $\alpha = 1, 2, 3, \ldots$ acting in $\mathcal{H}$. These satisfy:

$$P_\alpha(t)P_{\alpha'}(t) = \delta_{\alpha\alpha'}P_\alpha(t), \quad \sum_\alpha P_\alpha(t) = I. \tag{A.1}$$

Projections on bigger subspaces are more coarse-grained, projections on smaller subspaces are finer-grained.

Projection operators representing the same quantity at different times are connected by unitary evolution by the Hamiltonian H

$$P_\alpha(t') = e^{iH(t'-t)}P_\alpha(t)e^{-iH(t'-t)}. \tag{A.2}$$

For example, for the quasiclassical realm of the model box universe described in Sec. 3, the projections would be products of projections for each subvolume onto ranges of values of the quasiclassical variables — averages of energy, momentum and number.

A set of alternative coarse-grained histories is specified by a sequence of such sets of orthogonal projection operators at a series of times $t_1, t_2, \ldots, t_n$. An individual history corresponds to a particular sequence of events $\alpha \equiv (\alpha_1, \alpha_2, \ldots, \alpha_n)$ and is represented by the corresponding chain of projections:

$$C_\alpha \equiv P^n_{\alpha_n}(t_n) \ldots P^1_{\alpha_1}(t_1). \tag{A.3}$$

Branch state vectors corresponding to the individual histories can be defined by

$$|\Psi_\alpha\rangle \equiv C_\alpha|\Psi\rangle. \tag{A.4}$$

The set of histories (medium) decoheres when the quantum interference between branches is negligible

$$\langle\Psi_\alpha|\Psi_\beta\rangle \propto \delta_{\alpha\beta}. \tag{A.5}$$

As a consequence of decoherence, probabilities $p(\alpha)$ consistent with usual probability rules can be assigned to the individual histories that are

$$p(\alpha) = \||\Psi_\alpha\rangle\|^2 = \|C_\alpha|\Psi\rangle\|^2. \tag{A.6}$$

A decoherent set of alternative coarse-grained histories is called a "realm" for short.

This description of histories may seem similar to those for sequences of ideal measurements in Copenhagen quantum mechanics.[21] However, there are at least three crucial differences. First, there is no posited separate classical world as in the Copenhagen formulation. It is all quantum. Second, the alternatives represented by the P's are not restricted to measurement outcomes. They might, for example, refer to the orbit of the Moon when no one is looking at it, or to the magnitude of density fluctuations in the early universe when there were neither observers nor apparatus to measure them. Laboratory measurements can of course be described in terms of correlations between two particular kinds of subsystems of the universe — one being measured, the other doing the measuring. But laboratory measurements play no central role in formulating DH, and are just a small part of what it can predict.[g] Third, the P's can refer to alternatives describing observers and other IGUSes and to histories of how they formed and what they are doing: for instance, whether they are making measurements or not.

[g]Indeed, the Copenhagen quantum mechanics of measured subsystems is an approximation appropriate for measurement situations to the more general quantum mechanics of closed systems. See, e.g. Ref. 22, Sec. II.10.

References

1. A. Einstein, Physik und Realität, *J. Franklin Inst.* **221**, 313–382 (1936) including an English translation by J. Picard. The phrase has been translated in various ways. Picard's translation is "The eternal mystery of the world is its comprehensibility"; G. Holton[23] suggests the version used here. But there is also wide use of the version given as "the most incomprehensible thing about the world is its comprehensibility".
2. M. Gell-Mann, *The Quark and the Jaguar* (W. H. Freeman, New York, 1994).
3. R. B. Griffiths, *Consistent Quantum Theory* (Cambridge University Press, Cambridge, UK, 2002); R. Omnès, *Interpretation of Quantum Mechanics* (Princeton University Press, Princeton, 1994); M. Gell-Mann, *The Quark and the Jaguar* (W. H. Freeman, New York, 1994); M. Gell-Mann and J. B. Hartle, Quantum mechanics in the light of quantum cosmology, in *Complexity, Entropy, and the Physics of Information*, ed. W. Zurek (Addison Wesley, Reading, MA, 1990); P. C. Hohenberg, An introduction to consistent quantum theory, *Rev. Mod. Phys.* **82**, 2835–2844 (2010).
4. J. B. Hartle, The quantum mechanics of closed systems, in *Directions in General Relativity*, Vol. 1, eds. B.-L. Hu, M. P. Ryan and C. V. Vishveshwara (Cambridge University Press, Cambridge, 1993), arXiv:gr-qc/9210006.
5. H. Everett III, Relative state formulation of quantum mechanics, *Rev. Mod. Phys.* **29**, 454 (1957).
6. E. Joos and H. D. Zeh, The emergence of classical properties through interaction with the environment, *Zeit. Phys. B* **59**, 223 (1985).
7. J. B. Hartle, The quasiclassical realms of this quantum universe, *Found. Phys.* **41**, 982 (2011), arXiv:0806.3776; a slightly shorter version is published in *Many Worlds?* eds. S. Saunders, J. Barrett, A. Kent and D. Wallace (Oxford University Press, Oxford, 2010).
8. M. Gell-Mann and J. B. Hartle, Quasiclassical coarse graining and thermodynamic entropy, *Phys. Rev. A* **76**, 022104 (2007), arXiv:quant-ph/0609190.

9. M. Gell-Mann and J. B. Hartle, Decoherent histories quantum mechanics with one 'real' fine-grained history, *Phys. Rev. A* **85**, 062120 (2012), arXiv:1106.0767.
10. S. F. Elena and R. E. Lenski, Evolution experiments with microorganisms: the dynamics and genetic bases of adaption, *Nat. Rev. Genetics* **4**, 487 (2004).
11. L. M. Wahl and D. C. Krakauer, Models of experimental evolution: The role of genetic chance and selective necessity, *Genetics* **156**, 1437 (2000).
12. T. Brun and J. B. Hartle, Classical dynamics of the quantum harmonic chain, *Phys. Rev. D* **60**, 123503 (1999), arXiv:quant-ph/9905079.
13. S. Saunders, Decoherence and evolutionary adaption, *Phys. Lett. A* **184**, 1 (1993).
14. J. B. Hartle, *Measures for Classicality*, http://web.physics.ucsb.edu/~quniverse/papers/class-excpt.pdf.
15. J. B. Hartle and T. Hertog, *The Observer Strikes Back*, arXiv:1503.07295.
16. J. B. Hartle, S. W. Hawking and T. Hertog, No-boundary measure of the universe, *Phys. Rev. Lett.* **100**, 201301 (2008), arXiv:0711.4630.
17. J. B. Hartle, S. W. Hawking and T. Hertog, Classical universes of the no-boundary quantum state, *Phys. Rev. D* **77**, 123537 (2008), arXiv:0803.1663.
18. J. B. Hartle, S. W. Hawking and T. Hertog, Local observation in eternal inflation, *Phys. Rev. Lett.* **106**, 141302 (2011), arXiv:1009.2525.
19. J. B. Hartle and T. Hertog, Anthropic bounds on lambda from the no-boundary quantum state, *Phys. Rev. D* **88**, 123516 (2008), arXiv:1309.0493.
20. T. Hertog, Predicting a prior for Planck, *JCAP* **02**, 043 (2014), arXiv:1305.6135.
21. H. J. Groenewold, Information in quantum measurement, *Proc. Akad. van Wetenschappen, Amsterdam, Ser. B* **55**, 219 (1952); E. Wigner, The problem of measurement, *Am. J. Phys.* **31**, 6 (1963).

22. J. B. Hartle, The quantum mechanics of cosmology, in *Quantum Cosmology and Baby Universes: Proc. 1989 Jerusalem Winter School for Theoretical Physics*, eds. S. Coleman, J. B. Hartle, T. Piran and S. Weinberg (World Scientific, Singapore, 1991), pp. 65–157.
23. G. Holton, What precisely is "thinking"?... Einstein's answer, *Phys. Teach.* **17**, 157 (1979); Reprinted in *Einstein, A Centenary Volume*, ed. A. P. French (Heinemann, London, 1979).

CHAPTER 15

Quantum Cosmology: Problems for the 21st Century*

Two fundamental laws are needed for prediction in the universe: (1) a basic dynamical law and (2) a law for the cosmological initial condition. Quantum cosmology is the area of basic research concerned with the search for a theory of the initial cosmological state. The issues involved in this search are presented in the form of eight problems.

*Appeared in *Physics 2001*, ed. M. Kumar and in the *Proceedings of the 10th Yukawa-Nishinomiya Symposium*, November 7–8, 1996, Nishinomiya, Japan.
arXiv:gr-qc/9701022

1. What are the Fundamental Laws?

Physics, like other sciences, is concerned with explaining and predicting the regularities of specific physical systems. Stars, the solar system, high-temperature superconductors, fluid flows, atoms, and nuclei are just some of the many examples. Beyond particular systems, however, physics aims at finding laws that predict the regularities exhibited universally by *all* physical systems — without exception, without qualification, and without approximation. These are the *fundamental* laws of physics. This essay is concerned with the fundamental law for the initial condition of the universe.

Ideas for the nature of the fundamental laws have varied as new realms of phenomena have been explored experimentally. However, until recently, all of the various ideas for fundamental laws have had one feature in common: They were proposals for dynamical laws — laws that predicted regularities in time. The laws of Newtonian mechanics, electrodynamics, general relativity and quantum theory all have this character.

The Schrödinger equation is an example of a fundamental dynamical law:

$$i\hbar \frac{\partial \Psi}{\partial t} = H\Psi. \tag{1}$$

A fundamental theory of dynamics supplies the Hilbert space and the Hamiltonian operator H. However, a differential equation like (1) makes no predictions by itself. To solve (1), an initial condition — the state vector at one moment — must also be given. The Schrödinger equation then expresses the regularities in time that emerge from this initial state.

Where do the boundary conditions necessary to solve dynamical laws come from? In most of physics we study the evolution of subsystems of the universe and determine the boundary conditions by observation or experimental preparation. If we are interested in the evolution of the electromagnetic field in a room and observe no incoming radiation, we solve Maxwell's equations with no incoming radiation boundary conditions. To predict the probability for the decay of an atom prepared in an excited state, we solve the Schrödinger equation with that excited state as an initial condition at the time of preparation, and so on. Boundary conditions for the evolution of subsystems are obtained from *observations* of the rest of the universe outside the subsystem of interest.

Cosmology, however, presents us with an essentially different problem. The dynamical laws governing the evolution of the universe — the classical Einstein equation, for instance — require boundary conditions to yield solutions. But in cosmology, by definition, there is no "rest of the universe" to pass their specification off to. The cosmological boundary condition must be one of the fundamental laws of physics.

The inference is inescapable from the physics of the last sixty years that the fundamental laws are all quantum-mechanical. If that is assumed, a theory of the initial condition is a theory of the universe's initial quantum state. The search for a fundamental theory of this initial cosmological quantum state is the aim of that area of basic research which has come to be called quantum cosmology.

A view thus emerges that there are two fundamental laws of physics:

- A theory of the basic dynamics,
- A theory of the initial condition of the universe.

Were the universe governed by the Schrödinger equation (1), the basic theory of dynamics would specify the Hamiltonian H; a theory of the initial condition would be a law for the initial quantum state.

The search for the fundamental dynamical law has been seriously under way since the time of Newton. Classical mechanics, Newtonian gravity, electrodynamics, special relativity, general relativity, quantum mechanics, quantum electrodynamics, quantum chromodynamics, electro-weak theory, grand unified theories and superstring theories are but some of the important milestones in this search. By contrast, the search for a theory of the initial condition of the universe has been seriously under way for only a few decades. Why this difference? The answer lies in the empirical locality of the fundamental interactions on scales above the Planck length ($\sim 10^{-33}$ cm), or, put differently, the empirical fact that the fundamental interactions may be effectively described by a local quantum field theory on these scales. Assuming locality, the Hamiltonian of the whole universe can be deduced from experiments on familiar, laboratory, scales. However, typical ideas for the initial quantum state of the universe are non-local. They imply regularities *in space* that emerge mostly on large, cosmological scales. For example, the temperature of the cosmic microwave background is the same across the sky to one part in 10^5, a distance which corresponded to 10^{20} km at the time the radiation was emitted. It is only the recent progress in observational cosmology that has given us a picture of the universe on large enough scales of both space and time that is sufficiently detailed to confront with the predictions of a theory of the initial state of the universe.

2. Quantum Cosmology and The Everyday

Can those not interested in regularities on cosmological scales safely ignore the initial condition of the universe? Not if they seek a fundamental explanation of a number of its features we ordinarily take for granted. In this section we offer a few examples.

• Isolated Subsystems

In one way, we use a very weak theory of the initial condition every day. Many subsystems of the universe, in the laboratory and elsewhere, are approximately isolated for periods of time and can be approximately described by solving the Schrödinger equation for the subsystem alone. In effect, we assume that, for the purposes of making predictions about the subsystem, the wave function of the universe can be approximated by

$$\Psi(q^i, Q^A, t) \approx \psi(q^i, t)\Phi(Q^A, t) \tag{2}$$

where q^i and Q^A are coordinates referring to the subsystem and the rest of the universe, respectively, and ψ and Φ evolve *separately* under the Schrödinger equation. But what are the grounds for such an approximation? They do not lie in the nature of the Hamiltonian, because that generally specifies interactions between all the coordinates. Rather, the existence of isolated subsystems is a property of the quantum state. In discussing isolated subsystems, we are making weak quantum cosmological assumptions about the nature of this initial state.

• **The Quasiclassical Realm**
Classical deterministic laws govern a wide range of phenomena in the universe over a broad span of time, place, and scale. This quasi-classical realm is one of the most immediate facts of our experience. But indeterminacy and distributed probabilities are the characteristics of a quantum-mechanical universe. Classical deterministic dynamics can be but an approximation to the unitary evolution of the Schrödinger equation and the reduction of the state vector. To what do we owe the validity of this approximation? In part it arises from a coarse-grained description with positions and momenta specified to accuracies well above the limitations of the uncertainty principle, for instance. But coarse-graining is not enough; there must also be some restriction on the initial state. Ehrenfest's theorem is a simple illustration of why. For the motion of a particle in one dimension, Ehrenfest's theorem relates the acceleration of the expected position to the expected value of the force:

$$m\,\frac{d^2\langle x\rangle}{dt^2} = -\left\langle \frac{\partial V(x)}{\partial x} \right\rangle. \tag{3}$$

This is generally true, but for *certain* states, typically narrow wave packets, the right-hand side may be replaced by the force evaluated at the expected position to a good approximation, resulting in the deterministic classical equation of motion

$$m\,\frac{d^2\langle x\rangle}{dt^2} \approx -\frac{\partial V(\langle x\rangle)}{\partial x}. \tag{4}$$

Just as only certain states lead to classical behavior in this simple model, so also only certain cosmological initial conditions will lead to the quasiclassical realm of familiar experience.[a] That too is a feature of the universe that must ultimately be traced to the initial condition.

• **Homogeneity of the Thermodynamic Arrow of Time**
Isolated systems evolve toward equilibrium. That is a consequence of statistics. But in this universe, presently isolated systems are mostly evolving towards equilibrium in the *same* direction of time. That

[a] For a more quantitative discussion, see, e.g. Ref. 1.

is the homogeneity of the thermodynamic arrow of time. This is not a fact which can be explained by statistics or a property of the Hamiltonian alone, for that is approximately time-reversal invariant. The homogeneity of the thermodynamic arrow of time follows from a fundamental law of the initial condition which mandates that the progenitors of today's isolated systems were all far from equilibrium in the early universe. As Boltzmann put it: "The second law of thermodynamics can be proved from the [time-reversible] mechanical theory if one assumes that the present state of the universe ... started to evolve from an improbable state".[2]

• **History**

The reconstruction of history is useful for understanding the present in science as well as in human affairs. For example, we can best understand the character of biological species by understanding their evolution. We can best explain the present large scale distribution of galaxies by understanding how galaxies arose from tiny density fluctuations present shortly after the big bang. Such examples could be easily multiplied.

In physics, the reconstruction of history means using the fundamental laws to calculate the probabilities of alternative past events assuming the values of present records. Classically, present records *alone* are enough to calculate those probabilities by using them as the starting point for running the deterministic classical equations of motion backward in time. To reconstruct history in quantum mechanics, however, requires a theory of the initial condition *in addition* to present records.

The source of this difference between classical and quantum mechanics can be traced to the *arrow of time* in usual quantum theory.[b] Quantum mechanics treats the future differently from the past. To be sure, the Schrödinger equation (1) can be run backwards in time as well as forwards. But the Schrödinger equation is not the only law

---

[b]There are generalizations of quantum theory without an arrow of time in which the asymmetry of the usual theory may be understood as a difference between initial and final conditions, e.g. Ref. 3. We shall not consider these here.

of evolution in quantum theory. In the usual story, when a measurement is made, the wave function is "reduced" by the action of the projection operator P representing the outcome of the measurement, and then renormalized. This is a "second law of evolution":

$$\Psi \to \frac{P\Psi}{||P\Psi||}. \tag{5}$$

The evolution of the Schrödinger equation forward in time is interrupted by (5) on a measurement. While the Schrödinger equation can be run backward in time, the law (5) cannot, and that is a simple way of seeing the arrow of time in usual quantum mechanics. The same kind of arrow of time persists in more general quantum theories of closed systems where (5) is effectively used in the construction of the probabilities of histories which are not necessarily of the outcomes of measurements.

How then does one calculate the probabilities of past events assuming present records in quantum mechanics? The simple answer is that one works forward in time from the initial state. Evolving forward using (1) and (5), one calculates the joint probabilities of *both* alternative events of interest in the past *and* the alternative values of the present records that follow them. From these, one calculates the conditional probabilities of the past events given our particular present records in the usual way.

This process involves the initial state in an essential way. Strictly speaking, therefore, one cannot make any statements about the past without a theory of the universe's initial condition.

• **Phenomenology of the Initial Condition**

While the above four everyday features of the universe are fundamentally traceable to the universe's initial quantum state, there is a large set of initial states that would give rise to them. Put differently, the existence of isolated subsystems together with the applicability of classical physics, the second law of thermodynamics, and the possibility of historical explanation are not strong constraints on the initial quantum state. Neither are the observations of large scale features of the universe, such as its approximate homogeneity and isotropy or the fluctuations in the cosmic background radiation. The data are meager and the Hilbert space of the observable universe is vast.

It would be possible to investigate quantum cosmology *phenomenologically*, by asking for the constraints that present observations place on the initial state of the universe. A density matrix ρ is the way quantum mechanics represents the statistical distribution of states with associated probabilities that would be inferred. To investigate the initial condition phenomenologically is therefore to ask for the density matrices consistent with observed features of the universe.

The observed features of the universe may not uniquely fix an initial condition, but one should not exaggerate their weakness. The density matrix $\rho = I/Tr(I)$, where I is the unit matrix, is the unique representation of complete ignorance of the initial condition (i.e. no condition at all). But it also corresponds to infinite temperature in equilibrium ($\rho \propto \exp(-H/kT)$) — an initial condition whose implication of infinite temperature today is obviously inconsistent with present observations.

The entropy $S/k = -Tr(\rho \log \rho)$ is a measure of the missing information about the initial state in a density matrix ρ. Most of the entropy in the matter in the visible universe is in the cosmic background radiation, a number of order $S/k \approx 10^{80}$. As Penrose[4] has stressed, this is a large number, but infinitesimally small compared to the maximum possible value of $S/k \approx 10^{120}$ if all that matter composed a black hole.

This essay, however, is not concerned with phenomenology. *Rather, it is concerned with the fundamental law of the initial condition. We shall therefore assume that the universe has an initial state $|\Psi\rangle$ and discuss the issues involved in a search for the principles which determine it.*

3. Problems

Enumerating issues is one way of summarizing the present status of an area of science, and motivating future research. Certainly, setting problems is a more pleasant task than solving them, and quantum cosmology is such a young field that it is easier to summarize problems than to survey accomplishments. It is in this spirit that the author offers the following eight problems in quantum cosmology:

• Problem 1: What Principle Determines the Initial Condition of the Universe?

The evidence of the observations is that the universe was simpler earlier than it is now — more homogeneous, more isotropic, with matter more nearly in thermal equilibrium. This is evidence for a simple, discoverable initial condition of the universe. But what principles determine that initial state?

The most developed proposal for a principle determining the initial condition is the "no-boundary" wave function of Stephen Hawking and the author.[5] The idea is that the initial condition of a closed universe is the cosmological analog of a ground state. This does not mean the lowest eigenstate of some Hamiltonian. Intuitively, the total energy of a closed universe is zero, for there is no place outside from which to measure it. Correspondingly, the Hamiltonian vanishes.

But the lowest eigenstate of a Hamiltonian is not the only way to find the ground state even in the elementary case of a particle moving in a potential $V(x)$. In that case, the ground state wave function may be expressed directly as a sum over Euclidean paths, $x(\tau)$:

$$\psi_0(y) = \sum_{\text{paths } x(\tau)} \exp\Big(-I\big[x(\tau)\big]/\hbar\Big) \tag{6}$$

where $I = \int d\tau[m\dot{x}^2/2 + V(x)]$ is the Euclidean action. The sum is over paths $x(\tau)$ that have the argument of the wave function, y, as one end point, and a configuration of minimum action in the infinite past as another. Verify it for the harmonic oscillator, for example!

This construction of a ground state wave function generalizes to closed universes. For definiteness, suppose, for a moment, that the basic variables of the fundamental dynamical theory are the geometry of four-dimensional spacetime $\mathcal{G}$, represented by metrics on manifolds, together with matter fields such as the quark, lepton, gluon, and Higgs fields which we generically denote by $\phi(x)$. The arguments of cosmological wave functions are these basic variables restricted to spacelike surfaces, specifically the three-geometries of these surfaces, ${}^3\mathcal{G}$, and the field configurations on these surfaces, $\chi(\mathbf{x})$. The "no-boundary" wave function is of the form

$$\Psi_0[^3\mathcal{G}, \chi(\mathbf{x})] = \sum_{\mathcal{G},\phi(x)\in C} \exp\Big(-I\left[\mathcal{G}, \phi(x)\right]/\hbar\Big) \tag{7}$$

where $I[\mathcal{G}, \phi(\mathbf{x})]$ is the action for gravitation and matter. The "no-boundary" wave function is specified by giving the class C of four geometries $\mathcal{G}$, and matter fields $\phi(x)$ summed over in (7). So that the construction is analogous to (6), these geometries $\mathcal{G}$ should have Euclidean (signature $++++$) and have one boundary at which they match the three-geometry where the wave function is evaluated. The matter fields must similarly match their boundary value. The defining requirement is that the $\mathcal{G}$'s have no *other* boundary, hence the name "no-boundary" proposal.

Nothing goes on in a typical ground state in a fixed background spacetime. In field theory, the ground state is the time-translation invariant vacuum! However, this is not the context of the quantum cosmology of closed universes. Spacetime geometry is not fixed and there is therefore no notion of time-translation. Interesting histories therefore can happen; and the attractive nature of gravity makes things happen even in this cosmological analog of the ground state. In particular, initial, small, quantum, ground state fluctuations from homogeneity and isotropy that are predicted by this initial condition can grow by gravitational attraction to produce all the complexity in the universe that we see today.

This prescription for the "no-boundary" wave function is not complete. The reason is that the action $I[\mathcal{G}, \phi(x)]$ for gravitation coupled to matter is unbounded below. Were the sum in (7) extended over real, Euclidean geometries and fields, it would diverge! Rather, the sum must be taken over a class C of *complex* geometries and fields. A complex contour of summation is, in fact, essential for the "no-boundary" wave function to predict the nearly classical behavior of geometry we observe in the present epoch. But many different complex convergent contours are possibly available and correspondingly there are many different "no-boundary" wave functions. These do not differ in their semiclassical predictions; but we still lack a complete principle for fixing this wave function of the universe.

The "no-boundary" idea has been described in terms of an effective theory of dynamics in which spacetime and matter fields are treated as fundamental variables. If spacetime is not fundamental, as in string theory or non-perturbative quantum gravity, then extending the idea to such theories becomes an important question. The essentially topological nature of the idea gives some hope that such an extension is possible.

The "no-boundary" wave function is not the only idea for a theory of the initial condition. Other notable candidates are the "spontaneous nucleation from nothing wave function",[6] and the ideas associated with the "eternally self-reproducing inflationary universe".[7] Space does not permit a review of these and other theories, and the similarities and differences in their predictions. Discriminating between these and other ideas that may arise is certainly a problem for the 21st century.

• **Problem 2: How Can Quantum Gravity be Formulated for Cosmology?**

Gravity governs the evolution of the universe on the largest scales of space and time. That fact alone is enough to show that a quantum theory of gravity is required for a quantum theory of cosmology. Were the behavior of the universe on present cosmological scales all that was of interest, then a low-energy approximation to quantum gravity would be adequate. Indeed, most of the exploration of quantum cosmology has been carried out in such a low-energy approximation assuming spacetime geometry and quantum fields are the basic variables with Einstein's theory coupled to matter as the basic action. Any divergences that arise are truncated in one way or another.

It is a reasonable expectation that low-energy, large scale, features of the universe, such as the galaxy–galaxy correlation function, are insensitive to the nature of quantum gravity on very small scales. But in quantum cosmology we aim not only at an explanation of such large scale features, but also at a theory of the initial condition adequate to describe the probabilistic details of the earliest moments of the universe. The inevitability of an initial singularity in classical Einstein cosmologies strongly suggests that the earliest moments of the universe will exhibit curvatures of spacetime characterized by the

Planck length

$$\ell \equiv (\hbar G/c^3)^{1/2} \approx 10^{-33}\text{cm} \tag{8}$$

— the only combination of the three fundamental constants governing relativity, quantum mechanics and gravity that has the dimensions of length. By making similar combinations with the right dimensions, we can exhibit the Planck scales of energy and time. The universe at the epochs characterized by these scales will therefore depend on the detailed form of the fundamental quantum dynamical law for gravity.

There are a number of candidates for a finite, manageable quantum theory of gravity, notably superstring theory and non-perturbative canonical quantum gravity. However, neither of these theories is ready for application to quantum cosmology. String theory, for instance, exists in a practical sense as a set of rules for classical backgrounds and quantum perturbations away from them. Developing such theories to the point where they can be used for the non-perturbative quantum dynamics of closed cosmologies is thus an important problem.

The problem to be faced is not merely one of technique. Both of the approaches mentioned, and others as well, hint that spacetime geometry may not be a basic dynamical variable. If that is true, it becomes a conceptual issue — just how to frame cosmological questions in the variables of the fundamental dynamical theory.

• Problem 3: What is the Generalization of Quantum Mechanics Necessary for Quantum Gravity and Quantum Cosmology?

A generalization of usual quantum mechanics is needed for quantum gravity. That is because usual quantum mechanics relies in essential ways on a fixed, background spacetime geometry, in particular, to specify the notion of time that enters centrally into the formalism. This reliance on a fixed notion of time shows up in any of the various ways of formulating usual quantum theory — the idea of a state at a moment of time, the preferred role of time in the Schrödinger equation, the inner product at a moment of time, the reduction of the state vector at a moment of time, the commutation of fields at spacelike separated points, the equal time commutators of coordinates and momenta, etc.

But in quantum gravity, spacetime geometry is not fixed, rather it is a quantum dynamical variable, fluctuating and generally without definite value. A generalization of usual quantum theory that does not require a fixed spacetime geometry, but to which the usual theory is a good approximation in situations when the geometry is approximately fixed, is therefore needed for quantum gravity and quantum cosmology. What, therefore, do we mean more generally by a quantum-mechanical theory?

The most general objective of any quantum theory is the prediction of the probabilities of alternative, coarse-grained histories of the universe as a single, closed quantum-mechanical system. For example, one might be interested in predicting the probabilities of the set of possible orbits of the earth around the sun. Any orbit is possible, but a Keplerian ellipse has overwhelming probability. Such histories are said to be coarse-grained because they do not specify the coordinates of every particle in the universe, but only those of the center of mass of the earth and sun, and these only crudely and not at every time.

However, the characteristic feature of a quantum-mechanical theory is that consistent probabilities cannot be assigned to every set of alternative histories because of quantum-mechanical interference. Nowhere is this more clearly illustrated than in the famous two-slit experiment shown in Fig. 1. Electrons can proceed from an electron gun at left towards detection at a point y on a screen along one of two possible histories — the history passing through the upper slit, A, and the history passing through the lower slit, B. In the usual story, probabilities cannot be assigned to these two histories if we have not *measured* which slit the electron had passed through. It would be inconsistent to do so because the the probability to arrive at y would not be the sum of the probabilities to arrive there going through the upper and lower slits:

$$p(y) \neq p_A(y) + p_B(y) \tag{9}$$

because of quantum-mechanical interference. In quantum mechanics,

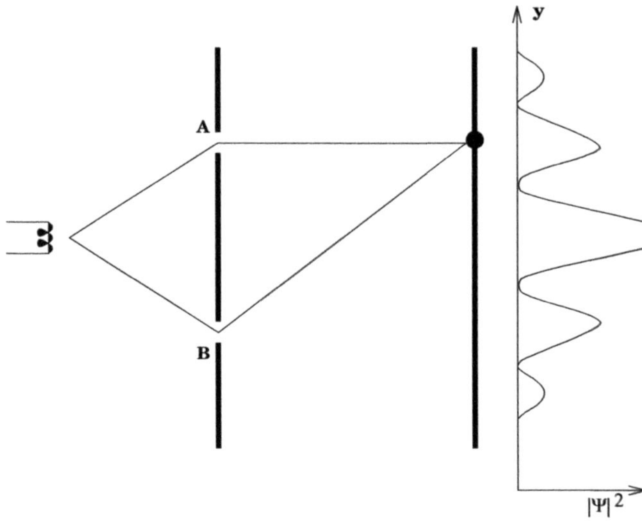

Fig. 1. The two-slit experiment.

probabilities are squares of amplitudes and, of course,

$$|\psi_A(y) + \psi_B(y)|^2 \neq |\psi_A(y)|^2 + |\psi_B(y)|^2. \tag{10}$$

A necessary consistency condition would not be satisfied.

A rule is thus needed in quantum theory to specify which sets of alternative histories may be assigned probabilities and which may not. In the usual, "Copenhagen" formulations of quantum mechanics presented in textbooks, probabilities can be assigned to the histories of alternatives of a subsystem that were "measured" by an "observer". But such formulations are not general enough for quantum cosmology which seeks to describe the early universe where there were neither measurements nor observers present.

In the more general quantum mechanics of closed systems[c] that rule is simple: probabilities can be assigned to just those sets of alternative histories for which there is vanishing interference between the individual members as a consequence of the system's initial state $|\Psi\rangle$. To make this quantitative, we need the measure of this interference.

[c]For expositions, see Refs. 8 and 9.

When there is a well-defined fixed notion of time, sequences of alternative sets of events at a series of times define a set of alternative histories. An individual history in such a set is a particular series of events, say, $\alpha \equiv (\alpha_1, \alpha_2, \ldots, \alpha_n)$ at times $t_1 < t_2 < \cdots < t_n$. In usual quantum mechanics, such a history is represented by a corresponding chain of (Heisenberg-picture) operators,

$$C_\alpha \equiv P^n_{\alpha_n}(t_n) \cdots P^2_{\alpha_2}(t_2) P^1_{\alpha_1}(t_1), \tag{11}$$

time-ordered from right to left. The application of the C_α to the initial state vector $|\Psi\rangle$ gives the branch state vector

$$C_\alpha |\Psi\rangle \tag{12}$$

corresponding to the history. Interference vanishes in a set of alternative histories when the branch state vectors corresponding to the different histories are mutually orthogonal. Sets of alternative histories with vanishing interference are said to *decohere*. The probabilities $p(\alpha)$ of the individual histories in a decoherent set are the squared lengths of the branch state vectors

$$p(\alpha) = ||C_\alpha |\Psi\rangle||^2. \tag{13}$$

Decoherence insures the consistency of these probabilities.

Interference is thus measured by the *decoherence functional*:

$$D(\alpha', \alpha) = \left\langle \Psi | C^\dagger_{\alpha'} C_\alpha | \Psi \right\rangle \tag{14}$$

which becomes the central element in the theory. The condition of decoherence and the resulting probabilities may be expressed by the single formula

$$D(\alpha', \alpha) = \delta_{\alpha'\alpha} p(\alpha). \tag{15}$$

The sets of possible coarse-grained histories, their decoherence functional, and (15) are the minimal elements of a quantum theory. A broad framework for quantum theories built on these elements, called generalized quantum mechanics, can be formulated in terms of decoherence functionals obeying general principles of Hermiticity, normalization, positivity, and the principle of superposition.[10,11]

Histories represented by strings of projections at definite moments of time (11) and a decoherence functional (14) are the ways that usual quantum theory implements the principles of generalized quantum theory. But there are many other ways, and among them are possibilities for generalizing usual quantum mechanics so that it works in the absence of a fixed spacetime geometry. Generalized sum-over-histories quantum theories have been discussed that put quantum theory into fully spacetime form with four-dimensional notions of histories, coarse-grainings, and decoherence.[11] But the principles of generalized quantum mechanics are only a minimal set of requirements for quantum theory. What further principles determine the correct quantum mechanics for quantum gravity and quantum cosmology?

• **Problem 4: What are the Definite Predictions of the Initial Condition for the Universe on Large Scales?**

Extracting the predictions of a theory of the initial condition and comparing them with observations is a central problem in quantum cosmology. Predictions take the form of probabilities for present observations. The theory stands or falls on those predictions with probabilities sufficiently close to one (or zero) being observed (or not observed). These are called the *definite* predictions of the theory. We expect few of them. A simple, comprehensible, discoverable theory of the initial condition cannot predict all the complexity observed in the present universe with probability near one.[12] Rather, most predictions, as of the stock market, the weather, or the number of moons of Jupiter, will have more distributed probabilities based on the initial condition alone. (The vast majority will be nearly uniformly distributed, which is no prediction at all.) In quantum cosmology one must search among the possible predictions for those which are predicted with probability near one. Interestingly, definite predictions may occur on all scales. For the purposes of simplicity, we have divided the problem of what are the definite predictions of a theory of the initial condition into problems concerning regularities on cosmological, familiar, and microscopic scales.

Quantum cosmologists expect that a number of the general large scale features of the universe will be definite predictions of a theory of its initial condition. These include an approximately classical cosmological spacetime geometry after the Planck epoch, the approximate homogeneity and isotropy of the geometry and matter on scales above several hundred megaparsecs,[d] the approximate spatial flatness of the universe (or what is the same thing as its vast age in Planck units), the initial spectrum of quantum fluctuations which grew to become the galaxies, a sufficiently long inflationary epoch, and the cosmological abundances of the matter and radiation species.

The probabilities for these features of the universe arising from various theories of the initial condition have been explored in highly simplified models valid only in limited regions of the configuration space of possible present universes. The output of some of these calculations, such as the prediction of the spectrum of initial quantum fluctuations,[13] are among the most successful achievements of quantum cosmology. But much more needs to be done to extend these calculations to the whole of the configuration space with greater accuracy, generality and a precise quantum mechanical interpretation. That is a practical and immediate problem for quantum cosmology.

• **Problem 5: What are the Definite Predictions of the Initial Condition for Features of the Universe on Familiar Scales?** We may treat this problem briefly because the obvious features of the universe on familiar scales that are traceable to the initial condition have been discussed qualitatively in Sec. 2. However, those qualitative conclusions raise quantitative questions:

What are the coarse-grained variables defining a quasiclassical realm governed by deterministic laws and how are these variables related to the principle that determines the initial condition? How much refining of a quasiclassical description of the universe is possible before decoherence is lost or determinism is overwhelmed by quantum noise? How far in space and time can a quasiclassical description be extended? How do the phenomenological equations of motion that exhibit the determinism of the quasiclassical realm follow from the fundamental dynamical law, an initial condition, and

[d]The megaparsec is a convenient unit for cosmology. One megaparsec is 3.3 million light years. The size of the visible universe is several thousand megaparsecs.

an appropriate coarse-grained description? What is the connection of the coarse-graining used to define a quasiclassical realm with that which is necessary to exhibit a second law of thermodynamics? How far out of equilibrium is the early universe in this coarse-graining?

In short, a theory of the initial condition presents the challenge of defining *quantitatively* those features of the universe on familiar scales which are traceable, in part, to the nature of the initial condition.

• Problem 6: What are the Definite Predictions of the Initial Condition on Microscopic Scales?

Our understanding of the world on microscopic scales above that are set by the Planck length is summarized by the effective field theories which govern phenomena on these scales — for example, the standard model of elementary particle physics. However, the forms of these effective field theories may be only distantly related to the form of the fundamental dynamical law. An analogous situation at a different scale may help to explain why: The form of the Navier–Stokes equation which governs the dynamics of much of the quasiclassical realm is not easily guessed from the Lagrangian of the standard model of particle physics. In particular, the Navier–Stokes equation incorporates dissipation and depends on constitutive relations between density, pressure, temperature, viscosity, etc. — relations not contained in the Lagrangian of the standard model.

Of course, we understand qualitatively the relation between the laws of the standard model and the Navier–Stokes equation. The Navier–Stokes equation applies, not generally, not exactly, but only approximately in particular circumstances. It is an *effective* equation with a limited range of approximate validity. In quantum mechanics, particular circumstances are represented by the quantum state and a coarse-grained description. It is the quantum state whose special properties allow a classical approximation, set up the conditions for dissipation, determine the constituents, and allow for the local equilibrium from which the constitutive relations follow.

But the standard model itself may be only an effective approximation to a more fundamental dynamical law, such as heterotic superstring theory or non-perturbative quantum gravity. We may therefore restate the problem of the definite predictions of the initial condition as follows: **What features of the effective dynamical laws that govern the elementary particle system at accessible energy scales are traceable to the cosmological initial condition and what, to the fundamental dynamical law?** For instance, what is the origin of the locality of the effective interactions in a theory of the quantum state that is intrinsically non-local?

The investigations of the effects of wormholes by Hawking, Coleman, Giddings and Strominger, and others indicate just how strong the effect of the initial condition on the effective interactions could be (for a review see, e.g. Ref. 14). Suppose that the sum over geometries defining the "no-boundary" wave function in (7) includes a sum over wormhole geometries — four-dimensional geometries with many "handles", rather like a teacup has a handle. Suppose that the Planck scale (8) is the characteristic size of these wormholes in the geometries that contribute the most to the sum. Fields propagating in such geometries can go down a wormhole and emerge from one. On the much larger scales accessible to us, we would see the effect of Planck scale wormholes as local interactions which create and destroy particles. The net effect is to add to any local Lagrangian an infinite series of local interactions with coupling constants that are not fixed once and for all by the fundamental dynamical law or even by a renormalization procedure, but rather vary probabilistically with a distribution determined by the initial condition. If the distribution was sharp (as was hoped for the cosmological constant), then the couplings would be predicted.

A similar decoupling between the observed coupling constants and the basic Lagrangian would hold if the initial condition predicted domains of space much larger than our visible universe, in which breaking of the symmetries of the fundamental dynamical law occurred in different ways and in different places leading to differing effective theories in different domains. The form of the effective theory governing our domain would then be only a probabilistic prediction of the fundamental dynamical law.

It has proved difficult to push such ideas very far, but their lesson is clear. The form and couplings of the effective interactions at accessible scales may be probabilistically distributed in a way which depends on the initial condition. Finding these distributions and how sharp they are, therefore important problems in quantum cosmology.

• **Problem 7: What Does Quantum Cosmology Predict for IGUSes?**

Most predictions of the initial condition that we have considered so far are described in terms of alternatives of the quasiclassical realm. But there are many sets of decohering histories of the universe arising from a theory of its initial condition and dynamics that have nothing to do with the usual quasiclassical realm. These sets may be quantum-mechanically incompatible with each other and with the usual quasiclassical realm in the sense that pairs of them cannot be combined into a common decohering set. Such incompatible sets are not contradictory; rather they are complementary ways of viewing the unfolding of the initial condition into alternative histories. The quantum mechanics of closed systems does not distinguish between such incompatible sets of alternative histories except by properties such as their classicality. All are in principle available for the process of prediction.

Yet, as observers, we describe the universe almost exclusively in terms of the familiar variables of classical physics. What is the reason for this narrow focus in the face of all the other non-quasiclassical decohering sets of alternative histories? Some see this disparity between the possibilities allowed by quantum theory and the possibilities utilized by us as grounds for augmenting quantum mechanics by a further fundamental principle that would single out one decohering set of histories from all others.[15] That is an interesting line of thought, but another is to seek an explanation within the existing quantum mechanics of closed systems.

Human beings, bacteria, and certain computers, are examples of information gathering and utilizing systems (IGUSes). Roughly, an IGUS is a subsystem of the universe that makes observations and thus acquires information, then makes predictions on the basis of that information using some approximation (typically very crude) to the quantum-mechanical laws of nature, and exhibits behavior based on these predictions.[e] To explain why IGUSes are exhibited by the universe, or why they behave the way they do, or to answer questions like "Why do we utilize quasiclassical variables?", one must seek to understand how IGUSes evolved as physical systems within the universe. In quantum cosmology, that means examining the probabilities of a set of histories that define alternative evolutionary tracks. For IGUSes that can be characterized in terms of alternatives of the usual quasiclassical realm, it is a plausible conjecture that they evolved to focus on the usual quasiclassical alternatives because these present enough regularity over time to permit prediction by relatively simple models (schemata). This would be one explanation of why we utilize the usual quasiclassical realm. However, we should not pretend that we are close to being able to carry out a calculation of the relevant probabilities, or even likely to be, in the early 21st century!

But what of sets of histories that are completely unrelated to the usual quasiclassical realm? Might some of these exhibit IGUSes with high probability that make predictions in terms of variables very different from the familiar quasiclassical ones? Or is the usual quasiclassical realm somehow distinguished with respect to exhibiting IGUSes? To answer such questions one would need a general characterization of IGUSes that is applicable to all kinds of histories — not just quasiclassical ones — and an ability to calculate the probabilities of various courses of the IGUSes' evolution. Such questions, while quite beyond our power to answer in the present, illustrate the range of predictions that are in principle possible in a quantum universe following from a fundamental theory of dynamics and the initial condition of the universe.

4. Unification

The universal laws that govern the regularities of every physical system are one goal of physics. A fundamental dynamical law is one objective. Quantum cosmology is concerned with the equally

---

[e]IGUSes are complex adaptive systems in the context of quantum mechanics. For more on the general characterization of complex adaptive systems, see Ref. 16.

necessary fundamental law specifying the initial condition of the universe.

Historically, many of the advances toward the fundamental laws have had in common that some idea that was previously thought to be universal, was subsequently seen to be only a feature of our special place in the universe and the limited range of our experience. With more data, the idea was seen to be a true physical fact, but one which is a special situation in a yet more general theory. The idea was a kind of "excess baggage" which had to be jettisoned to reach a more general, more comprehensive, and more fundamental perspective.[17]

It is not difficult to cite examples of such excess theoretical baggage in the history of physics: the idea that the earth was the center of the universe, the idea of Newtonian absolute time, the idea that the increased entropy was a basic dynamical law, the idea that spacetime geometry is fixed, the idea of a classical world separate from quantum mechanics, etc. Further, and more importantly for the present discussion, one can cite examples concerning the nature of the fundamental laws themselves: the idea that thermodynamics was separate from mechanics, the idea that electricity was separate from magnetism, and more recently the idea that there were separate weak and electromagnetic interactions. These seemingly distinct theories were eventually unified. Today, extrapolations of the standard model of the electro-weak and strong interactions suggest a unified theory of these forces characterized by an energy scale that is a little below the Planck scale.

Examples such as those just cited have led some physicists to speculate that the existing separation between the dynamical laws for the gravitational and other forces is also an example of excess baggage arising from the limitations of present experiments to energies well below the Planck scale, and to search for a unified fundamental law for dynamics of *all* the forces. Secure in the faith that fundamental laws are mathematically simple, heterotic superstring theory or its extensions have been the inspiring results.

However, such a unified *dynamical* law does not really deserve the common designation of "a theory of everything" or a "final theory". Quantum cosmology offers a further opportunity for unification beyond dynamical laws. Could it be that the apparent division of the fundamental laws into a law for dynamics and a law for an initial quantum state, is a kind of excess baggage similar to those described above? Gell-Mann[18] has stressed that there is already an element of unification in ideas such as the "no-boundary" proposal. In (7), the same action that determines fundamental dynamics also determines the quantum state of the universe. Despite this connection, the "no-boundary" proposal is a separate principle specifying one wave function out of many possible ones. Thus we have an eighth problem for quantum cosmology:

• **Problem 8: Is there a Fundamental Principle that would Single Out *Both* a Unified Dynamical Law *and* a Unique Initial Quantum State for the Universe? Could that same Principle Single Out the Form of Quantum Mechanics from Among Those Presented by Generalized Quantum Theory?** In such a unification of the law of dynamics, the cosmological boundary condition, and the principles of quantum mechanics, we would, at last, have a truly unified fundamental law of physics governing the universe as a whole and everything within it. That is truly a worthy problem for physics in the 21st century!

5. Further Reading

Reference 19 is a *Scientific American* article introducing quantum cosmology. An accessible but more advanced introductory review is Ref. 20. That article contains a nearly exhaustive list of references at the time and a guide to the literature. For an introduction to the quantum mechanics of closed systems, see e.g. Ref. 8. For an exposition on the applications of quantum mechanics to cosmology, see Ref. 11.

Acknowledgments

Thanks are due to G. Horowitz and J. West for critical readings of the manuscript. Preparation of this essay was supported in part by the US National Science Foundation under grants PHY95-07065 and PHY94-07194.

References

No attempt has been made to assemble a complete list of references to the topics touched on in this essay. Rather, the emphasis has been placed on reviews of topics that are most useful directly to the points considered. These are not always the earliest nor the latest references.

1. J. B. Hartle, Quasiclassical domains in a quantum universe, in *Proc. Cornelius Lanczos International Centenary Conference*, North Carolina State University, December 1992, eds. J. D. Brown, M. T. Chu, D. C. Ellison and R. J. Plemmons (SIAM, Philadelphia, 1994) arXiv: gr-qc/9404017.
2. L. Boltzmann, Zu Hrn. Zermelo's Abhandlung Über die mechanische Erklärung irreversibler Vorgange, *Ann. Physik* **60**, 392 (1897).
3. M. Gell-Mann and J. B. Hartle, Time symmetry and asymmetry in quantum mechanics and quantum cosmology, in *Proc. NATO Workshop on the Physical Origins of Time Asymmetry*, Mazagón, Spain, September 30–October 4, 1991, eds. J. Halliwell, J. Pérez-Mercader and W. Zurek (Cambridge University Press, Cambridge, 1993) arXiv: gr-qc/9309012.
4. R. Penrose, Singularities and time asymmetry, in *General Relativity: An Einstein Centenary Survey*, eds. S. W. Hawking and W. Israel (Cambridge University Press, Cambridge, 1979).
5. S. W. Hawking, The quantum state of the universe, *Nucl. Phys. B* **239**, 257 (1984).
6. A. Vilenkin, Predictions from quantum cosmology, in *String Gravity and Physics at the Planck Scale*, eds. N. Sanchez and A. Zichichi (Kluwer Academic, Dordrecht, 1996) and Approaches to Quantum Cosmology, *Phys. Rev. D* **50**, 2581 (1994).
7. A. D. Linde, *Particle Physics and Inflationary Cosmology* (Harwood Academic Publishers, Chur, Switzerland 1990); A. D. Linde, D. A. Linde and A. Mezhlumian, From the big bang theory to the theory of a stationary universe, *Phys. Rev. D* **49**, 1783 (1994).

8. J. B. Hartle, The quantum mechanics of closed systems, in *Directions in General Relativity, Vol. 1: A Symposium and Collection of Essays in honor of Professor Charles W. Misner's 60th Birthday*, eds. B.-L. Hu, M. P. Ryan and C. V. Vishveshwara (Cambridge University Press, Cambridge, 1993) arXiv: gr-qc/9210006.
9. R. Omnès, *The Interpretation of Quantum Mechanics* (Princeton University Press, Princeton, 1994).
10. C. J. Isham, Quantum logic and the histories approach to quantum theory, *J. Math. Phys.* **35**, 2157 (1994).
11. J. B. Hartle, Spacetime quantum mechanics and the quantum mechanics of spacetime, in *Gravitation and Quantizations, Proc. 1992 Les Houches Summer School*, eds. B. Julia and J. Zinn-Justin, Les Houches Summer School Proceedings, Vol. LVII (North Holland, Amsterdam, 1995) aXiv: gr-qc/9304006.
12. J. B. Hartle, Scientific knowledge from the perspectives of quantum cosmology, in *Boundaries and Barriers: On the Limits to Scientific Knowledge*, eds. J. L. Casti and A. Karlqvist (Addison-Wesley, Reading, Mass., 1996) arXiv: gr-qc/9601046.
13. J. Halliwell and S. W. Hawking, Origin of structure in the universe, *Phys. Rev. D* **31**, 1777 (1985).
14. A. Strominger, Baby universes, in *Quantum Cosmology and Baby Universes: Proc. 1989 Jerusalem Winter School for Theoretical Physics*, eds. S. Coleman, J. B. Hartle, T. Piran and S. Weinberg (World Scientific, Singapore, 1991).
15. H. F. Dowker and A. Kent, On the consistent histories approach to quantum mechanics, *J. Stat. Phys.* **82**, 1574 (1996), arXiv: gr-qc/9412067.
16. M. Gell-Mann, *The Quark and the Jaguar* (W. Freeman, San Francisco, 1994).
17. J. B. Hartle, Excess baggage, in *Elementary Particles and the Universe: Essays in Honor of Murray Gell-Mann*, ed. J. Schwarz (Cambridge University Press, Cambridge, 1990).
18. M. Gell-Mann, Dick Feynman — The guy in the office down the hall, *Phys. Today* **42**, 50 (1989).

19. J. Halliwell, Quantum cosmology and the creation of the universe, *Sci. Am.* **265**, 76 (1991).
20. J. Halliwell, Introductory lectures on quantum cosmology, in *Quantum Cosmology and Baby Universes: Proc. 1989 Jerusalem Winter School for Theoretical Physics*, eds. S. Coleman, J. B. Hartle, T. Piran and S. Weinberg (World Scientific, Singapore, 1991), pp. 65–157.

CHAPTER 16

Anthropic Reasoning and Quantum Cosmology*

Prediction in quantum cosmology requires a specification of the universe's quantum dynamics *and* its quantum state. We expect only a few general features of the universe to be predicted with probabilities near unity conditioned on the dynamics and quantum state alone. Most useful predictions are of conditional probabilities that assume additional information beyond the dynamics and quantum state. Anthropic reasoning utilizes probabilities conditioned on "us". This paper discusses the utility, limitations and theoretical uncertainty involved in using such probabilities. The predictions resulting from various levels of ignorance of the quantum state are discussed.

*This is a slightly abbreviated version of an article that appeared in *The New Cosmology: Proc. Conf. Strings and Cosmology*, College Station, Texas, March 14–17, 2004, eds. R. Allen, D. Nanopoulos and C. Pope, AIP Conference Proceedings, Vol. 743 (American Institute of Physics, Melville, NY, 2004).
arXiv:gr-qc/0406104

1. Introduction

If the universe is a quantum-mechanical system, then it has a quantum state. This state provides the initial condition for cosmology. A theory of this state is an essential part of any final theory summarizing the regularities exhibited universally by all physical systems and is the objective of the subject of quantum cosmology. This essay is concerned with the role that the state of the universe plays in anthropic reasoning — the process of explaining features of our universe from our existence in it.[1] The thesis will be that anthropic reasoning in a quantum-mechanical context depends crucially on assumptions about the universe's quantum state.

2. A Model Quantum Universe

Every prediction in a quantum-mechanical universe depends on its state, if only very weakly. Quantum mechanics predicts probabilities for alternative possibilities, most generally the probabilities for alternative histories of the universe. The computation of these probabilities requires both a theory of the quantum state as well as the theory of the dynamics specifying its evolution.

To make this idea concrete while keeping the discussion manageable, we consider a model quantum universe. The details of this model are not essential to the subsequent discussion of anthropic reasoning but help to fix the notation for probabilities and provide a specific example of what they mean. Particles and fields move in a large, perhaps expanding box, say, presently 20,000 Mpc on a side. Quantum gravity is neglected — an excellent approximation for accessible alternatives in our universe later than 10^{-43} s from the big bang. Spacetime geometry is thus fixed with a well-defined notion of time and the usual quantum apparatus of Hilbert space, states and their unitary evolution governed by a Hamiltonian can be applied.[a]

The Hamiltonian H and the state $|\Psi\rangle$ in the Heisenberg picture are the assumed theoretical inputs to the prediction of quantum-

[a]For a more detailed discussion of this model in the notation used here, see Ref. 2. For a quantum framework when spacetime geometry is not fixed, see e.g. Ref. 3.

mechanical probabilities. Alternative possibilities at one moment of time t can be reduced to yes/no alternatives represented by an exhaustive set of orthogonal projection operators $\{P_\alpha(t)\}$, $\alpha = 1, 2, \cdots$ in this Heisenberg picture. The operators representing the same alternatives at different times are connected by

$$P_\alpha(t) = e^{iHt/\hbar} P_\alpha(0)\, e^{-iHt/\hbar}. \tag{1}$$

For instance, the P's could be projections onto an exhaustive set of exclusive ranges of the center of mass position of the Earth labeled by α. The probabilities $p(\alpha)$ that the Earth is located in one or another of these regions at time t is

$$p\,(\alpha|H, \Psi) = \|P_\alpha(t)\,|\Psi\rangle\|^2. \tag{2}$$

The probabilities for the Earth's location at a different time is given by the same formula with different P's computed from the Hamiltonian by (1). The notation $p\,(\alpha|H, \Psi)$ departs from usual conventions (e.g. Ref. 2) to indicate explicitly that all probabilities are conditioned on the theory of the Hamiltonian H and quantum state $|\Psi\rangle$.

Most generally, quantum theory predicts the probabilities of sequences of alternatives at a series of times — that is, *histories.* An example is a sequence of ranges of center of mass position of the Earth at a series of times, giving a coarse-grained description of its orbit. Sequences of sets of alternatives $\{P^k_{\alpha_k}(t_k)\}$ at a series of times t_k, $k = 1, \ldots, n$, specify a set of alternative histories of the model universe. An individual history α in the set corresponds to a particular sequence of alternatives $\alpha \equiv (\alpha_1, \alpha_2, \ldots, \alpha_n)$, and is represented by the corresponding chain of projection operators C_α

$$C_\alpha \equiv P^n_{\alpha_n}(t_n) \cdots P^1_{\alpha_1}(t_1), \quad \alpha \equiv (\alpha_1, \ldots, \alpha_n). \tag{3}$$

The probabilities of the histories in the set are given by

$$p\,(\alpha|H, \Psi) \equiv p\,(\alpha_n, \ldots, \alpha_1|H, \Psi) = \|C_\alpha|\Psi\rangle\|^2 \tag{4}$$

provided the set decoheres, i.e. provided the branch state vectors $C_\alpha|\Psi\rangle$ are mutually orthogonal. Decoherence ensures the consistency of the probabilities (4) with the usual rules of probability theory.[b]

[b]For a short introduction to decoherence, see Ref. 2 or any of the classic expositions of decoherent (consistent) histories quantum theory.[4–6]

To use either (2) or (4) to make predictions, a theory of *both* H and $|\Psi\rangle$ is needed. No state; no predictions.

3. What is Predicted?

"If you know the wave function of the universe, why aren't you rich?" This question was once put to me by my colleague Murray Gell-Mann. The answer is that there are unlikely to be any alternatives relevant to making money that are predicted as sure bets conditioned just on the Hamiltonian and quantum state alone. A probability $p(\text{rise}|H,\Psi)$ for the stock market to rise tomorrow could be predicted from H and $|\Psi\rangle$ through (2) in principle. But it seems likely that the result would be a useless $p(\text{rise}|H,\Psi) \approx 1/2$ conditioned just on the "no-boundary" wave function[7] and M-theory.

It is plausible that this is the generic situation. To be manageable and discoverable, the theories of dynamics and the quantum state must be short — describable in terms of a few fundamental equations and the explanations of the symbols they contain. It is therefore unlikely that H and $|\Psi\rangle$ contain enough information to determine most of the interesting complexity of the present universe with significant probability.[8,9] We *hope* that the Hamiltonian and the quantum state are sufficient conditions to predict certain large scale features of the universe with significant probability. Approximately classical spacetime, the number of large spatial dimensions, the approximate homogeneity and isotropy on scales above several hundred Mpc, and the spectrum of density fluctuations that were the input to inflation are some examples of these. But even a simple feature like the time the Sun will rise tomorrow will not be usefully predicted by our present theories of dynamics and the quantum state *alone.*

The time of sunrise *does* become predictable with high probability if a few previous positions and orientations of the Earth in its orbit are supplied in addition to H and $|\Psi\rangle$. That is a particular case of a *conditional probability* of the form

$$p\,(\alpha|\beta, H, \Psi) = \frac{p\,(\alpha, \beta|H, \Psi)}{p\,(\beta|H, \Psi)} \tag{5}$$

for alternatives α (e.g. the times of sunrise) given $H, |\Psi\rangle$ and further alternatives β (e.g. a few earlier positions and orientations of the Earth). The joint probabilities on the right-hand side of (5) are computed using (4), as described in Sec. 2.

Conditioning probabilities on specific information can weaken their dependence on H and $|\Psi\rangle$, but does not eliminate it. That is because any specific information available to us as human observers (like a few positions of the Earth) is but a small part of that needed to specify the state of the universe. The P_β used to define the joint probabilities in (5) by (4) therefore spans a very large subspace of Hilbert space. As a consequence, $P_\beta|\Psi\rangle$ depends strongly on $|\Psi\rangle$. For example, to extrapolate present data on the Earth to its position 24 hours from now requires that the probability be high that it moves on a classical orbit in that time and that the probability be low that it is destroyed by a neutron star now racing across the galaxy at near light speed. Both of these probabilities depend crucially, if weakly, on the nature of the quantum state.[10]

Many useful predictions in physics are of conditional probabilities of the kind discussed in this section. We next turn to the question of whether we should be part of the conditions.

4. Anthropic Reasoning — Less is More

4.1. *Anthropic Probabilities*

In calculating the conditional probabilities for predicting some of *our* observations given others, there can be no objection of principle to including a description of "us" as part of the conditions,

$$p\,(\alpha|\beta, \text{`us'}, H, \Psi). \tag{6}$$

Drawing inferences using such probabilities is called *anthropic reasoning.* The motivation is the idea that probabilities for certain features of the universe might be sensitive to this inclusion.

The utility of anthropic reasoning depends on how sensitive probabilities like (6) are to the inclusion of "us". To make this concrete, consider the probabilities for a hypothetical cosmological parameter we will call Λ. We will assume that H and $|\Psi\rangle$ imply that Λ is constant over the visible universe, but only supply probabilities for the various constant values it might take through (4). We seek to compare $p(\Lambda|H, \Psi)$ with $p(\Lambda|\text{`us'}, H, \Psi)$. In principle, *both* are calculable from (4) and (5). Figure 1 shows three possible ways they might be related:

- $p(\Lambda|H, \Psi)$ is peaked around one value as in Fig. 1(a). The parameter Λ is determined either by H or $|\Psi\rangle$, or by both.[c] Anthropic reasoning is not necessary; the parameter is already determined by fundamental physics.
- $p(\Lambda|H, \Psi)$ is distributed and $p(\Lambda|\text{``us''}, H, \Psi)$ is also distributed as in Fig. 1(b). Anthropic reasoning is inconclusive. One might as well measure the value of Λ and use this as a condition for making further predictions[d] i.e. work with probabilities of the form $p(\alpha|\Lambda, H, \Psi)$.
- $p(\Lambda|H, \Psi)$ is distributed but $p(\Lambda|\text{``us''}, H, \Psi)$ is peaked. Anthropic reasoning helps to explain the value of Λ.

The important point to emphasize is that a *theoretical hypothesis for H and $|\Psi\rangle$ is needed to carry out anthropic reasoning.* Put differently, a theoretical context is needed to decide whether a parameter like Λ *can* vary, and to find out *how* it varies, before using anthropic reasoning to restrict its range. The Hamiltonian and quantum state provide this context. In Sec. 5, we will consider the situation where the state is imperfectly known.

4.2. *Less is More*

While there can be no objections of principle to including "us" as a condition for the probabilities of our observations, there are formidable obstacles of practice:

- We are complex physical systems requiring an extensive environment and a long evolutionary history, whose description in terms of the fundamental variables of H and $|\Psi\rangle$ may be uncertain, long and complicated.

---

[c] As, for example, in the as yet inconclusive discussions of baby universes.[12]
[d] As stressed by Hawking and Hertog.[13]

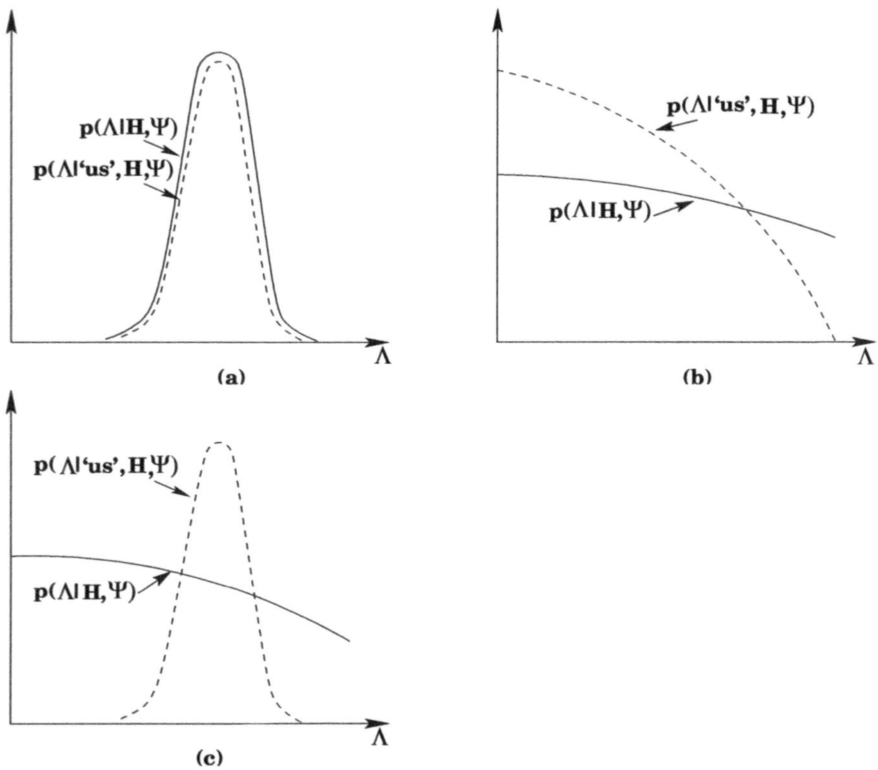

Fig. 1. Some possible behaviors for probabilities for the value of a cosmological parameter Λ with and without the condition "us" are illustrated. In the situation illustrated in (a), the value of Λ is fixed by H and $|\Psi\rangle$ and anthropic reasoning is not needed. In (b), anthropic probabilities are distributed so that anthropic reasoning is useless in fixing Λ. Anthropic reasoning is useful in the situation (c).

- The complexity of the description of a condition including "us" may make the calculation of the probabilities long or impossible as a practical matter.

In practice, therefore, anthropic probabilities (6) can only be estimated or guessed. Theoretical uncertainty in the results is thereby introduced.

The objectivity striven for in physics consists, at least in part, in using probabilities that are not too sensitive to "us". We would not have science if anthropic probabilities for observation depended significantly on which individual human being was part of the conditions. The existence of schizophrenic delusions shows that this is possible, so that the notion of "us" should be restricted to exclude such cases.

For these reasons it is prudent to condition probabilities, not on a detailed description of "us", but on the weakest condition consistent with "us" that plausibly provides useful results, like those illustrated in Fig. 1(c). A short list of conditions of roughly decreasing complexity might include:

- human beings;
- carbon-based life;
- information gathering and utilizing systems (IGUSes);
- at least one galaxy;
- a universe older than 5 Gyr;
- no condition at all.

For example, the probabilities used to bound the cosmological constant Λ,[1,14] make use of the fourth and fifth on this list under the assumption that including earlier ones will not much affect the anthropically-allowed range for Λ. To move down in the above list of conditions is to move in the direction of increasing theoretical certainty and decreasing computational complexity. With anthropic reasoning, less is more.

5. Ignorance is NOT Bliss

The quantum state of a single isolated subsystem generally cannot be determined from a measurement carried out on it. That is because the outcomes of measurements are distributed probabilistically and the outcome of a single trial does not determine the distribution. Neither can the state be determined from a series of measurements, because measurements disturb the state of the subsystem. The Hamiltonian cannot be inferred from a sequence of measurements on one subsystem for similar reasons. In the same way, we cannot generally determine either the Hamiltonian or the quantum state of the universe from our observations of it. Rather, these two parts of a final theory are theoretical proposals, inferred from partial data to be sure, but incorporating theoretical assumptions of simplicity, beauty, coherence, mathematical precision, etc. To test these proposals, we search among the conditional probabilities they imply for predictions of observations yet to be made with probabilities very near one. When such predictions occur we count it a success of the theory, when they do not we reject it and propose another.

Do we need a theory of the quantum state? To analyze this question, let us consider various degrees of theoretical uncertainty about it.

5.1. *Total Ignorance*

In the model cosmology in a box of Sec. 2, theoretical uncertainty about the quantum state can be represented by a density matrix ρ that specifies probabilities for its eigenstates to be $|\Psi\rangle$. Total ignorance of the quantum state is represented by a ρ proportional to the unit matrix. To illustrate this and the subsequent discussion, assume for the moment that the dimension of the Hilbert space is very large but finite. Then total ignorance of the quantum state is represented by

$$\rho_{\text{tot. ign.}} = \frac{I}{Tr(I)} \tag{7}$$

which assigns equal probability to any member of any complete set of orthogonal states.

The density matrix (7) predicts thermal equilibrium, infinite temperature, infinitely large field fluctuations, and maximum entropy.[9] In short, its predictions are inconsistent with observations. This is a more precise way of saying that every useful prediction depends in some way on a theory of the quantum state. Ignorance is not bliss.

5.2. *What We Know*

A more refined approach to avoiding theories of the quantum state is to assume that it is unknown except for reproducing our present observations of the universe. The relevant density matrix is

$$\rho_{\rm obs} = \frac{P_{\rm obs}}{Tr(P_{\rm obs})} \tag{8}$$

where $P_{\rm obs}$ is the projection on our current observations — "what we know". "Observations" in this context mean what we directly observe and record here on Earth, and not the inferences we draw from this data about the larger universe. That is because those inferences are based on assumptions about the very quantum state that (8) aims to ignore. For instance, we observed nebulae long before we understood what they were or where they are. The inference that the nebulae are distant clusters of stars and gas relies on assumptions about how the universe is structured on very large scales that are in effect weak assumptions on the quantum state.

Even if we made the overly generous assumption that we had somehow directly observed and recorded every detail of the volume 1 km above the surface of the Earth, say, at a 1 mm resolution, that is still a tiny fraction ($\sim 10^{-60}$) of the volume inside the present cosmological horizon. The projection operator $P_{\rm obs}$ therefore defines a very large subspace of Hilbert space. We can expect that the entropy of the density matrix (8) will therefore be near maximal, close to that of (7), and its predictions similarly inconsistent with further observations.

In the context of anthropic reasoning, these results show that conditioning probabilities on "us" alone is not enough to make useful predictions. Rather, a theory of H and $|\Psi\rangle$ are needed in addition, as described in the previous section.

6. A Final Theory

Let us hope that one day we will have a unified theory based on a *principle* that will specify *both* quantum dynamics (H) and a unique quantum state of the universe ($|\Psi\rangle$). That would truly be a final theory and a proper context for anthropic reasoning.

Acknowledgments

Appreciation is expressed to the Mitchell Institute for hospitality and to the National Science Foundation for partial support under grant NSF-PHY02-44764.

References

1. See, *e.g.* J. Barrow and F. Tipler, *The Anthropic Cosmological Principle* (Oxford University Press, Oxford, 1986).
2. J. B. Hartle, The quantum mechanics of closed systems, in *Directions in General Relativity, Vol. 1: A Symposium and Collection of Essays in Honor of Professor Charles W. Misner's 60th Birthday*, eds. B.-L. Hu, M. P. Ryan and C. V. Vishveshwara (Cambridge University Press, Cambridge, 1993) gr-qc/9210006.
3. J. B. Hartle, Spacetime quantum mechanics and the quantum mechanics of spacetime, in *Gravitation and Quantizations, Proc. 1992 Les Houches Summer School*, eds. B. Julia and J. Zinn-Justin, Les Houches Summer School Proc. Vol. LVII (North Holland, Amsterdam, 1995) gr-qc/9508023.
4. R. B. Griffiths, *Consistent Quantum Theory* (Cambridge University Press, Cambridge, 2002).
5. R. Omnès, *The Interpretation of Quantum Mechanics* (Princeton University Press, Princeton, 1994).
6. M. Gell-Mann, *The Quark and the Jaguar* (W. Freeman, San Francisco, 1994).
7. S. W. Hawking, The quantum state of the universe, *Nucl. Phys. B* **239**, 257 (1984).
8. J. B. Hartle, Scientific knowledge from the perspective of quantum cosmology, in *Boundaries and Barriers: On the Limits to Scientific Knowledge*, eds. J. L. Casti and A. Karlqvist (Addison-Wesley, Reading, MA, 1996) gr-qc/9601046.
9. J. B. Hartle, The state of the universe, in *The Future of Theoretical Physics and Cosmology: Stephen Hawking 60th Birthday Symposium*, eds. G. W. Gibbons, E. P. S. Shellard and S. J. Ranken (Cambridge University Press, UK, 2003) gr-qc/0209046.

10. J. B. Hartle, Quasiclassical domains in a quantum universe, in *Proc. Cornelius Lanczos International Centenary Conf.*, North Carolina State University, December 1992, eds. J. D. Brown, M. T. Chu, D. C. Ellison and R. J. Plemmons (SIAM, Philadelphia, 1994) gr-qc/9404017.
11. A. Strominger, Baby universes, in *Quantum Cosmology and Baby Universes: Proc. 1989 Jerusalem Winter School for Theoretical Physics*, eds. S. Coleman, J. B. Hartle, T. Piran and S. Weinberg (World Scientific, Singapore, 1991).
12. S. W. Hawking, The cosmological constant is probably zero, *Phys. Lett. B* **134**, 403 (1984).
13. S. W. Hawking and T. Hertog, to be published.
14. For recent reviews with references to earlier literature, see S. Weinberg, *The Cosmological Constant Problems*, astro-ph/0005265; A. Vilenkin, *Cosmological Constant Problems and Their Solutions*, hep-th/0106083.

CHAPTER 17

Are We Typical?*

Bayesian probability theory is used to analyze the oft-made assumption that humans are typical observers in the universe. Some theoretical calculations make the *selection fallacy* that we are randomly chosen from a class of objects by some physical process, despite the absence of any evidence for such a process or any observational evidence favoring our typicality. It is possible to favor theories in which we are typical by appropriately choosing their prior probabilities, but such assumptions should be made explicit to avoid confusion.

*With Mark Srednicki (Department of Physics, University of California Santa Barbara, CA 93106-9530, USA).
http://arxiv.org/abs/0704.2630v3

An increasingly common kind of reasoning in fundamental cosmology starts from an assumption that some property of human observers is typical in some class $\mathcal{C}$ of objects in the universe. This assumption is then used in conjunction with a physical theory of cosmology to draw conclusions about the properties of the universe that we should expect to observe. However, it is perfectly possible (and not necessarily unlikely) for us to live in a universe in which we are not typical. Indeed, there are many choices of the class $\mathcal{C}$ (such as all compact objects with mass between 10 and 100 kg) for which we have data that definitively establishes that humans are not typical. Any assumption of typicality or mediocrity should therefore be carefully scrutinized.

Such an examination seems especially timely because of recent discussions drawing conclusions from the possible existence (in a very large universe) of vacuum fluctuated brains (also known as "Boltzmann brains" or "freak observers"). Astounding conclusions have been drawn using arguments that involve these possibilities, notably those of Page,[1–4] that the universe must be limited in time and space in order to avoid having too many fluctuated brains that would make human observers untypical in any class that includes these fluctuated brains. Some other recent discussions of this issue include Refs. 5–9. Early discussions of the role of typicality in inflationary cosmology include Refs. 10 and 11. Previous criticism of typicality assumptions include Refs. 12 and 13.

Typicality assumptions involving human observers in a class containing other kinds of observers can be based on different types of information. These include (1) empirical extrapolations of data concerning extraterrestrial observers, (2) plausible guesses for the predictions of a more fundamental theory, and (3) mere personal preference for theories in which we are typical of something.

Since we currently have no data on extraterrestrial observers, there are no examples of type (1). An example of type (2) is the oft-made assumption in multiverse cosmology that the number of ordinary observers is proportional to the number of galaxies, or to the number of baryons, or to some other similar proxy. This paper does not criticize such assumptions when they are clearly stated. However, it will be argued that assumptions of type (3) are not well justified, and can lead to ridiculous conclusions that will be illustrated with simple examples.

To indicate where the paper is headed, some of its conclusions are summarized below.

- A theory is not incorrect merely because it predicts that we are atypical.
- Theories are tested using our data as a collectivity of human observers. What other observers might see, how many of them there are, and what properties they do or do not share with us are irrelevant for this process.
- No part of our data should be neglected in the process of discriminating between competing theories unless it can be demonstrated that the relevant probabilities are insensitive to it.
- Two theories that predict our data with equal probability are indistinguishable on the basis of that data. Cosmological models that predict that at least one instance of our data exists (with probability one) somewhere in spacetime are indistinguishable no matter how many other exact copies of these data exist.
- We have data that we exist in the universe, but we have no evidence that we have been selected by some random process. We should not calculate as though we were.
- In a fundamental theory of quantum cosmology, there is no need for any assumption of typicality to predict what we might see. However, typicality, and the measures to implement it, may be implied by the fundamental theory. In the absence of a manageable candidate for such a theory, it can be useful to try to guess such measures.

Bayesian probability theory is the basis for this analysis. In addition to supporting the conclusions above, it will be used to discuss some classes of mistakes that can be made in applying the Bayesian procedure. These include neglecting relevant parts of our data, and making the *selection fallacy* that we are randomly selected from some class $\mathcal{C}$.

1. Testing Theories Using Bayesian Probabilities

This section briefly reviews the elements of the Bayesian framework for testing theories. (For a compact introduction, see Ref. 14; for complete details, see Ref. 15; for an elegant presentation of the necessity of the Bayesian point of view, see Ref. 16.) Suppose we have a set of theories that we wish to discriminate between. For simplicity, assume the theories form a finite discrete set $\{T_1, \ldots, T_N\}$, with any continuous parameters binned. The specification of a theory is taken to include both the dynamics and the initial conditions, so that the theory contains enough information to make calculable predictions. For simplicity, we assume that the set of theories is complete, so that one of them must be true; if this is not the case, the probabilities to be calculated are relative rather than absolute.

Bayesian analysis has three inputs: First is a *prior probability* or *prior* $P(T_i)$ for each theory. These priors should be assigned before any data is obtained. Although the priors can in principle be chosen arbitrarily, standard practice is to try to make a choice that suitably reflects our ignorance; for example, by assigning equal probabilities to the discrete set of theories. In the terminology of Bayesian theory, the choice should be "non-informative". The exact choice of priors is ultimately unimportant in situations where there can be an arbitrarily large number of identical trials. However, in areas like cosmology, where data is hard to come by, priors can strongly influence the final conclusions.

Second, there is the specification of the data D with which we seek to discriminate between the theories. This is the data of the collectivity of human observers doing the Bayesian analysis.

The third input is the *likelihood* $P(D|T_i)$ that (upon suitable observation) we will obtain a particular data set D, given that the theory is T_i. An important point is that, for each theory, the likelihood should be *calculable*. If we can only estimate but not calculate a likelihood, then we should enlarge our set of theories by introducing new parameters that account for our ignorance of the likelihood, and allow these parameters to vary over the new, larger, set of theories.

Finally, there is the output — what we really want to know — the *posterior probability* $P(T_i|D)$, the probability of theory T_i, given the data D that we have obtained. The posterior probabilities for the various theories are related to the prior probabilities and the likelihoods by *Bayes' theorem*,

$$P(T_i|D) = \frac{P(D|T_i)P(T_i)}{\sum_i P(D|T_i)P(T_i)}. \tag{1}$$

This is the key formula of Bayesian analysis.

2. Priors Favoring Typicality

Our typicality can be enforced simply by choosing priors that favor theories that imply that we are typical. A specific example is useful to illustrate the dangers inherent in this kind of reasoning (type 3 above).

Consider two theories of the development of planet-based intelligent life, based on the appropriate physics, chemistry, biology, and ecology. Theory A predicts that there are likely to be intelligent beings living in the atmosphere of Jupiter; theory B predicts that there are no such beings. Because Jupiter is much larger than the Earth, theory A predicts that there are today many more jovians than humans.

Would we reject theory A solely because humans would not then be typical of intelligent beings in our solar system? Would we use this theory to predict that there are no jovians, because that is the only way we could be typical? Such a conclusion seems absurd.

We could nevertheless enforce an assumption of typicality by assigning a low prior probability to theory A, and a high prior probability to theory B. Alternatively, we could choose priors that favor theory B, so that the Earth is more typical of the planets in the solar system.[a] However, any *a priori* favoring of either theory is contrary

---

[a]This was the point of view taken by Huygens[17] in 1698: "A Man that is of Copernicus's Opinion, that this Earth of ours is a Planet, carry'd round and enlighten'd by the Sun, like the rest of them, cannot but sometimes have a fancy, that it is not improbable that the rest of the planets must have their Dress and Furniture, nay and their inhabitants too as well as this Earth of ours." Further he argues from the similarities among species from distant parts of the Earth (such as Europe and America) that we should be

to the standard scientific practice of assigning noninformative priors that do not preselect conclusions before relevant data is obtained.

3. Data

A discussion of what is meant by "our data" necessarily begins with a discussion of what is meant by "we", "us", and "our". In this paper, "we" refers to the information gathering and utilizing system (IGUS) that is engaged in the process of gathering data, constructing alternative theories, and using Bayesian analysis to discriminate between these theories. In this epoch, that is naturally the human scientific IGUS on Earth (HSI). If we ever have scientific exchanges with IGUSes elsewhere, the definition would have to be enlarged.

"Our data" D includes every scrap of information that the HSI possesses about the physical universe: every record of every experiment, every astronomical observation of distant galaxies, every available description of every leaf, etc., and necessarily every piece of information about the HSI itself, its members, and its history. The choices of coarse-graining used to define this information, the level of accessibility demanded, etc., are inevitably subjective but assumed here to be fixed. The data must all be expressed in physical terms in the language of the theories being tested, so that the likelihoods $P(D|T_i)$ can be computed.

All the information possessed by the HSI should be included in D unless it can be demonstrated that the posterior probabilities $P(T_i|D)$ are insensitive to it. Of course, we can practically test theories only if we can assume that this is the case for most of our data.

---

typical of the inhabitants of other planets: "'Tis more probable that all the difference there is between us and them, springs from the greater or less distance and influence from that Fountain of Heat and Life the Sun; which will cause a difference not so much in their Form and Shape, as in their Matter and Contexture." He then concludes that there must be substances on the other planets serving the role of water on this one: "I can't say that they are exactly of the same nature with our Water, but that they should be liquid, their use requires, as their beauty does that they should be clear." In this case, Huygens' assumption of the typicality of the Earth obviously did not lead him to correct conclusions.

The important point is that any such assumption can be checked by experiment or calculation. To arbitrarily neglect some pieces of data in favor of others is unscientific and risks contradiction when the neglected data is later considered.

What are the implications of this for typicality? It is *our data* that is used in a Bayesian analysis to discriminate between theories. What other hypothetical observers with data different from ours might see, how many of them there are, and what properties they might or might not share with us (defining some notion of typicality) are irrelevant for this process. In particular, the number of vacuum fluctuated brains with different data (for example, with disordered observations as discussed by Page[1,2]) is irrelevant. As far as Bayes' theorem is concerned, the only other observers that matter are ones that have exactly the same data D.

Observers with identical data may occur in multiple places in a large universe. For example, in contemporary inflationary models, we may know that we are 14 Gyr from the nucleation of our pocket universe. But we know neither the location of that pocket universe in spacetime or how many others like it with identical data may have been nucleated elsewhere. Vacuum or thermal fluctuations that produce identical data must also be considered. These unknown locations and origins must be summed over in calculating the likelihoods $P(D|T)$. *All we know is that there exists at least one such region containing our data.* This idea is illustrated with simple models in the following sections.

Beyond testing theories, questions sometimes arise like "What would be the probability to observe a cosmic microwave background (CMB) at a temperature of 3 K if we didn't already know the value of this temperature?" In such cases, the data D (which includes the observation of a 3 K CMB) must be divided into a part that could have varied and a notion of "us" that could not. There is additional ambiguity, and additional subjective choice, in making this division. At one extreme, the 3 K could be included in the meaning of "us" (no division), in which case "we" could never have observed anything else, by definition. At the other extreme, "us" could be defined to be an object less than 200 kg, in which case we could be talking about the CMB temperature that a rock might have received. Most physicists would make less extreme choices, but it is important to recognize that the answers to such questions depend on the choice.

4. Likelihoods and the Selection Fallacy

Return for the moment to the human/jovian model and assume that each theory is able to predict a probability for the number of humans H and jovians J alive today. The question then becomes, what is the data? We know that the actual number of humans today is $H^* = 6.57 \times 10^9$. (For simplicity, assume that this number is known exactly.) We know nothing at all about the actual number of jovians today; this data has not yet been obtained. Thus, the likelihood that a theory T_i predicts the data D that we actually have in this case, is given by summing over the mutually exclusive probabilities for different numbers of jovians, with the number of humans fixed to its known value,

$$P(D|T_i) = \sum_{J=0}^{\infty} P(H^*, J|T_i). \tag{2}$$

The likelihood for the data is independent of the number of jovians predicted by the theory, simply because we have not observed the number of jovians.

It is important to note that there is an alternative, incorrect computation of the likelihood that favors typicality. The invalid reasoning goes as follows. I observe that I am human. I might, however, have been jovian. If there are H humans and J jovians, the probability that I observe myself to be human is $H/(H+J)$. Therefore, I should multiply the probability $P(H, J|T_i)$ that there are H humans and J jovians by a factor of $H/(H+J)$ to account for the probability that I have observed myself to be human and not jovian. Then I compute the likelihood as

$$P(D|T_i) = \sum_{J=0}^{\infty} \frac{H^*}{H^*+J} P(H^*, J|T_i), \tag{3}$$

instead of (2). The factor of $H^*/(H^*+J)$ in (3) results in likelihoods that (when used in Bayes' theorem) disfavor theories that predict

$H \ll J$, and hence lead to the conclusion that humans should be typical. This is the kind of argument made in Refs. 5 and 6, and would seem to be a natural interpretation of Refs. 1–4.[b]

The factor of $H^*/(H^*+J)$ would be correct if we (or some outside agency) had performed an experiment that had (somehow) selected a single intelligent being in the solar system, with equal *a priori* probability for each, and that we or the outside agency knew the result of this selection. However, noticing that we are human is not equivalent to performing such an experiment. For example, it would certainly be incorrect to conclude that, since each of the two authors of this paper is human, the correct factor to multiply by is $(H^*/(H^* + J))^2$. This is because the probability that one of us is human is obviously not independent of the status of the other. Thus, whatever selection is performed when one of us notices that he is human, it is surely not equivalent to a random selection of a single intelligent being anywhere in the solar system.

In fact, there has been no selection at all. Intelligent beings on Earth notice that they exist, call themselves "humans", and count up how many there are. Whatever may or may not be happening on Jupiter is entirely irrelevant. This leads immediately to Eq. (2) for the likelihood, which does not favor or disfavor the typicality of humans. If we now set all priors equal, $P(T_i) = 1/N$ with N the number of theories, we immediately find $P(T_i|D) \propto P(D|T_i)$. That is, the probability that a theory is correct is directly proportional to the probability that it predicts the correct number of humans, independent of the number of jovians.

To compute likelihoods as though we had been randomly selected by some physical process, when there is no evidence for such a process, commits what might be called the *selection fallacy.* We are not a disembodied entity that was randomly selected to have a particular physical description; instead, we are the meaning of a transcription of "we" into the language of physical theory.

---

[b]However, in private discussions, Page says that the typicality factors arise naturally in his theory of quantum mechanics,[18] in which the basic quantities are operators representing the conscious perceptions of observers. These ideas are not discussed here.

As shown above, the correctly computed likelihood does not favor the typicality of humans. However, we can, if we wish to, implement an assumption of typicality through the choice of priors. We do this by identifying the set of theories that predicts that $P(H, J|T_i)$ is small whenever H is much less than J. Then we assign to these theories, small values of $P(T_i)$. For example, we could choose $P(T_i) \propto \bar{H}_i/(\bar{H}_i + \bar{J}_i)$ (or some power of that), where $\bar{H}_i$ and $\bar{J}_i$ are the mean number of humans and jovians that are predicted by theory T_i.

As was mentioned earlier, there is nothing wrong in principle with any choice of priors. However, choosing priors that favor one or another prediction (that we happen to like) goes against standard scientific practice, and simply represents an unsupported (by data) personal preference for a certain result.

5. A Simple Cosmological Model

A simple (highly idealized) cosmological model will further illustrate the key issues.

Consider a model universe which has N cycles in time, $k = 1, \ldots, N$. In each cycle, the universe may have one of two global properties: red (R) or blue (B), which could be thought of as (for example) two different possible values of the CMB temperature. To further simplify the discussion, the only relevant observables are assumed to be (1) the value of the property and (2) the existence of an observing system that is able to determine this value. In each cycle, the probability for such an observing system to exist is taken to be p_E; this probability is assumed to be independent of whether the universe is red or blue in that cycle. Furthermore, observations are assumed to be perfectly accurate, so that if red is observed in any cycle, then the universe is red in that cycle, and conversely.

Two competing theories of this model universe are proposed. One, *all red* or AR, in which all the cycles are red, and another, *some red* or SR, in which some number of particular cycles are red and the rest are blue. We (an idealized observing system) seek to discriminate between these two theories on the basis of our data.

Suppose that we (a particular observing system) observe red. Our data D is then (E, R), which in the context of the model could be more fully described as "there is at least one cycle in which an observing system exists and the universe is red". Using this data, we carry out a Bayesian analysis to discriminate between the two theories, assuming equal priors for simplicity. The probability that there is at least one cycle with (E, R) is the same as one minus the probability of the negation of this, which is the probability that no observing system exists in a cycle in which the universe is red. Since the probability for an observing system *not* to exist in any one cycle is $1 - p_E$, the likelihoods are

$$P(E, R|T) = 1 - (1 - p_E)^{N_R(T)}, \tag{4}$$

where $N_R(T)$ is the number of red cycles in theory T, equaling N when T is AR. Assuming equal priors, (1) yields the posterior probabilities

$$P(T|E, R) = \frac{P(E, R|T)}{P(E, R|AR) + P(E, R|SR)}, \tag{5}$$

where T is either AR or SR.

A number of limiting cases of the likelihoods (4) and their consequent posterior probabilities (5) are of interest.

If p_E is close to one, then $P(E, R|T) \approx 1$ for *both* theories. That is not very surprising, since the probability is high for there to be an observing system in every cycle. The probability that there is at least one red cycle with an observing system is therefore also high in both theories. Thus, our data do not discriminate between the two theories, and indeed $p(T|E, R) \approx 1/2$ for both.

Another case where our data does not discriminate between the two theories occurs when $p_E < 1$, and N_R is large in *both* theories to make $(1 - p_E)^{N_R(T)} \ll 1$, so that $P(E, R|T) \approx 1$ for both theories. Even though there may be many more red cycles in the AR theory than the SR theory, the probability that there is *at least one* red cycle with an observing subsystem approaches one for both theories when N_R becomes large in both. In this case, our data is not enough to discriminate between the two theories. This is important because inflationary theories can be expected to imply large numbers of any conceivable situation.

If $p_E \ll 1/N$, then $P(E, R|T) \approx N_R(T)p_E$, and the resulting posterior probabilities are

$$P(AR|E, R) = \frac{1}{1 + f_R}, \quad P(SR|E, R) = \frac{f_R}{1 + f_R}, \tag{6}$$

where $f_R \equiv N_R(SR)/N$ is the fraction of red cycles in the SR theory. In this case, if f_R is small, then the AR theory is strongly favored.

In the small p_E limit, the result (6) is the same as it would be if it was assumed that we were the unique observing system existing in the universe, but that we did not know the cycle k in which we were located. More specifically, were we unique, the probability that we were in cycle k observing R would be $1/N$, if cycle k is red in theory T, and zero if it is blue. Summing these probabilities over k yields (6).

Confusingly, this result is also the same under the assumption that our location is randomly selected from the N possible cycles. This notion presupposes that we exist separately from our physical description. But we are not separate from our physical description in our data; we *are* the physical system described, and no random selection has been made. In contemporary inflationary models, we are very unlikely to be unique as physical systems within the infinite multiverse. In this case, reasoning as if we had been randomly selected will not give correct results, as the above example shows.

The implications of this simple model for typicality can now be discussed. To emphasize that other observers with data different from ours are irrelevant for predicting what we observe, the likelihoods (4) can be calculated in a different way. First consider the probability $P(n_R, n_B|T)$ that there are n_R observing systems in red cycles and n_B observing systems in blue cycles,

$$P(n_R, n_B|T) = \binom{N_R(T)}{n_R}\binom{N_B(T)}{n_B} \times p_E^{n_R+n_B}(1 - p_E)^{N-n_R-n_B}, \tag{7}$$

where $N_B(T)$ and $N_R(T)$ are the number of blue and red cycles in each theory. Our data is "there is at least one red cycle containing an observing system". This is related to (7) by

$$P(E, R|T) = \sum_{n_R=1}^{N_R(T)} \sum_{n_B=0}^{N_B(T)} P(n_R, n_B|T), \tag{8}$$

which, as is easily shown, yields (4). The elementary but important point is that the number of observing systems measuring blue must be summed over. The value of that number therefore does not affect the probability for our data.

If the number of red cycles is infinite in both models, then the two theories are not distinguished by our data even if the number of the fraction of red cycles is small in SR. That is because our data is that there is *at least* one red cycle. As the number of red cycles approaches infinity, that becomes a certainty in both theories. Even though the typical observing system in the SR theory is observing blue, our data provides no evidence that we are typical.

In such cases it is tempting to reason as follows: In the SR theory with mostly blue cycles, we are more likely to observe blue than red. Since we observe red, we should reject this theory. This is the kind of argument made in Ref. 5 against eternal deSitter space. This, however, is an instance of the selection fallacy, and in more familiar situations leads to absurd conclusions such as those discussed earlier.

6. Deriving Measures

After a sufficient time, an isolated quantum system with (sufficiently complex interactions) will come to equilibrium, and the predictions for the probabilities of certain variables like total energy, momentum and number will be given by an equilibrium density matrix. In principle, there is no need for an equilibrium density matrix. Were the initial state known and powers of computation unlimited, the state could be evolved forward by the appropriate time and the probabilities calculated directly. However, the power, utility, and perhaps even the necessity of a notion of equilibrium and the density matrix that defines it are well known. The important point is that such a density matrix cannot be posited arbitrarily; it follows from the dynamics in principle, and its form can be checked (see, e.g. Ref. 19).

Similarly, in a complete quantum theory of the universe, there is in principle no need for any further assumption about the typicality of human observers, or a measure implementing that notion of typicality. The probabilities for our observations would follow from a theory of the dynamics, the initial quantum state, and an accurate quantum description of human beings, their knowledge, and their data.[20]

However, we presently lack even one example of such a complete theory, and, in practice, calculations of the history and observations of human observers are well beyond current and foreseeable abilities to compute. It then becomes useful to make plausible hypotheses of what such a complete theory might predict, and (as mentioned above) to expand the family of theories being tested to include these hypotheses. For example, inflationary cosmology frequently assumes a *classical* spacetime that models distributions of bubbles nucleated in quantum events. Our location in this spacetime is unknown. To calculate the probabilities of what we might see, it becomes convenient to posit a distribution of positions in space and time together with a measure to define it. This distribution can in principle be checked in a complete theory of quantum cosmology. It might even be possible to infer a distribution for where other observers are located in the universe and a measure to define that.

In a Bayesian analysis of such situations, priors must be assigned to candidates for a complete theory, and also to different hypotheses of what it will predict. In the example of inflationary cosmology, priors would be assigned to different measures and typicality assumptions. In Bayesian terms, debates about which measure is correct are discussions of these priors. The important thing is that these priors can be improved through the usual process of Bayesian updating after acquiring more data or by firming up predictions by better calculation. That is in contrast to priors which merely express personal preference for theories in which we are typical.

7. Conclusion

The course of physics, both theoretically and experimentally, is guided by prejudice as to the nature of the theory sought for.

We favor theories that are simple, beautiful, precisely formulable mathematically, economical in their assumptions, comprehensive, unifying, explanatory, accessible to existing intuition, etc. Most importantly, we favor theories that are successful in predicting new data beyond what we have at the moment. The bases for such prejudices do not lie in logic but, rather, previous experience with constructing successful theories.

Especially in areas far from immediate experiment, it is important to distinguish facts, logical deduction and prejudices. Bayesian analysis provides a framework for doing this. Data are the domain of facts, likelihoods are the domain of logical deduction, and the priors are the domain of theoretical prejudice.

This paper has analyzed typicality assumptions using the Bayesian framework. The main conclusions have already been given in the Introduction and we will not repeat them here. At present, there are no observational data supporting an assumption that we are typical in some class of observers, and our understanding of biological evolution is insufficient to supply a theoretical justification through the likelihoods. Many calculations that produce likelihoods that favor our typicality do so via the selection fallacy; this can lead to absurd conclusions, as the human/jovian model shows.

On the other hand, a preference for typicality can legitimately be made through a suitable choice of priors. However, such choices should be made explicit, so that others can properly evaluate the conclusions that are ultimately reached.

Acknowledgments

We are grateful to Don Page for voluminous and clarifying communications, and to Tom Banks, Andre Linde, Leonard Susskind, Alex Vilenkin, and colleagues in Santa Barbara, for useful remarks and helpful criticism. This work was supported by the National Science Foundation under grants PHY04-56556 and PHY05-55669.

References

1. D. N. Page, *Is our universe likely to decay within 20 billion years?* arXiv:hep-th/0610079.

2. D. N. Page, *Susskind's challenge to the Hartle–Hawking no-boundary proposal and possible resolutions*, arXiv:hep-th/0610199.
3. D. N. Page, *Return of the Boltzmann brains*, arXiv:hep-th/0611158.
4. D. N. Page, *Is our universe decaying at an astronomical rate?* arXiv:hep-th/0612137.
5. L. Dyson, M. Kleban and L. Susskind, Disturbing implications of a cosmological constant, *JHEP* **0210**, 011 (2002), arXiv:hep-th/0208013.
6. R. Bousso and B. Freivogel, *A paradox in the global description of the multiverse*, arXiv:hep-th/0610132.
7. A. Linde, *Sinks in the landscape and the invasion of Boltzmann brains*, arXiv:hep-th/0611043.
8. A. Vilenkin, *Freak observers and the measure of the multiverse*, arXiv:hep-th/0611271.
9. T. Banks, *Entropy and initial conditions in cosmology*, arXiv:hep-th/0701146.
10. J. Garcia-Bellido, A. D. Linde and D. A. Linde, Fluctuations of the gravitational constant in the inflationary Brans–Dicke cosmology, *Phys. Rev. D* **50**, 730 (1994), arXiv:astro-ph/9312039.
11. A. Vilenkin, Predictions from quantum cosmology, *Phys. Rev. Lett.* **74**, 846 (1995), arXiv:gr-qc/9507018.
12. L. Smolin, Scientific alternatives to the anthropic principle, in *Universe or Multiverse?* ed. B. Carr (Cambridge University Press, Cambridge, 2007), arXiv:hep-th/0407213.
13. R. M. Neal, *Puzzles of anthropic reasoning resolved by using full non-indexical conditioning*, arXiv:math.ST/0608592.
14. M. Srednicki, Subjective and objective probabilities in quantum mechanics, *Phys. Rev. A* **71**, 052107 (2005), arXiv:quant-ph/0501009.
15. E. T. Jaynes, *Probability Theory: The Logic of Science* (Cambridge University Press, Cambridge, 2003).
16. D. M. Appleby, Probabilities are single-case, or nothing, *Opt. Spectrosc.* **99**, 447 (2005), arXiv: quant-ph/0408058.

17. C. Huygens, *Cosmotheoros* (Adriaan Moetjens, The Hague, 1698); Eng. trans. from the Latin, *The Celestial Worlds Discovered: or, Conjectures on the Inhabitants, Plants, and Productions of the World in the Planets* (Timothy Childe at the White Hart at the west end of St. Paul's Churchyard, London, 1698) [www.phys.uu.nl/~huygens].
18. D. N. Page, Sensible quantum mechanics: Are probabilities only in the mind? *Int. J. Mod. Phys. D* **5**, 583 (1996), arXiv:gr-qc/9507024.
19. M. Srednicki, Chaos and quantum thermalization, *Phys. Rev. E* **50**, 888 (1994), arXiv:cond-mat/9406056; The approach to thermal equilibrium in quantized chaotic systems, *J. Phys. A* **32**, 1163 (1999), arXiv:cond-mat/9809360.
20. J. B. Hartle, Anthropic reasoning and quantum cosmology, in *The New Cosmology*, eds. R. Allen, D. Nanopoulos and C. Pope, AIP Conf. Proc. Vol. 743 (AIP, Melville, NY, 2004), arXiv:gr-qc/0406104.

CHAPTER 18

Science in a Very Large Universe*

As observers of the universe, we are quantum physical systems within it. If the universe is very large in space and/or time, the probability becomes significant that the data on which we base predictions, is replicated at other locations in spacetime. The physical conditions at these locations that are not specified by the data may differ. Predictions of our future observations therefore require an assumed probability distribution (the *xerographic distribution*) for our location among the possible ones. It is the combination of basic theory plus the xerographic distribution that can be predictive and testable by further observations.

*With Mark Srednicki (Department of Physics, University of California Santa Barbara, CA 93106-9530, USA).
http://arxiv.org/abs/0906.0042v3

1. Introduction

Theories of our universe are tested using the data that we acquire. When calculating predictions, we customarily make an implicit assumption that our data D_0 occur at a unique location in spacetime. However, there is a quantum probability for these data to exist in any spacetime volume. This probability is extremely small in the observable part of the universe. However, in the large (or infinite) universes considered in contemporary cosmology, the following predictions often hold:

1. The probability is near unity that our data D_0 exist somewhere.
2. The probability is near unity that our data D_0 is exactly replicated elsewhere many times. An assumption that we are unique is then false.

This paper is concerned with the implications of these two statements for science in a very large universe. Some implications of the first were discussed in Ref. 1. We shall return to these below, but we first focus on the implications of the second.

2. Third-Person and First-Person Prediction

It is useful to distinguish between first-person and third-person predictions in cosmology. Third-person predictions are made through the probabilities for alternative features of the universe that it may exhibit. Examples are the probabilities that the universe is homogeneous and isotropic, that it had a certain number of inflationary e-folds in the past, that it will end in a big crunch, or that it exhibits a certain number of instances of our data D_0. Familiar quantum theories of the universe make such third-person predictions by specifying a quantum state and a prescription for dynamics (see, e.g. Ref. 2); such theories are denoted by T. But to use and test theories we need predictions of what *we* will measure or predict. These are first-person predictions. Defining these is not trivial if there is more than one instance of our data D_0. We now discuss how to do it.

As observers of the universe, we are a subsystem within it that we call the human scientific IGUS[a] (HSI). Terms like "we", "us", "our" refer to this specific subsystem. The HSI can be described at various levels of coarse-graining. Here it is assumed that a description is fixed at a classical level.[b] This description can be divided into two parts: First, the data D_0 that the HSI has: every scrap of information that the HSI possesses about the physical universe, including every record of every experiment, every astronomical observation of distant galaxies, every available description of every leaf, etc., and necessarily every piece of information about the HSI itself, its members, and its history. Second, there are the quantities not included in the data but which are necessary for a complete physical description of the HSI. Location in the universe is one example; physical circumstances that have not yet been measured is another. (The HSI is assumed to exist at a unique location in the universe, but its data D_0 may be duplicated at many other locations in a very large universe.)

First-person predictions are through the probabilities of what our specific instance of an IGUS with data D_0 will observe or measure. But theories T of a quantum state and dynamics do not make such predictions directly if the chance is significant that our data D_0 are replicated at different places in the universe. Third-person probabilities computed from T make no reference to "us" and contain no information about which of several instances of D_0 is "us". A further assumption is therefore needed to connect the third-person probabilities of theory with the first-person probabilities for our observations. We call this assumption the *xerographic distribution* and describe it in the next section.

---

[a]IGUS is an acronym for Information Gathering and Utilizing System.

[b]More precisely, we assume that there is a description in terms of the quasiclassical variables that describes the quasiclassical realm of everyday experience in terms of a decoherent set of coarse-grained alternative histories defined in terms of these variables; see e.g. Ref. 3 for more details.

3. The Xerographic Distribution

Consider a universe that contains N copies of D_0 at different locations in spacetime x_A, $A = 1, \ldots, N$. We are one of these copies, but we have no information as to which, since our data is replicated at each of these locations. Therefore, to predict what we will observe in our future, we must choose a distribution that gives the probability ξ_A that we (the HSI) are the copy located at x_A. The probabilities ξ_A constitute the *xerographic distribution.* Location is used here as an illustration; the xerographic distribution could also refer to any other aspect of our physical situation that is not specified by the data D_0.

As mentioned above, the xerographic distribution is *not* fixed by usual candidates for the theory T, such as an initial quantum state and a prescription for its evolution. The xerographic distribution is a *further assumption* that *must* be made (explicitly or implicitly) in order to make first-person predictions. What is tested by observation is not just the theory T, but rather the combination (T, ξ); such a combination of a fundamental theory (including the initial state) T and the xerographic distribution ξ will be called a *theoretical framework* or *framework* for short.[c]

A natural assumption is that we are typical of the instances of D_0; this implies a uniform xerographic distribution $\xi_A = 1/N$. But it is also possible to assume that we are *a*typical. We argued in Ref. 1 that typicality was no more motivated by observation than atypicality.

---

[c]A good case can be made for calling the combination (T, ξ) the "theory", as it is the collection of assumptions from which testable predictions are made. But to do so would risk confusion with the usual notion of a theory as consisting of fundamental laws for dynamics and an initial state. Furthermore, the two parts of the combination (T, ξ) are of different characters. The theory T supplies third-person probabilities for features of the whole universe. The xerographic distribution refers to assumptions about a particular subsystem: the HSI. For these reasons, in this paper, "theory" is used in the usual way, and the combination (T, ξ) is called a "theoretical framework".

More generally, the xerographic distribution can be used to express typicality assumptions that involve data sets other than D_0. For instance, as we have defined it, D_0 includes the results of our observations as well as the conditions for them. But for some purposes it may be useful to assume that our results are typical of all instances of similar observational situations. Or it may be useful to assume that observations are typical of those made by any IGUS in the galaxy, or by any IGUS in the universe. Each of these typicality notions corresponds to a xerographic distribution that is uniform on the class involved. In the following we will show that the different assumptions about ξ can be testable, and use this to address the issues raised by the possibility of "Boltzmann brains".

It is important to note that the use of a xerographic distribution does not constitute a modification of the laws of the quantum mechanics of closed systems. The usual laws, the usual rules for implementing them (including Born's rule), and the usual interpretation apply to third-person probabilities. First-person probabilities are the new feature. These are made necessary because we, like other IGUSes, are quantum subsystems of the universe with a non-zero probability of being replicated exactly elsewhere. Quantum theory must be augmented by a prescription to calculate the first-person probabilities. There is nothing mysterious or even especially quantum about this prescription because it operates at the classical level.[d] Conversely, we cannot expect to derive the xerographic distribution from the rules of quantum theory for third-person probabilities.

4. Comparing Theories of Large Universes

Many different criteria can be used by physicists to discriminate between competing theoretical frameworks on the basis of the available data D_0. Frameworks are favored that are testable, simple, beautiful, precisely formulable mathematically, economical in their assumptions, comprehensive, unifying, explanatory, accessible to existing intuition, etc. Most importantly, scientists favor frameworks that are successful in predicting new data beyond what they have

[d]Indeed, it would be equally necessary if the theory T were classical.[4]

accumulated at the moment. That is, they favor frameworks that are *predictive.* We will discuss this criterion more fully in Sec. 6. In this section we discuss the simple idea that theoretical frameworks can be distinguished by the probabilities (likelihoods) that they give for the data D_0.

The process of distinguishing between frameworks by likelihoods is formalized in the widely familiar Bayesian schema for testing theories. While seldom applied in practice, this schema is useful to state the assumptions made in the process clearly and quantitively. The process involves computing posterior probabilities for frameworks from the likelihoods for the data D_0 and a set of prior probabilities for the competing frameworks.[e]

For a very large universe, there are different possibilities for a Bayesian comparison of theoretical frameworks allowed by the distinction between first-person and third-person probabilities. We can seek to distinguish between theories T on the basis of their third-person likelihoods for the data D_0, independent of any typicality assumption represented by a xerographic distribution. Or, we can seek to distinguish between different frameworks (T, ξ) on the basis of first-person probabilities that follow from an assumption of a xerographic distribution. The question of which to use is moot in a small universe where the data D_0 are unique to a good approximation, and first-person and third-person probabilities agree. However, the question becomes important in a large universe where the data D_0 may be replicated many times.

First, consider applying the Bayes procedure to distinguish between theories just on the basis of their third-person predictions for D_0. As we have emphasized in Ref. 1, all we know for certain from our data D_0 is that the universe exhibits at least one instance of it. We do not know how many times it is replicated or how frequently it occurs. A theory T is consistent with our data if the probability for the existence of at least one instance of our data is nonzero. But, as has been stressed by many, in a large universe, the likelihood that at least one instance of our data exists somewhere approaches unity

[e]A brief review consistent with present notation is given in Ref. 1.

for any theory that is consistent with our data. The third-person Bayes procedure is therefore not effective for discriminating between theories in a very large universe.

We therefore consider applying the Bayes schema to frameworks (T, ξ). This involves the following elements: First, prior probabilities $P(T, \xi)$ must be chosen for the different frameworks. Next, the first-person likelihoods $P^{(1p)}(D_0|T, \xi)$ must be computed. Finally, the first-person posterior probabilities are given by

$$P^{(1p)}(T, \xi|D_0) = \frac{P^{(1p)}(D_0|T, \xi)P(T, \xi)}{\sum_{(T,\xi)} P^{(1p)}(D_0|T, \xi)P(T, \xi)} \qquad (1)$$

The larger these are, the more favored are the corresponding frameworks.

As already noted, the theory T and the xerographic distribution ξ correspond to very different kinds of assumptions. The theory consists of a fundamental quantum state and a prescription for the dynamics of the universe. These are the quantities that summarize the universal regularities exhibited by all physical systems within the universe. By contrast, the xerographic distribution ξ concerns a particular subsystem of the universe — the HSI — and its relation to other subsystems with the same data D_0. Priors representing independent assumptions about T and ξ are therefore natural; this implies a factorization

$$P(T, \xi) = P_{\mathrm{xd}}(\xi)P_{\mathrm{th}}(T). \qquad (2)$$

With these kind of priors we can compete different theories T with the same typicality assumption ξ and also compete different typicality assumptions ξ with the same theory T.

There are those who are secure in the faith that the HSI must be typical of all other IGUSes in the universe, despite the absence of any experimental evidence either for or against the supposition. They will choose a prior $P_{\mathrm{xd}}(\xi)$ that is unity for the ξ that is uniform over the class of IGUSes, and zero elsewhere. Others, like the authors, who see ξ as an assumption much like any other, will allow different typicality assumptions to be competed against one another in the search for a successful framework for the universe. The essence of science is to concede that there is at least some chance that any assumption is incorrect, and then check for that with experiment and observation. The next section illustrates how this works in a simple model.

5. A Simple Cosmological Model

The ideas in the preceding section can be illustrated by the simple red-blue model that we employed for illustrative purposes in Ref. 1. Consider a model universe which consists of N cycles in time, $k = 1, \ldots, N$. In each cycle the universe may have one of two global properties: red (R) or blue (B), which could, for example, be thought of as two different possible values of the CMB temperature. In each cycle, there is a probability p_E for a physical system to exist (E) that is able to observe this global property. The model assumes that the observations are perfectly accurate, so that if red is observed in any cycle, then the universe is red in that cycle. It is further assumed that whether the universe is red or blue does not affect whether an observing system exists or not.

Two competing theories of this model universe are proposed. One, *all red* (AR), in which all the cycles are red, and another, *some red* (SR), in which some number of particular cycles are red and the rest are blue. Suppose that we (the HSI) observe red. Our data D_0 is then (E, R), and we seek to discriminate between the two theories on the basis of the likelihoods for this data. But, as described above, there are several choices for these likelihoods.

The use of third-person likelihoods based on the theory T alone was already discussed above and in Ref. 1. All we know about this universe in third-person terms is that it exhibits at least one instance of a cycle with (E, R) — the one we are in. The third-person probability that there is at least one cycle with (E, R) is the same as one minus the probability of the negation of this, which is the probability that no observing system exists in a cycle in which the universe is red. Since the probability for an observing system *not* to exist in any one cycle is $1 - p_E$, the likelihoods are

$$P(E, R|T) = 1 - (1 - p_E)^{N_R(T)} \tag{3}$$

where $N_R(T)$ is the number of red cycles in theory T, equal to N when T is AR. Our data do not discriminate between the two theories when N_R is large enough in *both* theories to make $(1 - p_E)^{N_R(T)} \ll 1$. Then $P(E, R|T) \approx 1$ for both theories. Even though there may be more many more red cycles in the AR theory than the SR theory, the probability that there is *at least one* red cycle with an observing subsystem approaches one for both theories when N_R becomes large in both. The likelihoods are the same. Our little bit of data is not enough to discriminate between the two theories on the basis of third-person probabilities.

The situation is different when we use first-person likelihoods to discriminate between different frameworks (T, ξ). To illustrate this, consider a variety of typicality assumptions expressed as different xerographic distributions.

The simplest assumption we can make is that we are typical in the class of other instances of our data D_0. In the context of the present model, that is the assumption that we are equally likely to be any of the instances of (R, E) — observers that exist in a red cycle. If there are n_R such systems, then the probability that we are the Ath instance is $1/n_R$ and the corresponding xerographic distribution is

$$\xi^{\text{typD}} = \frac{1}{n_R}, \tag{4}$$

where the superscript "typD" means typical in the class with data D_0. The first-person likelihoods for our data are then denoted by $P^{(1p)}(E, R|T, \xi^{\text{typD}})$.

In Appendix B, we work through the transition from third- to first-person probabilities using an explicit example of a typicality assumption and its associated xerographic distribution. That is instructive, but this model is so simple and symmetric that the results for the likelihoods follow from a few simple arguments.

The xerographic distribution ξ^{typD} is non-zero only on instances of D_0 in red cycles. The probability that we see red, $P^{(1p)}(R|E, T, \xi^{\text{typD}})$, is thus unity trivially, and the probability that we see blue is zero. But then, from the definition of conditional probabilities,

$$1 = P^{(1p)}(R|E,T,\xi^{\text{typD}}) \equiv \frac{P^{(1p)}(E,R|T,\xi^{\text{typD}})}{P^{(1p)}(E|T,\xi^{\text{typD}})}. \tag{5}$$

The first-person probability that we exist given that we are in a red cycle is the same as the third-person probability that at least one observing system (us) exists in a red cycle. That is, from (3)

$$P^{(1p)}(E|T,\xi^{\text{typD}}) = 1 - (1 - p_E)^{N_R(T)}. \tag{6}$$

Thus the first-person likelihoods are given by

$$P^{(1p)}(E,R|T,\xi^{\text{typD}}) = 1 - (1 - p_E)^{N_R(T)}. \tag{7}$$

The likelihoods for R and B correctly sum to unity. These first-person likelihoods are unchanged from the third-person ones (3), and no more able to discriminate between theories than they were.

We now turn our attention to other possible typicality assumptions represented by different xerographic distributions. First, let us consider the assumption that we are typical of *all* the observing systems that exist (E) in the model universe — not just the ones that have our data D_0. Equivalently, the assumption is that we are equally likely to be any of the observing systems that the universe exhibits. If there are n_O observing systems, then the probability that we are the Ath instance is $1/n_O$, and that defines the xerographic distribution

$$\xi_A^{\text{typO}} = \frac{1}{n_O}. \tag{8}$$

The first-person likelihoods for our data are then denoted by $P^{(1p)}(E,R|\xi^{\text{typO}},T)$.

We are equally likely to exist in any cycle since they are all the same. The probability that we see red, is therefore the probability that our cycle is red. This is $N_R(T)/N$. Thus

$$P^{(1p)}(R|E,\xi^{\text{typO}},T) = N_R(T)/N. \tag{9a}$$

Similarly, the probability that we see blue is

$$P^{(1p)}(B|E,\xi^{\text{typO}},T) = N_B(T)/N. \tag{9b}$$

We can now proceed as we did above from (5) to (7). The result for the first-person likelihoods is

$$P^{(1p)}(E, R|\xi^{\text{typO}}, T) = \frac{N_R(T)}{N}[1 - (1 - p_E)^N]. \tag{10}$$

The ratio of the likelihoods for the two theories is

$$\frac{P^{(1p)}(E, R|\xi^{\text{typO}}, AR)}{P^{(1p)}(E, R|\xi^{\text{typO}}, SR)} = \frac{N}{N_R(SR)} > 1. \tag{11}$$

Thus, assuming equal priors for the two theories, AR is always favored even if $N_R(SR)$ becomes arbitrarily large, provided there are at least some blue cycles.

In the above examples, the two theories AR and SR compete with the same typicality assumption. But it is also possible to compete different typicality assumptions for the same theory. Suppose we assign unit prior probability to the theory SR and equal priors to ξ^{typO} and ξ^{typD}. From (10) and (7) we find

$$\frac{P^{(1p)}(E, R|\xi^{\text{typO}}, SR)}{P^{(1p)}(E, R|\xi^{\text{typD}}, SR)} = \frac{N_R(SR)}{N}\frac{1 - (1 - p_E)^N}{1 - (1 - p_E)^{N_R(SR)}} < 1. \tag{12}$$

Thus, an assumption of typicality in the class of our data D_0 does a better job of explaining our data (trivially).

Certain theoretical models may imply that the number of instances of our data N is infinite. An example is provided by the infinite number of Hubble volumes on a surface of homogeneity inside a Coleman–De Luccia bubble of false vacuum.[5] The results above are well defined provided that the fractions $f_R(T)$ and $f_B(T)$ of red and blue Hubble volumes are well defined in the competing theories T. For instance, (9) becomes

$$P^{(1p)}(R|E, \xi^{\text{typO}}, T) = f_R(T), \quad P^{(1p)}(B|E, \xi^{\text{typO}}, T) = f_B(T) \tag{13}$$

and similarly with the other formulae. Indeed, the expressions are generally simpler than for the finite case since the probability that there is at least one instance of our data (6) becomes exactly unity.

This set of examples shows that frameworks (T, ξ) are an umbrella formalism for organizing, comparing, and (most importantly) making explicit a number of different assumptions that are commonly made in the quantum cosmology of a very large universe. Some of these possible assumptions will be discussed in more detail in Sec. 8 and Appendix A.

6. Predictivity

First, let us consider theories T where there is at most one copy of our data D_0. No xerographic distribution is needed. In this setting, a theory is *predictive* if the likelihood is high for some piece of data that we might acquire in the future, given the theory and the data that we have acquired in the past. To make this more concrete, consider a simplified situation in which a stream of data $d_{-n}, \ldots, d_{-1}, d_0$ is acquired at a sequence of times $t_{-n}, \ldots, t_{-1}, t_0$, where t_0 denotes the present time, defined relative to the local clocks provided by each particular occurrence of D_0. Assuming that all this data is accessible now, our present data is the union, $D_0 = d_{-n} \cup \ldots \cup d_0$. Let $P^{(1p)}(q_1|T, D_0)$ be the first-person likelihood that at some future time t_1, we will acquire some piece of data q_1; q_1 is a subset of all the data d_1 that we acquire at that time. A theory is *predictive* if, for *some* kinds of data q_1, the likelihood $P^{(1p)}(q_1|T, D_0)$ is sharply peaked around particular values $\bar{q}_1$.

There is obvious motivation to favor theories that are predictive and provide a coherent story connecting past data that we currently have, to future data that we may acquire. The utility of physical theory lies in its predictive power. Theories that are predictive are also testable in the sense that they are falsifiable when new data disagree with that predicted. Indeed, the whole history of science can be read as the search for predictive theories.

We turn now to theories of large universes with replication of our data D_0. As already noted, in this case we can only make predictions if we specify a xerographic distribution ξ in addition to a conventional theory T. Once the xerographic distribution is specified, we can (at least in principle) compute the likelihood $P^{(1p)}(q_1|T, \xi, D_0)$ that we will acquire a piece of data q_1 at a future time t_1. A framework (T, ξ) is *predictive* if, for *some* kinds of data q_1, the likelihood $P^{(1p)}(q_1|T, \xi, D_0)$ is sharply peaked around particular values $\bar{q}_1$.

To compute $P^{(1p)}(q_1|T, \xi, D_0)$, we first compute the third-person likelihood that the data subset q_1 is found at time t_1 at the Ath location of the data D_0; for this we need only the theory T and the data D_0. Denote this likelihood by $P(q_1@A|T, D_0@A)$, where $x@A$ means that data x occurs at location A. Then, the likelihood that "we" obtain q_1 is

$$P^{(1p)}(q_1|T, \xi, D_0) = \sum_A \xi_A P(q_1@A|\xi, T, D_0@A). \tag{14}$$

Without an assumed xerographic distribution, no prediction whatsoever can be made about what "we" will see in the future. A physical theory that is considered "complete" in the usual sense (such as a specified quantum state and a rule for its evolution, or even a fully deterministic classical theory plus initial data) is insufficient to determine the xerographic distribution, which must therefore be chosen as an additional ingredient of the theoretical framework.

7. Making Theories Predictive with Typicality Assumptions

Frameworks (T, ξ) with the same theory T but different assumptions for the xerographic distribution ξ can be compared and tested. Scientists favor frameworks that are predictive, that is, those that generate a stream of future predictions and hence are testable.

In this section, we give two examples of how the predictivity of frameworks (T, ξ) with a fixed theory T can be affected by different choices of ξ that reflect different assumptions about the typicality of our data D_0 in two different situations.

7.1. *Boltzmann Brains*

Thermal or vacuum fluctuations could replicate our data.[6,7] In one simplified model, a spatially closed universe originates in a Big Bang and eventually enters a de Sitter phase that persists forever. It is assumed that there is a very tiny probability per unit spacetime volume that a "brain" could fluctuate into existence. Such fluctuations are called "Boltzmann brains" (BBs) or "freak observers".[7] Since the spacetime volume is infinite in this model (even though the spatial volume is finite), it is assumed that an infinite number of BBs will be fluctuated into existence. A tiny fraction of this infinite number have the same data D_0 that we do. How do we know that we are not one of them? This is the Boltzmann brain problem.

Let us accept this scenario uncritically, and ask only if there is an assumption for the xerographic distribution ξ for which the framework is predictive.

An assumption that we are typical, $\xi_A = 1/N$, does not result in a predictive framework. In that case, we are much more likely to be a BB than an ordinary observer (OO). The BBs are deluded (e.g. Ref. 8); their data suggest that they are 13.7 Gyr from a Big Bang, but typically they are much further away. In contrast to ordinary observers that have 13.7 Gyr of history, the subsequent observations of any particular BB is overwhelmingly likely to be disordered, and inconsistent with its apparent history. Future data is thus uncorrelated with D_0, and firm predictions cannot be made.

Now let us consider a nonuniform ξ; that is, we assume that we are atypical. In particular, we suppose that ξ_A is non-zero only for locations sufficiently close to a Big Bang to make it much more likely that we are OOs rather than BBs. In this case, the framework is predictive in the usual way.

The answer to the question "How do we know that we are not BBs?" is this: we do not know. But if we assume a xerographic distribution ξ such that we are not likely to be BBs, then we get a predictive, testable framework (T, ξ). Confirmation of the predictions of this framework by a series of observations then supports the original assumption of atypicality.[f]

7.2. *Laboratory Experiment*

Consider a laboratory experiment to determine the value of a fundamental constant α by a sequence of measurements. We consider a class of theories that predict that α is constant throughout spacetime, but predict different values for it. In a very large universe, the experiment will be replicated in many different locations x_A in

[f]Others have expressed related ideas, e.g. Ref. 8.

all essential details. For the vast majority of these experiments, the mean value $\bar{\alpha}$ of the sequence of outcomes will be close to the true value of the constant to the accuracy of the experiment. But for a tiny fraction of these experiments, the sequence of measurement outcomes are mistaken — their mean value $\bar{\alpha}$ is far from the true value — just by the chances of statistics. How do we know that our particular sequence of results is not one of these?

The answer lies in the evidence from further measurements. The probabilities for further measurements can be predicted from the previous ones, once a xerographic distribution ξ is specified. An assumption that our experiment is typical of all the others results in a theoretical framework $(T(\bar{\alpha}), \xi_A{=}1/N)$, where $T(\bar{\alpha})$ is the theory that predicts that the value of the constant is the mean value of the previous outcomes, $\bar{\alpha}$. The prediction of this framework is that the next measurement should yield $\bar{\alpha}$ to within statistical error. Confirmation of this prediction by further measurements supports the framework $(T(\bar{\alpha}), \xi_A{=}1/N)$, both as to the value of α and the assumption of typicality.

In the way the problem has been set up, there is no evident variable with which to make an assumption of atypicality. The physical situations of all the experiments have been assumed to be the same, in contrast to the different situations represented by BBs and OOs. As emphasized correctly by a number of authors (e.g. Ref. 9), some kind of typicality is assumed implicitly in the analysis of every laboratory experiment. Here we have made that assumption explicit in terms of the xerographic distribution.

8. Measures for Cosmology

In models of eternally inflating cosmologies, relative probabilities for different kinds of physical situations are defined in terms of the ratios of the number of times they occur. Examples are the occurrence of different kinds of "bubble universes" and the ratio of the number of BBs to OOs. Since the numbers are typically infinite, a "measure" is required to define these fractions; without a measure, third-person probabilities are ill-defined.

Ideally, such a measure would emerge unambiguously from an underlying theory (such as string theory). In this case, the "measure problem" would be solved, and third-person probabilities could be computed.

However, a xerographic distribution would still be needed in order to define and compute first-person probabilities. Solving the measure problem does not remove the need to choose a xerographic distribution, but does make the choice of a uniform xerographic distribution well defined.

In some models (e.g. Ref. 2), there is a natural choice for the measure. However, in more general contexts there is as yet no consensus for how to determine the measure from the underlying theory, or even whether this is possible (see, e.g. Refs. 9 and 10 for discussions). If it turns out that the measure is not computable from the underlying theory, then whatever freedom remains in the choice of the measure can and should be incorporated into the choice of the xerographic distribution.

Starting from a different perspective, Page[11] has also argued that the quantum wave function alone contains insufficient information for the calculation of probabilities of subsequent observations by a particular observer when there are multiple copies of that observer. His work is discussed in Appendix A.

9. Conclusion

The possibility that our data may be replicated exactly elsewhere in a very large universe profoundly affects the way science must be done.

Central to cosmology are the third-person probabilities for properties of the universe, given a theory T of its quantum dynamics and quantum state. But of even greater interest are the first-person probabilities for the results of observations carried out by us — a particular subsystem of the universe — conditioned on our existing data. These probabilities are the means to test any prescription for prediction. In a very large universe, where our data will be replicated elsewhere with significant probability, predicting these probabilities requires not only a theory T but also a probability distribution on the sets of copies of us. This xerographic distribution cannot, even in principle, be determined from a theory of the dynamics and the quantum state, since such a theory has nothing to say about which copy is "us". It is only the theoretical framework consisting of both a theory and an assumed xerographic distribution that is predictive and testable by observation.

In cosmology, we should favor theoretical frameworks that generate a stream of predictions from our data that are confirmed by subsequent observation. The authors believe that choices of both the theory T and the xerographic distribution ξ should be competed against other alternatives, and that no element of the theoretical framework should be assigned a unit probability.

Ideas that imply particular notions of typicality (such as the "Copernican principle", "anthropic principle", or the "principle of mediocrity") cannot be universal laws of nature if only because they refer to a negligibly minor subsystem of the universe: "us". In the present context, these are simply notions that can motivate a particular choice of the xerographic distribution.

It is no surprise that information about us, is required to make predictions for our observations. Our data suggest that we are located some 13.7 Gyr from a Big Bang. To make a reliable prediction from that information, we have to assume that it truly describes our physical situation.

If the universe is rife with delusion, we must assume that we are atypical in order to have predictive and testable scientific theories. Indeed, it is only by making such assumptions that we are able to do science in a very large universe. We imagine that even Copernicus would have agreed that it was necessary to assume that Ptolemy was not deluded in his observations of the planets.

Acknowledgments

We thank Raphael Bousso, Brandon Carter, Sean Carroll, Ben Freivogel, Steve Giddings, Alan Guth, Thomas Hertog, Matthew Kleban, Don Page, Steve Shenker and Alex Vilenkin for numerous helpful conversations. This work was supported in part by the National Science Foundation under grants PHY05-55669 and PHY07-57035.

Appendix A

The Work of Page as an Example

As mentioned above, the schema for quantum cosmological prediction developed here provides a common framework for the discussion of different prescriptions for science in a very large universe. Prescriptions for which notions of first- and third-person probabilities can be distinguished, and a xerographic distribution identified.

Notable among these different presciptions are those found in the extensive contributions of Don Page.[11–13] In this appendix, we attempt to fit at least some parts of his ideas into the present context, in part to address his criticisms of our earlier work.[1]

Page reaches the conclusion that Born's rule of usual quantum theory must be modified to apply it to a very large universe where our data have a significant chance of being replicated elsewhere. We disagree with this conclusion, as we now explain.

We begin with Page's discussion of Born's rule. In Ref. 13, he says that "a goal of science is to produce complete theories T_i that each predict normalized probabilities $P_j(i)$ for observations O_j". These are the probabilities for observations made by "the observer". We interpret these as first-person probabilities for alternative outcomes of our observations. For example, they might be the probabilities for alternative CMB temperature maps, or the alternatives, red and blue in the R/B model. The sum of the probabilities of an exhaustive set of exclusive outcomes is, of course, one. This is illustrated explicitly by (9).

We agree with Page that first-person probabilities are not specified by the quantum state of the universe alone. As we have argued, a xerographic distribution is needed to connect them to the third-person probabilities that are specified by the quantum state. However this does not mean that "Born's rule dies." Born's rule is alive, well, and essential for constructing third-person probabilities.

Page suggests various candidates for replacing Born's rule. In his nomenclature,[12] our xerographic distribution for the assumption that we are typical of all the observing systems that exist, appears to be equivalent to Page's rule 5, which he calls "observational averaging".

Constructions such as those of Page or the theoretical frameworks discussed in this paper are tested by comparing observations with the first-person predictions for them. The third-person probabilities for features of the universe are essential for computing these. For instance, both Page's candidates for replacing Born's rule and the examples of xerographic distributions in this paper rely on an assumption of classical spacetime to make notions of location meaningful. But classical spacetime is neither fundamental nor inevitable in a quantum theory of gravity. Whether a quantum state predicts an ensemble of alternative classical spacetimes is a question of whether the third-person probabilities are high for correlations in time governed by the Einstein equation for suitably coarse-grained histories of geometry and matter fields (e.g. Ref. 2). Third-person probabilities are thus not dispensable; they are essential for the understanding of our quantum universe.

Appendix B

Third to First in the R/B Model with a Xerographic Distribution

In this appendix, the first-person likelihoods (10) in the red-blue model of Sec. 5 are derived by explicitly considering the form of the xerographic distribution ξ^{typO} without invoking the symmetry of the model directly.

We start with the third-person probability $P(n_O, n_R|T)$, that of the N total cycles, n_O are occupied by observing systems, with n_R of these in red cycles. The probability that there are n_O cycles occupied by observing systems is $p_E^{n_O}(1-p_E)^{N-n_O}$ multiplied by the number of ways of arranging n_R observing systems in N_R red cycles and the number of ways of arranging $n_B = n_O - n_R$ observing systems in $N_B = N - N_R$ blue cycles. This is

$$P(n_O, n_R|T) = \binom{N_R}{n_R}\binom{N-N_R}{n-n_R} p_E^{n_O}(1-p_E)^{N-n_O}. \tag{B.1}$$

The theoretical framework consists of the theory T and a xerographic distribution that assumes we are typical in the class of observing systems. The n observing systems can be labeled by an index A

that runs from 1 to n. The xerographic distribution that says we are equally likely to be any one of these systems is

$$\xi_A^{\text{typO}} = \frac{1}{n_O}, \tag{B.2}$$

which satisfies

$$\sum_{A=1}^{n_O} \xi_A^{\text{typO}} = 1. \tag{B.3}$$

With these assumptions, the first-person likelihood $P^{(1p)}(E, R|\xi^{\text{typO}}, T)$ that we exist and observe red is ξ_A^{typO} times the third-person probability $P(n_O, n_R|T)$ summed over A and the alternative configurations specified by (n_O, n_R). In general, the third-person probabilities would depend on which cycles are occupied and hence on A. But in the present case, where they are all identical, there is no such dependence. The result is

$$P^{(1p)}(E, R|\xi^{\text{typO}}, T) = \sum_{n_O=1}^{N} \sum_{n_R=1}^{n_O} \sum_{A=1}^{n_R} \xi_A^{\text{typO}} P(n_O, n_R|T). \tag{B.4}$$

Substituting in (B.1) and (B.2) and performing the sum over A, we get

$$\begin{aligned}
&P^{(1p)}(E, R|\xi^{\text{typO}}, T) \\
&\quad = \sum_{n_O=1}^{N} \sum_{n_R=1}^{n_O} \frac{n_R}{n_O} \binom{N_R}{n_R} \binom{N-N_R}{n_O-n_R} p_E^{n_O} (1-p_E)^{N-n_O} \\
&\quad = \sum_{n_O=1}^{N} \frac{N_R}{n_O} \sum_{n_R=1}^{n_O} \binom{N_R-1}{n_R-1} \binom{N-N_R}{n_O-n_R} p_E^{n_O} (1-p_E)^{N-n_O} \\
&\quad = \sum_{n_O=1}^{N} \frac{N_R}{n_O} \binom{N-1}{n_O-1} p_E^{n_O} (1-p_E)^{N-n_O} \\
&\quad = \frac{N_R}{N} \sum_{n_O=1}^{N} \binom{N}{n_O} p_E^{n_O} (1-p_E)^{N-n_O} \\
&\quad = \frac{N_R}{N} [1-(1-p_E)^N],
\end{aligned} \tag{B.5}$$

where the third line follows from Vandermonde's identity.

We see that (B.5) is the same as (10). The argument based on the symmetry given in Sec. 5 is evidently a more efficient way of getting this result. A general case without symmetry would be even more complicated. Suppose, for example, that there was a different value of p_E for each cycle. Then the third-person probabilities would depend on which cycles were occupied, and not just on the total number of them as given here. The prescription for computation, however, would be essentially the same: calculate the third-person probabilities for a configuration of occupied and unoccupied cycles; multiply by the xerographic distribution; sum over the possible configurations.

References

1. J. B. Hartle and M. Srednicki, Are we typical? *Phys. Rev. D* **75**, 123523 (2007), arXiv:0704:2530.
2. J. B. Hartle, S. W. Hawking and T. Hertog, The no-boundary measure of the universe, *Phys. Rev. Lett.* **100**, 202301 (2008), arXiv:0711:4630; Classical universes of the no-boundary quantum state, *Phys. Rev. D* **77**, 123537 (2008), arXiv:0803:1663; The no-boundary measure in the realm of eternal inflation, arXiv:1001.0262; J. B. Hartle and T. Hertog, Replication regulates volume weighting in quantum cosmology, *Phys. Rev. D* **80**, 063531 (2009), arXiv:0905.3877.
3. J. B. Hartle, *The quasiclassical realms of this quantum universe*, arXiv:0806.3776.
4. M. Srednicki and J. B. Hartle, *The xerographic distribution: Scientific reasoning in a large universe*, arXiv:1004.3816.
5. S. Coleman and F. DeLuccia, Gravitational effects on and of false vacuum decay, *Phys. Rev. D* **21**, 3305 (1980).
6. D. N. Page, Is our universe decaying at an astronomical rate? *Phys. Lett. B* **669**, 197 (2008), arXiv:hep-th/0612137.
7. A. Albrecht and L. Sorbo, Can the universe afford inflation? *Phys. Rev. D* **70**, 063528 (2004), arXiv:hep-th/0405270. There is an extended set of references in D. N. Page, Cosmological measures without volume weighting, *JCAP* **0810**, 025 (2008), arXiv:0808:0351.

8. J. R. Gott, *Boltzmann brains — I'd rather see than be one*, arXiv:0802.0233.
9. J. Garriga and A. Vilenkin, Prediction and explanation in the multiverse, *Phys. Rev. D* **77**, 043526 (2008), arXiv:0711.2559; R. Bousso, B. Freivogel and I.-S. Yang, Boltzmann babies and the proper time measure, *Phys. Rev. D* **77**, 103514 (2008), arXiv:0712.3324.
10. R. Bousso, B. Freivogel and I.-S. Yang, Properties of the scale factor measure, *Phys. Rev. D* **79**, 063513 (2009), arXiv:0808.3770; A. De Simone, A. H. Guth, A. Linde, M. Noorbala, M. P. Salem and A. Vilenkin, *Boltzmann brains and the scale-factor cutoff measure of the multiverse*, arXiv:0808.3778; and references therein.
11. D. N. Page, Typicality derived, *Phys. Rev. D* **78**, 103517 (2008), arXiv:0804.3592; Insufficiency of the quantum state for deducing observational probabilities, *Phys. Lett. B* **678**, 41 (2009), arXiv:0808.0722.
12. D. N. Page, The Born rule fails in cosmology, *JCAP* **07**, 008 (2009), arXiv:0903.4888.
13. D. N. Page, *Born again*, arXiv:0907.4152.

CHAPTER 19

The Observer Strikes Back*†

In the modern quantum mechanics of cosmology, observers are physical systems within the universe. They have no preferred role in the formulation of the theory nor in its predictions of third-person probabilities of what occurs. However, observers return to the importance for the prediction of first-person probabilities for what we observe of the universe: What is most probable to be observed is not necessarily what is most probable to occur. This essay reviews the basic framework for the computation of first-person probabilities in quantum cosmology starting with an analysis of very simple models. It is shown that anthropic selection is automatic in this framework, because there is no probability for us to observe what is where we cannot exist. First-person probabilities generally favor larger universes resulting from inflation, where there are more places for us to be. In very large universes, it is probable that our observational situation is duplicated elsewhere.

(*continued overleaf*)

---

*Based on a talk given by the author at the conference on the *Philosophy of Cosmology*, Tenerife, Spain, September 13–15, 2014.

†With Thomas Hertog (Institute for Theoretical Physics, KU Leuven, 3001 Leuven, Belgium).

arXiv:1503.07205

The calculation of first-person probabilities then requires a specification of whether our particular situation is assumed to be typical of all the others. It is the combination of the model of the observational situation, including this typicality assumption, and the third-person theory, which is tested by observation. We conclude with a discussion of the first-person predictions of cosmological observables, such as the cosmological constant and features of the primordial density fluctuations, in the no-boundary quantum state of the universe, and a dynamical theory in which these are allowed to vary.

1. Introduction

The context for this essay is our universe — the whole closed system at all times, containing all the galaxies, stars, planets, biota, human societies, you, us, etc. There is nothing outside. Two kinds of description of the universe can be distinguished:

Third-Person Descriptions: Descriptions of what the universe contains and how that evolves — histories of what occurs.
First-Person Descriptions: Descriptions of what we as the collection of human scientists observe of the universe and use to test cosmological models.
The connection between these two kinds of description is the subject of this essay.

Quantum mechanical theories, and also classical ones, provide probabilities for different descriptions. Correspondingly, we can distinguish two different kinds of probabilities for any observable $\mathcal{O}$: Third-person probabilities[a]

$$p(\mathcal{O}) \tag{1a}$$

for what values of $\mathcal{O}$ occur, and first-person probabilities

$$p^{(1p)}(\mathcal{O}) \tag{1b}$$

for what values of $\mathcal{O}$ we observe. The connection between these two kinds of probabilities is the focus of this paper.

[a]Following Ref. 1, in most of our previous work we have used "bottom-up probabilities" and "top-down probabilities" for what are called here, "third-person probabilities" and "first-person probabilities".

We will consider theories whose direct outputs are third-person probabilities for histories of what occurs in the universe. These include probabilities for the existence, evolution and functioning of any observing subsystems, such as ourselves. Such subsystems play no special role in formulating the theory — they are just one kind of subsystem among many. We refer to such theories as "third-person" theories. Our most successful theories of cosmology are of this kind. Classical physics in general is an example of a third-person theory. In quantum mechanics, the extensions of the ideas of Everett[2] are third-person theories, including those used in this paper.

Observers do play a preferred role in calculating first-person probabilities for observations from third-person ones. First-person probabilities, are third-person ones conditioned on a description of the observational situation. We shall use a variety of specific models to infer the following general conclusions:

- What is most probable to occur is not necessarily what is most probable to be observed.
- Anthropic selection is an automatic consequence of first-person probabilities.
- In universes large enough that we may be duplicated as physical subsystems elsewhere, the description of the observational situation needed to compute first-person probabilities must also specify whether our particular situation is typical of all the others. It is the combination of the model of the observer, including this typicality assumption, and the third-person theory which is tested by observation.

Observers and their observations are of central importance in the formulation of Copenhagen quantum mechanics. In the quantum mechanics of closed systems, observers might seem to have been demoted to the status of one subsystem among many. Indeed, they have little effect on third-person probabilities. But, as a consequence of the conclusions above, the observer returns to the importance in the calculation of first-person probabilities for observations by which the theory is used and tested. The observer strikes back.

Section 2 sketches the framework of the third-person theory we will employ. Section 3 discusses issues involved with first-person probabilities. Section 4 uses a model universe to make more concrete the notions of first- and third-person descriptions of the universe and their associated probabilities. Here, we also describe the connection between third- and first-person probabilities in a set of simple models. Section 5 describes how anthropic selection emerges automatically as a feature of a certain class of first-person probabilities. Section 6 shows how first-person probabilities can sometimes be calculated directly from the theory with an appropriate coarse-graining. Section 7 discusses the first-person predictions of a number of cosmological observables, such as the cosmological constant, in cosmological models based on the no-boundary quantum state of the universe and a dynamical landscape theory in which these observables can take a range of values. Section 8 deals with eternal inflation and anthropic selection. We then try to set our results in a more general view of physical theories.

2. Third-Person Quantum Cosmology

This section briefly describes the elements of the theory that predicts third-person probabilities for histories of the universe which we use to reach the general conclusions mentioned in the introduction.[b]

We view the universe as a closed quantum-mechanical system. It contains everything. Galaxies, stars, planets, their biota, observers (including us!), etc. are physical subsystems of the universe subject to its quantum-mechanical laws consistent with an Everettian point of view.[2] The basic variables describing the universe and its contents are four-dimensional cosmological spacetime geometries and four-dimensional configurations of matter fields. The basic ingredients of the theory are an action I describing the dynamics of geometry

---

[b]For more details, the reader can consult the authors' papers on which this essay is implicitly based and through them find further references. For the quantum mechanics of closed systems, see e.g. Refs. 3 and 4. For quantum cosmology, see e.g. Refs. 1, 5–9. There is a little more detail about the no-boundary quantum state of the universe[10] in Appendix B.

coupled to matter fields, and a quantum state of the universe Ψ. We denote the theory as (I, Ψ).

The theory (I, Ψ) predicts third-person probabilities for the individual members of sets of alternative four-dimensional histories of the universe, including those histories that describe its classical evolution. In this way, (I, Ψ) can supply third-person probabilities for such large-scale features of the universe as the amount of inflation, the approximate homogeneity and isotropy, the pattern of cosmic microwave background (CMB) variations, and the formation and evolution of the distribution of galaxies. In principle, (I, Ψ) also supplies third-person probabilities for the accidents of biological evolution, the existence of observers like ourselves, etc. that are well beyond our power to compute or even estimate. For the examples in this paper, we will mostly use histories in which we are at a single moment of time. The time is approximately 14 Gyr after the big bang when our observations of the universe are made. The theory (I, Ψ) is an example of what we will call a third-person theory.

We test and utilize a theory not by its third-person probabilities for what occurs, but by its first-person probabilities for what we observe. To compute first-person probabilities, we first need to model the observational situation, which we do next.

3. First-Person Quantum Cosmology

We begin by recalling the definition of a Hubble volume. We cannot see further in the universe than the distance that light travels to us from the big bang, roughly 14 Gyr ago. A present volume of the same size as this distance is called a "Hubble volume". In order of magnitude, the distance is c/H_0 where H_0 is the present Hubble constant. This size is approximately 4,000 Megaparsecs or 10^{23} km. This is the largest scale over which we can currently observe the universe.

As observers, we are physical systems within the universe with only a probability to have evolved in any one Hubble volume and a probability to be replicated in other Hubble volumes if the universe has a very large number of them. An observer is a very special kind of fluctuation D_0 in the universe. It is a fluctuation that is not singled out by quantum theory from, say, density fluctuations that produced the CMB. But the probability for observers is very difficult to compute. We will therefore employ a highly simplified model of observers: All observers are alike (copies of us) and either exist in any Hubble volume with a third-person probability $p_E(D_0)$ or do not exist with a probability $1 - p_E$. Realistically, the probability p_E incorporates the probability of the accidents of several billion years of biological evolution. Therefore, whatever its value is, it is very, very, very small. This is a very crude model of complex observers, but better than many treatments where the probability that observing systems evolved as part of the universe's evolution is not considered at all.

Since both we and what we observe are part of the universe, first-person probabilities can be computed from the third-person ones. If we are unique as physical systems within the universe, the first-person probabilities are simply third-person ones for what is observed, conditioned on a description of the observational situation like the one given above, in terms of probabilities p_E for data D_0. But if we are not unique, then a more careful specification of the observer is required, which includes an assumption about which instance of D_0 is us, or, more generally, a probability distribution on the set of copies called a xerographic distribution.[11] In this essay we will make the minimal assumption that we are equally likely to be any of the incidences of D_0 that the third-person theory (I, Ψ) predicts. The way to make other assumptions is discussed in Appendix A.

It is a common intuition that the presence or absence of observers is unimportant for the behavior of the universe on cosmological scales because observers are generally small subsystems. That intuition is correct for third-person probabilities, but it is not correct for first-person ones. As physical systems, we have only a tiny probability to exist in any one Hubble volume. That is why third-person probabilities are little affected by the presence or absence of us. But this also means that we have a greater probability to live in a larger universe than a smaller one, because in the larger, there are more Hubble volumes in which to be. Therefore, even if the third-person probabilities favor smaller universes, the first-person probabilities may favor larger ones. As we will show by example, what is most probable to occur (third-person) is not necessarily what is most probable to be observed (first-person). It is in this way that the observer returns to importance in cosmology.

4. Toy Model Universes

This section derives the connection between third- and first-person probabilities for a very simple class of models that provide elementary examples showing that what is most probable to be observed is not necessarily what is most probable to occur. In the trade, these are called "box models".[12]

4.1. *Third-Person Description and Probabilities*

Histories of the universe at one time are modeled as a collection of N boxes representing Hubble volumes at that time. Each box has a color — say red or blue — modeling different CMB maps. Each box may or may not contain an observer. A third-person description of a history specifies the number of boxes, their color, and whether each box contains an observer or not. A third-person quantum theory specifies probabilities for these histories — a probability for the number of boxes, a probability for the color of each box, and the probability p_E for whether an observer exists in each.

We illustrate this with the very simple set of single time histories[12] shown in Fig. 1. At one time, only two possible histories of N and color are possible — N_1 boxes, all red, that occur with a third-person probability p_1, and N_2 boxes, all blue, that occur with third-person probability $p_2 = 1 - p_1$. The probability that there is an observer in any box is p_E — the same for all boxes. The probability that there is no observer in a box is $1 - p_E$. The third-person probability of a history with a specific set of n_E of N_k boxes occupied by observers is

$$p_k p_E^{n_E} (1 - p_E)^{N_k - n_E} \tag{2}$$

where $k = 1$ (all red) or $k = 2$ (all blue). The complete set of histories consists of all red and all blue histories, with different numbers

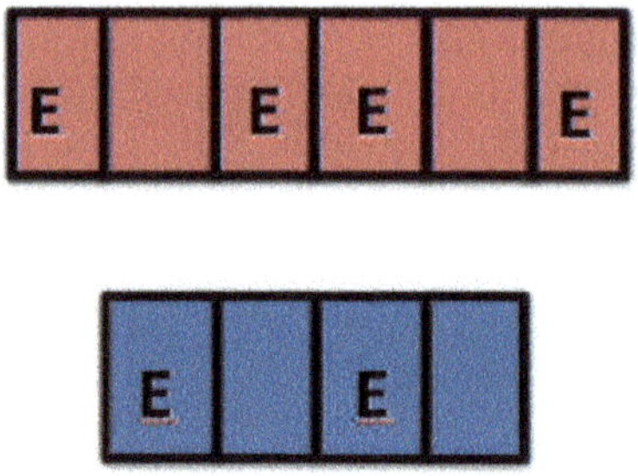

$$p_1 p_E^4 (1 - p_E)^2$$

$$p_2 p_E^2 (1 - p_E)^2$$

Fig. 1. Two histories of the simple box model universe described in the text. The boxes model Hubble volumes. Their color models an observable like the CMB. An "E" means that an observer is in the box observing its color. A blank means there is no observer in the box. The third-person probabilities for these histories to occur are at the right.

of boxes in each, and the various possible ways the boxes can be occupied by observers.

4.2. *First-Person Probabilities for Observation*

We are one of the observers in one of the histories. We now ask for the theory's prediction for the first-person probability that we observe red (WOR). To calculate that, assume that in either history we are equally likely to be any one of the occurrences of E. Then the probability that we observe red (WOR) is evidently the probability that we are in the history with all red boxes ($k = 1$).

The probability that WOR is *not* the probability p_1 that the history with all red boxes occurs, because that could happen with no boxes occupied by observers. Rather, the probability for WOR is the probability that the history $k = 1$ occurs with *at least one observer.* Denoting "at least one E" by $E^{\geq}$ we have

$$\begin{aligned} p(E^{\geq} \text{ in } 1) &= 1 - p(\text{no E in } 1) \\ &= 1 - (1 - p_E)^{N_1}. \end{aligned} \tag{3}$$

The normalized probability that we observe red (WOR) is then

$$p^{(1p)}(\text{WOR}) = \frac{p_1[1-(1-p_E)^{N_1}]}{\sum_k p_k[1-(1-p_E)^{N_k}]}. \tag{4a}$$

Similarly, the probability that we observe blue is

$$p^{(1p)}(\text{WOB}) = \frac{p_2[1-(1-p_E)^{N_2}]}{\sum_k p_k[1-(1-p_E)^{N_k}]}. \tag{4b}$$

We now discuss important limiting cases.

Rare in All Histories: When both $p_E N_1 \ll 1$ and $p_E N_2 \ll 1$, physical systems like us occur only rarely in each of the two possible histories. Therefore, we can assume that as a physical subsystem we are unique in the universe. The probabilities for color observation (4) then become

$$p^{(1p)}(\text{WOR}) \approx \frac{N_1 p_1}{N_1 p_1 + N_2 p_2}, \tag{5a}$$

$$p^{(1p)}(\text{WOB}) \approx \frac{N_2 p_2}{N_1 p_1 + N_2 p_2}. \tag{5b}$$

The first-person probabilities for our observation of red or blue are the third-person probabilities of the all red and all blue histories *weighted* by the number of Hubble volumes in each. This is called "volume weighting". It favors larger universes where there are more places for us to occur, as has been extensively discussed in cosmology (e.g. Refs. 5, 13 and 14).

It is important to emphasize that volume weighting is not an extra assumption in addition to the theory of the histories. Rather, it is a straightforward consequence of that theory in models where we are rare in all histories.

In a third-person description of this model, blue is more likely to occur if $p_2 > p_1$. But red is more likely to be observed if $N_1 p_1 > N_2 p_2$. Thus, we have an elementary example of what is most probable to be observed is not necessarily what is most likely to occur. That is the return of the observer.

Other limiting cases are also interesting:

Common in Both Histories: When both $p_E N_1 \gg 1$ and $p_E N_2 \gg 1$, copies of us as physical systems are common in both histories. Then the probabilities (4) are

$$p^{(1p)}(\text{WOR}) \approx p_1, \tag{6a}$$

$$p^{(1p)}(\text{WOB}) \approx p_2. \tag{6b}$$

Thus, when all histories in the ensemble are very large universes, what is predicted to be observed is also what is predicted to occur.

Rare in One History, Common in the Other: When, say, $p_E N_1 \gg 1$ but $p_E N_2 \ll 1$, copies of us as physical systems are common in the all red history and rare in the blue one. We have $[1-(1-p_E)^{N_1}] \approx 1$ and $[1-(1-p_E)^{N_2}] \approx N_2 p_E \approx 0$ since p_E is very, very small. Then,

$$p^{(1p)}(\text{WOR}) \approx 1, \tag{7a}$$

$$p^{(1p)}(\text{WOB}) \approx 0. \tag{7b}$$

This is the case where the return of the observer has its most striking effect. The first-person probabilities select large histories over small ones even when the latter have larger third-person probabilities. We will see a concrete example of this in a more realistic model in Sec. 8.

4.3. *A Crisis of Computability?*

Box models divide the theory for predicting first-person probabilities for what we observe into two parts. First, there is the specification of the third-person probabilities p_k for the large scale features of the models — the number of Hubble volumes and the color of each. Second, there are the third-person probabilities for the occurrence of observers inside each Hubble volume, summarized by the one number p_E, used to describe and to condition on the observational situation.

As we will see in Sec. 7, it is possible to make computationally tractable calculations of p_k in simple models. But the probability p_E would naturally include the probability that human observers evolved in a Hubble volume. To calculate this, or even estimate it, would involve considering several billion years of the chance accidents of biological evolution. This is well beyond our power to compute even assuming that we have a theory that is formulated well enough to define the task.

It is therefore fortunate that in all of the interesting limiting cases discussed above, the probability p_E cancels out. Thus, it is in the regime of universes so small that we are unique, or universes so large that we are common, that observations are easily and objectively calculated.

5. Anthropic Selection is Automatic in Quantum Cosmology

Consider again the simple box model in Sec. 4.2, but suppose that the probability for an observer to occupy a red box p_E^R is different from the probability p_E^B to occupy a blue box. Suppose further that, for some reason, red is necessary for observers, so that the probability p_E^B to occupy a blue box is exceedingly small. It is then obvious, and easily worked out, that the first-person probability that we observe red is very near unity. This is a very simple example of anthropic selection: The all red history is selected because observers are rare in the alternative.

The key point here is that anthropic selection is automatic. No additional assumptions or principles were needed beyond the probabilities p_1, p_2, p_E^R and p_E^B, which are all third-person probabilities following from the underlying theory. Anthropic selection emerges as an intrinsic feature of the first-person probabilities for the observer's observations.[15]

This is the case also in more general and more realistic cosmological models. As an example, consider an ensemble of single time histories which we take to be at the present age of the universe $t_0 \approx 14$ Gyr. Assume that the theory (I, Ψ) predicts third-person probabilities for what we may call a set of background histories, each with the same number of Hubble volumes but differing in the value of the cosmological constant Λ, which is assumed positive and the same in all Hubble volumes of one history. The theory thus allows the cosmological constant to vary. The histories can be labeled by the value of Λ and their third-person probabilities written as $p(\Lambda)$. As a fluctuation on these backgrounds, the theory (I, Ψ) also predicts a probability that in any Hubble volume there occur data D_0 that describe our observational situation (but not including any record we might have of the value of Λ). These probabilities depend on Λ. The complete set of histories is thus labeled by the backgrounds and which Hubble volumes in them are occupied by D_0.

Assume that we are typical of the incidences of D_0 in any one history. The first-person probability that we observe a value of Λ (WOΛ) is the third-person probability for the history Λ conditioned on the existence of at least one instance of D_0 [cf. (3) *et seq.*]. Using the Bayes identity we have

$$p^{(1p)}(WO\Lambda) \equiv p(\Lambda|D_0^{\geq 1}) = \frac{p(D_0^{\geq 1}|\Lambda)p(\Lambda)}{\sum_\Lambda p(D_0^{\geq 1}|\Lambda)p(\Lambda)}. \tag{8}$$

For values of Λ for which $p(D_0^{\geq 1}|\Lambda)$ is negligibly small, the probability that we observe that value will also be negligibly small. Hence, anthropic selection of values of Λ that are consistent with observers is automatic. This kind of argument assumes that the third-person theory allows Λ to vary over an anthropically allowed range — a range consistent with the evolution of D_0. We give an example of such a model in Sec. 8. If, however, the theory determines a unique value of Λ, then either that must be consistent with $D_0^{\geq 1}$ or the theory is incorrect (e.g. Ref. 21).

Barrow and Tipler,[16] and Weinberg[17] have argued that the observed value of the cosmological constant could not be much larger than $\Lambda \sim 10^{-122}$ in Planck units, not far from its observed value. Were Λ larger, the universe would expand too rapidly for galaxies to have formed by the present age $t_0 \sim 14$ Gyr, and human observers would not be here. In our scheme, the probability $p(D_0^{\geq 1}|\Lambda)$ would be near zero.

This kind of argument is an example of what one could call traditional anthropic reasoning. In this, the anthropically allowed range of values of an observable like Λ is determined from classical arguments like those involving galaxy formation mentioned above. It is then assumed that there is some unknown mechanism for Λ to vary over this range. In the absence of a specific model of this mechanism, a uniform distribution over the range is often assumed and the most probable Λ predicted on the basis of purely anthropic arguments. Impressively detailed calculations were carried out in this way with mixed results, see e.g. Ref. 19. But this program suffers from several uncertainties. For example, different results are obtained if different combinations of constants were assumed to vary.[20] In addition, traditional anthropic reasoning is not part of any theoretical framework. Rather, anthropic selection arises from an additional assumption or "principle".

The chief difference with anthropic selection in quantum cosmology is that: (1) There the theory (I, Ψ) provides a mechanism for what constants vary and how they vary, (2) the observer is a physical subsystem with a certain probability predicted by the theory to evolve in any Hubble volume and (3) anthropic selection emerges automatically as a property of the first-person probabilities by which we use and test the theory. To summarize, anthropic selection in quantum cosmology:

- Is not a choice.
- Does not require invoking some "anthropic principle".
- Does not change the objective nature of the underlying third-person theory.
- Does require a typicality assumption if there is a chance we are replicated.

The observer is important for first-person probabilities for observations through anthropic selection. In Sec. 8.3, we will describe a more precise calculation of $p^{(1p)}(WO\Lambda)$ in which the probabilities $p(\Lambda)$ in (8) are obtained from a concrete model of the universe's quantum state Ψ in combination with a dynamical theory I in which Λ can vary.

6. A Remark on Coarse-Graining

Our observations of the universe extend at most over a Hubble volume. But this may only be a tiny region in a vastly larger inflationary universe of the kind contemplated in contemporary cosmology. Indeed, some calculations[24] suggest that the universe typically becomes spatially infinite as a consequence of eternal inflation. Much of the third-person information about what occurs in the universe on very large scales is irrelevant for the first-person predictions for our observations in our Hubble volume, and perhaps not even well defined. In this section we illustrate how, in certain models, first-person probabilities for our observations can be calculated directly using coarse-grainings that ignore most of the structure outside our Hubble volume. Coarse-graining is not an *ad hoc* assumption. It is central and inevitable in quantum mechanics, statistical physics, complexity, and many other areas of science (e.g. Ref. 28).

We illustrate how coarse-graining works with a simple box model of the kind used in Sec. 4. Consider a universe with an infinite set of boxes as illustrated in Fig. 2. Each box has a color. This is either yellow with probability p_Y, or green with probability $p_G = 1 - p_Y$. There is a probability p_E for an observer to be in any box observing its color. Thus, all the boxes are statistically identical. In this sense, this universe has a discrete translation symmetry.

Now we ask for the first-person probability that we observe yellow. The answer is obvious. We are in one box. All the boxes are

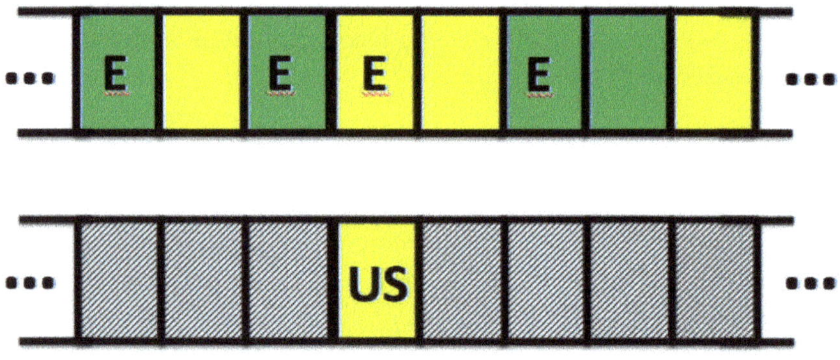

Fig. 2. Fine- and coarse-grained histories of a box model discussed in the text. The boxes model Hubble volumes. Their color models an observable like the CMB, in this case, either yellow or green. An "E" means that an observer is in the box observing its color. The top history is fine-grained with a color and either E or no E in every box. The bottom history is coarse-grained. The possibilities have been summed over for every box except one — our box. That enables a straightforward calculation of the first-person probability that we see one color or the other. The details outside our box are irrelevant for this.

statistically the same. The first-person probability $p^{(1p)}(Y)$ for observing yellow is the same as the probability that any of the boxes is yellow, viz.

$$p^{(1p)}(Y) = p_Y. \tag{9}$$

Although obvious, it is instructive to see how this result follows from our general framework through coarse-graining starting from third-person probabilities. A fine-grained history would specify the color (Y, G) of each box, and whether there exists an observer in it or not $(E, \bar{E})$ (top in Fig. 2). The third-person probability for one particular history having n_Y yellow boxes, n_G green boxes, n_E boxes with observers, and $n_{\bar{E}}$ without, is

$$(p_Y)^{n_Y}(1-p_Y)^{n_G}(p_E)^{n_E}(1-p_E)^{n_{\bar{E}}}. \tag{10}$$

But these probabilities tend to zero in a universe with an infinite number of boxes. The probabilities for these fine-grained histories are not well defined.

Finite probabilities can be obtained by coarse-graining. That is, they can be obtained by summing (10) over what is irrelevant for our observations. To get first-person probabilities for our observations, we can coarse-grain over the alternatives in every box but ours. That means to sum the probabilities (10) over the alternatives (Y, G) and $(E, \bar{E})$, giving a factor of unity for every box but ours (bottom in Fig. 2). That gives the first-person probabilities.

This result generalizes straightforwardly to any finite number of kinds (colors) of Hubble volumes, to models where the probabilities p_E depend on the color, and to more than one configuration (history) of boxes with third-person probabilities, like the $p(k)$ of Sec. 4.1. In Sec. 8, we review an example of coarse-graining used in realistic cosmology (see also Ref. 9).

To summarize, in infinite or just very large universes, focusing on our observations in our Hubble volume motivates coarse-grainings that directly lead to well defined first-person probabilities for observations. It is an intriguing open question whether such a local framework for prediction can be achieved more generally in quantum cosmology. Such a framework would truly return the observer to importance.

7. Inflation in Quantum Cosmology

We now turn from illustrative but artificial toy models to more realistic cosmological models. In this and the next section, we consider two examples in which the return of the observer is important — where the probabilities for what we observe are significantly different from the probabilities for what occurs.

The two examples share a common theoretical framework (I, Ψ). For the dynamics, we assume a spatially closed cosmological spacetime metric g coupled to a number of scalar fields $\vec{\phi}$. The dynamics are specified by an (Euclidean) action $I[g, \vec{\phi}]$ consisting of the action for general relativity plus the action for the scalar fields $\vec{\phi}$ coupled to the metric g. For the state Ψ we assume the no-boundary wave function of the universe (NBWF).[10] This is the natural analog of the notion of "ground state" for closed cosmologies.

Many of our large scale observations are of properties of our universe's classical history. The rate of the universe's expansion and the distribution of galaxies in our Hubble volume are examples. A history of the universe behaves classically when the quantum probability is high that it exhibits correlations in time governed by the Lorentzian Einstein equation and the classical field equations. The NBWF predicts an ensemble of alternative classical histories along with third-person probabilities for which history in the ensemble occurs. For a little more on what the NBWF is, and how it predicts probabilities, see Appendix B. For much more, see Ref. 6.

The first example is concerned with the classical cosmological histories predicted by the NBWF when the matter consists of a single scalar field moving in a quadratic potential

$$V(\phi) = \frac{1}{2}m^2\phi^2 \tag{11}$$

with $m^2 \ll 1$ and zero cosmological constant. As in the rest of this paper, we are using Planck units where $\hbar = 8\pi G = c = 1$. Geometry and field are restricted to be homogeneous and isotropic, thus defining a minisuperspace model. Lorentz signatured homogeneous and isotropic spacetime geometries can be described by metrics of the form

$$ds^2 = -dt^2 + a^2(t)d\Omega_3^2. \qquad (12)$$

Here, $d\Omega_3^2$ is the metric on a unit round three-sphere. The time-dependence of the scale factor $a(t)$ describes how this closed universe expands and contracts. Standard closed FLRW cosmological models have metrics of this form that satisfy the Einstein equation. The homogeneous field is a function only of time, viz. $\phi = \phi(t)$. A quantum history of this model universe is therefore specified by $(a(t), \phi(t))$. Classical histories are described by a pair $(a(t), \phi(t))$ that obey the Einstein equation and the classical equations of motion for the field.

As sketched in Appendix B, the NBWF in this minisuperspace model predicts a one-parameter ensemble of possible classical histories. These classical histories can be labeled by a parameter ϕ_0 that can be roughly thought of as the value of the scalar field from which it starts to roll down to the bottom of the potential (11). By examining any one classical $a(t)$, we can find the number of efolds N_e of slow roll inflation it has. Remarkably, all histories in the NBWF classical ensemble turn out to have some inflation at early times.[6] Inflation and the emergence of a classical universe in the NBWF are therefore profoundly connected.[6,27]

Does our theory (I, Ψ) predict a significant probability for an extended period of inflation in the early universe? As stated, this question is ambiguous. It could mean "Are the third-person probabilities from (I, Ψ) high for classical histories with an early period of inflation?" But it could also mean "Are the first-person probabilities high that the classical history we observe has a significant period of early inflation?" We will display the answers to both questions in our model and find that they are significantly different.

To evaluate the first-person probabilities for the amount of inflation, we can assume that we are rare physical systems in any of the classical histories predicted by the NBWF. This is because the number of Hubble volumes in all histories of the classical ensemble with matter densities below the Planck density is much smaller than any realistic value of p_E^{-1}.[7] Hence, volume weighting (5) connects the first-person probabilities for our observations with the third-person probabilities for what occurs.

The third-person probabilities $p = p(\phi_0)$ for the histories predicted by the NBWF are given approximately by [(B.4)]

$$p(\phi_0) \propto \exp[3\pi/V(\phi_0)]. \tag{13}$$

It turns out in this model that roughly, $N_e \approx (3/2)\phi_0^2$ starting at $N_e \approx 1$. The third-person NBWF probabilities for classical histories (13) thus imply probabilities $p(N_e)$ for the number of efolds that occur, starting roughly at $N_e \approx 1$,

$$p(N_e) \propto \exp\left(\frac{9\pi}{m^2 N_e}\right). \tag{14}$$

Third-person probabilities are therefore much larger for a small number of efolds than for the minimal number ~ 60 required for agreement with observation, as illustrated in Fig. 3.

The situation is much different for first-person probabilities for the amount of inflation in *our* history — the one we observe. Since it is assumed that we are rare, the first-person probabilities are the volume weighted third-person probabilities as in (5). During a period of inflation, the volume of the universe expands by a factor $\exp(3N_e)$. The first-person probabilities are then approximately

$$p^{(1p)}(N_e) \propto \exp\left(3N_e + \frac{9\pi}{m^2 N_e}\right). \tag{15}$$

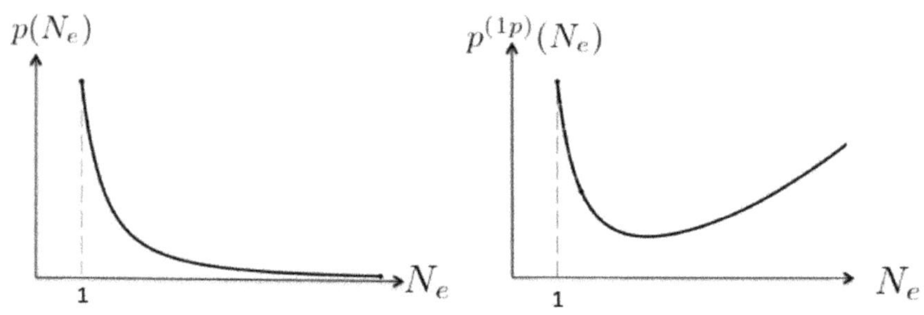

Fig. 3. Third- (left) and first-person (right) probabilities for the number of efolds N_e of matter-driven inflation in the early universe in the no-boundary quantum state Ψ. The first-person probabilities favor universes with a large amount of inflation because, in the larger universes that result from an extended period of early inflation, there are more Hubble volumes for us to be in.

Thus the NBWF predicts a significant probability for us to find ourselves in a universe with a large number of efolds (Fig. 3). This implies that there is a significant probability for us to observe the consequences of inflation — approximate spatial flatness, a scale invariant spectrum of density fluctuations, etc.

There is a higher probability for us to live in a larger universe because we are a physical subsystem within the universe with a very small probability p_E to have evolved in any Hubble volume. In the larger universes that result from an extended period of early inflation there are more Hubble volumes for us to be in. Thus the observer returns to importance in the probabilities for observing the consequences of inflation.

8. Eternal Inflation and Anthropic Selection

We now extend the above model in three crucial ways to arrive, at last, at a realistic model for the early universe:

(1) *Fluctuations:* We include fluctuations away from homogeneity and isotropy. These fluctuations provide the necessary degrees of freedom to describe, e.g. the pattern of temperature variations in the CMB. A consequence is that a range of classical histories exhibit eternal inflation, becoming spatially very large and highly inhomogeneous on the largest scales.
(2) *Landscape Potential:* We no longer assume one scalar field in a potential with a single minimum like (11). Rather, we assume many scalar fields in a multi-field potential that has many different minima with different directions of approach, as a toy model for the string landscape.[26] As a consequence, the dynamical theory provides a mechanism for the observable parameters of the histories to vary. Automatic anthropic selection can then be explicitly illustrated.
(3) *Not Rare but Common:* The above assumption that we are rare in all histories is no longer tenable with eternal inflation. The probability p_E for us to exist in any Hubble volume is very small. Nevertheless, we will be common in a range of histories where the universe becomes sufficiently large. In that range, the connection between third- and first-person probabilities will no longer involve volume weighting but instead be given by (6).

We now consider the meaning and implications of these extensions in more detail.

8.1. *Eternal Inflation — Histories where Observers are Common*

Cosmological perturbation theory extends the homogeneous and isotropic models of Sec. 7 straightforwardly to include linear fluctuations away from these symmetries for each classical history. The early period of slow roll inflation in the classical NBWF histories stretches and amplifies quantum vacuum fluctuations and generates a pattern of classical perturbations on scales larger than the horizon, which much later produce the small temperature fluctuations we observe in the microwave sky. However, it turns out that very long-wavelength fluctuations, those which leave the horizon and become classical at values of ϕ where

$$V^3 \geq |dV/d\phi|^2, \tag{16}$$

have a large expected amplitude. There is non-perturbative evidence that histories in which ϕ is initially in this regime reach very large (or even infinite) spatial volumes, because these large, very long-wavelength fluctuations tend to make them highly inhomogeneous on scales much larger than our Hubble volume.[22–24] This dynamical process is known as eternal inflation,[25] and (16) defines the regime of field values where eternal inflation occurs.

For quadratic potentials like (11), the condition for eternal inflation (16) is met for sufficiently large $\phi_0 > \phi_{\rm ei}$ where

$$\phi_{\rm ei} \sim 1/\sqrt{m}, \tag{17}$$

well below the Planck scale $\phi_{\rm pl} \sim 1/m$ for realistic values $m \sim 10^{-5}$.

Therefore, when we include fluctuations, the set of classical NBWF histories divides into two parts: Those with $\phi_0 \gtrsim \phi_{\rm ei}$ which are inhomogeneous on the largest scales and have a great many Hubble volumes N due to eternal inflation, and those with $\phi_0 \lesssim \phi_{\rm ei}$ which are much smaller and approximately homogeneous.

This division has significant implications for the first-person probabilities for observation. We are now effectively in the box model case

(7) where (assuming typicality)

$$[1-(1-p_E)^N] \approx 1, \quad \phi_0 > \phi_{\rm ei}, \tag{18a}$$

$$[1-(1-p_E)^N] \approx p_E N \ll 1, \quad \phi_0 < \phi_{\rm ei}. \tag{18b}$$

Eternally inflating histories are thus strongly selected, whereas histories with slow roll inflation only are strongly suppressed by the very small value of p_E. In selecting for eternally inflating universes as the ones we observe, the observer has returned in force.

8.2. *Landscapes — A Mechanism for Cosmological Parameters to Vary*

As discussed in Sec. 5, the anthropic selection of observed cosmological parameters requires a third-person theory (I, Ψ) that allows the parameters to vary. Theories with landscape potentials are a very simple example.

To illustrate what landscape potentials are, consider a third-person theory with two scalar fields ϕ_1 and ϕ_2 moving in a potential $V(\phi_1, \phi_2)$. A three-dimensional plot of this potential could be made using ϕ_1 and ϕ_2 as the x- and y-axes and plotting V along the z-axis. The plot might resemble a mountainous landscape on Earth, hence the name "landscape potential".

Suppose the potential has a number of different minima, each surrounded by a number of different valleys leading to it.[c] In our past history, the fields "rolled down" a particular valley ("our" valley) to a particular minimum ("our" minimum). The value of the potential at our minimum is the value of the cosmological constant we would observe. The shape of our valley near our minimum determines the spectrum of density fluctuations in the CMB we would see. A third-person theory that predicts probabilities for which of the possible histories occurs is thus a starting point for calculating the first-person probabilities for the values of these parameters we observe.

The notion of a landscape potential is easily extended to many scalar fields $\vec{\phi}$ so that $V = V(\vec{\phi})$. Assume that each minimum in $V(\vec{\phi})$ is surrounded by effectively one-dimensional valleys. Suppose

---

[c]Valleys were called "channels" in Ref. 9 and other places.

that these valleys are separated by large barriers so that transitions between valleys are negligible. Then we have, in effect, an ensemble of one-dimensional potentials $V_K(\phi_K)$, $K = 1, 2, \cdots$ — one for each valley.

The NBWF predicts the probabilities for classical inflationary histories for each one-dimensional potential in this landscape. The total classical ensemble is the union of all these. An individual history can therefore be labeled by (K, ϕ_{0K}) where, as explained before, ϕ_{0K} is roughly the value of the field ϕ_K at the start of its roll down the potential V_K. Thus, the NBWF predicts third-person probabilities $p(K, \phi_{0K})$ that our universe rolled down the potential from ϕ_{0K} in the valley K to its minimum.

Generically, a landscape of the kind under discussion contains some valleys K, where the potential $V_K(\phi_K)$ has a regime of eternal inflation. Equation (18) implies that histories in these valleys will dominate the first-person probabilities. In the presence of eternal inflation, the first-person probabilities are

$$p^{(1p)}(K, \phi_{K0}) \approx p(K, \phi_{K0}) \approx \exp[3\pi/V_K(\phi_{K0})], \quad \phi_{k0} \gtrsim \phi_{K\mathrm{ei}}$$
$$\approx 0, \quad \phi_{k0} \lesssim \phi_{K\mathrm{ei}}. \tag{19}$$

When the potentials are increasing with field, the most probable NBWF history in a given valley will be that where the field starts around the exit of eternal inflation, i.e. $\phi_{K0} = \phi_{K\mathrm{ei}}$. Thus, to good approximation, the probability that we rolled down in valley K is

$$p^{(1p)}(K) \approx p(K) \approx \exp[3\pi/V_K(\phi_{K\mathrm{ei}})]. \tag{20}$$

We now apply this result in a concrete model landscape.

8.3. *First-Person Predictions of the No-Boundary State in a Landscape Model*

We now discuss an example of a first-person prediction in a landscape model with valleys where the eternal inflation condition (16) is satisfied. This example relates directly to the historical effort described briefly in Sec. 5 to determine the anthropically allowed ranges of cosmological parameters that are consistent with our existence as observers (see, e.g. Refs. 16, 17, 19 and 20).

Specifically, we calculate the NBWF predictions for the first-person probability of a correlation between three observed numbers: First is the observed value of the cosmological constant $\Lambda \sim 10^{-123}$ (in Planck units). Second is the observed value of the amplitude of primordial density fluctuations $Q \sim 10^{-5}$ in the CMB. And third is the part of our data $D_{\rm loc}$ on the scales of our galaxy and nearby ones which, together with Λ, determine the present age of the universe $t_0(\Lambda, D_{\rm loc}) \sim 14$ Gyr. Thus, we are interested in $p^{(1p)}(\Lambda, Q, D_{\rm loc})$ — the first-person probability for this correlation to be observed. Many other such correlations could be investigated to test a given theory.[d] But this example will nicely illustrate several implications of the previous discussion.

We consider a landscape potential in which the parameters Λ and Q vary, so the theory (I, Ψ) will predict probabilities for their values. Specifically, we assume a landscape of different one-dimensional valleys with potentials of the form[e]

$$V_K(\phi) = \Lambda_K + \frac{1}{2}m_K^2\phi_K^2, \quad K = 1, 2, \cdots . \tag{21}$$

A valley in this landscape is therefore specified by the values (Λ, m). (From now on we will drop the subscripts K to simplify the notation.)

The NBWF predicts an ensemble of inflationary universes in this landscape. The observed classical history of our universe rolled down one of the valleys — "our valley". The values of Λ and m that characterize our valley can be determined by observation. Measurement of the expansion history of the universe determines Λ and CMB measurements determine m, because the amplitude of the primordial temperature fluctuations is given by[8,15]

$$Q \approx N_*^2 m^2. \tag{22}$$

Here $N_* \sim \mathcal{O}(60)$ is the number of efolds before the end of inflation such that the COBE scale left the horizon during inflation. Observations indicate that $Q \sim 10^{-5}$.

[d]For a recent discussion of observables associated with features of the CMB fluctuations, see e.g. Ref. 15.
[e]For examples of more general landscapes, see Ref. 27.

The first-person probability that we observe specific values of Λ and Q, is the first-person probability that our past history rolled down the valley which has those values. Since each of the valleys (21) has a regime of eternal inflation for $\phi > \phi_{\rm ei} \sim 1/\sqrt{m}$, the first-person probabilities (20) that we observe values of Λ and Q can be written as

$$p^{(1p)}(\Lambda, Q) \approx p(\Lambda, Q) \propto \exp[3\pi/(\Lambda + c\sqrt{Q}/N_*)] \tag{23}$$

where $c \sim \mathcal{O}(1)$. For fixed Λ this distribution favors small Q. As mentioned above, we are interested in the joint probability $p^{(1p)}(\Lambda, Q, D_{\rm loc}) \approx p(\Lambda, Q, D_{\rm loc})$ (first equals third in a regime of eternal inflation). The latter can be rewritten[f]

$$p(\Lambda, Q, D_{\rm loc}) = p(D_{\rm loc}|\Lambda, Q)p(\Lambda, Q). \tag{24}$$

The second factor in the expression above is given by (23). The first will be proportional to the number of habitable galaxies that have formed in our Hubble volume by the present age $t_0(\Lambda, D_{\rm loc})$. This is plausibly proportional to the fraction of baryons in the form of galaxies by the present age t_0, assuming that is bigger than the collapse time to form a galaxy. Denoting this by $f(\Lambda, Q, t_0)$, we have from (23)

$$\begin{aligned} &p^{(1p)}(\Lambda, Q, D_{\rm loc}) \\ &\qquad \approx p(\Lambda, Q, D_{\rm loc}) \propto f(\Lambda, Q, t_0) \exp[3\pi/(\Lambda + c\sqrt{Q}/N_*)]. \end{aligned} \tag{25}$$

All the quantities in (25) are determinable in one Hubble volume — ours. The rest of the calculation can therefore be carried out in that volume. We do not need to ask in which of the vast number of Hubble volumes in the third-person eternally inflating universe we dwell. In this model they are all statistically the same as far as the quantities in (25) go. We are effectively in the situation of the yellow-green box model in Sec. 6. We can coarse-grain over all other

[f]In traditional anthropic reasoning, the first factor in (24) is called the "selection probability" which can be calculated; the second is the "prior" which is assumed. See, e.g. Eq. (1) in Ref. 19. The "prior" is typically assumed to be uniform in the anthropically allowed range reflecting ignorance of part of the theory. Here both factors follow from the same theory (I, Ψ).

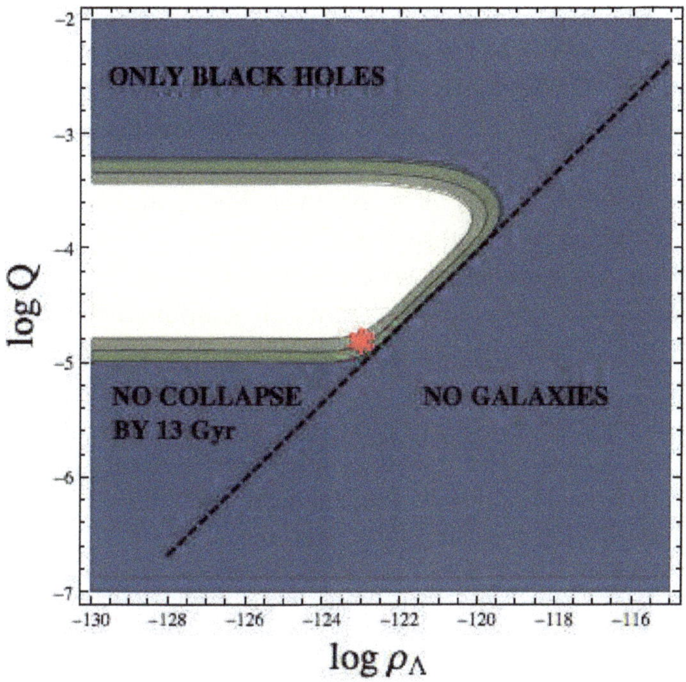

Fig. 4. A contour plot of the fraction $f(\Lambda, Q, D_{\rm loc})$ of baryons in the form of galaxies by the time $t_0 \sim 11$ Gyr, adapted from the calculation in Ref. 19. As discussed in the text, the fraction is negligible in the dark (blue) region either because gravitationally bound systems do not collapse or because they collapse to black holes.

Hubble volumes but ours. In this way, we are able to make contact — and use — calculations in traditional anthropic reasoning.

Figure 4 is a contour plot of f at $t_0 \sim 11$ Gyr — earlier than the present age 14 Gyr, because galaxies have been around for a while. This was adapted from the detailed astrophysical calculations in Ref. 19. For values of (Λ, Q) to the right of the diagonal dotted line, the universe accelerates too quickly for pre-galactic halos to collapse. Fluctuations have to be large enough to collapse into galaxies a bit before t_0 and produce $D_{\rm loc}$. That determines the bottom boundary. If the fluctuations are too large (top boundary) the bound systems are mostly large black holes inconsistent with $D_{\rm loc}$. The central region where $f \geq 0.6$ is the anthropically allowed region.

The cosmological constant Λ is negligible compared to Q in the anthropically allowed (white) region of Fig. 4. The exponential dependence $\exp[3\pi N_*/c\sqrt{Q}]$ implied by the NBWF means that the probabilities (25) are sharply confined to the smallest allowed values of Q consistent with galaxies by t_0, i.e. $Q \sim 10^{-5}$. The resulting marginal distribution for Λ is shown in Fig. 5 and is peaked about $\Lambda \sim 10^{-123}$ close to the observed value.

However, the agreement with observation is not the most important conclusion. The model is still too simplified for that. What is important is how the example illustrates the previous discussion. Specifically, how anthropic reasoning emerges automatically in quantum cosmology, how it can be sharpened by a theory of the universe's quantum state, how first-person probabilities select for large eternally inflating universes, and how what is most probable to be observed is not necessarily what is most probable to occur.

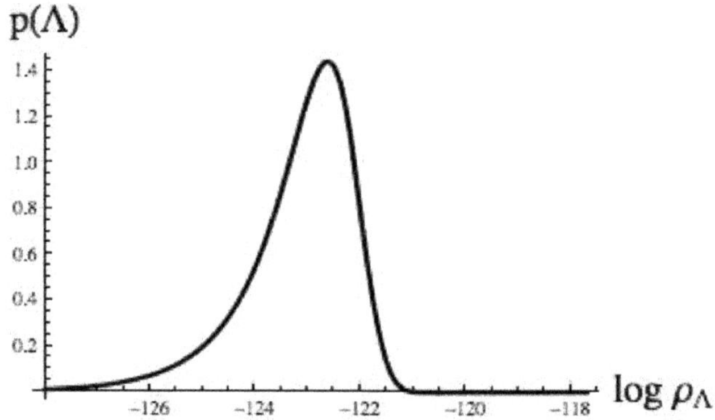

Fig. 5. The marginal distribution for the cosmological constant Λ obtained by integrating (25) over Q.

9. Conclusion

Fundamental physical theories have generally been organized into a third-person theory of what occurs and a first-person theory of what we observe. In quantum mechanics, the output of the two parts are third-person probabilities and first-person probabilities.

The character of these two parts, their relative importance, and the relation between them has changed over time as new data on new scales of observation required new theory and as the understanding of our observational situation in the universe evolved. The history of the transitions from classical physics, to Copenhagen quantum mechanics, and then to the quantum mechanics of closed systems that was briefly sketched in the Introduction, is part of this evolution.

As we discussed in this paper, modern cosmology implies new requirements on both the third-person theory of what occurs in the universe and the first-person theory of what we observe of the universe.

A third-person theory of the whole universe requires a quantum mechanics of closed systems including quantum spacetime, as in Ref. 4, for example. It requires a quantum mechanics that predicts probabilities for what happened in the early universe when there were no observers around and no measurements being made. It requires a quantum mechanics that predicts probabilities for the emergence of classical spacetime in a quantum theory of gravity. It requires a quantum mechanics that can explain the origin of the rest of classical predictability in distant realms of the universe that we will never visit.

Quantum cosmology also has implications for first-person theory. As observers, individually and collectively, we are physical systems within the universe with only a probability to have evolved in any one Hubble volume, and, a probability to be replicated in many, if the universe is very large. What we have seen in this essay is that this implies that the observer returns to importance in at least the following ways:

- By generally showing that what is most probable to be observed is not necessarily the most probable to occur.
- By favoring larger universes over small ones if we are rare or if we are common, and thus favoring observations of a significant amount of inflation.
- By making anthropic selection automatic. We will not observe what is where we cannot exist.
- By requiring the addition of an assumption of typicality to the

theory of what occurs, made concrete by the xerographic distribution — an addition that can be tested and used to improve prediction.

- By leading to an understanding of how to compute probabilities for our observations of fundamental constants in a landscape that allows them to vary.

This list is a brief summary of the results of this essay, but these lead naturally to a number of questions that we discuss here.

A natural question, is why the observer was not important in classical cosmology when it is in quantum cosmology. An answer can be traced to differences in starting points and objectives. The starting point for classical cosmology was the assumption that the universe had a single spacetime geometry. The goal was to infer the geometry of that spacetime from large-scale observation. Is it approximately homogeneous and isotropic on large scales, open or closed...? What are the values of the cosmological parameters that characterize it — the cosmological constant, the Hubble constant, the amount of radiation, the amount of baryons, the amplitude of the density fluctuations, etc.? Is there evidence that the spacetime had an early period of inflation? Observers were presumed to exist but had a negligible influence on the answers to these questions.

Quantum cosmology does not start by assuming classical spacetime. Rather, it starts from a theory of the universe's quantum state and dynamics. From that it seeks to explain when spacetime is classical and predict probabilities for what the different possible classical spacetimes there are — questions that cannot even be asked in the context of classical cosmology. It therefore answers to the classical questions with probabilities about large scale geometry.

A second natural question is *why* do fundamental physical theories encompass both a third-person theory of what occurs and a first-person theory of what is observed? The historical success of theories organized in this way is indisputable. That success tells us something about the world. It supports the idea of some form of realism, perhaps along the lines of what Putnam called "realism with a human face".[30] To paraphrase J. A. Wheeler, "In a quantum world, the universe is a grand synthesis, putting itself together all the time as a whole. Its history is ... a totality which includes us and in which what happens now gives reality to what happened then".[29]

A third natural question concerns the scale on which we have to know something of the universe to make first-person predictions. An IGUS[g] like us is confined to a very local region inside of a vast Hubble volume, which itself is typically but a small part of a much larger universe. Yet the present formulation of the first-person theory requires information beyond our Hubble volume to determine whether we are rare, or common, or other. It remains to be seen whether a more local computation of probabilities for observation in quantum cosmology can be found. That would be a further way that the observer and universe are unified.

Everett's insight was that, as observers of the universe, we are physical systems within it, not outside it. We are subject to the laws of quantum mechanics but play no special role in its formulation. We are negligible perturbations on a third-person description of the universe. But, as shown in this essay in several different ways, we return to importance for calculating first-person probabilities for our observations precisely because we are physical systems in the universe. We may have only begun to appreciate the ramifications of that insight.

Acknowledgments

The authors have had the benefit of collaboration and discussions with a great many scientists in this work. The cited papers have those acknowledgments. However, our collaborators on the papers that underlie this work should be thanked. They are Murray Gell-Mann, Stephen Hawking, and Mark Srednicki. Gregory Benford suggested the title. Comments of Don Page and Simon Saunders were useful. The work of JH was supported in part by the US NSF grant PHY12-05500. The work of TH was supported in part by the National Science Foundation of Belgium (FWO) grant G.001.12 Odysseus and by the European Research Council grant no. ERC-2013-CoG 616732 HoloQosmos.

[g]Information gathering and utilizing system.

Appendix A

Typicality — The Xerographic Distribution

Consider, for a moment, a third-person theory — classical or quantum — that predicts a large number of Hubble volumes, some with one kind of observable property, some with another. A box model like those in Sec. 4 with one history and different colored volumes would be a very simple example.

Suppose that the data D_0 that describe our observational situation (including us) occur in many different Hubble volumes. One of them is ours, but in other volumes, the result of the observation specified by D_0 could be different. A third-person theory (I, Ψ) would predict what is observed in all of the instances of D_0. But it does not predict which one is ours. Indeed, the theory has no notion of "we" or "us".

We do not know which of these copies of D_0 are us. It could be any one of them. To make first-person predictions for our observations, the third-person theory must therefore be augmented by an assumption about which instance of D_0 is us, or more generally with a probability distribution on the set of copies. If there is no such assumption, there are no predictions. Put differently, predictions for our observations require a statement on what exactly we mean by "us" — a specification of how we think our particular situation relates to other instances of D_0 in the universe.

In the body of this essay, we have consistently assumed that we are equally likely to be any of the incidences of our data D_0 that the third-person theory (I, Ψ) predicts. This is the simplest and least informative assumption, but it is not the only possible one. Other, more informative assumptions may lead to better agreement between theory and observation, be more justified by fact, and be more predictive.[h]

We call a distribution that gives the probability that we are any of the incidences of D_0 or a subset of it, a *xerographic distribution.*[11]

---

[h]As for the Boltzmann brain problem.[12] Boltzmann brains are not a problem if we assume that our observations are not typical of deluded observers who only imagine that they have the data D_0.

It is usually written ξ_i, where the index ranges over all incidences of D_0. A xerographic distribution is effectively a formal expression of the assumption about how typical we are in the universe in the set of all other incidences of D_0.

It is the theoretical structure consisting of (I, Ψ) and ξ that yields first-person predictions for observation. Each of the elements in this combination is therefore testable by experiment and observation. Just as we can compete different (I, Ψ) by their predictions for observation, we can also compete different ξ.

Appendix B

The No-Boundary Wave Function in Homogeneous and Isotropic Minisuperspace

The example in Sec. 7 assumed the no-boundary wave function (NBWF) for the quantum state Ψ in the third-person theory (I, Ψ). The NBWF is the analog of the ground state for closed cosmologies, and therefore a natural candidate for the wave function of our universe. Its predictions for observations are in good agreement with observation. For example, it predicts that fluctuations away from homogeneity and isotropy start out near the big bang in their ground state.[32] Combined with its prediction that our universe underwent an early period of inflation,[5] that leads to good agreement with the observed fluctuations in the CMB. This Appendix provides the reader a little more explanation of what that wave function is and how its consequences used in the examples are derived. For more details, see Ref. 10.

A quantum state of the universe like the NBWF is represented by a wave function on a configuration space of three geometries and matter field configurations on a spacelike surface Σ. On the minisuperspace of homogeneous and isotropic geometries (12) and a single homogeneous scalar field, we write

$$\Psi = \Psi(b, \chi). \tag{B.1}$$

Here, b is the scale factor of the homogeneous, isotropic metric (12) on Σ and χ is the value of the homogeneous scalar field. The no-boundary wave function is a particular wave function of this form.

The NBWF is formally defined by a sum over homogeneous and isotropic Euclidean geometries that are regular on a topological four-disk and match b and χ on its boundary,[10] weighted by $\exp(-I)$ where I is the Euclidean action of the configurations. We will not need to go into this construction. We will only need its leading order in $\hbar$ semiclassical approximation. That is given by the saddle points (extrema) of the action $I[a(\tau), \phi(\tau)]$ on this disk for metric coupled to scalar field. There is one dominant saddle point for each (b, χ). Generally these saddle points will have complex values of both metric and field. The semiclassical NBWF is a sum over saddle point terms of the form (in units where $\hbar = 1$)

$$\Psi(b, \chi) \propto \exp[-I(b, \chi)] \propto \exp\left[-I_R(b, \chi) + iS(b, \chi)\right] \tag{B.2}$$

where $I(b, \chi)$ is the action evaluated at the saddle point and I_R and $-S$ are its real and imaginary parts.

The wave function (B.2) has a standard WKB semiclassical form. As in non-relativistic quantum mechanics, an ensemble of classical histories is predicted in regions of configuration space where S varies rapidly in comparison with I_R. The histories are the integral curves of S defined by solving the Hamilton–Jacobi relations relating the momenta p_b and p_χ conjugate to a and ϕ to the gradients of S

$$p_b = \nabla_b S, \qquad p_\chi = \nabla_\chi S. \tag{B.3}$$

There is a one-parameter family of classical histories — one for each saddle point. It is convenient to label them by the magnitude of the scalar field ϕ_0 at the center of the saddle point. One can think crudely of ϕ_0 as the value of the scalar field at which it starts to roll down to the bottom of the potential in a classical history.

The third-person probabilities for these histories are proportional to the absolute square of the wave function (B.2)

$$p(\phi_0) \propto \exp[-2I_R(\phi_0)] \approx \exp[-3\pi/V(\phi_0)]. \tag{B.4}$$

The last term is a crude approximation to the action which is useful in rough estimates.[18]

This is a very quick summary of a lot of work. For more details, see Ref. 6. In short, in the leading semiclassical approximation, the

NBWF predicts an ensemble of possible classical spacetimes obeying (B.3) with third-person probabilities (13).

References

1. S. W. Hawking and T. Hertog, Populating the landscape: A top down approach, *Phys. Rev. D* **73**, 123527 (2006), arXiv:hep-th/0602091.
2. S. Saunders, J. Barrett, A. Kent and D. Wallace (eds.) *Many Worlds?* (Oxford University Press, Oxford, 2010).
3. J. B. Hartle, The quantum mechanics of closed systems, in *Directions in General Relativity*, Vol. 1, eds. B.-L. Hu, M. P. Ryan and C. V. Vishveshwara (Cambridge University Press, Cambridge, 1993), arXiv:gr-qc/9210006.
4. J. B. Hartle, Spacetime quantum mechanics and the quantum mechanics of spacetime, in *Gravitation and Quantizations, Proc. 1992 Les Houches Summer School*, eds. B. Julia and J. Zinn-Justin (North Holland, Amsterdam, 1995) gr-qc/9304006.
5. J. B. Hartle, S. W. Hawking and T. Hertog, The no-boundary measure of the universe, *Phys. Rev. Lett.* **100**, 202301 (2008), arXiv:0711:4630.
6. J. B. Hartle, S. W. Hawking and T. Hertog, Classical universes of the no-boundary quantum state, *Phys. Rev. D* **77**, 123537 (2008), arXiv:0803.1663.
7. J. B. Hartle and T. Hertog, Replication regulates volume weighting in quantum cosmology, *Phys. Rev. D* **80**, 063531 (2009), arXiv:0906.0042.
8. J. B. Hartle, S. W. Hawking and T. Hertog, The no-boundary measure in the regime of eternal inflation, *Phys. Rev. D* **82**, 063510 (2010), arXiv:1001.0262.
9. J. B. Hartle, S. W. Hawking and T. Hertog, Local observation in eternal inflation, *Phys. Rev. Lett.* **106**, 141302 (2011), arXiv:1009.252.
10. J. B. Hartle and S. W. Hawking, The wave function of the universe, *Phys. Rev. D* **28**, 2960–2975 (1983).
11. M. Srednicki and J. B. Hartle, Science in a very large universe, *Phys. Rev. D* **81**, 123524 (2010), arXiv:0906.0042, a summary

of this work with different emphases is found in *The Xerographic Distribution: Scientific Reasoning in a Large Universe*, arXiv:1004.3816.
12. J. B. Hartle and M. Srednicki, Are we typical? *Phys. Rev. D* **75**, 123523 (2007), arXiv:0704.2630.
13. D. Page, Space for both no-boundary and tunneling quantum states of the universe, *Phys. Rev. D* **56**, 2065 (1997).
14. S. W. Hawking, *Volume Weighting in the No Boundary Proposal*, arXiv:0710.2029.
15. J. B. Hartle and T. Hertog, Anthropic bounds on lambda from the no-boundary quantum state, *Phys. Rev. D* **88**, 123516 (2013).
16. J. Barrow and F. Tipler, *The Anthropic Cosmological Principle* (Oxford University Press, Oxford, 1986).
17. S. Weinberg, Anthropic bound on the cosmological constant, *Phys. Rev. Lett.* **62**, 485 (1989).
18. G. W. Lyons, Complex solutions for the scalar field model of the universe, *Phys. Rev. D* **46**, 1546–1550 (1992).
19. M. Tegmark, A. Aguirre, M. J. Rees and F. Wilczek, Dimensionless constants, cosmology, and other dark matters, *Phys. Rev. D* **73**, 023505 (2006).
20. M. Livio and M. J. Rees, Anthropic reasoning, *Science* **309**, 1022 (2005).
21. J. B. Hartle, Anthropic reasoning and quantum cosmology, *AIP Conf. Proc.* **743**, 298 (2005), arXiv:gr-qc/0406104.
22. A. Starobinsky, Stochastic de Sitter (Inflationary) state in the early universe, in *Field Theory, Quantum Gravity and Strings*, eds. H. de Vega and N. Sanchez (Springer-Verlag, Berlin, 1986).
23. A. D. Linde, D. A. Linde and A. Mezhlumian, From the big bang theory to the theory of a stationary universe, *Phys. Rev. D* **49**, 1783 (1986).
24. P. Creminelli, S. Dubovsky, A. Nicolis, L. Senatore and M. Zaldarriaga, The phase transition to slow-roll eternal inflation, *JHEP* **0809**, 036 (2008), arXiv:0802.1067.

25. A. Vilenkin, Birth of inflationary universes, *Phys. Rev. D* **27**, 2848 (1983); A. D. Linde, Eternally existing self-reproducing chaotic inflationary universe, *Phys. Lett. B* **175**, 395 (1986).
26. L. Susskind, The anthropic landscape of string theory, in *Universe or Multiverse?* ed. B. Carr (Cambridge University Press, Cambridge, 2007), pp. 247–266.
27. T. Hertog, Predicting a prior for Planck, *JCAP* **02**, 043 (2014), arXiv:1305.6135.
28. M. Gell-Mann and J. B. Hartle, Quasiclassical coarse graining and thermodynamic entropy, *Phys. Rev. A* **76**, 022104 (2007), arXiv:quant-ph/0609190.
29. J. A. Wheeler and K. Ford, *Geons, Black Holes and Quantum Foam* (Norton & Company, New York, 1998).
30. H. Putnam, *Realism with a Human Face* (Harvard University Press, Cambridge, 1990).
31. H. Everett III, Relative state formulation of quantum mechanics, *Rev. Mod. Phys.* **29**, 454 (1957).
32. J. J. Halliwell and S. W. Hawking, Origin of structure in the universe, *Phys. Rev. D* **31**, 1777 (1985).

CHAPTER 20

Quantum Multiverses

A quantum theory of the universe consists of a theory of its quantum dynamics (H) and a theory of its quantum state (Ψ). The theory (H, Ψ) predicts quantum multiverses in the form of decoherent sets of alternative histories describing the evolution of the universe's spacetime geometry and matter content. A small part of one of these histories is observed by us. These consequences follow: (a) The universe generally exhibits different quantum multiverses at different levels and kinds of coarse-graining. (b) Quantum multiverses are not a choice or an assumption, but are consequences of (H, Ψ) or not. (c) Quantum multiverses are generic for simple (H, Ψ). (d) Anthropic selection is automatic because observers are physical systems within the universe, not somehow outside it. (e) Quantum multiverses can provide different mechanisms for the variation of constants in effective theories (like the cosmological constant) enabling anthropic selection. (f) Different levels of coarse-grained multiverses provide different routes to calculation as

(*continued overleaf*)

arXiv:1801.08631

a consequence of decoherence. We support these conclusions by analyzing the quantum multiverses of a variety of quantum cosmological models aimed at the prediction of observable properties of our universe. In particular, we show how the example of a multiverse consisting of a vast classical spacetime containing many pocket universes having different values of the fundamental constants, arises automatically as part of a quantum multiverse describing an eternally inflating false vacuum that decays by the quantum nucleation of true vacuum bubbles. In an FAQ section in Appendix A, we argue that the quantum multiverses of the universe are scientific, real, testable, falsifiable, and similar to those in other areas of science, even if they are not directly observable on arbitrarily large scales.

1. Introduction

The universe may present us with an ensemble of alternative possible situations, only one of which is observed by us. In this paper, we call such an ensemble a "multiverse".

A much discussed classical example of a multiverse is a single, vast cosmological spacetime containing "pocket universes" at different locations. Physics inside different pockets is assumed to be governed by different low energy effective theories. For example, the effective theories could differ in the value of the cosmological constant Λ. This is a multiverse of pockets with different values of Λ, only one, which is observed by us — the one we live in. For short, we call this a *pocket multiverse.* Only small values $\Lambda \lesssim 10^{-120}$ are consistent with the rest of our cosmological data, including a description of us as physical systems within the universe and the formation of galaxies by the present age (e.g. Refs. 1–3). Had we not yet measured Λ, we would predict that our pocket has a value in this small range.[a] This is a simple example of *anthropic selection.* We will not observe properties of a universe where we cannot exist. We will discuss this essentially classical example in a quantum-mechanical context in Sec. 5.

Quantum theories of a closed system like the universe provide multiverses in the form of decoherent sets of alternative coarse-grained

[a]Some might restrict the term "multiverse" to just this example, but in our opinion there is clarity and simplicity in our more general definition.

histories of the universe.[b] Such sets are quantum multiverses in the sense of the first sentence in this paper — an ensemble of possible histories, only one of which is observed by us. Quantum mechanics predicts probabilities for which of the individual members of the ensemble of histories happens, starting from theories of the universe's dynamics (H) and quantum state (Ψ). When these probabilities are conditioned on a description of our observational situation (including us), we get probabilities for what we observe of the universe.

Quantum multiverses have been discussed extensively in connection with the fundamentals of quantum mechanics.[c] This essay is devoted to discussing quantum multiverses in the context of quantum cosmology. We will illustrate the idea in simple, concrete, calculable models based on the author's joint work with Stephen Hawking and Thomas Hertog, calculating the predictions for observations of the no-boundary wave function proposal[7] for Ψ (mainly Refs. 3, 8–10). We shall show how predicting the results of our observations is enabled by quantum mechanics. In particular, we shall show how a pocket multiverse emerges from an appropriate (H, Ψ).

These models support the following conclusions about quantum multiverses in general:

(a) Many Quantum Multiverses. The theory (H, Ψ) does not predict just one multiverse of alternative histories. It predicts many different multiverses at different levels and kinds of coarse-graining. All are available for prediction.

(b) Quantum Multiverses are Not a Choice. Quantum multiverses are not a choice or an assumption separate from the theory (H, Ψ). Rather they follow or do not follow from (H, Ψ).

(c) Quantum Multiverses are Generic. Simple, manageable, discoverable theories (H, Ψ) generically predict quantum multiverses with many histories. To predict just one history with certainty, all of the present complexity would have to be encoded (H, Ψ). Rather, we

---

[b]In many other works, we have called decoherent sets of alternative coarse-grained histories, "realms". In this paper we use "multiverse".
[c]See, for example the discussions in Refs. 4–6.

expect that the present complexity arose not just from (H, Ψ), but through a multiverse of histories that describe frozen accidents that occurred over the course of the universe's history — chance events that could have gone one way or the other, for which the consequences of the way it did go proliferated. The accidents of biological evolution are an example.

(d) Anthropic Selection is Automatic. Anthropic selection is an automatic consequence of quantum-mechanical probabilities for observations. This is because observers are physical systems within the universe and not somehow outside it. Probabilities for our observations are conditioned on a description of our observational situation, and we won't observe what is where we cannot exist. Anthropic reasoning does not rely on some anthropic principle and is not a choice to be made or not made. It is an automatic consequence of calculating probabilities for observations.

(e) Quantum Multiverses Provide Several Mechanisms for the Constants of Effective Theories to Vary. The pocket universe considered above concerns one history, with Λ varying from place to place. But there are (H, Ψ) leading to multiverses of spatially homogeneous histories with Λ the same at all places in each history, but differing from history to history, hence also enabling anthropic selection. We will provide examples of both mechanisms in Secs. 5 and 7.

(f) Two Routes to Coarse-Graining. Quantum multiverses are restricted to sets of alternative coarse-grained histories that are decoherent — that is, restricted to sets that have negligible quantum interference between the individual histories in the set as a consequence of (H, Ψ). This ensures that the probabilities are consistent with the rules of probability theory.[11] Further coarse-graining can be carried out either by summing probabilities or by summing quantum amplitudes. That can be a considerable computational advantage, as we illustrate by example in Sec. 5.

Understanding multiverses as decoherent sets of alternative coarse-grained histories of a quantum universe can help address some of the concerns and objections, that have been raised about multiverses. Multiverses arise naturally and inevitably in quantum theory, as well as in many other areas of science. The author's views on some of the concerns are given in the FAQ in Appendix A.

The paper is structured as follows: Sec. 2 uses two-slit model closed systems to describe the decoherent (or consistent) histories formulation of quantum mechanics (DH) which we use throughout. Sections 3–8 present the five model quantum cosmologies and their quantum multiverses on which we base our conclusions above. Conclusions are in Sec. 9. The FAQ in Appendix A briefly addresses some of the concerns and objections that have been raised concerning multiverses. A further Appendix B gives more details about DH beyond those in Sec. 2.

2. Quantum Mechanics for the Universe Illustrated in Two-Slit Models

This section uses three models based on the two-slit experiment to illustrate how different quantum multiverses can arise in the same physical situation, illustrating points (a)–(d) of the Introduction. We assume the decoherent or consistent histories formulation of the quantum mechanics of a closed system such as the universe.[d] Decoherent histories quantum theory (DH) is logically consistent, consistent with experiment as far as is known, consistent with the rest of modern physics, such as special relativity and quantum field theory, general enough for histories, general enough for cosmology, and generalizable to include semiclassical quantum gravity. Copenhagen quantum mechanics is contained within DH as an approximation appropriate for measurement situations. DH can be thought of as an extension, clarification, and, to some extent, a completion of the program started by Everett.[13] DH may not be the only formulation of quantum mechanics with these properties, but it is the only one we have at present.

The basic ideas of DH that will be needed in this paper can be introduced with three model two-slit situations, each in a closed box — three very simple model universes. We will illustrate a variety of

[d]For classic references, see Ref. 11. For a tutorial see, e.g. Ref. 12.

quantum multiverses with these simple situations. The three models are illustrated in Fig. 1. We discuss these informally with a minimum of equations in this section. The same discussion with more equations can be found in Appendix B.

TSS: In the first model the box contains an electron gun on the left that emits an electron which moves through a screen with two slits $S = (U, L)$ to arrive at a further screen on the right in one of a set of position intervals $Y = (1, 2, 3 \cdots)$. We call this the "simple two-slit model (TSS)".
TSG: In the second model, in addition to the contents of TSS, there is a gas of particles near the slits. These scatter off the electron weakly enough not to disturb its motion but strongly enough to carry away phases. The gas particles constitute an *environment* for the electron in the sense discussed in Refs. 14–16 and many others. We call this the "two-slit with gas model (TSG)".
TSGO: In the third model the gas particles in TSG have an observable color — red near the upper slit and blue near the lower one. The electron is replaced by an observer moving through the slits — so the box must be very large! In these respects, the model is closer to the real universe, where observers are physical systems within the universe and not somehow outside. We call this the "two-slit with gas and observer model (TSGO)". Figure 2 is a more evocative image of this.

We stress that in all three examples the boxes are closed — like the universe. There are no observers outside the box looking at the inside, or measuring what goes on there, or otherwise meddling with the inside. Any observers are physical systems within a box, as in the TSGO model.

The inputs to the prediction of quantum multiverses are the box's Hamiltonian H and its quantum state $|\Psi(t)\rangle$. The latter is a function of time in the Schrödinger picture in which we work. The state can be defined by specifying a wave function of the electron at the initial time when it leaves the gun. This state evolves to later times by the Schrödinger equation as the electron moves through the box. Starting in this way, branch state vectors can be constructed for the individual members of sets of alternative histories of the

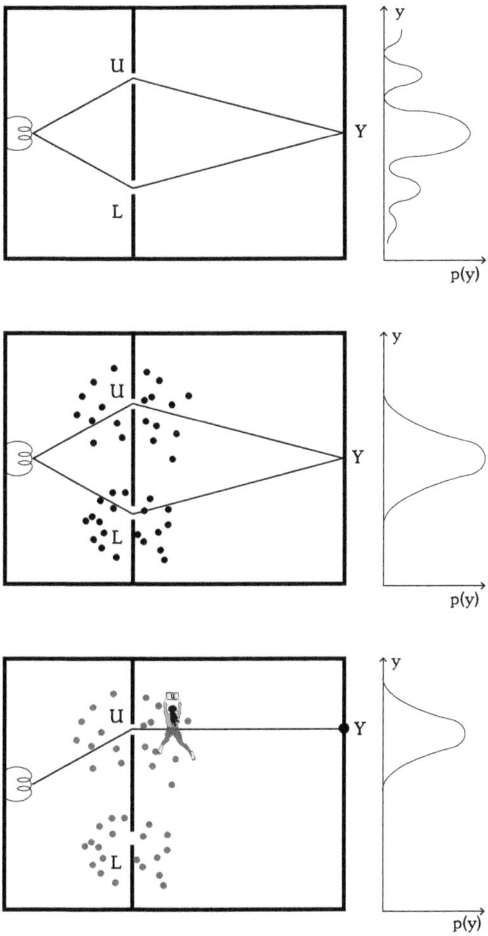

Fig. 1. Three idealized model closed systems based on the two-slit setup. The top model (TSS) is just the familiar two-slit example. The second model (TSG) includes a gas of particles in the vicinity of the slits that scatter weakly off the electron as it passes through. The third model (TSGO) considers a very large box of this kind with an observer who can pass through the slits (Fig. 2). The gas particles have a distinguishable color detectable by the observer — red near the upper slit and blue near the lower one. The probabilities for detection in an interval Y on the far screen is indicated schematically in the graphs on the right.

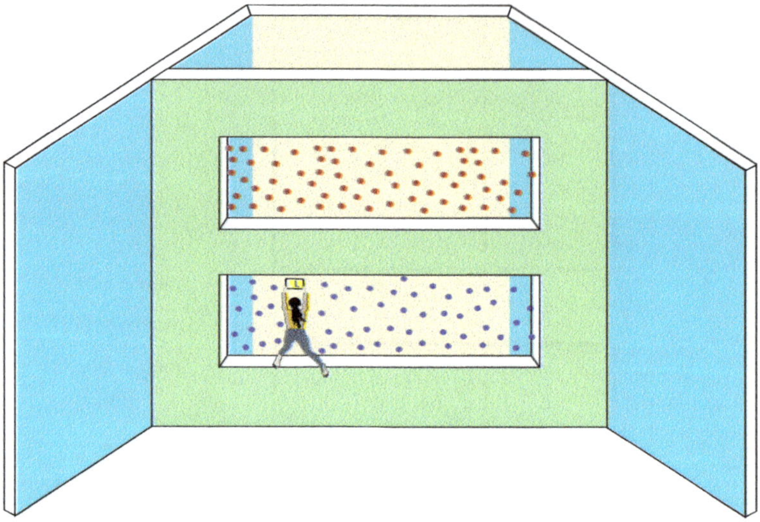

Fig. 2. A more evocative figure of an observer going through a two-slit experiment. The observer is passing through the lower slit and detecting particles of the blue gas which are in her vicinity.

electron's motion through the box. The set is decoherent when all the branches are approximately mutually orthogonal. Probabilities for histories are then the norms of the branch state vectors. The computation of these will be described qualitatively in this section; more mathematical details are described in Appendix B. We move back and forth between wave functions like (B.1) and bras and kets like $|\Psi(t)\rangle$ as convenient.

2.1. *Simple Two-slit Model (TSS)*

Two sets of coarse-grained histories are readily identified. There is the set of alternative histories defined by the position interval Y that the electron arrives at the screen, at time t_d. A finer-grained set of histories is defined by also specifying the slit S that the electron passes through on the way to Y. An even finer-grained set of alternative histories would specify the path the electron takes at each moment of time, but just the first two sets will be sufficient to illustrate what we need.

Do these two sets of histories define different quantum multiverses that TSS exhibits? Not yet! We still have to check that the sets are decoherent, that is, that the squares of their amplitudes give probabilities that are consistent with the usual rules of probability theory.

A branch state vector corresponds to each of the histories in a set of alternative ones. Consider, for example, the set where the histories are labeled only by the different values Y at the screen. The branch state vectors for this set are

$$|\Psi_Y(t_d)\rangle \equiv P_Y|\Psi(t_d)\rangle, \quad Y = 1, 2, 3, \cdots \tag{1}$$

where P_Y is the (Schrödinger picture) projection operator onto the range of positions corresponding to arrival at Y. The collection of projection operators for different Y obey

$$P_Y P_{Y'} = \delta_{YY'} P_Y, \quad \sum\nolimits_Y P_Y = I \tag{2}$$

capturing the idea that the electron must arrive at one or another of the intervals. As a consequence, the branch state vectors are orthogonal and sum to the state

$$\langle\Psi_Y(t_d)|\Psi_{Y'}(t_d)\rangle = 0, \; Y \neq Y', \qquad \sum\nolimits_Y |\Psi_Y\rangle = |\Psi\rangle. \tag{3}$$

The orthogonality means that there is vanishing quantum interference between the branch state vectors. This set of histories is therefore *decoherent*. The probabilities $p(Y)$ for arrival at an interval Y are

$$p(Y) = \||\Psi_Y(t_d)\rangle\|^2 = \||P_Y|\Psi(t_d)\rangle\|^2. \tag{4}$$

Now, let us consider the finer-grained set of histories which specify which slit the electron passes through in addition to where it arrives at the far screen. The branch state vectors for these histories can be constructed in the same way as in the above example. At the time t_s when the electron passes through the slits, define

$$|\Psi_S(t_s)\rangle \equiv P_S|\Psi(t_s)\rangle \tag{5}$$

where P_S is a projection on a region around slit S, either U or L. Evolve these vectors to the time t_d and project on the different regions Y to find the branch state vectors

$$|\Psi_{YS}(t_d)\rangle \equiv P_Y|\Psi_S(t_d)\rangle \tag{6}$$

for the histories that went through slit S and arrived at interval Y. Evidently

$$\sum\nolimits_S |\Psi_{YS}\rangle = |\Psi_Y\rangle. \tag{7}$$

Unlike the previous example, this set of histories does *not* decohere. Vanishing overlap does not follow from the orthogonality of the projections expressed in (2). Indeed, the quantum interference between these branches is directly responsible in the classic two-slit interference pattern.

It would be inconsistent to try to assign probabilities to this set of alternative histories. The probability to arrive at Y should be the sum of the probability to go through U and arrive at Y and the probability to go through L and arrive at Y. But in quantum mechanics, probabilities are squares of amplitudes and

$$\begin{aligned} \||\Psi_Y\rangle\|^2 &= \||\Psi_{YU}(t_d)\rangle + |\Psi_{YL}(t_d)\rangle\|^2 \\ &\neq \||\Psi_{YU}(t_d)\rangle\|^2 + \||\Psi_{YL}(t_d)\rangle\|^2 \end{aligned} \tag{8}$$

because of quantum interference.

Thus, the finer-grained set following both Y and S does not give us a further quantum multiverse because it is not decoherent. In Copenhagen quantum mechanics we would have said that there are no probabilities for S because probabilities are assigned only to the outcomes of measurements and we didn't measure which slit the electron went through. The Copenhagen approximation is consistent with DH because measured alternatives decohere (e.g. Refs. 17 and 18). But decoherence is a more general, more observer independent criterion for assigning probabilities that apply to histories of the universe as a whole.

2.2. *Two-slit Model with Gas (TSG)*

TSG differs from TSS only in the presence of the weakly interacting gas. The weak interaction with the gas means the set of histories coarse-grained only by different values Y will decohere and have nearly identical probabilities to the same set of histories in TSS. But there will be a significant difference between TSS and TSG for a set of histories that follows S as well as Y. The branch state vectors (6) must now include the degrees of freedom of the scattering particles. Scattering from the upper slit leads to a different state of the gas than scattering from the lower slit. If enough gas particles scatter, these states can be nearly orthogonal leading to decoherence of the set of histories that follows both S and Y. This argument is given more quantitatively in Appendix B

Thus for TSG we have exhibited two multiverses at different levels of coarse-graining. The first one (just Y) is a coarse-graining of the second (Y, S). The first ignores the gas, the second follows it.

Which of these two multiverses really describes the box model universe? Both of them. They are descriptions of the same system at different levels of coarse-graining. Which of the two descriptions should be used to calculate the probability $p(Y)$ that the electron arrives at interval Y on the far screen? Either of them, because they both supply probabilities that are consistent with one another.

Useful descriptions of a physical system at different levels of coarse-graining are familiar from statistical mechanics. A fine-grained description of a box of gas would specify the position and momentum of every particle in the box. A much coarser but more useful description specifies only the total energy, angular momentum, and number of particles in the box. These differ greatly in utility and computational complexity.

This model illustrates two of the conclusions in the Introduction. First, (a) even for simple systems like TSG, the theory (H, Ψ) will predict many different multiverses at different levels and kinds of coarse-graining. The two illustrated here are compatible in the sense that one is a coarse-graining of the other. But that does not have to be the case. The universe may exhibit incompatible coarse-grainings for which there is no finer-grained decoherent set of which they are both coarse-grainings (think position and momentum.) (See, e.g. Ref. 17.)

The model also illustrates conclusion (b) in the Introduction that multiverses are not a choice or an assumption. The possible multiverses follow from the basic theory of the two-slit experiment. You can choose one multiverse or another to calculate with, but you cannot choose whether the theory exhibits them or not.

The model also illustrates the conclusion (f) in the Introduction. The equivalence of coarse-graining by summing quantum amplitudes, with coarse-graining by summing probabilities is one way of stating the consistency between different sets that is a consequence of decoherence. The TSG model gives a specific example. Suppose that we are interested in the probability $p(Y)$ that the electron arrives at interval Y. This can be written in two different ways as a consequence of decoherence

$$p(Y) = \sum_S p(Y, S) = \sum_S \| |\Psi_{YS}\rangle \|^2, \qquad \text{(sum probabilities)} \quad (9a)$$

$$p(Y) = \| |\Psi_Y\rangle \|^2 = \| \sum_S |\Psi_{YS}\rangle \|^2, \qquad \text{(sum amplitudes)}. \quad (9b)$$

The first is coarse-graining by summing probabilities for histories, the second is coarse-graining by summing amplitudes for histories. Which of these formulae should be used to calculate $p(Y)$? It doesn't make any difference; they both give the same answer because the probabilities are consistent as a consequence of decoherence. Summing amplitudes is usually easier than summing probabilities because less computation is involved. This is a triviality in this example, but, as we will illustrate in Sec. 6, this simplicity is a considerable advantage in physically complex situations.

2.3. *Two-slit Model with Gas and Observer (TSGO)*

2.3.1. *Third- and first-person probabilities*

TSS and TSG have quantum multiverses that provide probabilities for which of a set of alternative histories happens in the box whether or not they are subject to the attention of observers. These probabilities for which history happens are called *third-person probabilities* and are derivable just from (H, Ψ). As observers of the universe, we are interested in the probabilities for what we observe. To discuss this, TSGO includes a model observer as a physical system within the box and assumes that the gas particles of TSS have a detectable color — red near the upper slit and blue near the lower slit. See Fig. 2.

An observer moves through the two-slit setup (which must be very large) equipped with a detector that can measure the color of any gas detected. If the detector registers "red" the observer knows that she is passing through the upper slit. Given this data *red*, or what is the same thing, given that she passed through U, what does she predict for the probability that she will arrive at the interval Y at the further screen? This is an example of a *first-person probability* — a probability for the result of an observation.[e] First-person probabilities are third-person probabilities for what happens conditioned on data that describes the observational situation. In this case, the observer's data is that she passed through the upper slit U. The first-person probability that she observes Y at the further screen is then

$$p^{(1p)}(Y) = p(Y|U) = \frac{p(Y,U)}{p(U)} = \frac{\|P_Y|\Psi_U(t_d)\rangle\|^2}{\||\Psi_U(t_d)\rangle\|^2}. \tag{10}$$

The distribution is illustrated in the bottom image in Fig. 1. As should be clear from comparing that with the figure immediately above, what is most probable to occur is not the most probable to be observed.

2.3.2. *Anthropic selection*

Up to now, we have tacitly assumed in TSGO that a live observer (us) exists (E) in the box for all times in all cases. Since an observer is a physical system within the box, there is a third-person probability for it to exist (E) or not exist ($\bar{E}$). Our data as observers trivially includes E.

As an example of the consequences of this, consider histories specified by both Y and S. But suppose that the red radiation is lethal to any observer passing through the upper slit. When we arrive at an interval Y at the far screen, we have data (Y, E). What are the probabilities, given that data, that we got there[f] by going through

---

[e]In other works, we have called first-person probabilities "bottom-up probabilities" and third-person probabilities "top-down probabilities", see Ref. 19 and, especially, Ref. 20.

[f]This is thus a very simple example of a probability for our past history of the kind that is unavailable in Copenhagen quantum mechanics but which is essential in cosmology, as we will see in Sec. 7. See, e.g. Ref. 21.

U or L? Evidently (and trivially) these probabilities are

$$p^{(1p)}(U) = p(U|Y,E) = 0, \quad p^{(1p)}(L) = p(L|Y,E) = 1. \qquad (11)$$

This is an example of automatic anthropic selection — point (d) in the Introduction. We would not arrive at Y if we had gone through the lower slit. No "anthropic principle" had to be invoked, no extra ingredient added to (H, Ψ). We just calculated the probabilities for what we observe. For more discussion of living in a superposition like the observer in this model, see Ref. 22.

3. Five Exemplary Quantum Cosmological Multiverses

The most striking observable feature of our quantum universe is its classical spacetime. At some level and kind of coarse-graining, this extends over the whole of the visible universe from near the big bang to the distant future. What is the origin of this realm of classical predictability in a quantum theory characterized fundamentally by indeterminacy and distributed probabilities? How does it emerge from theory (H, Ψ) that includes quantum gravity where spacetime geometry is generally fluctuating and without definite value? Many of our cosmological observations are of properties of the universe's classical spacetime and its contents — the expansion, the approximate homogeneity and isotropy on large scales, the distribution of galaxies, the CMB, the value of the cosmological constant, etc.

Classical behavior is not a given in a quantum universe. It is a matter of quantum probabilities. A quantum system behaves classically when, in a suitable quantum multiverse of alternative histories, the probabilities are high for those histories exhibiting correlations in time governed by deterministic laws. The relevant probabilities follow from (H, Ψ). Classical spacetime emerges when the probabilities are high for spacetime geometries correlated in time by the Einstein equation.[g]

---

[g]The rest of the domain of applicability of classical physics constituting the quasiclassical realm of everyday experience is a consequence of this.[23]

Subsequent Secs. 3 through 8 exhibit five models of a quantum multiverse of classical spacetimes that illustrate conclusions (a)–(f) of the Introduction. The models are based on the author's joint work with Stephen Hawking and Thomas Hertog (mainly Refs. 3, 8–10). We do not pretend to present these models in the depth and precision that they are discussed in those papers. Rather, we present a qualitative and intuitive description of the models.

3.1. *Common Elements*

The five models have the following elements in common:

3.1.1. *General theory*

Semiclassical Quantum Gravity: Quantum multiverses of the universe necessarily include alternative histories of cosmological spacetime geometry and therefore involve quantum gravity at some approximate level. Phenomena like the emergence of classical spacetime from the quantum big bang, the generation of large scale structure from quantum fluctuations away from homogeneity and isotropy, the nucleation of bubbles of true vacuum in a false vacuum, and eternal inflation are all fundamentally quantum spacetime phenomena. We discuss these, treating quantum gravity in semiclassical approximation.

Quantum State: For the quantum state Ψ we adopt the no-boundary wave function of the universe (NBWF) in its semiclassical approximation.[7]

Dynamics: For a theory of dynamics (H), we assume Einstein gravity coupled to a single homogeneous scalar field $\phi(t)$ moving in a potential $V(\phi)$. Different models have different $V(\phi)$.

Coarse-Graining: The theory (H, Ψ) predicts different quantum multiverses defined by coarse-grainings that follow different variables on very different scales. Coarse-grainings relevant for laboratory experiment follow the outcomes of the experiment and ignore cosmological scale features of the universe if these do not affect the outcomes. Coarse-grainings relevant for cosmology follow the large-scale features of the universe and ignore small-scale fluctuations like planets, biota, human observers, and their laboratory experiments whose presence or absence has little effect on the large scale behavior of the universe.

A Model of Observers and Observation: By itself, the theory (H, Ψ) predicts third-person probabilities for which of a set of alternative histories of classical spacetime geometries and the matter within happens. First-person probabilities for our observations are third-person probabilities for what happened conditioned on the data D describing our observational situation. TSGO provided a very simple example of this, leading to (10) and (11). There, the data D was that an observer existed, E. Naturally, probabilities for the results of her observations were conditioned in E.

As observers of the universe we, and the apparatus we use, are quantum physical systems within it, not somehow outside it. Our observations of the universe are limited to a spatial volume of rough size $c/(\text{Hubble constant}) \sim 4,000$ Mpc at the present time — our Hubble volume. This is just one Hubble volume in a universe that may have a great many similar volumes. We have only a very, very small probability, denoted by $p_E(D)$, to have evolved in any Hubble volume. Incorporating, as it does, the probabilities for accidents of biological evolution, the probability $p_E(D)$ is a very, very, very small number, much beyond the ability of present day physics to compute. This is a very crude model of an observing situation, but still better than many treatments where the evolution of observing systems in the universe is not considered at all.[h]

Despite the small value of $p_E(D)$, there can be a significant probability that in a very large universe the data D is replicated in many Hubble volumes. First-person probabilities for our observations are third-person probabilities for what happened conditioned on the existence of at least one instance of our observational situation — our instance. That's all we know for sure about instances of D. We

[h]For details on observers, observations of first- and third-person probabilities and what is included in D, etc., see e.g. Refs. 19 and 24.

abbreviate (at least one instance of D) by $D^{\geq 1}$. TSGO provided a very simple example of this, leading to (10) and (11).

The bottom line is that first-person probabilities for an observable feature of the universe $\mathcal{O}$ are third-person probabilities for $\mathcal{O}$ to happen from (H, Ψ) conditioned on the existence of at least one instance of our observational situation D, viz.

$$p^{(1p)}(\mathcal{O}) \equiv p(\mathcal{O}|D^{\geq 1}). \tag{12}$$

We describe how to calculate this in the various models, beginning with the one in the next section.

4. A Quantum Multiverse of Homogeneous, Isotropic, Classical Histories

This is a very simple model that illustrates how a quantum multiverse of classical spacetime geometries emerges from (H, Ψ).

The model assumes a minisuperspace based on homogeneous, isotropic, spatially closed spacetime geometries with metrics of the form

$$ds^2 = -dt^2 + a^2(t)d\Omega_3^2. \tag{13}$$

Here, $a(t)$ is the scale factor and $d\Omega_3^2$ is the metric on a unit round 3-sphere.

Dynamics: For a theory of dynamics (H) we assume Einstein gravity and a single homogeneous scalar field $\phi(t)$ moving in a potential $V(\phi)$ with the form:

$$V(\phi) = \Lambda + \frac{1}{2}m^2\phi^2 \tag{14}$$

in the Planck units ($\hbar = G = c = 1$) that we use throughout. An action $I[(a(t), \phi(t)]$ summarizes this dynamics, but we won't need its explicit form.

Quantum State: The NBWF is a function(al) of the geometry and field configurations on a spacelike three-surface. In this minisuperspace model, this means that the wave function depends on the scale factor of the three-surface and the homogeneous value of the scalar field there. When used as arguments of the wave function, we denote these by b and χ, respectively. Thus,

$$\Psi = \Psi(b, \chi). \tag{15}$$

In the semiclassical approximation, the NBWF is a sum of terms of the form

$$\Psi(b, \chi) \propto \exp[-I(b, \chi)/\hbar] \tag{16}$$

where $I(b, \chi)$ is the action at a saddle point of the action $I[(a(t), \phi(t)]$ that is regular on a four-disk and for which $(a(t), \phi(t))$ assume the values (b, χ) on its boundary. These saddle points are generally complex, so we can write

$$\Psi(b, \chi) \propto \exp\{-[I_R(b, \chi) + iS(b, \chi)]/\hbar\} \tag{17}$$

where I_R and $-S$ are the real and imaginary parts, respectively, of the saddle point action.

In regions of (b, χ) where S varies rapidly compared to I_R, this wave function takes a WKB form and predicts a multiverse of Lorentzian, classical histories obeying the Einstein equation.[25] This classical multiverse consists of the integral curves of S. That is, the histories are the solutions of the Hamilton–Jacobi expressions for the momenta π_b and π_χ conjugate to b and χ, viz.

$$\pi_b = \frac{\partial S}{\partial b}, \quad \pi_\chi = \frac{\partial S}{\partial \chi}. \tag{18}$$

There turns out to be a one-parameter family of such curves conveniently labeled by the magnitude of the scalar field ϕ_0 at the center of the saddle point geometry. The label ϕ_0 turns out to be approximately equal to the initial value at which the field starts rolling down the potential in the classical histories labeled by ϕ_0. The probabilities of these histories are approximately

$$p(\phi_0) \propto \exp[-2I_R(\phi_0)/\hbar], \tag{19}$$

which, for this model, is approximately

$$p(\phi_0) \propto \exp\left[\frac{3\pi}{\Lambda + (1/2)m^2\phi_0^2}\right]. \tag{20}$$

Thus, the theory (H, Ψ) predicts a quantum multiverse of classical spacetime geometries and matter fields labeled by ϕ_0 with probabilities (20). We can say that these are the third-person probabilities that the history labeled by ϕ_0, was the history of the universe that happened or occurred.

This multiverse does not provide a mechanism for the variation of Λ. There is only one minimum characterized by one value of Λ, and all the possible histories roll down to that. The same is true for m, which turns out to govern the size of the primordial density fluctuations observed in the CMB.[26]

4.1. *Probabilities for Inflation*

The number of efolds of scalar field driven inflation $N_e(\phi_0)$ is a simple example of a quantity which, if not directly observable, has significant observable consequences for our universe. Numerical solution of the equations (18) for the individual members of the multiverse of classical histories shows that approximately,[27]

$$N_e(\phi_0) \approx 3\phi_0^2/2 \tag{21}$$

for $\phi_0 \gtrsim 1$ in the Planck units used throughout. The third-person probabilities for (20) show that histories with a low amount of inflation are the most probable to occur. But what is most probable to occur is not necessarily the most probable for us to observe. Evaluating (12) gives those probabilities.

From the definition of conditional probabilities (Bayes theorem), we have

$$p(\phi_0|D^{\geq 1}) \propto p(D^{\geq 1}|\phi_0)p(\phi_0). \tag{22}$$

Normalizing the right-hand side gives a formula with an equality. In this simple model of the observing situation, third-person probabilities for what occurs are converted to first-person probabilities for what is observed by multiplying by the factor $p(D^{\geq 1}|\phi_0)$, called the "top-down" factor, and then renormalizing.

The probability that there is at least one instance of our observational situation somewhere in the universe, $p(D^{\geq 1}|\phi_0)$, is bigger in a larger universe where there are more Hubble volumes for an instance of D to have evolved than it is in a smaller universe where there are fewer Hubble volumes. This turns out to mean that we are more likely to observe universes with more inflation.[28]

We can understand this result more quantitatively with a small-scale model where $p(D^{\geq 1}|\phi_0)$ can be explicitly evaluated. Suppose that our data D locate us somewhere on a homogeneous isotropic spacelike surface with $N_h(\phi_0)$ Hubble volumes in each classical history. The probability $p(D^{\geq 1}|\phi_0)$ is 1 minus the probability that there are no instances of D on the surface. This, in turn, is the product of the probabilities $1 - p_E(D)$ that there are no instances of D in any particular Hubble volume on the surface. The result is the formula:

$$p(D^{\geq 1}|\phi_0) = 1 - [1 - p_E(D)]^{N_h(\phi_0)}. \tag{23}$$

Equation (23) is an explicit formula for the "top-down factor" that converts third-person probabilities to first-person ones, as in (22). Larger number of Hubble volumes N_h makes this factor larger and closer to 1. However small $p_E(D)$ is, in a sufficiently large universe, the probability is 1 that an instance of D occurs somewhere. When $N_h \gg 1/p_E(D)$, first-person probabilities are equal to third-person ones.

First-person probabilities thus favor larger universes, larger ϕ_0, with more efolds for inflation (21). In the quantum multiverse considered in this section, significant inflation is anthropically selected to be observed.

4.2. *How this Model Supports the Conclusions*

The main contribution of this model to our exposition is to illustrate, in a simple way, how quantum multiverses of classical histories (including spacetime geometry) are predicted by (H, Ψ). We will assume this for subsequent models. But the model does illustrate points (b) and (c) of the Introduction, namely, that quantum multiverses are a consequence of (H, Ψ) and typically consist of many different histories, not just one.

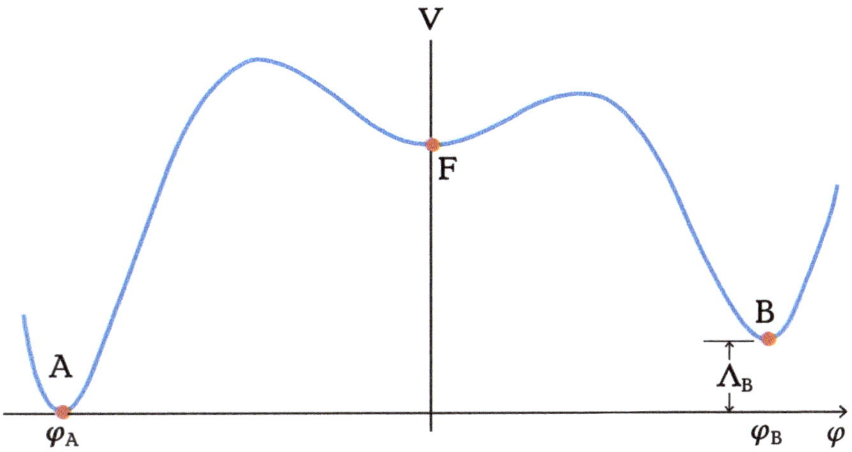

Fig. 3. A potential for the scalar field with one false vacuum F and two true vacua A and B. The false vacuum is separated from either true vacua by potential barriers and relatively flat patches where the conditions for slow roll inflation are satisfied. The different shapes of the barriers and of the potential in the two slow roll regimes leading to the true vacua result in different false vacuum decay rates and different predictions for CMB-related observables in universes ending up in A or B.

5. Quantum Multiverses of Pocket Universes

This section shows how pocket multiverses of the kind mentioned in the Introduction arise at various levels of coarse-graining from the NBWF (Ψ) and a dynamical theory (H) assuming a particular potential $V(\phi)$, like the one in Fig. 3. This potential has three minima (vacua) — two true vacua A and B and one false vacuum F. We can say that the potential defines a *landscape of vacua*, although there are only three here. As described in Sec. 4, (H, Ψ) predicts a one-parameter ensemble of classical histories labeled by the value at which they start to roll down.

The classical history that starts to roll down at F in Fig. 3 describes a universe that inflates with an effective cosmological constant $3\pi/V(0)$. Classically, this inflation is eternal — it goes on forever. But quantum-mechanically, regions of spacetime can tunnel through the barriers on either side of F and roll down classically to one of

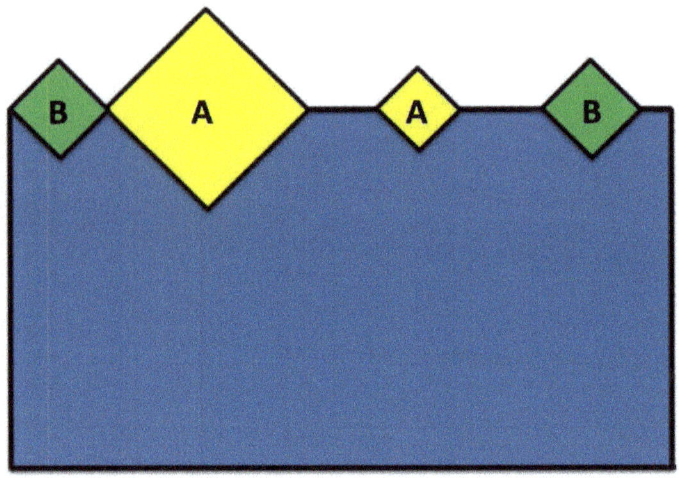

Fig. 4. The Penrose diagram for the pocket multiverse model when the bubble nucleation rate is low enough that collisions between bubbles are negligible. The blue (dark) region in the diagram represents the false vacuum, classical deSitter expansion. That deterministic classical evolution is interrupted by the nucleation of bubbles of true vacua in the false vacuum by quantum tunneling, which then expand at the speed of light. The bubbles nucleate at a point in the false vacuum and their walls expand at the speed of light. There are two kinds of true vacua, A (yellow) and B (green), corresponding to the two minima in the potential in Fig. 3.

the true vacua A or B. That is, the inflating background can nucleate bubbles of true vacua A or B. The result is a multiverse whose Penrose diagram is like the one in Fig. 4 for low nucleation rates. Different histories will have different numbers of true vacua of either kind in different places in the eternally inflating universe.

We don't live in the false vacuum devoid of matter. Rather, we live in one of the bubbles of true vacuum, either one of type A or of type B. The kind of bubble we live in can be determined from observation. The value of the potential at the minimum determines the local value of the cosmological constant, either $\Lambda_A = V(\phi_A)$ or $\Lambda_B = V(\phi_B)$. Observations of the expansion, can in principle, determine the cosmological constant in our bubble, and therefore determine which kind of bubble we are in. The curvatures of the potential at the minima are related to the amplitude of primordial density fluctuations which leave an imprint on the CMB.[26] Different curvatures at the minima of V produce observationally distinct CMB_A and CMB_B. By observing the CMB, we can determine which kind of bubble we are in.

This model supports several of the conclusions in the Introduction. It supplies a concrete example of how certain (H, Ψ) imply a pocket multiverse that is realized as a false vacuum eternally inflating universe that nucleates bubbles of true vacuum (the pockets) with different observational productions. In particular, for a suitable potential, it shows how the cosmological constant can vary from bubble to bubble in space. This pocket universe is not a choice or postulate once an appropriate (H, Ψ) is fixed. Were Λ_B too large to form galaxies as described in the Introduction, we would immediately predict we are living in a bubble of type A. That is a simple example of anthropic selection, but we will have more to say about this in the next section.

6. A Quantum Multiverse of the Observable Properties of Our Bubble

In the TSG model in Sec. 2, we showed how one quantum system can exhibit different quantum multiverses at different levels of coarse-graining. In this section, we show how the model cosmology defined by the potential like that in Fig. 3 can exhibit more than one quantum multiverse at different levels of coarse-graining.

A relatively fine-grained multiverse would consist of histories that describe all possible configurations of bubble nucleation and non-nucleation at all places and all times. The history illustrated in Fig. 4 is just one example. But there are an infinite number of other histories at this level of coarse-graining in a false vacuum deSitter expansion that extend to the infinite future.

The (third-person) probabilities per unit four-volume p_A and p_B for nucleating bubbles of different kinds were calculated by Coleman and De Lucia (CDL).[29] The probability per unit four-volume to remain in the false vacuum is $p_F = 1 - p_A - p_B$. With these probabilities, it is possible to imagine calculating the third-person probabilities for an entire set of histories of different mosaics of volumes of false and true vacua. Some kind of cutoff or "measure" would be required to deal with the infinite volume. Even then it would be a formidable calculation.

We do not observe entire four-dimensional histories of roiling seas of bubble nucleation extending to the infinite future. Neither do we know the location of our bubble. First-person probabilities for our observations, say of the CMB, are the probabilities that we are in a bubble of type A or B at some unknown location in the eternally inflating false vacuum spacetime. Such probabilities could in principle be calculated by summing up the probabilities for the fine-grained histories over all the alternative structures outside our bubble. Quantum mechanics provides a more direct route to the answer: Make predictions for our observations using a quantum multiverse based on a much coarser-grained multiverse that follows what goes on inside our bubble and ignores everything outside — a multiverse based on our bubble.

A history in our bubble multiverse is illustrated in Fig. 5. Our bubble occupies one part of the spacetime with (CDL) probabilities p_A or p_B, such that it is of type A or B. Since the coarse-graining doesn't specify what is going on in the volumes outside our bubble, their contribution to the probability of this coarse-grained history is $p_A + p_B + p_F = 1$.

Suppose the data D describing our observational situation locate us on the spacelike reheating surface inside one of these kinds of bubbles. There are an infinite number of Hubble volumes on these surfaces in both kinds of bubbles. The top-down factor (23) connecting first- and third-person probabilities is then unity. First-person probabilities equal third-person probabilities. Thus, the probabilities that we observe A, $p(WOA)$ or B, $p(WOB)$ are

$$p(WOA) = \frac{p_A}{p_A + p_B}, \quad p(WOB) = \frac{p_B}{p_A + p_B}. \tag{24}$$

It doesn't matter where in spacetime our bubble is located. CDL probabilities respect the symmetries of deSitter space and are the same at all locations.

This example illustrates that a quantum system like our universe is not described by only one quantum multiverse. It is described by

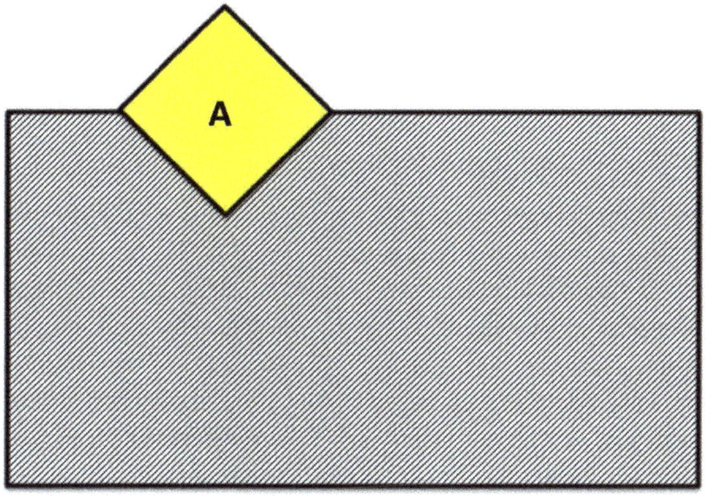

Fig. 5. A history of the universe that is much coarser-grained than the one in Fig. 4. There is only one bubble — our bubble. The mosaic of false vacua A and B bubbles that are outside our bubble suggested in Fig. 4 is ignored and shaded gray. We stress that this does not mean that there are no regions of A, B or F outside. Only that the coarse-graining does not distinguish whether a given region outside our bubble is A, B or F.

many different ones at different levels of coarse-graining within the same theory (H, Ψ) with the same potential as in Fig. 3. For a given question, it is generally best to use the coarsest-grained description that supplies an answer to it.

This model supports all of the conclusions (a)–(f) mentioned in the Introduction: (a) Different quantum multiverses follow from (H, Ψ) at different levels and kinds of coarse-graining. (b) They emerged by calculation from (H, Ψ) with the potential in Fig. 3. They were not an assumption beyond assuming (H, Ψ). (c) The coarse-graining following the alternatives A, B or F for all volumes in spacetime leads to quantum multiverses with a truly vast number of histories to compute third-person probabilities. The coarse-graining following only the inside of our bubble and ignoring everything outside has only alternatives A or B. Either way there is a multiplicity of alternative histories. (d) Anthropic selection automatically ruled out an observation of a false vacuum F devoid of matter where we cannot exist. (f) The simplicity of coarse-graining multiverses by summing up amplitudes rather than summing up probabilities.

6.1. *Different Dynamics, Different Multiverses*

Our story about pocket universes forming by the decay of an eternally inflating false vacuum depended crucially on a dynamical theory (H) incorporating something like the potential in Fig. 3. This dependence on dynamical theory is illustrated in the work by Hawking and Hertog.[30] They used the no-boundary quantum state (Ψ) together with dynamics (H) specified by an effective dual field theory defined on the exit surface of eternal inflation. This dual field theory provides a description of the transition between the essentially quantum realm of eternal inflation to the ensemble of possible classical universes, one of which we observe. It provides a quantum probabilistic measure on this ensemble that is different from the one in Sec. 6. Hawking and Hertog found that this measure predicts a smooth exit from eternal inflation with no bubbles and no pocket universes. Evidently, this dynamics is not represented by an effective dynamical theory incorporating a potential like that in Fig. 3.

7. A Quantum Multiverse of Homogeneous and Isotropic Classical Histories with Different Physical Constants

The potential on which this model is based is shown in Fig. 6. It has many minima (vacua) $K = 1, 2, \cdots$ at values ϕ_K near which it is approximately

$$V(\phi) \approx \Lambda_K + \frac{1}{2} m_K^2 (\phi - \phi_K)^2 + \cdots \qquad (25)$$

for constants Λ_K and m_K — defining a landscape of vacua.

This potential defines a one-parameter ensemble of alternative homogenous and isotropic classical histories that may roll down to one minimum or another. The theory (H, Ψ) supplies third-person probabilities for which the minimum is reached. The ensemble is thus a quantum multiverse in which the constants Λ and m are the

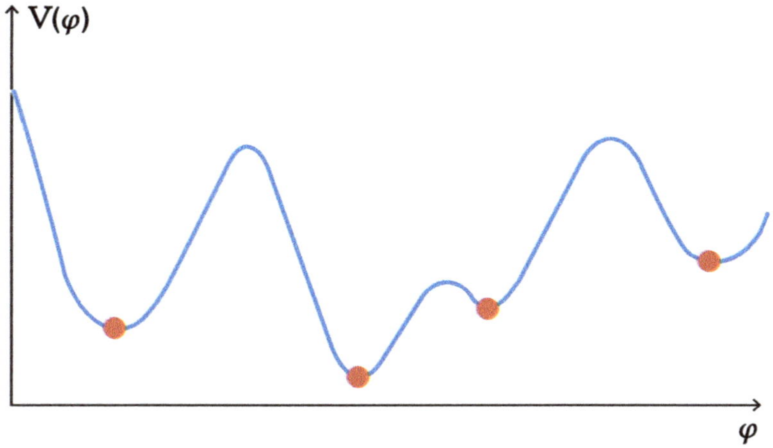

Fig. 6. A potential $V(\phi)$ for the scalar field with many minima at field values $\phi_K, K = 1, 2, \cdots$. The values of the field at the minima define different values of the cosmological constant $\Lambda_K = V(\phi_K)$. The first-person probability that we observe one value Λ or another, is the probability that our past history rolled down to one minimum or another.

same everywhere in space but vary from history to history. Their value is thus not fixed by the action but by the quantum accident of which one of the minima is reached.

As observers of the universe, we are a special kind of late time fluctuation away from homogeneity and isotropy living in one of the true vacua after reheating has produced ordinary matter. The values of Λ and m that we observe are the values for our minima. The parameter m is related to the amplitude Q of primordial density fluctuations when they leave their horizon. The relation turns out to be roughly $Q \approx m$ in the Planck units employed throughout. The first-person probability that we observe particular values of Λ and Q is a probability that our past history rolled down to a minimum with these values.[i]

---

[i]The approximate Copenhagen quantum mechanics of measured subsystems would not be able to give us these probabilities because it cannot generally retrodict the past (e.g. Ref. 21). A generalization like DH is necessary to do that.

In Sec. 4.1, we described how the transition between third-person probabilities for which history occurs and first-person probabilities for which history is observed favors larger universes where there are more places for us to be and disfavors smaller universes. The history that dominates the first-person probabilities for observation turns out to be the one with the lowest ϕ_0 consistent with significant inflation. In more technical terms, first-person probabilities are dominated by the lowest exit from eternal inflation. As explained in Sec. 4.1, for large universes with many Hubble volumes, the top-down factor (23) is unity and first-person and third-person probabilities are equal.

As mentioned before, it turns out that $m \approx Q$ so the first-person probabilities for Λ and Q then are approximately from (20)

$$p^{(1p)}(\Lambda, Q) \approx \exp\left(\frac{3\pi}{\Lambda + cQ}\right) \tag{26}$$

where c is a calculable, dimensionless constant of order unity.

Conditioning on at least one instance of a small amount of our data on the scales of the Earth turns out to be sufficient to derive (26). We can then use this result to look for predictions of correlations between the rest of our data on scales of our Hubble volume. To illustrate this, we look for correlations among three pieces of our data: the value of Λ, the value of Q, and the fact that we have a Hubble volume full of galaxies by the present age of approximately 14 Gyr. The range of values of Λ and Q for which these conditions can be satisfied is as calculated in Ref. 31 and displayed in Fig. 7. As noted in Refs. 1 and 2, too large a Λ would prevent proto-galaxies from collapsing. Too large a Q would result in a universe where most collapses produce black holes. Were Q too small, the fluctuations would not grow enough for galaxies to form by the present age. These anthropic constraints mean that the probability is negligible that we are outside the white region in Fig. 7.

Inside the white region, the NBWF first-person probabilities (26) favor small values of Q and larger values of $\log_{10} \Lambda$. Thus we predict the values indicated by the red $*$

$$Q \sim 10^{-5}, \qquad \Lambda \sim 10^{-123}. \tag{27}$$

Figure 8 shows the marginal distribution for Λ.

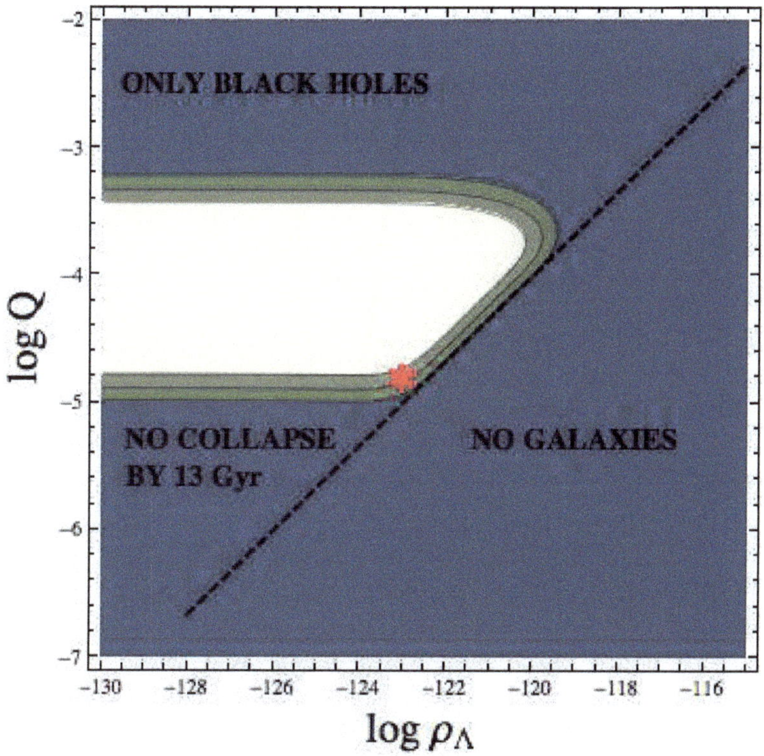

Fig. 7. Anthropic constraints on the values of Q and Λ in a figure from Ref. 3, based on calculations in Tegmark *et al.*[31] given the present age of the universe.

The few orders of magnitude agreement of the values in (27) with observations is not the main point. After all, this is just one example out of many possible ones. Rather, the main things the reader should take away is the following: First, the universe does not have to have pockets for the fundamental constants to vary. As discussed here, there can be quantum multiverses of homogeneous and isotropic histories in which the constants do not vary in space but over a multiverse of histories. Second, anthropic selection is automatic in quantum cosmology through first-person probabilities for observations. And finally, a theory of the quantum state has a significant impact on what the probabilities are and for our predictions of the first-person probabilities of the constants that we observe.

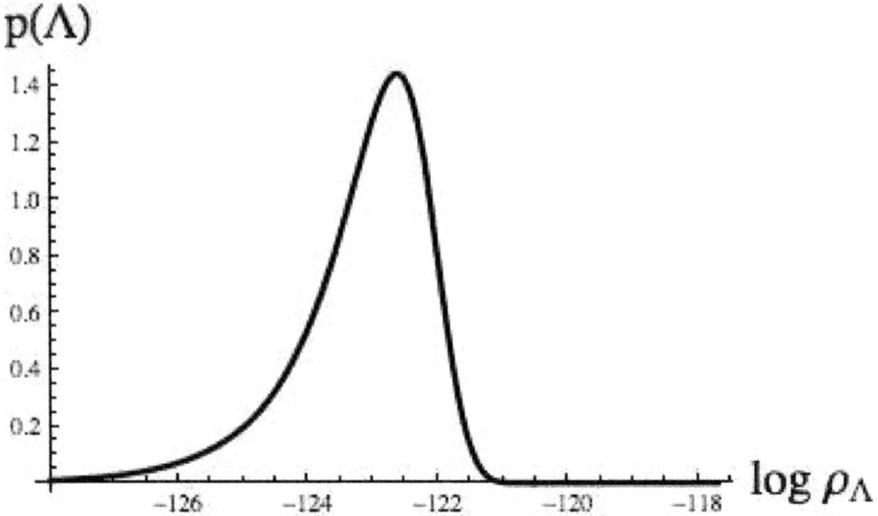

Fig. 8. The marginal probability distribution for observations of the cosmological constant obtained by integrating the first-person probabilities (26) over Q.

8. A Quantum Multiverse of Histories with Different CMBs

In the preceding section's calculation of first-person probabilities for observation, only values of the potential in Fig. 3 below the lowest exits from eternal inflation contributed to the final result. The potential is thus effectively equivalent to an ensemble of one-dimensional potentials, each with one minimum.

String theory has a vast landscape of possible vacua.[32] A landscape with many fields moving in an ensemble of one-dimensional potentials, each with one minimum, is probably more analogous to the string landscape than one potential with many minima of the kind in Fig. 3. Thomas Hertog used such a model landscape consisting of a number of one-dimensional potentials for (H) together with the no-boundary wave function for (Ψ) to estimate that theory's prediction for the tensor/scalar ratio in the CMB.[10] For a model landscape he took the scalar field potentials that were used to reduce data from the Planck satellite on the CMB.[33] That landscape included power law potentials of the form $V(\phi) = \lambda\phi^n$, plateau potentials of the form $V(\phi) = V_0(1 - \phi^n/\mu)$ and "R^2 inflation potentials". The quantum multiverse consists of all the histories in all the potentials. The most probable first-person history is the one with the lowest exit from eternal inflation among all these potentials.

Hertog's results are shown in Fig. 9. The important point about the model is not the specific numerical value for the tensor/scalar ratio, but the fact that it is a prediction for observations yet to be done. If further observations yield different numbers, then either the NBWF is ruled out or the *ad hoc* landscape of potentials would be ruled out.

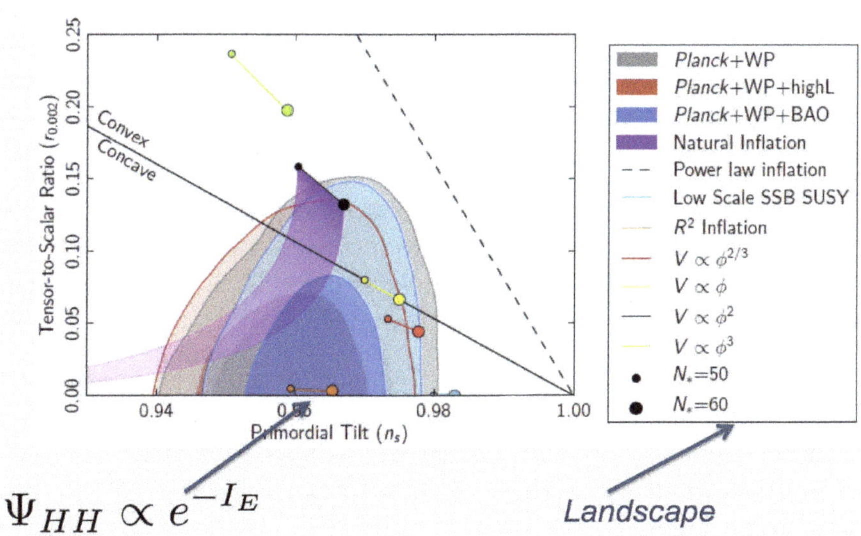

Fig. 9. Planck satellite results for the tensor/scalar ratio r and the scalar spectral tilt n_s of the observed CMB as compared with various theories. The figure was adapted by Thomas Hertog from Fig. 1 of Ref. 33. The shaded regions show the values consistent with the data when combined with other astronomical measurements. The predicted curves are predictions from various kinds of models. The NBWF plus the landscape of potentials discussed in the text and labeled at the right predicts the values indicated roughly by the arrow.

9. Conclusions

At the start of this paper, we defined a multiverse as an ensemble of alternative possible situations only one of which is observed by us. We have seen how decoherent histories quantum mechanics (DH) predicts quantum cosmological multiverses in the form of decoherent sets of alternative coarse-grained histories of the universe. In each set one history occurs, or happens, with a probability predicted by a fundamental quantum theory of cosmology consisting of theories of the universe's quantum state (Ψ) and dynamics (H). Quantum multiverses are not some posited speculative idea to be grafted onto the basic theory. They are the output of that theory.

9.1. *General Conclusions*

Conclusions that follow from the general perspective of this paper have been stated at several points, but we repeat them here in abbreviated form.[j]

(a) Many Quantum Multiverses. The theory (H, Ψ) does not just predict one multiverse of alternative histories. It predicts many different multiverses at different levels and kinds of coarse-graining. The pocket multiverse is just one example.

(b) Multiverses are Not a Choice. Quantum multiverses are not a choice or an assumption separate from the theory (H, Ψ). They follow or do not follow from (H, Ψ). If you have some prior objection to multiverses of some kind, then, restrict to theories (H, Ψ) that don't imply them.

(c) Multiverses are Generic. Simple, manageable, discoverable (H, Ψ)'s generically predict quantum multiverses consisting of an ensemble of many possible histories together with probabilities for which one occurs. If the ensemble consisted of just one history, with certainty all of present complexity would have to be encoded (H, Ψ).

[j]Many years ago, John Wheeler explained to the young author that some people only read the introduction, some other people only read the conclusions, and still some other people only read the figure captions, so your message should be in all of those places.

(d) Anthropic Selection is Automatic. Anthropic selection is an automatic consequence of first-person quantum mechanical probabilities for observations (Sec. 2.3.1). Probabilities for our observations are conditioned on a description of our observational situation, and we won't observe what is where we cannot exist.

(e) Different Mechanisms for the Variation of Constants. In different quantum multiverses, constants like Λ can be constant in each history but vary from history to history, or can vary in space within every history. Automatic anthropic selection is enabled in either case.

(f) Two Routes to Coarse-Graining. Quantum multiverses are restricted to sets of alternative coarse-grained histories that are decoherent — that have negligible quantum interference between individual histories in a set. Further coarse-graining can be carried out either by summing up probabilities or by summing up quantum amplitudes. That is a considerable computational advantage.

Understanding quantum multiverses as decoherent sets of alternative coarse-grained histories of the universe provides a unified perspective on different notions of multiverse that have been discussed. As in the two-slit models in Sec. 2, it includes the multiverses that describe the possible outcomes of laboratory experiment. As in Secs. 5 and 6, it includes a quantum picture of pocket universes at different levels of coarse-graining. As in Secs. 7 and 8, it provides a notion of a multiverse in which constants vary in a way that is different from pocket multiverses.

9.2. *How Specific Models Support the General Conclusions*

Sections 5–8 exhibited five quantum cosmological models that illustrate the different kinds of quantum multiverses that follow from an (H, Ψ)'s consisting of the no-boundary quantum state of the universe Ψ and dynamical theories H based on general relativity coupled to a single scalar field moving in different potentials. These models are of course consistent with the general conclusions above as are the two-slit models in Sec. 2. But they also illustrate the following specific consequences:

• How pocket (bubble) multiverse universes emerge from some dynamical theories (H, Ψ), like the one in Sec. 5, but not from others like those in Secs. 4 and 6.

• How the same theory (H, Ψ) can predict multiverses at vastly different levels of coarse-graining, like the multiverse of true vacuum bubbles in Sec. 5 and the multiverse of possibilities for our bubble in Sec. 6.

• How to use appropriate coarse-grainings implemented by summing up amplitudes rather than probabilities to make manageable predictions for our observations even when the large scale structure of the universe is a complex roiling sea of false vacuum bubbles in an ever expanding sea of false vacuum, as in Sec. 6.

• How deterministic classical physics emerges from (H, Ψ) as a quantum multiverse of histories with high probabilities for those exhibiting correlations in time by deterministic laws like the Einstein equation as in Sec. 4.

• How a landscape potential with many minima leads to different possible values of the cosmological constant and how first-person probabilities implement anthropic restrictions for values which we will observe as in Sec. 7

• How a landscape of effective potentials leads to testable predictions for features of the CMB, as in Sec. 8.

Quantum multiverses are what a fundamental theory (H, Ψ) predicts. It is through them that we understand our universe. As Weinberg wrote: "Most advances in the history of science have been marked by discoveries about nature, but at certain turning points we have made discoveries about science itself". The quantum multiverses of the universe are one of these turning points.

Acknowledgments

The author thanks Murray Gell-Mann, Stephen Hawking, Thomas Hertog and Mark Srednicki for discussions on the quantum mechanics of the universe over many years. He thanks Bernard Carr, George Ellis and Thomas Hertog for useful conversations, for critical readings of the manuscript, and for supplying relevant references.

He thanks David Krakauer and Steven Benner for discussions on experimental evolution. He thanks Thomas Hertog for permission to reproduce Fig. 9. He thanks the Santa Fe Institute for supporting many productive visits there. This work was supported in part by the National Science Foundation under grant PHY15-04541.

Appendix A

A FAQ for Discussion

There has been much discussion of cosmological multiverses by many authors. Some works known to the author (but not necessarily read carefully by the author) are Refs. 34–46. This FAQ addresses concerns that have been raised in some of these articles about whether a quantum multiverse is falsifiable, testable, predictive, real and scientific. No claim is made to cover all of the issues that have been raised or to review all the discussion that has taken place. This FAQ is separated from the main discussion because it reflects the opinions of the author more than the results of calculation.

The quantum multiverse framework for prediction presented here is a synthesis of many elements. These include a quantum mechanics for the universe (DH), a model of observers, for example, the one in Sec. 4.1, a theory of the universe's quantum state (Ψ), and a dynamical theory (H) expressed in particular variables, perhaps defining a landscape of vacua, etc.

Is the notion of a quantum multiverse of the universe ***falsifiable?*** **Yes.** As in every other theory in physics, this one can be falsified by falsifying any of the elements mentioned above that went into its construction. Take, for example, DH. There is overwhelming evidence for quantum multiverses on laboratory scales. But there is little evidence that the same quantum mechanics (DH) can be extrapolated to the scales of cosmology, as has been assumed in this paper. Suppose quantum laboratory experiment shows that DH is false on some intermediate scale. Then DH would be falsified. Similarly with the other elements of the predictive framework: a dynamical theory, a quantum state, a landscape for the variation of constants, etc.

Are quantum multiverses of the universe ***testable?*** **Yes.** A theory like (H, Ψ) is testable by the (first-person) probabilities it predicts for the outcomes of observations. These are probabilities supplied by (H, Ψ) conditioned on a description of our observational situation.[k] Ideally we would test the theory with situations where the predicted probabilities are near 1. If a prediction with probability near 1 does not occur, the theory is falsified. But this luxury is seldom accessible in environmental sciences like geology, planet formation, biological evolution, human history and cosmology. Rather, we judge the success of a theory by its successful prediction of correlations among our present data, as in the example in Secs. 7 and 8. (For more on criteria for success see, e.g. Ref. 3.) Hertog's prediction of the tensor/scalar ratio in the CMB described in Sec. 8 provides a clear example. If the observed ratio is much different from the predicted one, then either the landscape is wrong, the state is wrong, or the framework of quantum prediction is wrong.

Something like cosmological multiverses might be directly testable by experiment if a very large spacetime volume could be prepared with an initial quantum state from which galaxies, stars, life, etc. would emerge over billions of years. This is quite beyond human powers at present. But the laws of the universe do not have to be such as to make it easy for some negligible bits of protoplasm to test them directly on the scales that happen to be accessible to them at the moment.

Are ***other histories in the multiverse of alternatives observable?*** **No.** The different histories are exclusive alternatives. We are not outside the universe observing the whole ensemble, but rather inside it participating in the superposition. A live Schrödinger cat does not observe a dead cat. There are six alternative outcomes to the roll of a die, each with a probability of 1/6. We observe the one face that comes up. We do not observe the alternative outcomes of that one roll.

---

[k]In various works we have called these *top-down* or *first-person* probabilities. For more on this, see Ref. 8.

Is the quantum multiverse of the universe ***real?*** **Yes.** It is a tenet of many Everettian formulations of quantum theory for closed systems, that all the histories in a given multiverse are real.[4,6] Further, all the other multiverses based on different variables with different levels of coarse-graining are real. There are also formulations of DH in which only one history occurs in a multiverse of possible real ones.[47] For more of the author's more nuanced views on what is real, see Ref. 48.

Are there ***other areas of science*** *with similar issues?* **Yes.** Biological evolution is concerned with the multiverse of different histories of the origin of species. Individual histories can be illustrated by familiar tree diagrams like the example in Fig. A1. Different evolutionary histories correspond to different branching trees. In principle, the theory (H, Ψ) predicts probabilities for these various trees to occur, but these are much beyond our ability to compute at present. We do not observe directly ("see") any of the histories of this multiverse. They are in our past. Yet we believe that biological evolution happened because it explains regularities today. For example, plausible reconstructions of the evolutionary multiverse explain the similarities between species today. Those similarities are consequences of frozen accidents of evolution — chance events whose consequences proliferated. The theory of this multiverse is based on a number of elements: Genetic variation by mutation, genetic drift, or recombination; a fitness landscape of ecological niches; an initial condition of primordial DNA synthesis. This is testable in a limited way by controlled laboratory experiment.[49,50] Other historical sciences such as geology, planet formation, human history, etc. provide similar examples of multiverses. In cosmology we do not observe big bang nucleosynthesis directly, but we believe that it happened in the past because of its successful prediction of the correlations in the abundance of the elements observed today.

Are quantum multiverses a ***departure from laboratory physics?*** **No.** Laboratory experiments inside the universe are not somehow outside it. Laboratory experiments can therefore be described by highly coarse-grained alternative histories of the universe.

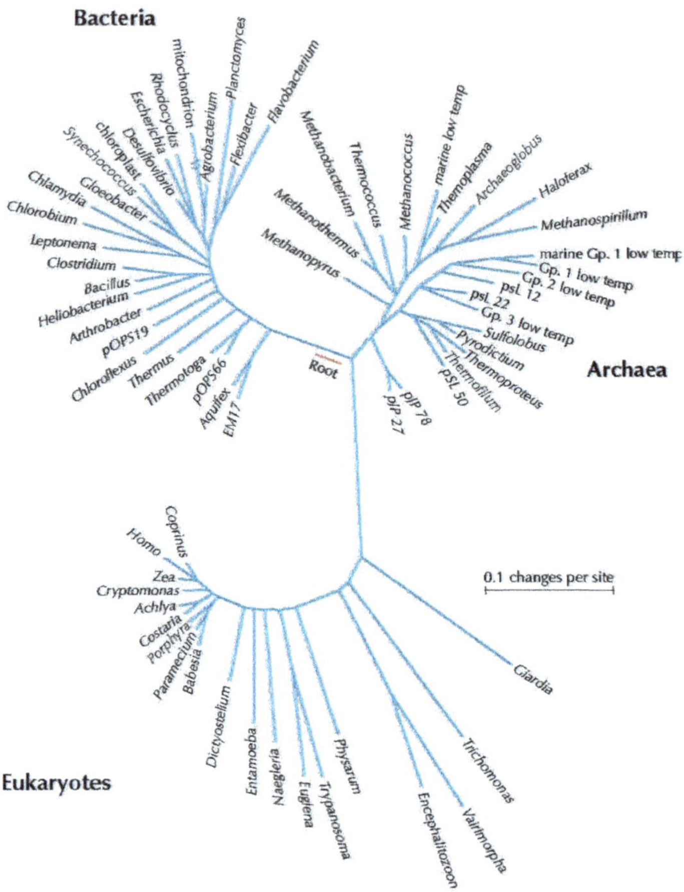

Fig. A1. A single history of biological evolution in a coarse-graining that only distinguishes certain genera.[51] In this representation, time roughly starts at "root" and proceeds outward. Different genera branch into others due to the accidents of biological evolution — mutation, recombination, genetic drift, extinction, etc. This is just *one* history of how the genera might have been formed. The evolutionary multiverse consists of all the different possible trees like this one, representing different ways that the branching structure could have happened. We do not directly observe any of this history. It is in the past. Yet we believe in the multiverse of biological evolution because it explains regularities between species that we observe today.

The highly successful Copenhagen formulation of quantum mechanics is not an alternative to DH, but an approximation to it that is appropriate for measurement situations.[18] Quantum cosmology is not in conflict with the successes in the laboratory.

But Copenhagen quantum mechanics must be generalized beyond the laboratory to be applicable to the early universe where no measurements were being made and no observers were around to carry them out. DH, the generalization used here, is not a departure from Copenhagen quantum mechanics where it applies, but rather a generalization of it to new domains of applicability. Some generalization is inevitable for cosmology, and various ones have been studied since the time of Everett.

Cosmology is a historical science like geology, biological evolution, and human history. Its aim is to use data that exists now to reconstruct the quantum past in order to simplify the prediction of the future (e.g. Ref. 21). We should not be surprised that the extension of observation to new domains of phenomena requires new extensions of existing theory.

Are quantum multiverses ***scientific?*** **Yes, in the author's opinion:** As sketched above, in many areas of science one finds multiverses of the kind mentioned in the introduction — an ensemble of possible situations, only one of which is observed by us.

Appendix B

A Little More DH

In this appendix, we repeat much of the qualitative discussion of the two-slit models in Sec. 2 with more equations. That may be helpful to some readers.

The inputs to the prediction of a quantum multiverse are the Hamiltonian H and quantum state $|\Psi(t)\rangle$ of the particles in the boxes. This is a function of time in the Schrödinger picture in which we work. The state of the electron can also be described by a wave function in configuration space, viz.

$$\Psi = \Psi(x, y, t) \tag{B.1}$$

using a coordinate x for the horizontal direction and y for the vertical direction in the three boxes as shown in Fig. 1 and assuming symmetry in the perpendicular direction. We move back and forth between wave functions like (B.1) and bras and kets like $|\Psi(t)\rangle$ as convenient.

Histories can be represented by quantum branch state vectors constructed from (H, Ψ). For an example, take the TSS model. Denote the initial state at time t_0 by $\Psi(x, y, t_0)$. This is a product of a wave packet in the x-direction $\phi(x, t_0)$ and a wave function $\psi_0(y, t_0)$ localized at the gun, viz.

$$\Psi(x, y, t_0) = \psi(y, t_0)\phi(x, t_0). \tag{B.2}$$

This wave function evolves in time by the Schrödinger equation

$$i\hbar\frac{\partial\Psi}{\partial t} = H\Psi. \tag{B.3}$$

We assume that $\phi(x, t)$ is a narrow wave packet peaked to the left of the slits, but moving to the right so as to reach the slits at time t_s and the detecting screen at t_d. Thus, its progress in x recapitulates evolution in time. After passing through the slits, the wave function has the approximate form

$$\Psi(x, y, t) = \psi_U(y, t)\phi(x, t) + \psi_L(y, t)\phi(x, t), \quad t_s < t < t_d \tag{B.4a}$$

$$\equiv \Psi_U(x, y, t) + \Psi_L(x, y, t). \tag{B.4b}$$

Here, in the first term, $uc\Psi_U(y, t)$ is localized near the upper slit at time t_s and spreads over a larger region of y by the time t_d such that the electron hits the detecting screen. Similarly for the second term. The last line defines branch wave functions $\Psi_U(x, y, t)$ and $\Psi_L(x, y, t)$ for the two histories in the set.

Simple two slit model (TSS): The electron starts localized at the gun. The alternative position intervals Y at the further screen define a coarse-grained set of alternative coarse-grained histories of the electron between the times t_0 and t_d. If $\{P_Y\}$ are a complete set of orthogonal projections on these intervals, the branch state vectors for these histories are $|\Psi_Y(t_d)\rangle \equiv P_Y|\Psi(t_d)\rangle$. The probability to arrive at Y predicted from (H, Ψ) is

$$p(Y) = \|P_Y|\Psi(t_d)\rangle\|^2 = \||\Psi_Y(t_d)\rangle\|^2. \tag{B.5}$$

A finer-grained set of histories for TSS would also specify whether the electron passed through the the upper slit U or the lower slit L on its way to a given interval Y. Naively, one might expect that the probabilities $p_U(Y)$ and $p_L(Y)$ that the electron exits the upper or lower slit respectively and arrived at Y would be, from (B.4b),

$$p_U(Y) = \|P_Y|\Psi_U(t_d)\rangle\|^2, \quad p_L(Y) = \|P_Y|\Psi_L(t_d)\rangle\|^2. \tag{B.6}$$

Then we would have from (B.4b) and (B.5)

$$\begin{aligned} p(Y) &= \|P_Y|\Psi_U(t_d)\rangle + P_Y|\Psi_L(t_d)\rangle\|^2 \\ &= \|P_Y|\Psi_U(t_d)\rangle\|^2 + \|P_Y|\Psi_U(t_d)\rangle\|^2. \end{aligned} \tag{B.7}$$

This is false unless the two branches $P_Y|\Psi_U(t_d)\rangle$ and $P_Y|\Psi_L(t_d)\rangle$ are orthogonal so that the quantum interference between them vanishes. That is, the set is inconsistent unless the set of histories decoheres.

$$\langle\Psi_U(t_d)|P_Y|\Psi_L(t_d)\rangle \approx 0. \tag{B.8}$$

That won't be the case in TSS. The two histories interfere as shown by an interference pattern that is a characteristic feature of the two-slit experiment. TSS does not have a multiverse of this kind.

Two-slit model with gas (TSG): The TSG model contains a gas of particles that interact weakly with the electron — an example of an *environment*. The initial wave function is

$$\Psi(x, y, t_0) = \psi(y, t_0)\phi(x, t_0)\chi(t_0), \quad t_0 < t < t_s \tag{B.9}$$

where $\chi(t_0)$ is the initial state of all the particles in the gas. After passing through the slits this becomes

$$\Psi(x, y, t) = \psi_U(y, t)\phi(x, t)\chi_U(t) + \psi_L(y, t)\phi(x, t)\chi_L(t), \quad t_s < t < t_d \tag{B.10a}$$

$$\equiv \Psi_U(x, y, t) + \Psi_L(x, y, t) \tag{B.10b}$$

where $\chi_U(t)$ and $\chi_L(t)$ denote the state of the gas particles that have scattered from the region of the upper and lower slits respectively. Then $\Psi_U(x, y, t_d)$ and $\Psi_L(x, y, t_d)$ are the branch wave functions for the two histories that the electron arrived in Y at time t_d after passing through either the upper or lower slit.

In the simple two-slit model (TSS), the set of alternative histories describing which slit (U or L) the electron went through on its way to arrive at Y on the further screen did not decohere as required for a quantum multiverse. That was because the states representing the two histories were not orthogonal. However, because of the gas in (TSG), the two histories do decohere. The decoherence condition (B.8) is satisfied because the state of the gas scattered from the upper slit $|\chi_U\rangle$ is orthogonal to the state of the gas scattered from the lower slit $|\chi_L\rangle$ if enough particles scatter.

To see that gas ensures this orthogonality, consider that a single gas particle a whose initial state is $|\chi_a\rangle$ scatters near the upper slit and will be in a final state $S_U|\chi_a\rangle$, where S_U is the S-matrix for scattering near the upper slit. Similarly for the lower slit with $S_L|\chi_a\rangle$; if N gas particles scatter, the overlap between $|\chi_U\rangle$ and $|\chi_L\rangle$ will be proportional to N inner products of the form

$$\prod_{a=1}^{N} |\langle\chi_a|S_U^\dagger S_L|\chi_a\rangle|. \tag{B.11}$$

Since the two final state vectors $S_U|\chi_a\rangle$ and $S_L|\chi_a\rangle$ are different, each individual product has a magnitude less than 1. The product of a very large number of such products will be near zero, implying decoherence.

Thus for TSG we have exhibited two quantum multiverses at different levels of coarse-graining. The first one (just Y) is a coarse-graining of the second (Y and S). The first ignores the gas, the second follows it.

References

1. J. Barrow and F. Tipler, *The Anthropic Cosmological Principle* (OUP, Oxford, 1985).
2. S. Weinberg, Anthropic bound on the cosmological constant, *Phys. Rev. Lett.* **62**, 485 (1989).
3. J. B. Hartle and T. Hertog, Anthropic bounds on Lambda from the no-boundary quantum state, *Phys. Rev. D* **88**, 123516 (2013), arXiv:1309.0493.
4. S. Saunders, J. Barrett, A. Kent and D. Wallace (eds.), *Many Worlds? Everett, Quantum Theory and Reality* (OUP, Oxford, 2010).

5. D. Deutsch, *The Beginning of Infinity* (Viking Penguin, New York, 2011).
6. D. Wallace, *The Emergent Multiverse: Quantum Theory According to the Everett Interpretation* (OUP, Oxford, 2012).
7. J. B. Hartle and S. W. Hawking, The wave function of the universe, *Phys. Rev. D* **28**, 2960–2975 (1983); S. W. Hawking, The quantum state of the universe, *Nucl. Phys. B* **239**, 257–276 (1984).
8. J. B. Hartle and T. Hertog, *One bubble to rule them all*, arXiv:1604.03580.
9. J. B. Hartle, S. W. Hawking and T. Hertog, Local observation in eternal inflation, *Phys. Rev. Lett.* **106**, 141302 (2011), arXiv:1009.252.
10. T. Hertog, Predicting a prior for Planck, *JCAP* **1402**, 043 (2014), arXiv:1305.6135.
11. R. B. Griffiths, *Consistent Quantum Theory* (Cambridge University Press, Cambridge, UK, 2002); R. Omnès, *Interpretation of Quantum Mechanics* (Princeton University Press, Princeton, 1994); M. Gell-Mann, *The Quark and the Jaguar* (W. H. Freeman, New York, 1994); M. Gell-Mann and J. B. Hartle, Quantum mechanics in the light of quantum cosmology, in *Complexity, Entropy, and the Physics of Information*, ed. W. Zurek (Addison Wesley, Reading, MA, 1990); P. C. Hohenberg, An introduction to consistent quantum theory, *Rev. Mod. Phys.* **82**, 2835–2844 (2010).
12. J. B. Hartle, The quantum mechanics of closed systems, in *Directions in General Relativity*, Vol. 1, eds. B.-L. Hu, M. P. Ryan and C. V. Vishveshwara (CUP, Cambridge, 1993), arXiv:gr-qc/9210006.
13. H. Everett III, Relative state formulation of quantum mechanics, *Rev. Mod. Phys.* **29**, 454 (1957).
14. E. Joos and H. D. Zeh, The emergence of classical properties through interaction with the environment, *Zeit. Phys. B* **59**, 223 (1985).
15. W. Zurek, Decoherence, einselection, and the quantum origins of the classical, *Rev. Mod. Phys.* **75**, 715 (2003).

16. M. Gell-Mann and J. B. Hartle, Classical equations for quantum systems, *Phys. Rev. D* **47**, 334 (1993), arXiv: gr-qc/9210010.
17. M. Gell-Mann and J. B. Hartle, Quantum mechanics in the light of quantum cosmology, in *Complexity, Entropy, and the Physics of Information*, ed. W. Zurek (Addison Wesley, Reading, MA, 1990).
18. J. B. Hartle, The quantum mechanics of cosmology, in *Quantum Cosmology and Baby Universes: Proc. 1989 Jerusalem Winter School for Theoretical Physics*, eds. S. Coleman, J. B. Hartle, T. Piran and S. Weinberg (World Scientific, Singapore, 1991), Section II.10.
19. J. B. Hartle and T. Hertog, The observer strikes back, in *The Philosophy of Cosmology*, eds. K. Chamcham, J. Silk, J. D. Barrow and S. Saunders (Cambridge University Press, Cambridge, 2017), arXiv:1503.07295.
20. S. W. Hawking and T. Hertog, Populating the landscape: A top down approach, *Phys. Rev. D* **73**, 123527 (2006), arXiv:hep-th/0602091.
21. J. B. Hartle, Quantum pasts and the utility of history, *Physica Scripta* **T76**, 67–77 (1998), arXiv:gr-qc/9712001.
22. J. B. Hartle, *Living in a Superposition*, arXiv:1511.01550.
23. J. B. Hartle, *The quasiclassical realms of this quantum universe*, arXiv:0806.3776. A slightly shorter version is published in *Many Worlds?* eds. S. Saunders, J. Barrett, A. Kent and D. Wallace (Oxford University Press, Oxford, 2010), the longer version was reprinted in *Foundations of Physics* **41**, 982 (2011).
24. J. B. Hartle and M. Srednicki, Are we typical? *Phys. Rev. D* **75**, 123523 (2007), arXiv:0704.2530.
25. J. B. Hartle and T. Hertog, *Classical Prediction in Quantum Cosmology* (to be finished).
26. J. B. Hartle, S. W. Hawking and T. Hertog, The no-boundary measure in the realm of eternal inflation, *Phys. Rev. D* **82**, 063510 (2010), arXiv:1001.0262.
27. J. B. Hartle, S. W. Hawking and T. Hertog, Classical universes of the no-boundary quantum state, *Phys. Rev. D* **77**, 123537 (2008), arXiv:0803.1663.

28. J. B. Hartle, S. W. Hawking and T. Hertog, The no-boundary measure of the universe, *Phys. Rev. Lett.* **100**, 202301 (2008), arXiv:0711.4630; Classical universes of the no-boundary quantum state, *Phys. Rev. D* **77**, 123537 (2008), arXiv:0803.1663; J. B. Hartle and T. Hertog, Replication regulates volume weighting in quantum cosmology, *Phys. Rev. D* **80**, 063531 (2009), arXiv:0905.3877.
29. S. Coleman and F. De Luccia, Gravitational effects on and of vacuum decay, *Phys. Rev. D* **21**, 3305 (1980).
30. S. W. Hawking and T. Hertog, *A smooth exit from eternal inflation*, arXiv:1707.7702.
31. M. Tegmark, A. Aguirre, M. J. Rees and F. Wilczek, Dimensionless constants, cosmology, and other dark matters, *Phys. Rev. D* **73**, 023505 (2006).
32. F. Denef and M. R. Douglas, Distributions of flux vacua, *JHEP* **2004**, 072 (2004).
33. P. A. R. Ade *et al.*, Planck 2013 results, XXII constraints on inflation, *Astron. Astrophys.* **571**, A22 (2014), arXiv:1303.5082.
34. B. J. Carr and M. J. Rees, The anthropic principle and the structure of the physical world, *Nature* **278**, 605–612 (1979).
35. S. Weinberg, *Living in the Multiverse*, published in Ref. 37, arXiv:hepth/0511037.
36. A. Vilenkin, *Many Worlds in One: The Search for Other Universes* (Hill and Wang, 2006).
37. The many fine articles in B. J. Carr, *Universe or Multiverse?* (Cambridge University Press, Cambridge, 2007).
38. B. J. Carr and G. F. R. Ellis, Universe or multiverse? *Astron. Geophys.* **49**(2), 2.29–2.33 (2008).
39. L. Mersini-Houghten, *Thoughts on Defining the Multiverse*, arXiv:0804.3419.
40. P. Steinhardt, Theories of anything, in *What Scientific Idea is Ready for Retirement*, https://www.edge.org/response-detail/25405
41. G. F. R. Ellis, Does the multiverse really exist? *Scientific American*, August 2011, p. 38ff.

42. A. Vilenkin and M. Tegmark, The case for parallel universes, *Scientific American*, July 19, 2011, p. 123.
43. F. Wilczek, Multiversality, *Class. Quant. Grav.* **30**, 193001 (2013), arXiv:2307.7376.
44. G. F. R. Ellis and J. Silk, Scientific method: Defend the integrity of physics, *Nature* **516**, 321–323 (2014).
45. G. F. R. Ellis, On the philosophy of cosmology, *Stud. Hist. Philos. Sci. B: Stud. Hist. Philos. Mod. Phys.* **46**, 5–23 (2014).
46. A. Linde, *A Brief History of the Multiverse*, arXiv:1512.01203.
47. M. Gell-Mann and J. B. Hartle, Decoherent histories quantum mechanics with one 'real' fine-grained history, *Phys. Rev. A* **85**, 062120 (2012), arXiv:1106.0767.
48. J. B. Hartle, Quantum physics and human language, *J. Phys. A: Math. Theor.* **40**, 3101–3121 (2007), arXiv:quant-ph/0610131.
49. S. F. Elena and R. E. Lenski, Evolution experiments with microorganisms: The dynamics and genetic bases of adaption, *Nat. Rev. (Genet.)* **4**, 487 (2004).
50. L. M. Wahl and D. C. Krakauer, Models of experimental evolution: The role of genetic chance and selective necessity, *Genetics* **156**, 1437 (2000).
51. Norman Pace, from http://pacelab.colorado.edu
52. J. B. Hartle and T. Hertog, Replication regulates volume weighting in quantum cosmology, *Phys. Rev. D* **80**, 063531 (2009), arXiv:0905.3877.

CHAPTER 21

Quantum Mechanics at the Planck Scale*

Usual quantum mechanics requires a fixed, background, spacetime geometry and its associated causal structure. A generalization of the usual theory may therefore be needed at the Planck scale for quantum theories of gravity in which spacetime geometry is a quantum variable. The elements of generalized quantum theory are briefly reviewed and illustrated by generalizations of usual quantum theory that incorporate spacetime alternatives, gauge degrees of freedom, and histories that move forward and backward in time. A generalized quantum framework for cosmological spacetime geometry is sketched. This theory is in fully four-dimensional form and free from the need for a fixed causal structure. Usual quantum mechanics is recovered as an approximation to this more general framework that is appropriate in those situations where spacetime geometry behaves classically.

*Talk given at the *Workshop on Physics at the Planck Scale*, Puri, India, December 12–21, 1994. This talk is a précis of the author's 1992 Les Houches Lectures *Spacetime Quantum Mechanics and the Quantum Mechanics of Spacetime.*[1]
http://arxiv.org/abs/gr-qc/9508023v2

1. Introduction

Generalizations of usual quantum theory are required for physics at the Planck scale. That is because the usual framework depends strongly on an assumed fixed, background, spacetime geometry. States are defined on spacelike surfaces in this geometry. States evolve unitarily in between such surfaces in the absence of measurement and by state vector reduction on them when a measurement occurs. The inner product between states is defined by integrals over fields on a spacelike surface. These are just some of the ways that a fixed spacetime geometry is central to the usual formulation of quantum mechanics. However, at the Planck scale, spacetime geometry is not fixed, but a quantum dynamical variable — fluctuating and without definite value. Given two nearby points on a spacetime manifold, it is not possible to say whether they are spacelike separated or not. Rather, the amplitudes for prediction are sums over *different* metrics on the manifold. Points separated by a spacelike interval in one metric may be timelike separated in another that contributes just as significantly to the sum. For this reason, usual quantum mechanics needs to be generalized to accommodate quantum spacetime. There are many approaches to this generalization whose difficulties have been lucidly reviewed in Refs. 2–5. This lecture describes another approach — the effort to provide a generalization of usual quantum mechanics that is in fully spacetime form and does not require a fixed spacetime geometry, but which yields the usual formulation approximately in those physical situations where spacetime geometry *is* approximately fixed.

The rules for the calculation of S-matrix elements in field theory may be summarized without reference to states on spacelike surfaces except in asymptotic regions. It is thus possible to carry out investigations of the dynamics of quantum gravity as expressed by S-matrix elements without addressing the issues of generalization raised above. However, one of the principal applications of quantum gravity is to the very early universe where the physics of the Planck scale becomes important. In quantum cosmology, we cannot evade the challenge of generalizing quantum theory. We live in the middle of this particular experiment.

The application of quantum mechanics to cosmology also requires another kind of generalization of the usual formulation. Usual quantum mechanics predicts the outcomes of "measurements" carried out on a system by another system outside it. But in cosmology there is no system outside. Cosmology requires a quantum mechanics of closed systems that is a generalization of the usual theory. Recent years have seen the emergence of such a generalization built on the works of Everett, Zeh, Zurek, Griffiths, Omnès, Gell-Mann, and many others.[a] The most general predictions of this formulation of quantum mechanics are the probabilities of the individual members of sets of alternative histories of the closed system. Consistency of probability sum rules is the criterion determining the sets of histories which may be assigned probabilities rather than any notion of measurement. The absence of quantum-mechanical interference between histories, or decoherence, is the sufficient condition for this consistency. The initial condition of the closed system and Hamiltonian determine which sets of histories decohere, rather than the action of any external observer.

There is a relation between these two required generalizations. By abstracting the principles of the quantum mechanics of closed systems, one arrives at a more general framework for quantum prediction — generalized quantum mechanics. Within generalized quantum mechanics one can construct fully spacetime quantum theories which do not require a fixed background spacetime. In this lecture, we shall review some simple examples of generalized quantum theories and sketch a generalized quantum mechanics of spacetime geometry. Our discussion is necessarily brief; for a more extensive one, see Ref. 1.[b]

2. Generalized Quantum Theory

Not every set of histories that may be described can be consistently assigned probabilities in quantum theory. In the two-slit experiment,

[a]For an introduction from the author's perspective, see Ref. 6.

[b]This article may be thought of as an abridgement of the author's lectures in Ref. 1.

it would be inconsistent to assign probabilities to the alternatives that the electron went through slit A or slit B on its way to detection at a point y on the screen in the absence of a measurement detecting which happened. The probability to arrive at y would not be the sum of the probability to go through A and arrive at y, and the probability to go through B and arrive at y, because of quantum interference. In quantum mechanics, probabilities are squares of amplitudes, and

$$\left|\psi_A(y) + \psi_B(y)\right|^2 \neq \left|\psi_A(y)\right|^2 + \left|\psi_B(y)\right|^2. \tag{1}$$

A criterion is therefore needed to specify which sets of histories may be consistently assigned probabilities. In the familiar quantum mechanics, that criterion is *measurement* — probabilities may be consistently assigned to histories of *measured* alternatives and not, in general, otherwise. In the two-slit experiment, if we measure which slit the electron passes through, then interference is destroyed, the probability sum rules are obeyed, and probabilities are predicted for the two alternative histories of the electron.

However, a criterion based on measurements or observers cannot be fundamental in a quantum theory that seeks to explain the early universe where neither existed. A more general criterion for closed systems assigns probabilities to just those sets of histories for which there is vanishing interference between its individual members as a consequence of the system's initial quantum state.[7–9] Such sets of histories are said to *decohere*. Decoherent sets of histories are what may be used for prediction and retrodiction in quantum cosmology, for they may be assigned probabilities.

The usual Hamiltonian formulation of quantum mechanics in a fixed background spacetime was the context for the treatment of the decoherent sets of histories of a closed system in Refs. 7–9. However, it is possible to abstract from these discussions, principles for a wider class of quantum-mechanical theories called *generalized quantum theories*. Within that framework, generalized quantum theories of spacetime geometry may be constructed which do not assume a fixed background spacetime.

The principles of generalized quantum mechanics were introduced in Ref. 10 and developed more fully in Refs. 11 and 1. The principles have been axiomatized in a rigorous mathematical setting by Isham.[12] Three elements are needed to specify a generalized quantum theory:

(1) The sets of *fine-grained histories.* These are the most refined possible description of a closed system.
(2) The allowed *coarse-grainings.* A coarse-graining of a set of histories is generally a partition of that set into mutually exclusive classes $\{c_\alpha\}, \alpha = 1, 2, \cdots$ called *coarse-grained histories.* Each coarse-grained history is a set of fine-grained histories, and the set of classes constitutes a set of coarse-grained histories, with each history labeled by the discrete index α.
(3) A *decoherence functional* defined for each allowed set of coarse-grained histories which incorporates a theory of the initial condition and dynamics of the closed system and measures the quantum-mechanical interference between pairs of histories in the set. A decoherence functional $D(\alpha', \alpha)$ must satisfy the following properties.

(i) *Hermiticity:*

$$D(\alpha', \alpha) = D^*(\alpha, \alpha'). \tag{2a}$$

(ii) *Positivity:*

$$D(\alpha, \alpha) \geq 0. \tag{2b}$$

(iii) *Normalization:*

$$\sum\nolimits_{\alpha'\alpha} D(\alpha', \alpha) = 1. \tag{2c}$$

(iv) *The Principle of Superposition:*
If $\{\bar{c}_{\bar{\alpha}}\}$ is a coarse-graining of a set of histories $\{c_\alpha\}$, that is, a further partition into classes $\{\bar{c}_{\bar{\alpha}}\}$, then

$$D(\bar{\alpha}', \bar{\alpha}) = \sum_{\alpha' \in \bar{\alpha}'} \sum_{\alpha \in \bar{\alpha}} D(\alpha', \alpha). \tag{2d}$$

Once these three elements are specified, the process of prediction proceeds as follows: A set of histories is said to (medium) decohere if all the "off-diagonal" elements of $D(\alpha', \alpha)$ are sufficiently small. The diagonal elements are the probabilities $p(\alpha)$ of the individual histories in a decoherent set. These two definitions are summarized in the one relation

$$D(\alpha', \alpha) \approx \delta_{\alpha'\alpha} p(\alpha). \tag{3}$$

As a consequence of (3) and the four properties of (2a)–(2d), the numbers $p(\alpha)$ lie between zero and one, sum to one, and satisfy the most general form of the probability sum rules

$$p(\bar{\alpha}) = \sum_{\alpha \in \bar{\alpha}} p(\alpha) \tag{4}$$

for any coarse-graining $\{\bar{c}_{\bar{\alpha}}\}$ of the set $\{c_\alpha\}$. The $p(\alpha)$ are therefore probabilities. They are the predictions of generalized quantum mechanics for the possible coarse-grained histories of the closed system that arise from the theory of its initial condition and dynamics incorporated in the construction of D.

In the following, we shall illustrate this general framework with examples designed to realize it in concrete form but also designed to show how to cast quantum theories into fully spacetime form. For these illustrative purposes, we shall confine ourselves to sum-over-histories formulations which posit a unique set of fine-grained histories. Relations to operator formulations are discussed in Ref. 1.

3. Non-Relativistic Quantum Mechanics

The non-relativistic quantum mechanics of a particle moving in one dimension is the simplest example of a generalized quantum theory when its three elements are identified as follows:

(1) *Fine-grained Histories:* The fine-grained histories are particle paths $x(t)$ which are *single-valued* functions of t on a fixed time interval, say, $[0, T]$.

(2) *Allowed Coarse-grainings:* The fine-grained paths may be partitioned by their behavior with respect to regions of x at definite moments of time. For instance, consider a division of the real line into intervals $\{\Delta^1_{\alpha_1}\}$ at time t_1 and $\{\Delta^2_{\alpha_2}\}$ at time t_2 where α_1 and α_2 run over a discrete set of labels. One coarse-grained history consists of all the paths which pass through, say, Δ^1_8 and Δ^2_{17}. This is the coarse-grained history in which the particle is

localized in region Δ^1_8 at time t_1 and Δ^2_{17} at time t_2. An exhaustive set of coarse-grained histories in this example is specified by considering all possible sequences of two regions in the sets $\{\Delta^1_{\alpha_1}\}$ and $\{\Delta^2_{\alpha_2}\}$.

(3) *Decoherence Functional:* Given a partition of the set of fine-grained paths into classes $\{c_\alpha\}, \alpha = 1, 2, \cdots$ we may define a class operator for each class (coarse-grained history) by a sum over paths in the class, viz.

$$\langle x''|C_\alpha|x'\rangle = \int_{c_\alpha} \delta x \, \exp\bigl(iS[x(\tau)]\bigr). \tag{5}$$

The functional $S[x(\tau)]$ is the classical action which defines quantum dynamics in path integrals. The sum is over paths which start at x' at $t = 0$, proceed to x'' at $t = T$, and lie in the class c_α. A sum of the form (5) over the class of *all* paths is just an expression for the propagator $\langle x''|\exp(-iHT)|x'\rangle$. Thus,

$$\sum_\alpha C_\alpha = e^{-iHT}. \tag{6}$$

With the class operators in hand, the decoherence functional for non-relativistic quantum mechanics is

$$D(\alpha', \alpha) = \mathrm{Tr}\bigl[C_{\alpha'} \rho \, C^\dagger_\alpha\bigr] \tag{7}$$

where ρ is the density matrix representing the initial condition of the closed system. It is not difficult to see from (5), (6), and the properties of density matrices, that the four requirements for decoherence functionals (2a)–(2d) are satisfied.

The class operators are simply expressed for the coarse-grainings by exhaustive sets of ranges of positions at definite movements of time that we have taken to *define* the allowed coarse-grainings of non-relativistic quantum mechanics above. If $\{P^k_{\alpha_k}(t_k)\}$ are a set of Heisenberg-picture projections onto ranges $\{\Delta^k_{\alpha_k}\}$ at time t_k, then the class operator for the paths that pass through intervals $\Delta^1_{\alpha_1}, \ldots, \Delta^n_{\alpha_n}$ at $t_1, \ldots, t_n$ is

$$C_{\alpha_n, \ldots, \alpha_1} = e^{-iHT} P^n_{\alpha_n}(t_n) \cdots P^1_{\alpha_1}(t_1). \tag{8}$$

In (8) one can see the two forms of evolution in usual quantum mechanics. Applied to a state, (8) evolves it unitarily in between times $t_1, \ldots, t_n$ (Heisenberg-picture evolution of the P's) and by state vector reduction (action of the P's) at the times $t_1, \ldots, t_n$.

This formulation of non-relativistic quantum mechanics is not in fully spacetime form. Quantum dynamics has been expressed in spacetime form through the use of Feynman path integrals over histories. But the alternatives are restricted to sequences of sets of alternative ranges of position at *definite moments of time*. However, this usual framework of non-relativistic theory is easily generalized to alternatives that are in spacetime form. Simply allow arbitrary partitions as coarse-grainings and replace (2) by

(2′) *Allowed Coarse-grainings:* Arbitrary partitions of the fine-grained paths.

This allows new types of *spacetime* alternatives which are not at definite moments of time. For example, given a region R with extent in both space and time, one could partition the paths into the class c_0 of paths that never cross R and the class c_1 that cross R sometimes. A way to see that this is a genuine generalization is to note that the resulting class operators are generally not unitary, and neither are they projections or products of projections as in (8). Thus, this generalized quantum mechanics of a non-relativistic particle cannot be reformulated in terms of states on surfaces of constant time which evolve unitarily or by reduction. Even without states on spacelike surfaces, the theory is predictive as described above, and this generalization is now in fully spacetime form with respect to both dynamics and alternatives.

4. Abelian Gauge Fields

The three elements of a generalized quantum mechanics of the free electromagnetic field are:

(1) *Fine-grained Histories:* Configurations of the potential $A_\mu(x)$ on a spacetime region between two constant time surfaces, say, those at $t = 0$ and $t = T$.

(2) *Allowed Coarse-grainings:* Partitions of the fine-grained histories of the potential into *gauge-invariant* classes $\{c_\alpha\}$.

(3) *Decoherence Functional:* Class operators on the physical Hilbert space of transverse degrees of freedom of the vector potential are defined by

$$\langle \vec{A}^{T''} | C_\alpha | \vec{A}^{T'} \rangle = \int_{c_\alpha} \delta A \, \Delta_\Phi[A] \, \delta[\Phi(A)] \, \exp\big(iS[A]\big). \tag{9}$$

The functional $S[A]$ is the action for the free electromagnetic field. $\Phi(A)$ is a function such that $\Phi(A) = 0$ is a gauge-fixing condition, and $\Delta_\Phi[A]$ is the associated Faddeev–Popov determinant. The sum is over all potentials $A(x)$ which match $\vec{A}^{T'}(\mathbf{x})$ on the initial surface $t = 0$ and $\vec{A}^{T''}(\mathbf{x})$ on the final surface $t = T$. In particular, the values of the time component $A^t(\mathbf{x})$ and the longitudinal component of $\vec{A}(\mathbf{x})$ on these surfaces are summed over. The decoherence functional is

$$D(\alpha', \alpha) = Tr\big[C_{\alpha'} \rho \, C_\alpha^\dagger\big] \tag{10}$$

where all operators and operations are defined in the Hilbert space of the transverse components of the vector potential.

When restricted to partitions by values of $\vec{A}^T(x)$ on spacelike surfaces, the generalized quantum mechanics specified above coincides with the familiar Hamiltonian quantum mechanics of the electromagnetic field. But more general alternatives are possible. One could consider partitions by the values of any gauge invariant functional, for instance, the average of a magnetic field component over a spacetime region R

$$F[A] = \frac{1}{\Delta V} \int_R d^4x B_z(x). \tag{11}$$

More specifically, let $\{\Delta_\alpha\}$ be a set of mutually exclusive intervals making up the real line. The coarse-grained history c_α consists of all those potentials $A_\mu(x)$ for which $F[A] \in \Delta_\alpha$. Such field averages are familiar from Bohr and Rosenfeld's discussion of the measurability of the electromagnetic field.

Beyond functionals of the type (11) which depend only on the transverse degrees of freedom, one can also consider partitions by values of averages of quantities like $\vec{\nabla} \cdot \vec{E}$ which are gauge invariant but involve non-transverse degrees of freedom. In this way the question of whether the constraint $\vec{\nabla} \cdot \vec{E} = 0$ is satisfied becomes a question of the probability of its value rather than a matter of the absence of the relevant degrees of freedom.

5. The Relativistic World Line

Classical dynamics in general relativity may be thought of as the evolution of the spatial geometry of a family of spacelike surfaces that foliate spacetime. This dynamics may be exhibited by decomposing the metric in $3+1$ form with respect to these surfaces:

$$ds^2 = -N^2 dt^2 + h_{ij}\big(dx^i + N^i dt\big)\big(dx^i + N^i dt\big). \tag{12}$$

Here, t is the label of a spacelike surface and the x^i are three coordinates in it.

Two types of invariance of the action of general relativity may be distinguished when it is expressed in terms of the variables defined by (12). First, there are transformations of the coordinates x^i in the spacelike surfaces

$$x^i \to \bar{x}^i(x^j), \tag{13}$$

which for infinitesimal transformations $x^i \to x^i + \xi^i(x^j)$ lead to

$$h_{ij}(x) \to h_{ij}(x) + D_{(i}\xi_{j)}(x) \tag{14}$$

where D_i is the derivative in the surface. The transformation rule (14) is not unlike that of gauge transformations in electromagnetism. It can be treated in a way analogous to that of Sec. 4 in the construction of a generalized quantum framework for general relativity.

A class of invariances of a different character are the reparameterizations of the time

$$t \to \bar{t}(t), \tag{15}$$

and these have their own special implications for the construction of a generalized quantum mechanics for gravitation. A simple model theory which exhibits time reparameterization invariance classically, is the relativistic world line. Its action may be taken to be

$$S\left[x^{\mu}, N\right]=\frac{m}{2} \int_{0}^{1} d\lambda\, N(\lambda)\left[\left(\frac{\dot{x}^{\mu}(\lambda)}{N(\lambda)}\right)^{2}-1\right]. \tag{16}$$

Here, $x^{\mu}(\lambda)$ are the coordinates of the world line as functions of a parameter λ along it, a dot denotes a derivative with respect to λ, and $N(\lambda)$ is a multiplier enforcing the constraint $(p^{\mu})^2=-m^2$. This action is invariant under reparameterizations $\lambda \rightarrow \bar{\lambda}(\lambda)$ that leave the endpoints fixed, with an appropriate transformation law for N.

It is not possible to construct a quantum theory of a *single* relativistic particle interacting with an external field within the framework of usual Hamiltonian quantum mechanics. That is because the external field will produce pairs so that the only consistent theory is a *many*-particle theory. However, it *is* possible to construct a generalized quantum theory of a single relativistic world line even interacting with an external field. This is not a theory of realistic relativistic particles such as the proton and electron, for that is provided by field theory. However, in many ways this is a useful model for the quantum cosmology of a universe with a single fixed topology. We now sketch the elements of a generalized quantum theory of a single relativistic world line, confining ourselves, for simplicity, to the non-interacting case.

(1) *Fine-grained Histories:* The fine-grained histories are paths $(x^{\mu}(\lambda), N(\lambda))$ in the extended configuration space of paths in spacetime $x^{\mu}(\lambda)$ and multiplier $N(\lambda)$. The paths in spacetime are not single-valued in the time of any Lorentz frame, but move both forward and backward in the time of any Lorentz frame.
(2) *Allowed Coarse-grainings:* Any partition of the fine-grained histories into *reparametrization invariant* classes c_{α} is an allowed coarse-graining. For example, given a fixed spacetime region R, one could partition the paths $x^{\mu}(\lambda)$ into those which never cross R and the class which cross R at least once. One could partition the fine-grained histories between two such regions by the value of the reparameterization invariant "proper time"

$$\int N(\lambda) d\lambda \tag{17}$$

between the last passage of the first region and the first passage of the second. More generally, one could partition the fine-grained histories by the values of any reparameterization invariant functional $F[x^\mu(\lambda), N(\lambda)]$.

It is *not* possible to partition this class of paths by the position that the world line intersects a spacelike surface. The world line may cross a given surface, not just at one position, but at an arbitrarily large number of them.

(3) *Decoherence Functional:* For each class of fine-grained histories c_α in an allowed coarse-graining, the matrix elements of a class operator may be defined by

$$\left\langle x'' \left\| C_\alpha \right\| x' \right\rangle = \int_{c_\alpha} \delta x \, \delta N \, \Delta_\Phi[x, N] \, \delta\left[\Phi(x, N)\right] \exp(iS[x, N]). \tag{18}$$

Here, the sum is over all fine-grained histories whose paths start at the spacetime point x''^μ, end at the point x'^μ, and lie in the class c_α. The functional S is the action (16). The condition $\Phi = 0$ fixes a parameterization and Δ_Φ is the associated Faddeev–Popov determinant. The measure for the integration is that induced by the Liouville "*dpdq*" measure on phase space.

Initial and final conditions are represented by wave functions $\psi(x)$ and $\phi(x)$ on initial and final spacelike surfaces in spacetime, labeled σ' and σ'' respectively. We define class operators in linear spaces of such wave functions $\{\psi_j(x)\}$ and $\{\phi_i(x)\}$ by "attaching" wave functions to (18) using a bilinear (but not necessarily positive) inner product $\circ$, viz.

$$\langle \phi_i \left| C_\alpha \right| \psi_j \rangle = \phi_i^*(x'') \circ \left\langle x'' \left\| C_\alpha \right\| x' \right\rangle \circ \psi_j(x'). \tag{19}$$

The Klein–Gordon inner product is an appropriate choice for $\circ$ for the relativistic world line. Specifically, in the case that σ' and σ'' are surfaces of constant time in some Lorentz frame, we define

$$\langle \phi_i \left| C_\alpha \right| \psi_j \rangle = -\int_{\sigma''} d^3x'' \int_{\sigma'} d^3x' \phi_i^*(x'') \frac{\overleftrightarrow{\partial}}{\partial t''} \left\langle x'' \left\| C_\alpha \right\| x' \right\rangle \frac{\overleftrightarrow{\partial}}{\partial t'} \psi_j(x'). \tag{20}$$

If the wave functions defined off the surfaces σ' and σ'' are consistent with the operator form of the constraint $(p^\mu)^2 = -m^2$, i.e. satisfy the Klein–Gordon equation:

$$(-\nabla^2 + m^2)\phi_i(x) = 0, \tag{21}$$

then the construction (20) is independent of deformations of σ' and σ'' provided they do not intersect.

For an initial condition represented by a single wave function $\psi(x)$, and a final condition of equal probability for a set of wave functions $\{\phi_i(x)\}$, the decoherence functional is

$$D(\alpha', \alpha) = \mathcal{N} \sum_i \langle \phi_i | C_{\alpha'} | \psi \rangle \langle \psi | C_\alpha | \phi_i \rangle \tag{22a}$$

where $\mathcal{N}$ is a normalizing factor

$$\mathcal{N}^{-1} = \sum_i |\langle \phi_i | C_u | \psi \rangle|^2 , \tag{22b}$$

with C_u being the class operator for *all* fine-grained histories. It is not difficult to verify that this decoherence functional satisfies the four general requirements (2a)–(2d).

With the decoherence functional (22a), sets of alternative histories which decohere may be identified and the probabilities of the individual alternative histories computed. This generalized quantum-mechanical framework cannot be reformulated in terms of states on spacelike surfaces, their unitary evolution and reduction. The inclusion of paths that move both forward and backward in time implies it is not meaningful to consider alternative values of position at moments of time so that identities like (8) can no longer hold. Thus, by moving beyond the strictures of usual Hamiltonian quantum mechanics, to a more general quantum framework, we have exhibited a quantum theory of a single relativistic world line in fully spacetime form.

6. General Relativity

The question of whether there is a consistent manageable quantum theory of Einstein's general relativity is still open at the time of writing. Whatever the answer, a generalized quantum theory of general relativity provides a model of the kind of conceptual issues that must be faced in any quantum theory of gravity. Further, such a generalized quantum theory, made finite by truncating its ultraviolet divergences, may be a useful tool in quantum cosmology for investigating the predictions of a theory of the initial condition for the very low-energy phenomena of the universe like the galaxy–galaxy correlation function.

As mentioned above, the classical theory of general relativity exhibits symmetries analogous both to gauge symmetries and reparameterizations of time. Building on the treatments of gauge theories in Sec. 4 and the relativistic world line in Sec. 5, it is possible to sketch the elements of a generalized quantum theory for quantum spacetime. For simplicity, we restrict attention to spatially closed cosmological four-geometries.

(1) *Fine-grained Histories:* A class of metrics on four-dimensional manifolds are the fine-grained histories of a generalized quantum theory of spacetime geometry. The framework is broad enough to allow for different manifolds, and thus we can discuss topology change, but for simplicity in this discussion we restrict attention to the case where the manifold is fixed and of form $\mathbf{R}\times M^3$ where M^3 is a closed three-manifold that supports spatially closed cosmological geometries. What behavior is permitted in the fine-grained histories on very small scales, and what singularities are allowed on large scales, are two issues related to the ultraviolet behavior of the theory that we shall not discuss here. The geometries should at least be such that the action for general relativity is finite.

(2) *Allowed Coarse-grainings:* The general idea is a partition of the fine-grained metrics into four-dimensional *diffeomorphism-invariant* classes $\{c_\alpha\}$. For example, we could partition all cosmological four metrics into the class that contains no three-surface with a volume bigger than V_0 and the class that contains at least one such three-surface. The probability of the second of these alternatives can be thought of as the probability that the universe expands to a volume bigger than V_0. The set of

fine-grained histories could be partitioned into the class of geometries which is homogeneous and isotropic to some standard at large volumes and the class of geometries which is not homogeneous and isotropic to that standard. The probability of the first class is the probability that the universe becomes homogeneous and isotropic far from its singularities. The set of fine-grained histories could be partitioned into geometries which obey the Einstein equations to some accuracy at large volumes and the class which do not obey the Einstein equation. The probability of the first class is the probability that the universe becomes classical at large size. With further partitions of this class, the probabilities of individual classical histories could be calculated.

In general, when we ask for the probability of *any* property of the universe which can be expressed in terms of spacetime geometry and matter fields, there is a corresponding partition of the fine-grained histories into the class which has this property and the class which does not have it. If it is not possible to tell which four-dimensional geometries have the property, and which do not, then the property is not well defined. Coarse-grainings by partitions of the fine-grained histories into diffeomorphism-invariant classes is thus the most general notion in quantum theory of a set of alternatives expressible in terms of spacetime geometry.

(3) *Decoherence Functional:* The construction of a decoherence functional for general relativity parallels the construction of that for the relativistic particle. We sketch it here, referring the reader to Ref. 1 for more details. Consider histories on a manifold $M = \mathbf{I} \times M^3$ with two end boundaries $\partial M'$ and $\partial M''$. Any four-dimensional metric and matter field configuration on M induces a three-metric $h'_{ij}(\mathbf{x})$ and a field configuration $\chi'(\mathbf{x})$ on $\partial M'$. Similarly, three-metrics and fields are induced on the other end of the history $\partial M''$. Class operator matrix elements for a class c_α may be defined by

$$\left\langle h''_{ij}, \chi'' \left\| C_\alpha \right\| h'_{ij}, \chi' \right\rangle = \int_{c_\alpha} \delta g \, \delta\phi \, \Delta_\Phi[g] \, \delta\left[\Phi(g)\right] \exp\left(iS[g, \phi]\right). \tag{23}$$

Here the sum is over four-dimensional metrics g and field

configurations that induce the assigned three-metrics and fields on the boundaries of M and lie in the class c_α. The functional S is the action for metric coupled to matter, and Φ is a suitable gauge fixing condition.

Pure initial or final conditions are represented by wave functions on the superspace of three-metrics and spatial matter field configurations that satisfy operator versions of the four constraints of general relativity

$$\mathcal{H}_\mu(\mathbf{x})\Psi[h_{ij}(\mathbf{x}),\chi(\mathbf{x})]=0. \tag{24}$$

One of these is the Wheeler–DeWitt equation. The action of the class operators on such wave functions is specified by a linear (but not necessarily positive) inner product $\circ$. Thus,

$$\langle\Phi|C_\alpha|\Psi\rangle=\Phi^*\left[h''_{ij},\chi''\right]\circ\left\langle h''_{ij},\chi''\|C_\alpha\|h'_{ij},\chi'\right\rangle\circ\Psi\left[h'_{ij},\chi'\right]. \tag{25}$$

There are various candidates for this $\circ$ product. The most immediate is the analog of the Klein–Gordon inner product — the DeWitt inner product on hypersurfaces in superspace.

Density matrices representing non-pure initial and final conditions may be constructed in linear spaces of wave functions $\{\Psi_k[h_{ij},\chi]\}$ satisfying the constraints (24). For example

$$\left\langle h''_{ij},\chi''\|\rho^i\|h'_{ij},\chi'\right\rangle=\sum\nolimits_k\Psi_k\left[h''_{ij},\chi''\right]p'_k\Psi^*_k\left[h'_{ij},\chi'\right] \tag{26}$$

and similarly for ρ^f. The decoherence functional is then constructed in the obvious way on these linear spaces

$$\begin{aligned}D(\alpha',\alpha)&=\mathcal{N}\,Tr\left[\rho^f C_{\alpha'}\rho^i C^\dagger_\alpha\right]\\&\equiv\mathcal{N}\sum\nolimits_{ij}p''_i\,\langle\Phi_i\,|C_{\alpha'}|\,\Psi_j\rangle\;p'_i\,\langle\Psi_j\,|C_\alpha|\,\Phi_i\rangle\end{aligned} \tag{27}$$

where $\mathcal{N}$ is a normalizing factor ensuring $\sum_{\alpha'\alpha}D(\alpha',\alpha)=1$.

The generalized quantum mechanics defined by the above three elements does not require a fixed background spacetime or its associated causal structure, at least formally. No special set of spacelike surfaces has been singled out in the specification of sets of fine-grained histories, the sets of possible alternatives, or the amplitudes that define the decoherence functional. This theory of four-dimensional quantum spacetimes is in fully four-dimensional form.

One cannot expect to reformulate this theory of quantum spacetime in terms of states on spacelike surfaces, not least for the reasons given in the Introduction. More specifically, this follows from the nature of the fine-grained histories. Spacelike surfaces may be defined intrinsically in a cosmological four-geometry by their three-geometries, including such properties as the total spatial volume. However, there will be fine-grained quantum geometries in which surfaces of a given three-geometry occur an arbitrary number of times. In this sense, the histories of quantum general relativity move both "forward" and "backward" in any time defined by three-geometry. The factorization of histories represented by relations like (8) cannot be expected to hold, and states on spacelike surfaces in superspace cannot be defined.

Yet, the usual formalism of quantum mechanics with states on spacelike surfaces in spacetime must hold in some approximation when geometry is approximately fixed and classical. To see how this comes about, consider a partition of the fine-grained histories into coarse-grained classes $\{c_\alpha\}$ representing different possible classical and non-classical evolutions defined to distance accuracies well above the Planck scale. Denote the possible classical evolutions by $\{c_\gamma\}$, and suppose these are further refined by partitions defining different behaviors of the matter fields in these geometries, into classes $\{c_{\gamma\beta}\}$. Suppose further that the initial and final conditions are such that, when these classes are sufficiently coarse-grained, the integral over metrics from (23) in (25) may be done by the method of steepest descents, with but a single classical metric g_γ contributing in each classical class and no metrics at all in the non-classical classes. The class operator matrix elements in this approximation vanish for the non-classical classes and are determined by functional integrals of the form

$$\int_{c_{\gamma\beta}} \delta\phi \exp\bigl(iS[g_\gamma, \phi]\bigr) \tag{28}$$

for those classes $c_{\gamma\beta}$ that define classical behavior of geometry. Functional integrals like (28) define a matter field theory in a fixed background spacetime g_γ. These amplitudes can be equivalently constructed from states of the matter field on spacelike surfaces and their evolution. The classical metric g_γ fixes the meaning of "spacelike" and defines the notion(s) of time.

In this way the usual formulation of quantum mechanics is recovered as an approximation to the more general theory for those coarse-grainings and initial and final conditions in which spacetime geometry is approximately fixed, behaves classically, and can define the necessary fixed causal structure.

7. Conclusions

The accompanying table shows where we have been. We have used the framework of generalized quantum theory to exhibit a series of generalizations of usual quantum theory. In each case we have specified the three elements: fine-grained histories, allowed coarse-grainings, and decoherence functional. The successive generalizations have incorporated spacetime alternatives, gauge symmetries, and histories that move forward and backward in time — all expected features of a theory of quantum spacetime. We were able to build on these generalizations to formally sketch a quantum theory of geometry and matter fields that does not require a fixed background spacetime geometry or causal structure, as familiar quantum theory does.

From this perspective, the usual framework, with its states on spacelike surfaces, with their unitary evolution and reduction, is not the most fundamental formulation of quantum theory. Rather, it is an approximation to a more general theory sketched above that is free from the prerequisite of a fixed background spacetime and therefore applicable at the Planck scale.

Acknowledgment

The preparation of this report as well as the work described in it were supported in part by the US National Science Foundation under grant PHY90-08502.

GENERALIZED QUANTUM THEORIES

	Non-Relativistic Quantum Mechanics	Gauge Field Theory	Single Relativistic World Line	General Relativity
Fine-Grained Histories $\{f\}$	Paths $x(t)$ that move forward in time.	Four-dimensional single-valued configurations of the potential $A^{\mu}(x)$.	Paths in spacetime $x^{\alpha}(\lambda)$ that move both forward and backward in time and multiplier $N(\lambda)$.	Four-dimensional manifolds, M with metrics $g_{\alpha\beta}(x)$, and matter fields $\phi(x)$.
Coarse Grainings $\{c_{\alpha}\}$	Partitions of the paths into classes.	Partitions of the potential into *gauge invariant* classes.	Partitions of the paths into *reparametrization invariant* classes.	Partitions of manifolds, metrics, and fields into *diffeomorphism invariant* classes.
	e.g., (i) By the position of crossing a surface of constant time, (ii) By whether a path crosses a given spacetime region or does not.	*e.g.*, By the values of the field averaged over a given spacetime region, or any other gauge-invariant functional.	*e.g.*, (i) By whether the path crosses a given spacetime region or does not, (ii) By values of the "proper time" $\int N d\lambda$.	*e.g.*, By whether a 4-geometry contains a spacelike surface with a given 3-geometry or or does not.
Decoherence Functional	$\langle \text{end}''\lvert C_{\alpha}\rvert \text{end}'\rangle = \Sigma_{f\epsilon c_{\alpha}} \exp(iS[f])$ $\langle \phi_i\lvert C_{\alpha}\rvert\psi_j\rangle = \phi_i \circ \langle \text{end}''\lvert C_{\alpha}\rvert \text{end}'\rangle \circ \psi_j$ $D(\alpha', \alpha) = \mathcal{N}\Sigma_{ij} p_i'' \langle \phi_i\lvert C_{\alpha'}\rvert\psi_j\rangle \langle \phi_i\lvert C_{\alpha}\rvert\psi_j\rangle^{*} p_j'$			
	$\circ = L_2$ inner product on functions of x.	$\circ = L_2$ inner product on "true degrees of freedom".	$\circ =$ Klein-Gordon inner product on a surface. in spacetime.	$\circ =$ DeWitt inner product on a surface in superspace.

References

1. J. B. Hartle, Spacetime Quantum Mechanics and the Quantum Mechanics of Spacetime, in *Gravitation and Quantizations*, eds. B. Julia and J. Zinn-Justin, Les Houches Summer School Proceedings Vol. LVII (North Holland, Amsterdam, 1995), UCSBTH-92-21, gr-qc/9304006.
2. C. Isham, in *Recent Aspects of Quantum Fields*, eds. H. Mitter and H. Gausterer (Springer-Verlag, Berlin, 1992).
3. C. Isham, in *Integrable Systems, Quantum Groups, and Quantum Field Theories*, eds. L. A. Ibort and M. A. Rodriguez (Kluwer Academic Publishers, London, 1993).
4. K. Kuchař, in *Proc. 4th Canadian Conf. General Relativity and Relativistic Astrophysics*, eds. G. Kunstatter, D. Vincent and J. Williams (World Scientific, Singapore, 1992).
5. W. Unruh, in *Gravitation: A Banff Summer Institute*, eds. R. Mann and P. Wesson (World Scientific, Singapore, 1991).
6. J. B. Hartle, in *Directions in General Relativity, Vol. 1: A Symposium and Collection of Essays in Honor of Professor Charles W. Misner's 60th Birthday*, eds. B.-L. Hu, M. P. Ryan and C. V. Vishveshwara (Cambridge University Press, Cambridge, 1993); and in *Gravitation and Relativity 1989: Proc. 12th Int. Conf. General Relativity and Gravitation*, eds. N. Ashby, D. F. Bartlett and W. Wyss (Cambridge University Press, Cambridge, 1990), gr-qc/9210006.
7. R. Griffiths, *J. Stat. Phys.* **36**, 219 (1984).
8. R. Omnès, *J. Stat. Phys.* **53**, 893 (1988); *ibid* **53**, 933 (1988); *ibid* **53**, 957 (1988); *ibid* **57**, 357 (1989); *Rev. Mod. Phys.* **64**, 339 (1992); *Interpretation of Quantum Mechanics* (Princeton University Press, Princeton, 1994).
9. M. Gell-Mann and J. B. Hartle, in *Complexity, Entropy, and the Physics of Information*, SFI Studies in the Sciences of Complexity, Vol. VIII, ed. W. Zurek (Addison Wesley, Reading, MA) or in *Proc. 3rd Int. Symp. Foundations of Quantum Mechanics in the Light of New Technology*, eds. S. Kobayashi, H. Ezawa, Y. Murayama and S. Nomura (Physical Society of Japan, Tokyo, 1990).

10. J. B. Hartle, The Quantum Mechanics of Cosmology, in *Quantum Cosmology and Baby Universes: Proc. 1989 Jerusalem Winter School for Theoretical Physics*, eds. S. Coleman, J. B. Hartle, T. Piran and S. Weinberg (World Scientific, Singapore, 1991), pp. 65–157.
11. J. B. Hartle, *Phys. Rev. D* **44**, 3173 (1991).
12. C. J. Isham, *J. Math. Phys.* **35**, 2157 (1994).

CHAPTER 22

The State of the Universe*

What is the quantum state of the universe? That is the central question of quantum cosmology. This essay describes the place of that quantum state in a final theory governing the regularities exhibited universally by all physical systems in the universe. It is possible that this final theory consists of two parts: (1) a dynamical theory such as superstring theory, and (2) a state of the universe such as the no-boundary wave function. Both are necessary because prediction in quantum mechanics requires both a Hamiltonian and a state. The simplicity observed in the early universe gives hope that there is a simple, discoverable quantum state of the universe. It may be that the predictions of the quantum state for late time, low-energy observations can be summarized by an effective cosmological theory. That should not obscure the need to provide a fundamental basis for such an effective theory. It could be that there is one principle that determines *both* the dynamical theory and the quantum state. That would be a truly unified final theory.

---

*Published in *The Future of Theoretical Physics and Cosmology: Stephen Hawking 60th Birthday Symposium* (Cambridge University Press). http://arxiv.org/abs/gr-qc/0209046v1

1. Introduction

The universe has a quantum state. What is it? That is the central question of quantum cosmology — the subject to which Stephen Hawking has contributed so many seminal ideas.

To ask this question is to assume that the universe is a quantum-mechanical system. We perhaps have little direct evidence of peculiarly quantum-mechanical phenomena on large and even familiar scales, but there is no evidence that the phenomena that we do see cannot be described in quantum-mechanical terms and explained by quantum-mechanical laws. Further, every major candidate for a fundamental dynamical law, from the standard model to M-theory, conforms to the quantum-mechanical framework for prediction. If this framework applies to the whole thing, there must be a quantum state of the universe.

It would be even more interesting if quantum mechanics broke down on cosmological scales. But there is not a shred of evidence for that, and my guess is that, even if it does, we will only find out by pursuing the assumption that quantum mechanics is the framework for a final theory of cosmology.

My talk will not review any of the current ideas for a quantum state of the universe — even Stephen's no-boundary wave function.[1] The articles of Don Page and Alex Vilenkin in this volume do that. Rather, in the limited time available, I want to concentrate on explaining why a theory of the quantum state of the universe must be part of any final theory. I am also not going to discuss the generalizations of usual quantum theory that are required for quantum cosmology.[2] For the essential points of this talk, you can just imagine that the universe is a vast number of particles in a very large expanding box.

2. Final Theories

The final theory (to use Steve Weinberg's term) predicts the regularities that are exhibited by all physical systems — without exception, without qualification and without approximation. Much of this conference has been concerned with the search for the final theory. A possible view at present is that it consists of two parts:

- A universal dynamical law, such as string theory or its successors;
- A law for the quantum state of the universe, such as Hawking's no-boundary wave function of the universe.

In a model universe in a box, these two parts are represented by the Hamiltonian specifying the form of the Schrödinger equation

$$i\hbar \frac{d|\Psi(t)\rangle}{dt} = H\,|\Psi(t)\rangle \tag{1}$$

and the initial quantum state

$$|\Psi(0)\rangle. \tag{2}$$

Both of these pieces are necessary for prediction. The Schrödinger equation makes no predictions by itself. The probabilities p_α predicted by quantum mechanics for a set of alternatives represented by projection operators $\{P_\alpha\}$ are

$$p_\alpha = \| P_\alpha |\Psi(t)\rangle \|^2. \tag{3}$$

To compute these, the quantum state is needed at least at one time. No state, no predictions.

To put this in a different way, if the state is arbitrary, the predictions are arbitrary. Pick any probabilities p_α that you would like for the alternatives P_α. There is some state that will reproduce them. For example, you can take

$$|\Psi(t)\rangle = \sum\nolimits_\alpha p_\alpha^{\frac{1}{2}} |\Psi_\alpha\rangle \tag{4}$$

where the $|\Psi_\alpha\rangle$ are any set of eigenstates of the P_α's

$$P_\alpha |\Psi_\beta\rangle = \delta_{\alpha\beta} |\Psi_\beta\rangle. \tag{5}$$

The $|\Psi(t)\rangle$ constructed according to (4) will reproduce the preassigned probabilities p_α.

Neither is ignorance bliss. If you assume that you know nothing about the state of the universe in a box, then you should make predictions with a density matrix proportional to unity

$$\rho = \frac{I}{Tr(I)} \tag{6}$$

reflecting that ignorance. But this density matrix corresponds to equilibrium at infinite temperature and its predictions are nothing like the universe we live in. In particular, there would be no evolution, since $[H, \rho] = 0$. There would be no second law of thermodynamics, since the entropy $-\mathrm{Tr}[\rho \log \rho]$ is already at its maximum possible value. There would be no classical behavior, since, although the expected value of a field averaged over a spacetime volume R might be finite, its fluctuations, $\langle \phi(R)^2 \rangle$, would be infinite.

The search for a unified fundamental dynamical law has been seriously under way at least since the time of Newton, with string theory or its generalizations being the most actively investigated direction today. By contrast, the search for a theory of the quantum state of the universe has only been actively underway since the time of Hawking, let us say on this occasion. Why this difference?

Dynamical laws govern regularities in time, and it is an empirical fact that the basic dynamical laws are local in space on scales above the Planck length. The laws that govern regularities in time across the whole universe are therefore discoverable and testable in laboratories on earth. By contrast, many of the regularities predicted with near-certainty by the quantum state of the universe are mostly in space on large cosmological scales. Only recently has there been enough data to confront theory with observation. That difference in the nature of the predicted regularities, or their difference in scales, should not obscure the fact that the state is just as much a part of the final theory as is its Hamiltonian.

Given these differences, what grounds do we have to hope that we can discover the quantum state of the universe? There are two: The first is the simplicity of the early universe revealed by observation — more homogeneous, more isotropic, more nearly in thermal equilibrium than the universe is today. It is therefore possible that the universe has a simple, discoverable, initial quantum state and that all of the complex universe of galaxies, stars, planets, and life today arose from quantum accidents that have happened since and the action of gravitational attraction. The second reason is the idea that the quantum state and the dynamical theory may be naturally connected, as in Hawking's no-boundary theory.

3. Effective Theories

We are used to the idea of effective dynamical theories that accurately describe limited ranges of phenomena. The Navier–Stokes equations, non-relativistic quantum mechanics, general relativity, quantum electrodynamics, and the standard model of particle physics are all familiar examples. To construct an effective theory, we typically assume a coarse-grained description (restricting attention to energies below the Planck scale, for instance) and assume some simple property that the state might predict there (classical spacetime, for example).

Cosmology, too, has its effective theory and its standard model. This is summarized neatly by the following list due to Martin Rees.[3] I reproduce it here with unauthorized additions.

- spacetime is classical, governed by the Einstein equation,
- our universe is expanding,
- from a hot big bang,
- in which light elements were synthesized,
- there was a period of inflation,
- which led to a flat universe today,
- structure was seeded by Gaussian irregularities,
- which are relics of quantum fluctuations,
- the dominant matter is cold and dark,
- but a cosmological constant (or quintessence) is dynamically dominant.

Possibly all current observations in cosmology, or at least the large scale ones, can be compressed into an effective "standard model" based on this list of ten assumptions and a few cosmological parameters. That is not unlike the situation in particle physics where most observations can be compressed into the Lagrangian of the standard model and its eighteen or so parameters.

However, the success of such effective theories which operate in limited ranges of phenomena should not obscure the need to find fundamental ones which apply to all phenomena without qualification and without approximation. It would be inconsistent, I believe, to pursue a fundamental dynamical theory in the face of a successful, effective, standard dynamical model, and not pursue a fundamental theory of the state of the universe because of the success of *its* effective, standard cosmological model. That not least, because the fundamental theory could provide a unified explanation of its assumptions.

It must be said, however, that when the natural domains of fundamental theories are as far from controllable experiments as string theory and the quantum initial condition, the possibility of definitive tests seems to recede. It could be that the predictions of string theory are limited to general relativity, gauge theories, supersymmetry, and the parameters of the standard particle model. In a similar way, the predictions of the state of the universe could be limited to classical spacetime, the initial conditions for inflation, and the quantum fluctuations that satisfy large scale structure. Perhaps that is prediction enough.

4. Directions

The instructions of the organizers were to discuss "future directions in theoretical physics and cosmology". Continuing the search for a final theory incorporating dynamics and the initial quantum state is certainly one direction. But I would like to mention three questions that might lead to different approaches to the main one.

4.1. *What's Environmental?*

Which features of the observed universe follow entirely from the dynamical theory (H) and which follow entirely from the initial condition ($|\Psi(0)\rangle$), and which are the results of quantum accidents that occurred over the course of the universe's history with probabilities specified by the combination of H and $|\Psi(0)\rangle$? Those that depend significantly on $|\Psi(0)\rangle$ are called "environmental". Some version of this question was number one on the list of top ten questions for the next millennium prepared by string theorists at the Strings 2000 conference.[4]

Take the coupling constants in effective dynamical theories, for instance. The viscosities and equation-of-state in the Navier–Stokes equation are certainly environmental. They vary with system, place and time. But, at a given energy, do the coupling constants of the standard model of the elementary particle interactions vary with place and time or with the possible history of the universe? If so, then the initial quantum state is central to determining their probabilities.

4.2. *Why Quantum Mechanics?*

The founders of quantum mechanics thought that the inherent indeterminacy of quantum theory "reflected the unavoidable interference in measurement dictated by the magnitude of the quantum of the action" (Bohr). But why then do we live in a quantum-mechanical universe which, by definition, is never measured from the outside?

The most striking general feature of quantum mechanics is its exact linearity — the principle of superposition. But why should there be a principle of superposition in quantum cosmology, which has only a single quantum state?

4.3. *Why a Division into Dynamics and Initial Condition?*

The schema for a final theory which I have been describing posits a separate theory of dynamics and quantum state. Could they be connected? They already are, in Hawking's no-boundary wave function[1]

$$\Psi = \int \delta g \, \delta\phi \; e^{-I[g,\phi]} \tag{7}$$

where the action for metric $g_{\alpha\beta}(x)$ coupled to matter $\phi(x)$ determines both the state and quantum dynamics. Is there a principle that determines both? Is there a connection between superstring theory and its successors and a unique quantum state?

A unified quantum theory of state and dynamics would be truly a final theory. Pursuing that vision is surely a direction for theoretical physics and cosmology.

Acknowledgment

This paper was supported in part by NSF Grant PHY00-70895.

References

1. S. Hawking, *Nucl. Phys. B* **239**, 257 (1984).
2. C. J. Isham, *J. Math. Phys.* **35**, 2157 (1994); C. J. Isham and N. Linden, *J. Math. Phys.* **35**, 5452 (1994); *ibid.* (1995).
3. M. J. Rees, The state of modern cosmology, in *Critical Dialogues in Cosmology*, ed. N. G. Turok (World Scientific, Singapore, 1997).
4. M. J. Duff, J. T. Liu and J. Lu (eds.) *Strings: Proc. 2000 Int. Superstrings Conf.* (World Scientific, Singapore, 2001).

CHAPTER 23

The Impact of Cosmology on Quantum Mechanics*

When quantum mechanics was developed in the 1920s, another revolution in physics was just starting. It began with the discovery that the universe is expanding. For a long time, quantum mechanics and cosmology developed independently of one another. Yet the very discovery of the expansion would eventually draw the two subjects together because it implied the big bang where quantum mechanics was important for cosmology and for understanding and predicting our observations of the universe today. Textbook (Copenhagen) formulations of quantum mechanics are inadequate for cosmology for at least four reasons: (1) They predict the outcomes of measurements made by observers. But in the very early universe no measurements were being made and no observers were around to make them. (2) Observers were outside of the system being measured. But we are interested in a theory of the whole universe where everything, including observers, are

(*continued overleaf*)

*A pedagogical essay. Based in part on a talk given at the conference *90 Years of Quantum Mechanics*, Singapore, January, 2017.
arXiv:1901.03933

inside. (3) Copenhagen quantum mechanics could not retrodict the past. But retrodicting the past to understand how the universe began is the main task of cosmology. (4) Copenhagen quantum mechanics required a fixed classical spacetime geometry, not least to give meaning to time in the Schrödinger equation. But in the very early universe spacetime is fluctuating quantum-mechanically (quantum gravity) and without definite value. A formulation of quantum mechanics general enough for cosmology was started by Everett and developed by many. That effort has given us a more general framework that is adequate for cosmology — decoherent (or consistent) histories quantum theory in the context of semiclassical quantum gravity. Copenhagen quantum theory is an approximation to this more general quantum framework that is appropriate for measurement situations. We discuss whether further generalization may still be required.

1. Introduction

In the late '20s and early '30s of the last century, quantum mechanics was revolutionizing our understanding of the physics of the very small. At roughly the same time, another revolution in physics was beginning, that was to change our understanding of the physics of the very large. This started with the discovery that the universe was expanding, through the work of Lemaître, Hubble, and Slipher and others. For many years these two revolutions proceeded independently of one another. But in more recent times each has had a significant impact on the other.

Quantum mechanics is central to an understanding of important physical processes that take place in the very early universe. Well-known examples are big bang nucleosynthesis, recombination, and, most importantly, the generation of quantum fluctuations in matter density that grow under gravitational attraction to produce the large scale structure we see today in the distribution of the galaxies and in the temperature fluctuations of the cosmic background radiation (CMB).

What is perhaps less widely appreciated is that cosmology had a significant impact on our understanding of quantum mechanics. That is because the Copenhagen formulations that appear in many textbooks are inadequate for cosmology for several reasons. Quantum mechanics had to be generalized to apply to cosmology, and those generalizations led to a different understanding of the subject and what it is about.

This essay will describe very briefly the impact of cosmology on quantum mechanics and give an overview of the more general formulations of quantum theory that resulted. Throughout, we attempt to explain the basic ideas with a minimum of technical development.[a] We trade lack of precision for hope of being more widely understandable.

We should also stress that we will not address issues of the interpretation of quantum theory. Rather, we aim at describing how cosmology has affected what has to be interpreted.

We begin in Sec. 2 by describing the status of Copenhagen quantum mechanics (CQM) today. Section 3 explains why this Copenhagen quantum theory is inadequate for cosmology and why a generalization of it is needed. Section 4 describes a model quantum universe in a very large box. Section 5 describes the decoherent (or consistent) histories generalization of Copenhagen quantum theory that is adequate for closed systems like the universe. Section 6 describes further generalizations to deal with quantum spacetime geometry. Section 7.1 describes how to get probabilities for the results of our observations from probabilities for histories of the universe. Section 7.2 recovers Copenhagen quantum mechanics as an approximation appropriate for measurement situations in more general formulations of quantum mechanics applicable to the universe. Section 8 describes how various formulations of quantum theory can be regarded as features of the universe that emerge along with other features that are a prerequisite to the formulations. Section 9 concludes by speculating on whether further generalizations will be required.

[a]That is with a minimum of equations. For those who want the equations, there are a large number of references, many by the author, that offer different levels of explanation.

2. The Copenhagen Quantum Mechanics of Laboratory Measurements (CQM)

The Copenhagen quantum mechanics (CQM) found in standard textbooks[b] is arguably the most successful theoretical framework in the history of physics. It is central to our understanding of a vast range of physical phenomena, including atoms, molecules, chemistry, the solid state, how stars are formed, how they shine, evolve, and die, nuclear energy, thermonuclear explosions, how transistors work, and many, many, many, more phenomena. It has been claimed that quantum mechanics is responsible for a significant fraction of the US GDP.

Work continues today to better understand Copenhagen quantum mechanics, to make its central notions of measurement and state vector reduction more precise, and to resolve "problems" that it is alleged to have, like the "measurement problem" (Fig. 1). Even given these concerns, the author knows no mistake that was made because of them in correctly applying Copenhagen quantum mechanics in the century since it was first formulated.

Despite its experimental success, a striking number of our most distinguished scientists are of the opinion that even the quantum mechanics of laboratory experiment will need to be revised (e.g. Ref. 2) for these concerns or perhaps others.

None of these concerns will be discussed here directly. We do not aim to clarify or perfect Copenhagen quantum mechanics. Rather, we aim to move beyond it to a quantum mechanics general enough for cosmology to which Copenhagen quantum mechanics is an approximation appropriate for laboratory situations. The issues alluded to in Fig. 1 could have a very different form or be absent entirely in such a generalization (e.g. Ref. 3).

---

[b]By "Copenhagen quantum mechanics", or "textbook quantum mechanics", we mean the standard formulation that is found in many textbooks, including the assumption of the quasiclassical realm of everyday experience as described in Sec. 3.1. We do not necessarily mean that it conforms exactly to what the founders of quantum mechanics meant by the "Copenhagen interpretation of quantum mechanics".[1]

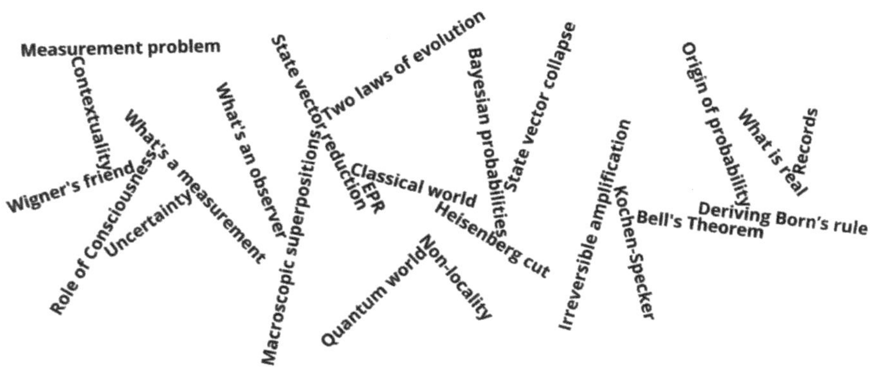

Fig. 1. Foundational issues sometimes raised for Copenhagen quantum mechanics. None of these issues are discussed directly in this essay. Rather, we aim to go beyond them with more general formulations of quantum theory.

3. Copenhagen Quantum Mechanics is Inadequate for Cosmology

There are at least four reasons why textbook quantum mechanics is inadequate for cosmology. We describe them in this section.

3.1. *No Quasiclassical Realm, No Observers, No Measurements, in the Early Universe*

Copenhagen quantum mechanics assumes the quasiclassical realm of everyday experience — the wide range of time, place and scale on which the deterministic laws of classical physics apply. Notable examples are the laws of gravity summarized by the Einstein equation and the laws of motion of continua summarized in the Navier–Stokes equation. These and other classical laws are needed just to describe measurement apparatus and observers and their operation in physical terms. Indeed, some formulations of Copenhagen quantum mechanics assumed the existence of entirely separate classical and quantum worlds with a kind of movable boundary between them (e.g. Ref. 4). There is no evidence for such a division in cosmology.

3.2. *Observers are Physical Systems within the Universe*

Copenhagen quantum mechanics predicts probabilities for the outcomes of measurements carried out on a subsystem of the universe by another subsystem called the "observer" or "measuring apparatus" that is outside the subsystem being measured.[c]

But in the very early universe there were no measurements being made and no observers around to make them. Are we then to believe that quantum mechanics does not apply to the early universe before the evolution of observers, approximately 10 Gyr after the big bang? Certainly not! But then we need a generalization of Copenhagen quantum mechanics in which observers are physical systems *inside* the universe. We need a generalization of textbook quantum mechanics in which observers and their measurements can be described in quantum-mechanical terms but play no special role in the theory's formulation and are not prerequisites for the theory's application.[d]

3.3. *No Retrodiction*

In his now classic book *The Mathematical Foundations of Quantum Mechanics*,[5] John von Neumann formulated quantum evolution in terms of two laws. First, unitary evolution of the state vector

---

[c]Individually and collectively, observers are examples of the general concept of IGUSes — information gathering and utilizing subsystems. It would be both more accurate and more accessible to present Copenhagen quantum mechanics in terms of IGUSes. But we aim at a discussion that is as broadly accessible as possible, and to that end we will stick with the more broadly familiar, if less nuanced, "observer" in this paper.

[d]Sometimes one meets the suggestion that no such generalization is needed. The idea is that Copenhagen quantum mechanics retrodicts the past when present observers measure present records of what went on in the past. But what exactly is a record? A record is a physical quantity today that is correlated with one or more past events with high probability. This probability is for a two-time history of the events and the formation of a later record of them. A generalization of Copenhagen quantum mechanics is needed to assign probabilities to such time histories and to define what is meant by records.

described by the Schrödinger equation. Second, the reduction of the state vector after an "ideal" measurement[e] that disturbs the measured subsystem as little as possible.

Starting from the state of an isolated subsystem at one time, and using these two laws, Copenhagen quantum theory can predict probabilities for the histories of later measurements that an observer might decide to carry out on the same subsystem. This is the sense in which Copenhagen quantum theory can predict the results of future measurements starting from the present state of the measured subsystem.

Copenhagen quantum mechanics cannot retrodict the past of a subsystem starting from its present quantum state. The Schrödinger differential equation can of course be evolved backward in time as well as forward. But state vector reduction runs only one way — forward in time, usually taken to be the direction in which entropy is increasing in the universe. Thus, retrodiction is impossible in Copenhagen quantum mechanics.[f]

But retrodiction is central to cosmology. We use present data and an initial quantum state to construct a model of what went on in our past to simplify our predictions of the future.[7] We are interested, in particular, in a quantum framework that can be used to construct a history of how the universe came to be the way it is today, starting from its initial condition roughly 14 Gyr ago, using its quantum state.

3.4. *Fixed Classical Spacetime*

Copenhagen quantum mechanics assumes a fixed spacetime geometry. This geometry defines the possible time directions for the "t" in the Schrödinger equation. States are defined on spacelike surfaces in this fixed spacetime and evolve by the Schrödinger equation through

[e]von Neumann had the opposite order of one and two but the one here is more used today.

[f]The sole consideration of the past in early Copenhagen quantum theory appears to be a paper by Einstein, Podolsky and Tolman,[6] which concluded, "quantum mechanics must involve an uncertainty in the description of past events which is analogous to the uncertainty in the prediction of future events". (Don Howard, private communication).

a foliating family of such surfaces in this fixed spacetime. After an ideal measurement, the state is "reduced" all across one of these spacelike surfaces that extend over the whole universe. Copenhagen quantum mechanics requires a fixed spacetime geometry just to make sense of these two modes of evolution.

But in the early universe near the big bang, energy scales above the Planck scale $(\hbar c^5/G)^{1/2} \sim 10^{19}$ Gev will be reached at which quantum gravity becomes important and spacetime geometry will fluctuate quantum-mechanically and be without the definiteness required by Copenhagen quantum theory.

Rather, we expect Copenhagen quantum mechanics to emerge in the early universe *along with* classical spacetime geometry when the latter is sufficiently slowly varying to provide the notions of time and spacelike surfaces that Copenhagen quantum mechanics assumes (e.g. Ref. 8).

The conclusion of the discussion in this section is that CQM has to be generalized for quantum cosmology. The rest of the paper discusses possible generalizations.

4. A Model Quantum Universe in a Closed Box

To better understand how quantum mechanics works for the universe, we first consider a model closed quantum system consisting of a large box, say 20,000 Mpc on a side, perhaps expanding, and containing particles and fields, as suggested in Fig. 2. We assume a fixed, flat, background spacetime inside, thus neglecting quantum gravity. This is an excellent approximation in the realistic universe for times later than a very short interval of $\sim 10^{-43}$ s after the big bang. There is then a well-defined notion of time in any particular Lorentz frame. The familiar apparatus of textbook quantum mechanics then applies — a Hilbert space, operators, states and their unitary evolution in time. We assume a quantum field theory in the flat spacetime for dynamics. We will return briefly to what happens when quantum gravity is not neglected, in Sec. 6.

The important thing is that everything is contained within the box — galaxies, planets, observers and observed, measured subsystems, and any apparatus that measures them, you and me.

There is nothing outside and no influence of the outside on the inside, or the inside on the outside.

The basic theoretical inputs for predicting what goes on in the box are the Hamiltonian H governing evolution and quantum state $|\Psi\rangle$, written here in the Heisenberg picture for convenience, and assumed pure for simplicity. Input theory is then (H, Ψ). The basic outputs are probabilities for what goes on in the box. More precisely, the outputs of (H, Ψ) are probabilities for the individual members of sets of coarse-grained alternative time histories of what goes on in the box. A few examples will make the important ideas introduced in this section more concrete.

4.1. *Coarse-grained Histories in General*

We might be interested in a set of histories that describe the positions of the Moon in its motion around the Earth at a series of times $t_1, \ldots, t_n$. We are then interested in the probabilities of the alternative orbits that the Moon could follow around the Earth. Each orbit is an example of a *history* — a sequence of events at a series of times. Relevant histories are *coarse-grained* because we are only interested in positions defined to an accuracy consistent with our observations of position, and further because positions are not specified at each and every time but only at a finite discrete sequence of times. Coarse-grained histories can be said to *follow* certain variables and *ignore* others. In the present example, the histories follow the center of mass of the Moon and ignore variables that describe the interior of the Moon and Earth. In quantum mechanics, there is no certainty that the coarse-grained history of the Moon's center of mass will follow a classical Keplerian orbit, but in the Moon's situation, the probability predicted by (H, Ψ) is vastly higher than for a non-classical orbit.

Coarse-graining is of the greatest importance in theoretical science, for example, in connection with the meaning of regularity and randomness and of simplicity and complexity.[9] Of course, it is also central to statistical mechanics, where a particular kind of coarse-graining leads to the usual physico-chemical or thermodynamic entropy. As we shall see, coarse-graining also plays a crucial role in the quantum mechanics of the universe, making possible the decoherence of alternative histories and enabling probabilities of those decoherent histories to be defined.

4.2. *Coarse-Grained Histories of the Universe*

In cosmology we are interested in the probabilities of sets of certain coarse-grained alternative classical histories that describe the universe's expansion, the primordial nucleosynthesis of light elements, the formation and evolution of the microwave background radiation, the formation of the galaxies, stars, planets, the evolution of biota, etc. Sets of alternative histories relevant for our observations are highly coarse-grained. They don't describe everything that goes on in the universe — every galaxy, star, planet, human history, etc., in all detail. Rather, they follow much coarser-grained histories of the universe. And, as we will see in quantum mechanics, it is only histories that are sufficiently coarse-grained for which the theory assigns probabilities at all!

In laboratory science, we are interested in histories that describe the preparation, progress, and outcomes of a measurement situation. In quantum cosmology, there are no measurements of the inside of the box by something outside it. Laboratory measurements are described realistically, as a correlation between one subsystem inside the box that includes the apparatus, observers, etc. and another subsystem inside that is thus measured. In this way, measurements can be described in the quantum mechanics of the universe but play no preferred role in the formulation of quantum mechanics as they do in Copenhagen quantum theory.

The measuring apparatus and the orbit of the Moon are examples of systems that behave according to the laws of classical physics at a suitably coarse-grained level. Indeed, most of our understanding of the universe on large scales is of its classical behavior. But classical behavior is not a given in a quantum universe. It is rather a matter of quantum probabilities. A quantum system behaves classically when, in a suitably coarse-grained set of alternative histories, the probabilities are high for histories exhibiting correlations in time governed by deterministic classical laws (e.g. Refs. 9–11).

Fig. 2. A simple model of a closed quantum system is a universe of quantum matter fields inside a large closed box (say, 20,000 Mpc on a side) with fixed flat spacetime inside. Everything is a physical system inside the box — galaxies, stars, planets, human beings, observers and observed, subsystems that are measured and subsystems that are measuring. The most general objectives for prediction are the probabilities of the individual members of decoherent sets of alternative coarse-grained histories that describe what goes on in the box. That includes histories describing any measurements that take place there. There is no observation of or other meddling with the inside from outside.

5. Decoherent Histories Quantum Theory (DH)

Decoherent histories quantum mechanics (DH) is a formulation of quantum theory that is general enough for cosmology. It has all the needed generalizations of Copenhagen quantum mechanics discussed in Sec. 3. It permits both prediction and retrodiction. Observers are physical systems within the closed universe. Measurements are physical processes within the universe. But neither observers nor their measurements play any preferred role in the formulation of DH.

As we will see, Copenhagen quantum mechanics is an approximation to this more general framework that is appropriate for measurement situations.

Decoherent histories quantum theory (DH) is the work of many;[12] the formulation we present here is the work of Murray Gell-Mann and the author.[13] On many essential points, it coincides with the earlier independent consistent histories (CH) formulation of quantum theory of Griffiths and Omnès.[12] DH can be seen as a generalization, clarification, and, to some extent, a completion of the program started by Everett.[14]

We will not develop the machinery necessary for an application of DH to cosmology in any detail here because we aim only at an exposition of the principles behind the theory with a minimum of technical complication.[g] For definiteness, we first focus this discussion on the universe in a box of the previous section with a fixed, flat, spacetime geometry. As described there, we aim at predicting probabilities for the individual members of sets of coarse-grained alternative histories of what goes on in the box. But quantum interference is an obstacle to assigning probabilities, sets of alternative histories for any closed system. Nowhere is this more simply illustrated than in the classic two-slit experiment shown in Fig. 3.

5.1. *Two-Slit Experiment*

In the two-slit setup shown in Fig. 3, suppose an electron arrives at a position Y on the detecting screen after passing through a screen with two slits U and L. There is an obvious set of two alternative histories of how it might have got to the detecting screen — one in which it went through the upper slit (U) to arrive at the position interval Y on the screen, and the other where it went through the lower slit (L) to arrive at Y. Quantum theory assigns amplitudes to the two individual histories set, for example, as in Feynman's sum-over-histories formulation of quantum theory.[16] Denote the quantum amplitudes of the two histories by $\Psi_U(Y)$ and $\Psi_L(Y)$. The amplitude

[g]For a pedagogical introduction with more equations than here see, e.g. Ref. 15.

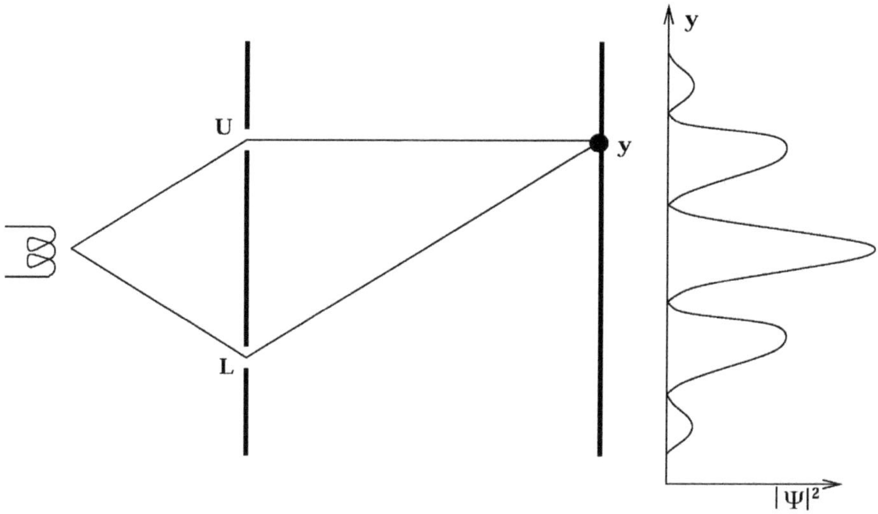

Fig. 3. The two-slit experiment. An electron gun at left emits an electron traveling towards a screen with two slits, its progress in space recapitulating its evolution in time. When precise detections are made of an ensemble of such electrons at the screen, the classic interference pattern emerges — evidence that the two histories interfere. Because of this quantum interference, it is not possible to assign a probability to the alternatives of an individual electron going through the upper slit or the lower slit. However, if the electron interacts with apparatus that measures which slit it passed through, then these alternatives decohere and probabilities can be assigned.

$\Psi(Y)$ to arrive at Y is then the sum

$$\Psi(Y) = \Psi_U(Y) + \Psi_L(Y). \tag{1}$$

It would be inconsistent to assign probabilities to these histories that were squares of these amplitudes. The probability to arrive at Y should be the sum of the probabilities to arrive there by going through the upper or lower slit. But in quantum mechanics, probabilities are squares of amplitudes and the sum of squares is not the square of the sum. It would be *inconsistent* with the standard rules of probability for a theory to predict probabilities for the two histories that were squares of amplitudes in this set. Specifically, we would have

$$p(Y) \neq p_U(Y) + p_L(Y) \tag{2a}$$

because

$$|\Psi(Y)|^2 = |\Psi_U(Y) + \psi_L(Y)|^2 \neq |\Psi_U(Y)|^2 + |\Psi_L(Y)|^2. \tag{2b}$$

Any formulation of quantum mechanics must therefore include a rule to specify which sets of alternative histories can be consistently assigned probabilities and which cannot. In Copenhagen quantum mechanics, the rule was simple: Probabilities are assigned only to sets of alternative histories that have been *measured*, and not otherwise. If we measured which slit the electron went through, then the interference pattern would be destroyed, the sum rule obeyed, and the probabilities consistent with the usual rules.

However, as discussed in Sec. 3 we can't have such a rule in a quantum theory of cosmology which seeks to describe early universe cosmology when no measurements were being made and there were no observers around to carry them out. The more general rule that defines DH is just this: *Probabilities can be assigned to just those sets of alternative histories of a closed system for which there is negligible interference between the individual histories in the set as a consequence of the Hamiltonian* (H) *and the state* (Ψ) *that the closed system has.* These probabilities are consistent with the rules of probability theory as a result of the absence of quantum interference. Such a set of histories for which the mutual interference between any pair of histories is negligible, is said to *decohere.*

5.2. *Calculating Amplitudes for Histories in the Two-Slit Experiment*

We can get an idea of how to calculate the amplitudes for histories in the two-slit experiment just from the Schrödinger equation defining the quantum evolution of the electron's wave function. To keep the discussion simple, we ignore the spin of the electron and consider only its position. The state of the electron in the two-slit experiment can then be described by a wave function of the form

$$\Psi = \Psi(x, y, t) \tag{3}$$

using a coordinate x for the horizontal direction and y for the vertical direction in Fig. 3 and assuming symmetry in the perpendicular direction.

Denote the initial state at time t_0 by $\Psi(x, y, t_0)$. This is a product of a wave packet in the x-direction $\phi(x, t_0)$ and a wave function $\psi_0(y, t_0)$ localized at the gun, viz.

$$\Psi(x, y, t_0) = \psi(y, t_0)\phi(x, t_0). \tag{4}$$

This wave function evolves in time by the Schrödinger equation

$$i\hbar\frac{\partial\Psi}{\partial t} = H\Psi \tag{5}$$

where H is the Hamiltonian of a free particle interacting with the screens. We assume that $\phi(x, t)$ is a narrow wave packet peaked to the left of the slits but moving to the right, so as to reach the slits at time t_s and the detecting screen at t_d. Thus, its progress in x recapitulates evolution in time. After passing through the slits, the wave function has the approximate form

$$\Psi(x, y, t) = \psi_U(y, t)\phi(x, t) + \Psi_L(y, t)\phi(x, t), \quad t_s < t < t_d \tag{6a}$$
$$\equiv \Psi_U(x, y, t) + \Psi_L(x, y, t). \tag{6b}$$

Here, in the first term, $\psi_U(y, t)$ is localized near the upper slit at time t_s and spreads over a larger region of y by the time t_d such that the wave packet hits the detecting screen. Similarly for the second term. When evaluated at the arrival position x_d and time t_d, and projected on the interval of arrival Y, these are the amplitudes $\Psi_U(Y)$ and $\Psi_L(Y)$ in (1). They are wave functions of two branch state vectors $|\Psi_U(Y)\rangle$ and $|\Psi_L(Y)\rangle$ for the two histories in the set. The set does not decohere because

$$\langle\Psi_U(Y)|\Psi_L(Y)\rangle \not\approx 0 \tag{7}$$

so that the decoherence condition is not satisfied. The two histories interfere. There will be no consistent set of probabilities for this set of histories.

5.3. *A Simple Model of Decoherence*

To see how decoherence can occur, suppose that near the slits there is a gas of particles (e.g. photons) that scatter off the electron but weakly enough not to disturb its trajectory. Even if the collisions do not affect the trajectories of the electrons very much, they can still carry away the phase correlations between the two histories. A coarse-graining that described only one of these two alternative histories of the electron would then approximately decohere as a consequence of the interactions with the gas, given adequate density, cross-section, etc. Interference is destroyed and consistent probabilities can be assigned to these alternative histories of the electron in a way that they could not be if the gas were absent (cf. Fig. 1). The lost phase information is still available in correlations between states of the gas and states of the electron. The alternative histories of the electron would not decohere in a coarse-graining that included both the histories of the electron *and* the operators that were sensitive to the correlations between the electrons and the gas.

Functioning in this way to dissipate relative phases between different histories, the gas is an example of an *environment* (e.g. Refs. 10, 17–19), and the example illustrates the widely occurring mechanism of *environmental decoherence.*

The initial wave function including the state of the gas is

$$\Psi(x,y,t_0) = \psi(y,t_0)\phi(x,t_0)\chi(t_0), \quad t_0 < t < t_s \tag{8}$$

where $\chi(t_0)$ is the initial state of all the particles in the gas. After passing through the slits this becomes

$$\Psi(x,y,t) = \psi_U(y,t)\phi(x,t)\chi_U(t) + \psi_L(y,t)\phi(x,t)\chi_L(t), \quad t_s < t < t_d \tag{9a}$$

$$\equiv \Psi_U(x,y,t) + \Psi_L(x,y,t) \tag{9b}$$

where $\chi_U(t)$ and $\chi_L(t)$ denote the state of the gas particles that have scattered from the region of the upper and lower slits, respectively. Then $\Psi_U(x,y,t_d)$ and $\Psi_L(x,y,t_d)$ are the branch wave functions for the two histories for the events that the electron arrived in the interval Y at time t_d after passing through either the upper or lower slit. The condition that this set of two histories decoheres is the absence

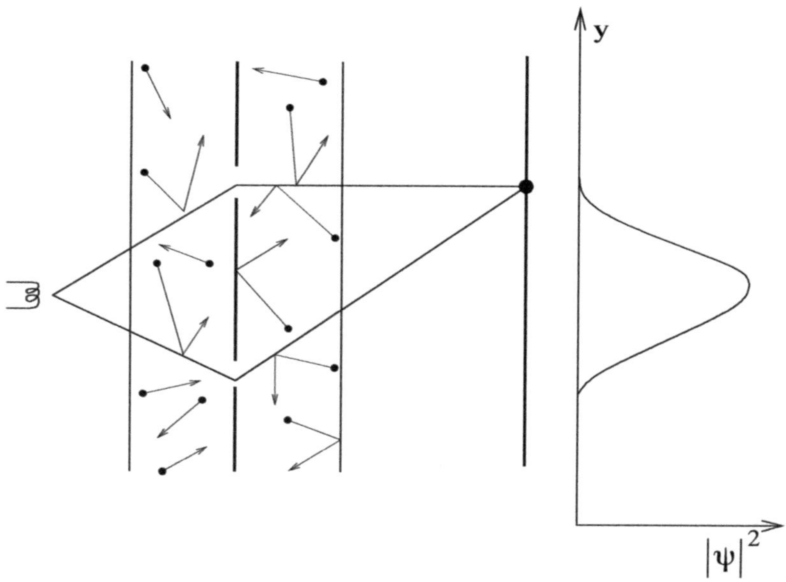

Fig. 4. The two-slit experiment with an interacting gas. Near the slits light particles of a gas collide with the electrons. Even if the collisions do not affect the trajectories of the electrons very much, they can still carry away the phase correlations between the histories in which the electron arrived at point y on the screen by passing through the upper slit and that in which it arrived at the same point by passing through the lower slit. A coarse-graining that described only one of these two alternative histories of the electron would approximately decohere as a consequence of the interactions with the gas, given adequate density, cross-section, etc. Interference is destroyed and probabilities can be assigned to these alternative histories of the electron in a way that they could not be, if the gas were not present (cf. Fig. 1).

of interference between these two branch wave functions, specifically

$$\langle \Psi_U(t_d) | \Psi_L(t_d) \rangle \approx 0. \tag{10}$$

In the simple two-slit model, the set of alternative histories describing which slit (U or L) the electron went through on its way to arrive at Y on the further screen did not decohere as required by DH for predictions of probabilities [cf. (7)]. That was because the branch states representing the two histories were not orthogonal. However, the set of two histories *does* decohere in this example because interactions with the gas dissipates relative phases. The decoherence condition (10) is satisfied because the state of the gas scattered from the upper slit $|\chi_U\rangle$ is orthogonal to the state of the gas scattered from the lower slit $|\chi_L\rangle$ if enough particles scatter.

To see that the gas ensures this orthogonality, consider a single gas particle a whose initial state is $|\chi_a\rangle$, that scatters near the upper slit and will be in a final state $S_U|\chi_a\rangle$, where S_U is the S-matrix for scattering near the upper slit. Similarly for the lower slit with $S_L|\chi_a\rangle$, if N gas particles scatter, the overlap between $|\chi_U\rangle$ and $|\chi_L\rangle$ will be proportional to N inner products of the form

$$\prod_{a=1}^{N} |\langle\chi_a|S_U^{\dagger}S_L|\chi_a\rangle|. \tag{11}$$

Since the two final state vectors $S_U|\chi_a\rangle$ and $S_L|\chi_a\rangle$ are different, each individual product has a magnitude less than one. The product of a very large number of such products will be near zero, implying decoherence as in (10).

5.4. *General Features of Quantum Mechanics Illustrated by the Two-Slit Experiment*

This two-slit example illustrates three general features of the quantum mechanics of histories that are central to its formulation.

First, it illustrates that sets of alternative histories must decohere in order to have probabilities consistent with the rules of probability theory.

Second, the two-slit experiment illustrates the crucial role that coarse-graining plays by making decoherence possible. Sets of fine-grained alternative histories, like the set of all possible Feynman paths of a particle, generally do not decohere except in trivial cases. Some degrees of freedom must be ignored to carry away the phase information between individual coarse-grained histories that can be followed. It is a remarkable fact that in quantum mechanics, some information must be lost to have any information at all.

Third, the two-slit experiment leads to understanding the central role that cosmology plays in implementing environmental decoherence. Mechanisms of decoherence are widespread in our epoch of the universe. Cosmology is not just important for formulating quantum mechanics, but is also important for implementing the decoherence that is necessary for its predictions.

6. Generalized Decoherent Histories Quantum Mechanics of Quantum Spacetime (GDH)

The exposition of decoherent histories quantum mechanics of closed systems (DH) in Sec. 5 assumed one fixed classical background spacetime. Histories describe the motion of particles and the evolution of fields *in* this background spacetime. But in a quantum theory of spacetime geometry, we do not expect just one classical spacetime geometry. Rather, we expect an ensemble of possible classical spacetimes for which quantum theory predicts probabilities for which we observe from the theories of dynamics and the quantum state[h] (H, Ψ).

It is conceptually simple but technically complicated to generalize the DH in Sec. 5 to include dynamical spacetime. Here we describe only the three simple conceptual changes from DH. We call the resulting generalization "Generalized Consistent Histories Quantum Mechanics" (GDH). The three important ingredients in GDH are:

- *Fine-grained histories:* The sets of possible alternative histories of spacetime geometry and matter field configurations that our universe might have are each specified by one spacetime geometry and one field history.
- *Coarse-grained histories:* These are defined by a partition of the set of fine-grained histories into exclusive bundles of fine-grained histories. The bundles are the coarse-grained histories.
- *A measure of quantum interference* between different coarse-grained histories constructed by Feynman path integrals $\int \exp\{-iS/\hbar\}$, over the fine-grained histories in the coarse-grained sets. The functional S is the action of the dynamical

[h]Theories of spacetime like general relativity are more straightforwardly summarized by an action than a Hamiltonian, but we will denote the dynamical theory by H however it is expressed.

theory H and the integral is weighted by a wave function of the universe representing Ψ like the no-boundary quantum state.[20,21]

Details can be found in Refs. 21 and 22. So formulated, GDH would predict probabilities for the individual members of decoherent sets of alternative coarse-grained cosmological classical histories of geometries and fields.[i]

7. Back to Observations, Measurements, Observers and Copenhagen Quantum Mechanics as an Approximation

7.1. *Predicting Probabilities for the Results of Observations*

When implemented in a quantum framework Q like DH or GDH, a theory (H, Ψ) predicts probabilities for which set of alternative coarse-grained histories of the universe occurs. A triple (H, Ψ, Q) is tested by its success in predicting the results of our observations of the universe. Predictions for our observations are through probabilities of histories supplied by (H, Ψ) conditioned on data D_{obs} describing our observational situation, including a description of any observers making the observation, and assuming here for simplicity that there is a unique instance of D_{obs} in the universe.[j]

By way of example, consider a prediction for the temperature of the CMB that will be observed in a particular experiment. Locally, the experiment would be described by data D_{obs} that would include a description of the detector, where it is pointing, whether it is shielded from radiation from the ground, and a description of the observers running the experiment, etc. But the theory (H, Ψ) by itself does not supply a probability for the one temperature observed. The

---

[i]At the time of writing, ways of implementing coarse-graining and decoherence for histories of spacetime geometry were at an early stage of development.

[j]In the vast universes contemplated by current theories of inflation, D_{obs} is not likely to be unique. Then a further assumption is needed about which of the instances of D_{obs} is us (e.g. Refs. 23–26).

temperature varies in time, getting lower as the universe expands. Rather, (H, Ψ) supplies probabilities for histories of how the temperature varies over time. But we don't observe entire four-dimensional histories of the CMB. We observe the CMB at a narrow range of times to about 14 Gyr after the big bang. Predictions for our observations of the CMB thus depend on data D_{obs} that includes the astronomical observations that determine the time from the big bang to the time when an observation is being made — the observed Hubble constant, the cosmological constant, and the mean mass density for instance.

7.2. *Copenhagen Quantum Mechanics Recovered as an Approximation*

The construction, operation, and outcomes of realistic measurement situations can be described by an appropriate set of four-dimensional histories. In a measurement situation, a subsystem — the apparatus — interacts over a short time with another subsystem — the measured subsystem — to produce records of the values of the measured quantities. DH supplies probabilities for the values of these records as the probabilities of appropriate histories.

Figure 5 shows a simple model of a measurement situation in which an observer is carrying out a two-slit experiment. The observer's apparatus measures which slit the electron went through, U or L, and the position interval of arrival at the screen Y. The results U or L and Y are recorded by the apparatus. All the ingredients for the application of Copenhagen quantum theory are there. There is a measured subsystem — the electron. There is a measurement apparatus which records the outcomes, and there is an observer. Given the initial state of the electron at the gun, the probabilities for these results can be calculated with Copenhagen quantum theory.

A measurement situation like this might have been constructed somewhere in the history of the universe. As such, the probabilities for the same recorded outcomes could be calculated with DH. When the CQM probabilities are close to those of DH for all the records, we have recovered CQM as an approximation to DH in suitable measurement situations. A fairly detailed discussion of how this works

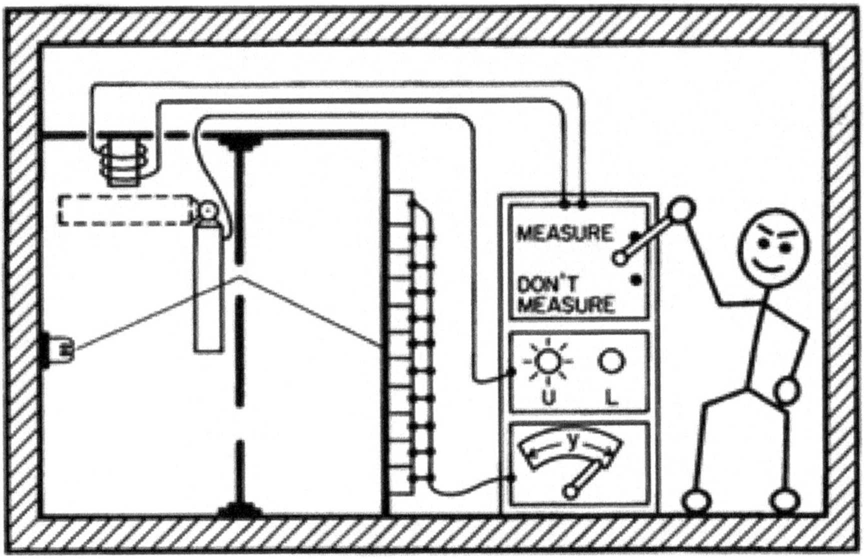

Fig. 5. A model closed quantum system containing an observer together with the necessary apparatus for carrying out a two-slit experiment. Alternatives for the system include whether the observer measured which slit the electron passed through or did not, whether the electron passed through the upper or lower slit, the alternative positions of arrival of the electron at the screen, the alternative arrival positions registered by the apparatus, the registration of these in the brain of the observer, etc. DH assigns probabilities to the individual alternative histories in such a set when there is negligible quantum-mechanical interference between them, that is, when the set of histories decoheres.

was given in Ref. 27, Sec. II.10. Other measurement models that could be used to analyze this connection can be found in Refs. 5, 28 and 29. Of course, we expect that measurement situations will occur only at certain epochs of the universe — not in the early universe, not when there were no IGUSes to construct them, not without classical spacetime, etc. These limitations are reasons why we need a generalization of CGM for cosmology.

8. Emergent Formulations of Quantum Mechanics for Cosmology

Sections 2–6 describe how the Copenhagen quantum mechanics of measurement situations (CQM) can be generalized to provide a quantum framework for prediction in quantum cosmology in two steps. The first is the construction of the decoherent histories quantum mechanics (DH) of a closed system like the universe with a fixed background spacetime geometry. The second step is a generalization of DH to a decoherent histories quantum mechanics *of* the spacetime geometries and quantum fields, namely GDH.

Each of these formulations assume the existence of particular features of our universe. Copenhagen quantum mechanics assumes both a classical world containing laboratories with measuring apparatus and observers that obey the laws of classical physics and a quantum world of alternatives that are measured. Decoherent histories quantum mechanics assumes a fixed background spacetime geometry. These features are examples of excess baggage.[k] They are true physical facts but describe special situations in the universe that occur only at particular times and circumstances. We can say that they *emerged* at those particular times and places in the evolution of our universe.

For example, in the early stages of the universe, spacetime geometry was fluctuating quantum-mechanically with no fixed value. As a consequence, there was no fixed notion of time and no fixed notion of spacelike or timelike surfaces. There then could be no notion of quantum states defined on a foliating family of spacelike surfaces and, perforce, no notion of the unitary evolution of states through such families of surfaces. There were no isolated subsystems, and no notion of measurement of one by another. There were no observers to exploit the regularities of the universe. The features of the universe

[k]In many advances in physics, some previously accepted general idea was found to be unnecessary and dispensable. The idea was not truly a general feature of the world, but only perceived to be general because of our special place in the universe and the limited range of our experience. It was excess baggage which had to be jettisoned to reach a more general perspective.[30]

assumed in the formulation of the Copenhagen quantum mechanics and DH were not there in the early universe.

Not much later the universe's classical spacetime emerged. With it emerged the quasiclassical realm — the wide range of time, place and scale also on which the deterministic laws of classical physics hold.[31] With it also emerged the notions of time, states on spacelike surfaces and their unitary evolution by the Schrödinger equation.[8] Eventually, galaxies, stars, planets and biota emerged many years later, and with them emerged the quasiclassical realm, and the observers, and measurements that are the subjects of Copenhagen quantum theory. It is then, and only then, that Copenhagen quantum mechanics is an appropriate approximation to the more general formulations.[l]

9. Conclusion — Are We Finished Generalizing Quantum Mechanics?

The impacts of cosmology on quantum mechanics are at least these:

- Cosmology requires a quantum mechanics of a closed system in which measurements and observers can be described but are not central in the formulation of the theory.
- Cosmology requires a formulation of quantum mechanics general enough to deal with quantum spacetime geometry, the big bang, and what emerged from it.
- Cosmology requires a formulation of quantum mechanics that can describe the emergence of classical spacetime in the early universe

---

[l]Thus only sufficiently late epochs of the universe will contain the observers necessary for Copenhagen quantum mechanics. But also we cannot expect observers (including ourselves) very late in the universe's history. The cosmological constant causes the universe to expand and cool exponentially quickly. Stars exhaust their thermonuclear fuel and die out. Black holes evaporate. The density of matter and the temperature approach zero. Mechanisms of environmental decoherence fail.[32] The universe becomes cold, dark and inhospitable to the formation and functioning of observers.[33]

and along with it the quasiclassical realm — the wide range of time, scale and place where the classical laws of physics apply.

- Cosmology requires a formulation of quantum mechanics to which the Copenhagen quantum mechanics is an approximation appropriate for measurement situations.

GDH potentially provides a generalization of CQM that can be applied to the prediction of the outcomes of our cosmological observations from theories of the universe's quantum state Ψ and dynamics H. GDH is therefore a formulation that is adequate for quantum cosmology. But it seems unlikely that it is the only possible adequate formulation. It is a large and daring extrapolation to go from quantum mechanics of laboratory situations to quantum mechanics of the whole universe. This is especially the case because the data testing of the extrapolation is very limited. Generalizations different from GDH are therefore likely to be possible. Indeed, different routes to generalizations have already been suggested.[34–39] A generalization of CQM should not just consist of a theory of dynamics (H) plus a theory of the quantum state (Ψ). It also must specify the quantum framework for prediction (Q). For the input theory in quantum cosmology we should write (H, Ψ, Q).

We hope to find generalizations Q that have less excess baggage than GDH, such as a principle of superposition for quantum states of the whole universe that can never be realized experimentally. Perhaps even the notion of spacetime geometry is excess baggage. We hope to find generalizations that are more deeply connected to ideas in contemporary fundamental physics such as duality (e.g. Ref. 40). We hope to find generalizations that more deeply unify the theory of the quantum state Ψ and the theory of dynamics H. But above all, we hope to find generalizations that lead to definitive tests either in experiments in the laboratory or by large scale observations. *We are not yet finished understanding the impact of cosmology on quantum mechanics.*

Cosmology, quantum mechanics, and fundamental physics are inextricably but fruitfully linked. The big bang is the place in the universe where the near-Planck energy scales that characterize much of today's fundamental theory are reached. The observable consequences of the big bang by which these theories might be tested are scattered over the large scales of space and time observable today. We should not imagine that predicting these consequences from (H, Ψ) is simply a matter of calculating in a fixed quantum framework established by low-energy experiment. Rather, as we have seen in this paper, our understanding of cosmology impacts our understanding of what is the quantum mechanics that applies to the universe as a whole.

We end with a quotation that is the final paragraph of the first paper on decoherent histories quantum mechanics[13]:

"We conclude that resolution of the problems of interpretation presented by quantum mechanics is not to be accomplished by further intense scrutiny of the subject as it applies to reproducible laboratory situations, but rather through an examination of the origin of the universe and its subsequent history. Quantum mechanics is best and most fundamentally understood in the context of quantum cosmology. The founders of quantum mechanics were right in pointing out that something external to the framework of wave function and Schrödinger equation *is* needed to interpret the theory. But it is not a postulated classical world to which quantum mechanics does not apply. Rather it is the initial condition of the universe that, together with the action function of the elementary particles and the throws of quantum dice since the beginning, explains the origin of quasiclassical domain(s) within quantum theory itself".

Acknowledgments

The author thanks Murray Gell-Mann, Thomas Hertog and Mark Srednicki for discussions of the quantum mechanics of the universe over a long period of time. He thanks the Santa Fe Institute for supporting many productive visits there. He thanks the organizers of the conference *90 Years of Quantum Mechanics* in Singapore, which was the occasion of a talk on which this work is partially based. This work was supported in part by the National Science Foundation under grants PHY15-04541 and PHY18-18018105.

References

1. M. Jammer, *The Conceptual Development of Quantum Mechanics* (McGraw-Hill, New York, 1966).
2. R. Penrose, Gravity's role in quantum state reduction, *Gen. Rel. Grav.* **28**, 582 (1996); A. J. Leggett, Testing the limits of quantum mechanics: Motivation, state-of-play, prospects, *J. Phys. Cond. Matt.* **14**, R415 (2002); P. Pearle and A. Valentini, *Generalizations of quantum mechanics*, arXiv:quant-ph/0506115; S. Weinberg, Collapse of the state vector, *Phys. Rev. A* **85**, 062116 (2012); S. Weinberg, *Lectures on Quantum Mechanics* (CUP, Cambridge, 2013), Chap. 3; S. Weinberg, The Trouble with Quantum Mechanics, (*New York Review of Books*, January 19, 2017); G. 't Hooft, *The cellular automaton interpretation of quantum mechanics*, Fundamental Theories of Physics, Vol. 185 (Springer International Publishing, 2016), arXiv:1405.1548.
3. S. Saunders, Decoherence and ontology, in *Many Worlds?* eds. S. Saunders, J. Barrett, A. Kent and D. Wallace (Oxford University Press, Oxford, 2010).
4. L. D. Landau and E. M. Lifshitz, *Quantum Mechanics* [Eng. trans.] (Pergamon Press, 1958).
5. J. von Neumann, *Mathematische Grundlagen der Quantenmechanik* (Springer, Berlin, 1932). [English trans. *Mathematical Foundations of Quantum Mechanics* (Princeton University Press, Princeton, 1955)].
6. A. Einstein, R. Tolman and B. Podolsky, Knowledge of the past and future in quantum mechanics, *Phys. Rev.* **37**, 780 (1931).
7. J. B. Hartle, Quantum pasts and the utility of history, in *The Proc. Nobel Symp.: Modern Studies of Basic Quantum Concepts and Phenomena*, Gimo, Sweden, June 13–17, 1997, *Phys. Scr.* **T76**, 67–77 (1998), arXiv:gr-qc/9712001.
8. J. B. Hartle, The spacetime approach to quantum mechanics, *Vistas Astron.* **37**, 569 (1993), arXiv:gr-qc/9304006.
9. M. Gell-Mann and J. B Hartle, Quasiclassical coarse graining and thermodynamic entropy, *Phys. Rev. A* **76**, 022104 (2007), arXiv:quant-ph/0609190.
10. M. Gell-Mann and J. B. Hartle, Classical equations for quantum systems, *Phys. Rev. D* **47**, 3345–3382 (1993), gr-qc/9210010.

11. J. B. Hartle, S. W. Hawking and T. Hertog, Classical universes of the no-boundary quantum state, *Phys. Rev. D* **77**, 123537 (2008), arXiv:0803.1663.
12. R. B. Griffiths, *Consistent Quantum Theory* (Cambridge University Press, Cambridge, UK, 2002); R. Omnès, *Interpretation of Quantum Mechanics* (Princeton University Press, Princeton, 1994); M. Gell-Mann, *The Quark and the Jaguar* (W. H. Freeman, New York, 1994); M. Gell-Mann and J. B. Hartle, Quantum mechanics in the light of quantum cosmology, in *Complexity, Entropy, and the Physics of Information*, ed. W. Zurek (Addison Wesley, Reading, MA, 1990); P. C. Hohenberg, An introduction to consistent quantum theory, *Rev. Mod. Phys.* **82**, 2835–2844 (2010).
13. M. Gell-Mann and J. B. Hartle, Quantum mechanics in the light of quantum cosmology, in *Complexity, Entropy, and the Physics of Information*, ed. W. Zurek (Addison Wesley, Reading, MA, 1990), arXiv:1803.04605.
14. H. Everett III, Relative state formulation of quantum mechanics, *Rev. Mod. Phys.* **29**, 454 (1957).
15. J. B. Hartle, The quantum mechanics of closed systems, in *Directions in General Relativity*, Vol. 1, eds. B.-L. Hu, M. P. Ryan and C. V. Vishveshwara (Cambridge University Press, Cambridge, 1993), arXiv:gr-qc/9210006.
16. R. P. Feynman and A. R. Hibbs, *Quantum Mechanics and Path Integrals* (McGraw-Hill, New York, 1964).
17. E. Joos and H. D. Zeh, The emergence of classical properties through interaction with the environment, *Zeit. Phys. B* **59**, 223 (1985).
18. W. Zurek, Decoherence, einselection, and the quantum origins of the classical, *Rev. Mod. Phys.* **75**, 715 (2003).
19. M. Gell-Mann and J. B. Hartle, Adaptive coarse graining, environments, strong decoherence, and quasiclassical realms, *Phys. Rev. A* **89**, 052125 (2014), arXiv:1312.7454.
20. J. B. Hartle and S. W. Hawking, The wave function of the universe, *Phys. Rev. D* **28**, 2960–2975 (1983).
21. J. J. Halliwell, J. B. Hartle and T. Hertog, *What is the no-boundary wave function of the universe*, arXiv:1812.01760.

22. J. B. Hartle, Spacetime quantum mechanics and the quantum mechanics of spacetime, in *Gravitation and Quantizations, Proc. 1992 Les Houches Summer School*, eds. B. Julia and J. Zinn-Justin, Les Houches Summer School Proceedings Vol. LVII (North Holland, Amsterdam, 1995), arXiv:gr-qc/9304006.
23. J. B Hartle and T. Hertog, The observer strikes back, in *The Philosophy of Cosmology*, eds. K. Chamcham, J. Silk, J. D. Barrow and S. Saunders (Cambridge University Press, Cambridge, 2017), pp. 181–205, arXiv:1503.07205.
24. S. W. Hawking and T. Hertog, Populating the landscape: A top down approach, *Phys. Rev. D* **73**, 123527 (2006), arXiv:hep-th/0602091.
25. J. B. Hartle and M. Srednicki, Are we typical? *Phys. Rev. D* **75**, 123523 (2007), arXiv:0704.2630.
26. M. Srednicki and J. B. Hartle, Science in a very large universe, *Phys. Rev. D* **81**, 123524 (2010), arXiv:0906.0042, a summary of this work with different emphases is in *The Xerographic Distribution: Scientific Reasoning in a Large Universe*, arXiv:1004.3816.
27. J. B. Hartle, The quantum mechanics of cosmology, in *Quantum Cosmology and Baby Universes: Proc. 1989 Jerusalem Winter School for Theoretical Physics*, Section II.10, eds. S. Coleman, J. B. Hartle, T. Piran and S. Weinberg (World Scientific, Singapore, 1991), pp. 65–157, arXiv:1805.12246.
28. F. London and E. Bauer, *La Théorie de l'Observation en Mécanique Quantique* (Hermann, Paris, 1939).
29. E. Wigner, The problem of measurement, *Am. J. Phys.* **31**, 6 (1963).
30. J. B. Hartle, Excess baggage, in *Elementary Particles and the Universe: Essays in Honor of Murray Gell-Mann*, ed. J. Schwarz (Cambridge University Press, Cambridge, 1990), arXiv:gr-qc/0508001. Chapter 2, p. 23, this volume.
31. J. B. Hartle, *The quasiclassical realms of this quantum universe*, arXiv:0806.3776. A slightly shorter version is published in *Many Worlds?* eds. S. Saunders, J. Barrett, A. Kent and D. Wallace (Oxford University Press, Oxford, 2010), the longer version was published in *Found. Phys.* **41**, 982 (2011). Chapter 11, p. 261, this volume.

32. J. B. Hartle and T. Hertog, Does the Universe Have a Predictable Future, forthcoming.
33. J. B. Hartle, *Why Our Universe is Comprehensible*, arXiv:1612. 01952. Chapter 14, p. 349, this volume.
34. J. B. Hartle, Linear positivity and virtual probability, *Phys. Rev. A* **70**, 022104 (2004), arXiv:quant-ph/0401108.
35. J. B. Hartle, Generalizing quantum mechanics for quantum spacetime, in *The Quantum Structure of Space and Time: Proc. 23rd Solvay Conf. Physics*, eds. D. Gross, M. Henneaux and A. Sevrin (World Scientific, Singapore, 2007), arXiv:gr-qc/0602013.
36. J. B. Hartle, Quantum mechanics with extended probabilities, *Phys. Rev. A* **78**, 012108 (2008), arXiv:0801.0688.
37. M. Gell-Mann and J. B. Hartle, Decoherent histories quantum mechanics with one real fine-grained history, *Phys. Rev. A* **85**, 062120 (2012), arXiv:1106.0767.
38. J. B. Hartle, *Decoherent Histories Quantum Mechanics Starting with Records of What Happens*, arXiv:1608.04145.
39. S. B. Giddings, Universal quantum mechanics, *Phys. Rev. D* **78**, 084004, arXiv:0711.0757.
40. T. Hertog and J. B. Hartle, Holographic no-boundary measure, *JHEP* **2012**, 95 (2012), arXiv:1111.6090.

CHAPTER 24

Murray Gell-Mann: A Short Appreciation

On September 25, 2014, Murray Gell-Mann was presented with the Helmholz Medal of the Berlin Brandenburg Academy of Sciences and Humanities in a ceremony at the Santa Fe Institute. The author, among others, was asked to speak for fifteen minutes on Murray and his accomplishments. The following is an edited transcription of the author's speaking text.

arXiv:1510.08787

I am a student of Murray's, and this year is the 50th anniversary of my Ph.D. under his supervision. In recent years, he and I have been working together to understand how our everyday classical world with all its complexity emerges from a simple quantum theory of the origin of the universe.

I am not going to talk much about the universe or quantum mechanics. I am not even going to talk much about Murray's renowned achievements in the physics of the elementary particles — strangeness, quarks, the eightfold way symmetries, the renormalization group, the standard model of the strong interactions (QCD), just to mention some. It is for some of these works in particle physics that he received the Nobel prize. It is the works for which he is perhaps best known. I'm not going to do that because others here will tell you something about it and you can read about it in many places. But, more importantly, to focus only on his work in particle physics would not do justice to the rest of his achievements.

- His work on the evolution of human languages. Is there evidence in the many existing human languages to show that they all had a common origin? What does that tell you about human history? There is the Gell-Mann law for the evolution of the order of subject-object-verb, for example.
- His seminal work on complex adaptive systems. We have no trouble exhibiting complex adaptive systems. The Santa Fe Institute is one. But what is there in common among the huge variety of complex systems we find in the world? What do we even mean by complexity? One answer is supplied in his beautiful work with Seth Lloyd on a general characterization of complexity. This is the kind of work that SFI was set up to do. It is encapsulated in the title of his book — *The Quark and the Jaguar.*
- His work on building a sustainable environment. A complex system that is central to our everyday life.
- His work on the foundations of quantum mechanics and its application to cosmology. Quantum mechanics is the fundamental framework for the physics of the very small. It is usually formulated for laboratory measurements carried out by observers. But in the early universe there were no observers. Are we to believe

that quantum theory doesn't apply there? No. In the very early universe, large and small are one. At the big bang, the biggest possible system has the smallest possible size. It is obviously quantum-mechanical. How, then, can quantum mechanics be formulated so that it works at the big bang? The decoherent histories quantum theory that Murray and I worked on, is one answer.

- And, perhaps most importantly, there is his work as a scientific citizen, especially with the President's Science Advisory Committee, the MacArthur Foundation, and his central role in setting up SFI.

It is an impressive and varied list. And I've often wondered what more he might have done in physics if he were more focused on that.

What I hope to do is to speculate (at great risk) on how these disparate areas can be seen as part of a whole. And to talk a little about what we can discern of Murray's approach to science from those connections and what we can learn about him from that.

Erwin Schrödinger is often credited with the following injunction:

The task is not to see what no one else has seen, but to think what no one else has thought, about that which everyone else has seen.

That is Murray Gell-Mann.

- To find the pattern that no one else had noticed was there.
- To find the connections that no one else had sought.
- To see the symmetry that is hidden or only manifested approximately.
- To aim for the generality and the connections which no one else realized were possible.
- To have the courage to give up on what was accepted as secure, general and inescapable.
- To know what is fundamental and what is excess baggage.
- To have the guts to guess the answer.

It is a common popular misconception that scientists working at the frontier are seeking to answer some well-defined, fixed, famous great questions: "What is the unified theory of all the fundamental interactions?" "What was the origin of the universe?" But this is not how it seems in the trenches.

The frontier of theoretical physics is a noisy, chaotic place. At any one time, we have cherished old ideas, well confirmed in territory already mapped, that we are very reluctant to give up. We also have a whole variety of unconfirmed new ideas vying for the best route forward into new territory that *require* us to give up some of the old ideas.

Rather, the question is, "What *is* the right question?" Murray always knows what the right question is. He has a remarkable ability to see through all the clutter, to cut to the heart of matter, and to focus on the essentials. He also has the courage to discard the cherished old ideas that are an obstacle to progress. Later, when looked at in the right way, what he does seems inevitable. But that's the genius.

From this perspective I think one can see some commonality in the subjects on which Murray has worked — particle physics, human languages, complex adaptive systems, cosmology, and even the environment. They are all complicated with no, or non-obvious, or hidden regularities. But all are systems for which Murray supplied some regularity.

Even his avocations — bird watching, collecting southwestern ceramics, wine tasting, etc. — somehow fit in.

The paradigmatic example is his work on the eightfold way symmetry of the strong interactions. At the time physicists were giving up on field theory, so there was no real theory. But, Murray found the approximate symmetry relating the elementary particles *before* there was anything like a theory that had the symmetry. The symmetry predicted a previously unobserved particle, the particle was found, and a Nobel prize awarded.

That led to quarks and eventually to string theory, which is the focus of the efforts of many physicists today.

The power of searching for symmetries, connections and organizing principles that is characteristic of Murray's approach to science is well illustrated in this example. That approach has had a lasting effect on how we do physics at the frontier.

Fig. 1. The author and Murray working in Murray's office at SFI in 2015. Photo courtesy of SFI.

...to think what no one else has thought about that which everyone else has seen.

Working with Murray is a unique kind of experience, at least in my life. To work with him, it helps first of all not to have too fragile an ego. Murray was very accessible to his students. But I remember one occasion when I went to his office and said, "Professor Gell-Mann, I have a question." He replied somewhat irritably, "Questions, questions — you are always coming in here with questions! Why don't you try bringing in some answers sometime!" — Caltech wasn't called the Marine Corps of graduate schools for nothing. But despite the obvious differences in intellect and insight, Murray somehow makes you feel an equal partner when you are working with him.

Working with Murray, you have to accept certain rigorous standards. There is typically a list of things that Murray is sure are connected, and we go over and over them again and again, posing them, attacking them, calculating them until a unified, new, surprising perspective emerges. Never stopping short, never settling for a picture that is incomplete, it can take years.

Working with Murray is challenging, but it is also fun! To a good approximation, Murray knows everything. The conversation is not only about physics, but about languages, the history of science, the history of the Ottoman empire, politics, the lives of colleagues, the environment, philosophy, etc. You enter a world in which everything is connected on a much bigger scale and in a much deeper way than in your own narrow speciality. You make personal progress not only in science but in almost everything else. It is a true adventure of the mind. And the same thing could be said about Murray's life.

In the future, as physicists, Murray will always be in our minds, because he is a great scientist, because he created much of what we have to think about, and because he has set a certain style on how to do science. But for those of us who have had the privilege of working with him, he will also be — always in our hearts. Thank you.

Papers with Murray Gell-Mann

- *Quantum mechanics in the light of quantum cosmology*, in *Complexity, Entropy, and the Physics of Information*, Santa Fe Institute Studies in the Sciences of Complexity VIII, ed. W. H. Zurek (Addison-Wesley, Reading, MA, 1990) and in *Proc. 3rd Int. Symp. Foundations of Quantum Mechanics in the Light of New Technology*, eds. S. Kobayashi, H. Ezawa, Y. Murayama and S. Nomura (Physical Society of Japan, Tokyo, 1990), arXiv:1803.04605.
- *Alternative decohering histories in quantum mechanics*, in the *Proc. 25th Int. Conf. High Energy Physics*, Singapore, August, 2–8, 1990, eds. K. K. Phua and Y. Yamaguchi (South East Asia Theoretical Physics Association and Physical Society of Japan), distributed by (World Scientific, Singapore, 1990).
- *Time symmetry and asymmetry in quantum mechanics and quantum cosmology*, in the *Proc. 1st Int. Sakharov Conf. Physics*, Moscow, USSR, May 27–31, 1991, eds. L. V. Keldysh and V. Ya. Fainberg (Nova Science Publishers, New York, 1992);

and in *Physical Origins of Time Asymmetry: Proc. NATO Workshop*, Mazagon, Spain, September 30–October 4, 1991, eds. J. Halliwell, J. Perez-Mercader and W. Zurek (Cambridge University Press, Cambridge, 1994), pp. 311–345, arXiv:gr-qc/9304023.

- *Classical equations for quantum systems*, *Phys. Rev. D* **47**, 3345–3382 (1993), arXiv:gr-qc/9210010.
- *Equivalent Sets of Histories and Multiple Quasiclassical Domains*, arXiv:gr-qc/9404013.
- *Strong decoherence* in the *Proc. 4th (1994) Drexel Conf. Quantum Non-Integrability: Quantum-Classical Correspondence*, eds. D.-H. Feng and B.-L. Hu (International Press of Boston/Hong Kong, 1998), arXiv:gr-qc/9509054.
- *Quasiclassical coarse graining and thermodynamic entropy*, *Phys. Rev. A* **76**, 022104 (2007), arXiv:quant-ph/0609190.
- *Decoherent histories quantum mechanics with one 'real' fine-grained history*, *Phys. Rev. A* **85**, 062120 (2012), arXiv:1106.0767.
- *Adaptive coarse graining, environments, strong decoherence, and quasiclassical realms*, *Phys. Rev. A* **89**, 052125 (2014), arXiv:1312.7454.

CHAPTER 25

Working with Stephen Hawking

The banquet for the July 2017 conference in Cambridge, UK celebrating Stephen Hawking's 75th birthday was held in Trinity College on July 3rd. The organizers asked the author, among others, to give a ten-minute after-dinner talk on what it was like to work with Stephen. The following is an edited version of the author's speaking text.

arXiv:1711.09071

I would like to thank the organizers for inviting me to this celebration and for giving me an opportunity to speak at a wonderful occasion. For after-dinner amusement, they have asked me to give a few personal recollections addressing the question, "What is it like to work with Stephen". It remains to be seen whether I should thank them for that.

My association with Stephen began forty-six years ago during a long visit to Fred Hoyle's Institute of Theoretical Astronomy (as it was known then). In residence were people like Brandon Carter, Martin Rees, Paul Davies, and Stephen Hawking — colleagues with whom I have maintained lifelong personal and scientific contacts.

Stephen had guessed that there must be solutions of the Einstein–Maxwell equations that represented many equally charged and equally massive black holes held in equilibrium with the electrostatic repulsion between them balancing their gravitational attraction. As it happened, I knew where the relevant metric was to be found, and we were off and running.

Since that time I have always felt that we were on the same wavelength — not the same in ability or insight, of course — but rather similar in style and understanding of what is important. Ten more papers were to follow.

It is impossible to address the question of what is it like to work with Stephen without first answering the question of what is it that makes him so special as a scientist.

It is a common misconception that scientists working at the frontier are participants in an organized effort seeking to answer some well-defined fixed famous great questions: "What is the unified theory of all the fundamental interactions?" "What was the origin of the universe?" But this is not how it seems to those of us in the trenches.

The frontier of theoretical physics is a noisy, chaotic place. At any one time, we have cherished *old* ideas, well confirmed in territory already mapped, that we are very reluctant to give up. We also have a whole variety of competing, different, unconfirmed *new* ideas vying for the best route forward into new territory that *require* us to give up some of the old ideas. It is certainly not organized! I think that

Stephen likes situations like this and the surprises that upset the status quo.

Rather, the questions to be asked are: "What is the right question?" "What do we keep and what do we give up?" Stephen knows the answers. He has a remarkable ability to see through all this clutter, to cut to the heart of the matter, and to focus on the essentials. He also has the courage to discard the cherished old ideas that are an obstacle to progress. Later, when looked at in the right way these seem inevitable. But that's the genius.

Working with Stephen day-to-day is not so very different from other collaborations. Ideas are supplied and vetted by all. Contrary views are reconciled by argument. The important is distinguished from the unimportant by debate. The feasible is separated from the unfeasible by trial and error. Stephen is inclusive in all this and I think, cognizant of its value. Many of his statements start with "we", for instance: "We could say that... ." That inclusion extends to a larger loose group of scientists with similar viewpoints working on similar problems — Thomas Hertog, for instance, Stephen's many students, and many more. Working with Stephen broadens your horizons both scientifically and personally.

Working with Stephen is fun. You get to draw on his vast and deep understanding of physics broadly and of general relativity, in particular. You learn. You get to appreciate his ability to explain with concision and clarity. You get to enjoy Stephen's wonderful sense of humor. Right now, I can see the expression on his face when he is getting ready to tell me that something I've just said is inconsistent, or violates some basic principle of physics, or some other rubbish. But it is not just about physics. The conversation is often diverted to other issues in science and life on which Stephen often has strong, well-thought-out positions.

The consequences of working with Stephen extend far beyond the particular paper or calculation. I have often thought that the signature of a great problem in physics is that its solution generates more great problems. Certainly that is the case with the Hawking radiation. In my case, the work with Stephen on the no-boundary wave function of the universe led to numerous specific calculations,

The author and Stephen Hawking working.
(Photo courtesy Kathy Page)

many done with Thomas Hertog and Stephen, of what it predicts for our observations of the universe on the largest scales of space and time. But it also motivated a new vision, formulated with Murray Gell-Mann, of how usual textbook quantum mechanics can be generalized to apply to cosmology — decoherent histories quantum theory, in particular.

Working with Stephen brings out the best in you. You are stretched. You learn. You are confronted with new ideas, and new perspectives that are more powerful than the ones you had before. Important new routes are opened up and important questions asked that you didn't think existed before. Stephen encourages thinking outside the box while at the same time setting high standards for that thinking.

Late in life, I find that I am proud of my record in theoretical physics. Not, of course, because of any great discoveries or huge impacts. But, rather, in the sense of "personal best". I think that I did the best that I could have done with the talents and circumstances that I had. But I don't think that I could have done even that, without having worked with Stephen.

A toast concluded the proceedings: "For seventy-five years of achievement both in science and life: Stephen Hawking".

Papers with Stephen Hawking

- Solutions of the Einstein–Maxwell Equations with Many Black Holes, *Comm. Math Phys.* **26**, 87–101 (1972).
- Energy and Angular Momentum Flow into a Black Hole, *Comm. Math. Phys.* **27**, 283–290 (1972).
- Path Integral Derivation of Black Hole Radiance, *Phys. Rev. D* **13**, 2188–2203 (1976).
- Wave Function of the Universe, *Phys. Rev. D* **28**, 2960–2975 (1983).
- The No-Boundary Measure of the Universe (with T. Hertog), *Phys. Rev. Lett.* **100**, 202301 (2008), arXiv:0711:4630.
- Classical Universes of the No-Boundary Quantum State (with T. Hertog), *Phys. Rev. D* **77**, 123537 (2008), arXiv:0803:1663.
- The No-Boundary Measure in the Realm of Eternal Inflation (with T. Hertog), *Phys. Rev. D* **82**, 063510 (2010), arXiv:1001:0262.
- Local Observation and Eternal Inflation (with T. Hertog), *Phys. Rev. Lett.* **106**, 141302 (2011), arXiv:1009.2525. (An early arXiv version had the title 'Eternal Inflation without Metaphysics')
- Vector Fields in Holographic Cosmology (with T. Hertog), *JHEP11* (2013) 201, arXiv:1305.719.
- Quantum Probabilities for Inflation from Holography (with T. Hertog), *JCAP* 01(2014) 015, arXiv:1207.6653.
- Accelerated Expansion from Negative Λ (with T. Hertog), arXiv:1205.3807.

List of Essays with publication data and arXiv references

(1) Computability and Physical Theories (with R. Geroch), *Found. Phys.*, **16**, 533–550, 1986; and reprinted in *Between Quantum and Cosmos: Studies in Honor of John Archibald Wheeler* (eds. W.H. Zurek, A. van der Merwe, and W.A. Miller) Princeton University Press, Princeton (1988), arXiv:1806.09237.

(2) Excess Baggage, in *Elementary Particles and the Universe: Essays in Honor of Murray Gell-Mann*, ed. J.H. Schwarz, Cambridge University Press, Cambridge (1991) 1–16, arXiv:gr-qc/0508001.

(3) Sources of Predictability, in *Complexity*, **3**, 22–25, 1997; arXiv:gr-qc/9701027.

(4) Theories of Everything and Hawking's Wave Function of the Universe, in *The Future of Theoretical Physics and Cosmology: Celebrating Stephen Hawking's 60th Birthday*, eds. G.W. Gibbons, E.P.S. Shellard, and S.J. Ranken, (Cambridge University Press, Cambridge UK, 2003); arXiv:gr-qc/0209047.

(5) Scientific Knowledge from the Perspective of Quantum Cosmology, in *Boundaries and Barriers: On the Limits of Scientific Knowledge* (1995 Abisko Conference), eds. J.L. Casti and A. Karlqvist, Addison-Wesley Publishing Co., Reading, MA (1996); arXiv:gr-qc/9601046.

(6) The Physics of Now, *Am. J. Phys.*, **73**, 101–109 (2005), arXiv:gr-qc/0403001. (A popularization of this by M. Chown appeared in *New Scientist*, **182**, May 1, 2004, p. 34.
(7) The Quantum Mechanics of Closed Systems, in *General Relativity and Gravitation 1992*, (Proceedings of the 13th Conference on General Relativity and Gravitation, Cordoba, Argentina, June 28–July 4, 1992) eds. R.J. Gleiser, C.N. Kozameh, and O.M. Moreschi, Institute of Physics Publishing, Bristol (1993); and in *Directions in Relativity, Vol. 1, (Essays in Honor of Charles W. Misner's 60th Birthday)* eds. B.-L. Hu, M.P. Ryan, and C.V. Vishveshwara, Cambridge University Press, Cambridge (1993); arXiv:gr-qc/9210006.
(8) The Spacetime Approach to Quantum Mechanics, in *Vistas in Astronomy*, **37**, 569, 1993; and in *Topics on Quantum Gravity and Beyond (Essays in Honor of Louis Witten on His Retirement)* eds. F. Mansouri and J.J. Scanio, World Scientific, Singapore (1993); and in *Quantum Physics, Chaos Theory, and Cosmology*, eds. M. Namiki, I. Ohba, K. Maeda, and Y. Aizawa (AIP Press, Woodbury, NY, 1996); arXiv:gr-qc/9210004.
(9) Quantum Physics and Human Language, *J. Phys. A: Math. Theor.*, **40**, 3101–3121 (2007), arXiv:quant-ph/0610131.
(10) Quantum Pasts and the Utility of History, in *The Proceedings of the Nobel Symposium: Modern Studies of Basic Quantum Concepts and Phenomena*, Gimo, Sweden, June 13–17, 1997, *Physica Scripta*, **T76**, 67–77 (1998); arXiv:gr-qc/9712001.
(11) The quasiclassical realms of this quantum universe, arXiv:0806.3776. A slightly shorter version is published in *Many Worlds?* eds. S. Saunders, J. Barrett, A. Kent, and D. Wallace (Oxford University Press, Oxford, 2010), the longer version was published in *Foundations of Physics*, **41**, 982 (2011).
(12) Living in a Superposition, arXiv:1511.01550.
(13) The Quantum Mechanical Arrows of Time, in *Fundamental Aspects of Quantum Theory: A Two-Time Winner, Proceedings of the conference to honor Yakir Aharonov's 80th Birthday*, eds. D. Struppa and J. Tollaksen, (Springer, Milan, 2013); arXiv:1301.2844.

(14) Why Our Universe is Comprehensible, arXiv:1612.01952.
(15) Quantum Cosmology: Problems for the 21st Century, in *Physics in the 21st Century: Proceedings of the 11th Nishinomiya-Yukawa Symposium*, Nishinomiya, Hyogo, Japan, eds. K. Kikkawa, H. Kunitomo, and H. Ohtsubo, World Scientific, Singapore (1997); arXiv:gr-qc/9701022.
(16) Anthropic Reasoning and Quantum Cosmology, in *The New Cosmology: Proceedings of the Conference on Strings and Cosmology, College Station, Texas, March 14–17, 2004*, eds. R. Allen, D. Nanopoulos and C. Pope, AIP Conference Proceedings, v. 743 (American Institute of Physics, Melville, NY, 2004), arXiv:gr-qc/0406104; essentially the same article is to be reprinted in *Universe or Multiverse*, ed. B. Carr, (Cambridge University Press, Cambridge).
(17) Are We Typical? (with M. Srednicki), *Phys. Rev. D*, **75**, 123523 (2007), arXiv:0704.2630.
(18) Science in a Very Large Universe (with M. Srednicki), *Phys. Rev. D*, **81** 123524 (2010), arXiv:0906.0042, a summary of this work with different emphases is in The Xerographic Distribution: Scientific Reasoning in a Large Universe, arXiv:1004.3816.
(19) The Observer Strikes Back (with T. Hertog), in *The Philosophy of Cosmology* eds. K. Chamcham, J. Silk, J.D. Barrow, and S. Saunders, (Cambridge University Press, Cambridge, 2017), pp. 181–205, arXiv:1503.07205.
(20) Quantum Multiverses, arXiv:1801.08631.
(21) Quantum Mechanics at the Planck Scale, in *Proceedings of the Workshop on Physics at the Planck Scale* held at Puri, India, December 12–21, 1994; arXiv:gr-qc/9508023.
(22) The State of the Universe, in *The Future of Theoretical Physics and Cosmology: Celebrating Stephen Hawking's 60th Birthday*, eds. G.W. Gibbons, E.P.S. Shellard, and S.J. Ranken, (Cambridge University Press, Cambridge UK, 2003); arXiv:gr-qc/0209046.
(23) The Impact of Cosmology on Quantum Mechanics, arXiv:1901.03933.
(24) Murray Gell-Mann: A Short Appreciation, arXiv:1510.08787.
(25) Working with Stephen Hawking, arXiv:1711.09071.

www.ingramcontent.com/pod-product-compliance
Ingram Content Group UK Ltd.
Pitfield, Milton Keynes, MK11 3LW, UK
UKHW022107170926
8624IPUK00024B/179

9 789811 216398